Psychrophilic Microbiome:

Diversity, Adaptation Strategies, and Biotechnological Potential

Editors

Burhan Hamid

Centre of Research for Development (Microbiology)
University of Kashmir, Srinagar, India

Riyaz Sayyed

Department of Microbiology
PSGVP Mandal's ASC College
Shahada, India

CRC Press
Taylor & Francis Group
Boca Raton London New York

CRC Press is an imprint of the
Taylor & Francis Group, an **informa** business
A SCIENCE PUBLISHERS BOOK

First edition published 2026
by CRC Press
2385 NW Executive Center Drive, Suite 320, Boca Raton FL 33431

and by CRC Press
4 Park Square, Milton Park, Abingdon, Oxon, OX14 4RN

CRC Press is an imprint of Taylor & Francis Group, LLC

Library of Congress Cataloging-in-Publication Data (applied for)

ISBN: 978-1-041-14546-2 (hbk)
ISBN: 978-1-041-20568-5 (pbk)
ISBN: 978-1-003-71700-3 (ebk)

DOI: 10.1201/9781003717003

Typeset in Times New Roman
by Prime Publishing Services

Foreword

Microorganisms inhabit every part of the biosphere. They are found everywhere, including harsh environments like the poles, deserts, geysers, rocks, and the deep sea. They play vital roles in ecosystems and human well-being. Their rich biodiversity has attracted the scientific community to explore their potential applications.

Extremophiles that produce extremozymes are known as psychrophiles. Psychrophiles endure, flourish, and even thrive in extremely cold climates, and have numerous applications that need to be explored. Recent studies have identified novel processes and biomolecular structures with a wide range of potential uses in biotechnology, bioremediation, and waste management. They are also a valuable source of industrially significant cold-active enzymes called psychrozymes.

This book highlights the importance of psychrophiles and psychrozymes in cold weather agriculture and industries. Thriving in cold environments makes them important and applicable for multiple fields from agriculture to biotechnology. These microbes help in decomposing organic matter and recycling of nutrients in low temperatures soils.

Moreover, psychrophiles support soil fertility and some help in plant growth by producing anti-freeze proteins. This book possesses significant information on adaptation strategies adopted by cryophiles with authentic data presentation. It encompasses the psychrophilic microbiome, its biodiversity, and its biochemical, physiological, and molecular adaptation in extreme habitats such as Antarctica, polar regions, permafrost, and deep ocean waters. In addition, it contains significant information about the application of psychrozymes in different industries, along with the impact of climate change on the psychrophilic microbiome of polar regions.

This book is a valuable resource that will, I am confident, steer researchers and students toward a deeper understanding of psychrophiles and psychrozymes in cold weather agriculture and their biotechnological utility. It provides a comprehensive overview of recent discoveries and the challenges that lie ahead, offering both theoretical and practical insights. I believe it will serve as a catalyst for further research and exploration in this exciting field.

Luz E. Bashan
Auburn, Alabama, USA

Acknowledgement

In the journey of bringing this book project to completion, we've been fortunate to receive the kindness, blessings, and unwavering support of many individuals. First and foremost, we humbly acknowledge the guiding presence of the Almighty ALLAH, navigating us through life's twists and turns, culminating in the joy of the "mission accomplished" today.

Our profound appreciation goes to our families, whose blessings, endurance, and persistent sacrifices have made compilation of this project a reality.

We would like to gratefully acknowledge and express sincere thanks to all authors; for their time, talent, and interest in our work. We cordially appreciate their contribution in our book, and without their contribution it couldn't have been possible.

We owe a special and heartfelt debt of gratitude to the Director and all other faculty members of Centre of Research for Development (CORD), University of Kashmir, India, whose moral encouragement and intellectual guidance was an inspirational factor that uplifted us to embark on the journey.

Burhan Hamid
Riyaz Sayyed

Preface

The biosphere is fundamentally structured by microorganisms, which drive essential biogeochemical cycles and sustain life on Earth. Other life forms wouldn't have existed or couldn't have evolved without them. Microorganisms can be found all over the place. Nearly every livable square inch of the globe is home to them. All regions of the biosphere, including harsh ones like the poles, deserts, geysers, rocks, and the deep sea, are home to them and they flourish there. They are crucial to both the ecosystem and human well-being. The scientific community is interested in exploring their potential due to their diverseness. Among them one major group of microorganisms are extremophiles.

The remarkable group of extremophiles producing extremozymes known as psychrophiles and their enzymes as psychrozymes, which love to exist in extremely cold climatic conditions, are the main focus of this book. Despite living in extremely delicate conditions, these microorganisms are able to endure and even flourish. Psychrophiles play a vital role in agricultural sustainability at low temperature environments. Recent studies have identified novel processes and biomolecular structures with a wide range of possible uses in biotechnology, bioremediation, and waste management. They are also a valuable source of industrially significant cold active enzymes called psychrozymes. This book provides in-depth knowledge to students, research scholars, and scientists about psychrophilic microbiomes, their diversity, cold adaptation, and their utility in biotechnology, especially their role in cold weather agriculture. This book covers the current focused issues of cold adaptation by cryophiles; adaptation strategies adapted by cryophiles have been discussed with authentic data. Psychrophiles offer a potential role in agricultural sustainability under harsh environmental conditions. Alternative to chemical fertilizers, potent cold-active microbiomes have been used for supporting agricultural biodiversity and sustainability.

This book sums up the diversity of cold-loving bacterial, fungal, and archaeal groups and molecular adaptation of psychrophilic microbiome in different habitats including Antarctica, polar regions, permafrost as well as in deep ocean waters.

Moreover, it contains significant data about application of psychrozymes in different industries along with the impact of climate change on the psychrophilic microbiomes of polar regions; that's the need of the hour. The book has been framed to provide information about the multidimensional role of psychrophiles in addressing environmental issues like waste management in cold environments and sustainability of the agro-industry. This book will guide researchers as well as students to work on extremophiles, especially psychrophiles which possess potential applications.

Book chapters in this book have been planned in a clear chronology so that readers will easily understand the significance of chapters. For research scholars, teachers, and postgraduate students who believe in quality work, this book will provide them with an upright platform. This book possesses important data necessary for researchers, microbiologists, bio-technologists, and environmentalists. In general, this book is comprised of significant data about diversity, adaptation strategies, and biotechnological applications of psychrophiles and psychrozymes. This book covers the role of psychrophiles and psychrozymes in cold-weather agriculture as the application of psychrophilic microbiomes is immensely increasing for sustainable farming and biotechnology in cold climates.

I am sure that the book will be of vital importance to postgraduate students and researchers who have keen interest in knowing life in the cold and also application of cryophiles and their enzymes in different areas of life. This book will further help readers in attaining both theoretical as well as practical exposure about recent discoveries and further challenges in this area. Moreover, we believe it will make a good impact on readers, so that they will work in cold climatic regions and will explore new avenues.

Burhan Hamid
Riyaz Sayyed

List of Abbreviations

%	:	Percent
α	:	Alpha
°C	:	Degree Celsius
ACC	:	1-Aminocyclopropane-1-Carboxylate
ADH	:	Alcohol Dehydrogenase
AFPs	:	Antifreeze Protein
Ag-NPs	:	Silver Nanoparticles
AHA	:	α-amylase
AMP	:	Adenosine Monophosphate
ANME	:	ANaerobic MEthanotrophic
AP	:	Apurinic/Apyrimidinic
AP	:	Alkaline Phosphatase
APIs	:	Active Pharmaceutical Ingredients
ASVs	:	Amplicon Sequence Variations
ATP	:	Adenosine Triphosphate
BCA	:	Biological Control Agents
BER	:	Base Excision Repair
Bs	:	Baltic Sea Ice
BT	:	Biotechnological Aspects
BTEX	:	Benzene, Toluene, Ethylbenzene & o-, m-, p-Xylenes
C	:	Celsius
CA	:	Cold Active
Ca^{2+}	:	Calcium Cation
CAEs	:	Cold Active Enzymes
CAGR	:	Compound Annual Growth Rate

CAP	:	Cold Acclimation Proteins
CBAs	:	Chlorobenzoates
CFB	:	Cytophaga Flexibacter Bacteroides
CIPs	:	Cold Induced Proteins
CO_2	:	Carbon Dioxide
CRESS	:	Circular Replication-Associated Protein (Rep) Encoding Single Stranded
CSP	:	Cold Shock Proteins
DBD	:	Dielectric Barrier Discharge
DHA	:	Decosahexaenoic Acid
DHABs	:	Deep Hypersaline Anoxic Basins
DNA	:	Deoxy Ribonucleic Acid
dNTP	:	Deoxynucleotide Triphosphate
DOM	:	Dissolve Oxygen Matter
DOVEs	:	Dark Oligotrophic Volcanic Ecosystems
DPP	:	Dairy Processing
Dump	:	Deoxyuridine Monophosphate
DyP	:	Dye Decolorizing Peroxidase
EAA	:	Enzymatic Activity and Adaptation
EC	:	European Commission
ECM	:	Ecto Mycorrhizal
ECs	:	Emerging Contaminants
eDNA	:	Environmental DNA
EES	:	Enzyme Engineering Strategies
EOI	:	Enzyme Optimization and Immobilization
EPA	:	Eicosapentaenoic Acid
EPRS	:	Electron Paramagnetic Resonance Spectroscopy
EPS	:	Exopolymer Substances
EPS	:	Exopolysaccharide
EPS	:	Extracellular Polymeric Substances
ES	:	Enzyme Substrate
FAD	:	Fatty Acid Desaturases
FAO	:	Food and Agriculture Organization
FE	:	Flavour Enhancement
FISH	:	Fluorescence *In Situ* Hybridization
FPA	:	Food Processing Applications
FPI	:	Food Processing Infrastructures

FTIR	:	Fourier Transform Infrared
GC-MS	:	Gas Chromatography Mass Spectrometer
GrIS	:	Greenland Ice Sheet
GST	:	Glutathione-S-Transferase
H	:	Hydrogen
H_2O	:	Water
HAALs	:	High-altitude Andean Lakes
HCN	:	Hydrogen Cyanide
HGT	:	Horizontal Gene Transfer
HPA	:	Human Pancreas
HPLC	:	High-Performance Liquid Chromatography
HSP	:	Heat Shock Proteins
IAA	:	Indole Acetic Acid
INP	:	Ice Nucleating Proteins
IPCC	:	Intergovernmental Panel on Climate Change
IPTG	:	Isopropyl β-D-1-Thiogalactopyranoside
IRI	:	Ice Recrystallization Inhibitors
IRIP	:	Ice Recrystallization Inhibition Proteins
J&K	:	Jammu and Kashmir
Kcat	:	Catalytic Constant
KGy	:	KiloGrays
Km	:	Michaelis Constant
LCPUFAs	:	Long-Chain Polyunsaturated Fatty Acids
LPS	:	Lipopolysaccharides
LPUFA	:	Long Chain Polyunsaturated Fatty Acids
LTS	:	Low-temperature Stress
M&SBA	:	Metagenomic and Synthetic Biology Approaches
Map	:	Metagenome Derived AP
MAP	:	Modified Atmosphere Packaging
MASE	:	Mars Analogues for Space Exploration
MBP	:	Maltose Binding Protein
MCF	:	Micro-Colonial Fungi
MD	:	Molecular Dynamics
MDMA	:	Methylene-Dioxy-Methamphetamine
MiRNA	:	MicroRNA
MM	:	Molecular Mechanisms
Mmol	:	Milli molar

MMR	:	Mismatch Repair
MOFs	:	Metal Organic Frameworks
mp	:	Melt Ponds
MPa	:	Megapascal
mRNA	:	Messenger RNA
MSA	:	Multiple Sequence Alignment
MT	:	Meat Tenderization
MTBE	:	Methyl Tert Butyl Ether
NAD	:	Nicotinamide Adenine diphosphate
NER	:	Nucleotide Excision Repair
NGS	:	Next-Generation Sequencing
NMR	:	Nuclear Magnetic Resonance
OM	:	Organic Matter
ORFs	:	Open Reading Frames
OUT	:	Operational Taxonomic Unit
PAHs	:	Poly Aromatic Hydrocarbons
PBDEs	:	Polybrominated Diphenyl Ethers
PCR	:	Polymerase Chain Reaction
PFAS	:	Per- and Polyfluoroalkyl Substances
PFCs	:	Perfluorinated Compounds
PFO	:	Perfluoro Octane Sulfonate
PGP	:	Plant Growth Promotion
PGPM	:	Plant Growth Promoting Microorganisms
PGPR	:	Plant Growth Promoting Rhizobacteria
pH	:	Potential of Hydrogen
PMs	:	Psychrophilic Microorganisms
POPs	:	Persistent Organic Pollutants
PPIase	:	Peptidyl Prolyl Isomerases
PSB	:	Phosphate Solubilizing Bacteria
Psp.	:	*Psychrobacter* sp.
PUFA	:	Polyunsaturated Fatty Acids
Rbps	:	RNA Binding Proteins
rDNA	:	Recombinant Deoxyribonucleic Acid
RIF	:	Rock-Inhabiting Fungi
RMSD	:	Root Mean Square Deviation
RNA	:	Ribonucleic Acid
ROS	:	Reactive Oxygen Species

rRNA	:	Ribosomal RNA
RubisCO	:	Ribulose-1,5-bisphosphate Carboxylase/Oxygenase
SDE	:	Stochastic Differential Equations
SFP	:	Sustainable Food Production
SIGEX	:	Substrate-Induced Gene Expression
SIMCOs	:	Sea Ice Microbial Communities
S-layer	:	Surface Layer
SOC	:	Soil Organic Carbon
sp	:	Species
sRNAs	:	Small RNA regulators
SSO	:	Specialized Spoilage Organisms
ssRNA	:	Single Stranded RNA
SUMO	:	Small Ubiquitin Like Modifier
SWC	:	Soil Water Content
TCA	:	Tri Carboxylic Acid
TF	:	Trigger Factor
TH	:	Thermal Hysteresis
TMAU	:	Tri Methyl Aminuria
T_{max}	:	Maximum Growth Temperature
T_{opt}	:	Optimum Temperature
Trx	:	Thioredoxin
UDG	:	Uracil DNA Glycosylase
UFA	:	Unsaturated Fattyacids
UNG	:	Uracil-DNA N-Glycosylase
USD	:	United States Dollar
UV	:	Ultraviolet
VFA	:	Volatile Fatty Acid
WHO	:	World Health Organization
β	:	Beta
γ-HCH	:	Gamma-Hexachlorocyclohexane
ΔG*	:	Reducing Activation Energy

List of Contributors

Aamina Shahab: Protein Research Laboratory, Department of Bioengineering, Integral University, Lucknow, Uttar Pradesh, India. aaminashahab4@gmail.com.

Amar P. Garg: Swami Vivekanand Subharti University, Subhartipuram, NH-58, Delhi – Haridwar Highway, 250005, India. amarprakashgarg@yahoo.com.

Anisa Tariq: Genetic and Molecular Epidemiology Research Group, Department of Bioinformatics and Biosciences, Capital University of Science and Technology, Islamabad, Pakistan. anisatariq5151@gmail.com.

Ashiq Hussain Magrey: Centre for Scientific Research and Development Peoples University Bhopal, India. ashiqmagrey@gmail.com.

Asma Shakeel: Division of Soil Science, FOA, Wadura, SKUAST-Kashmir, India. asmasm1391@skuastkashmir.ac.in.

Ayesha Aftab: Genetic and Molecular Epidemiology Research Group, Department of Bioinformatics and Biosciences, Capital University of Science and Technology, Islamabad, Pakistan. ayesha.aftab@cust.edu.pk.

Ayse Feyza Tufan Dulger: Department of Agricultural Biotechnology, Faculty of Agriculture, Ondokuz Mayis University, Samsun, Turkey. feyzadulger@omu.edu.tr.

Bijender Kumar Bajaj: School of Biotechnology, University of Jammu, Jammu, India. bajajbijenderk@gmail.com.

Burhan Hamid: Centre of Research for Development, University of Kashmir, Srinagar – 190006, J&K, India. peerzada19@gmail.com.

Bilal Ahmed: Department of Agricultural & Biological Engineering, Purdue University, West Lafayette, United States. ahmed249@purdue.edu

Cheng-Di Dong: Department of Marine Environmental Engineering, National Kaohsiung University of Science and Technology, Kaohsiung City, Taiwan. cddong@nkust.edu.tw.

Dilfuza Jabborova: Institute of Genetics and Plant Experimental Biology, Kibray 111208, Uzbekistan. dilfuzajabborova@yahoo.com

Fatima Bibi: Department of Botany, Rawalpindi Women University, 6th Road Satellite Town Rawalpindi, Punjab 46300. fatimazia00@yahoo.com.

Fayaz A. Mohiddin: Section of Plant Pathology, MRCFC, Khudwani, SKUAST-Kashmir, J&K – 192101, India. famohiddin@rediffmail.com

Garima Kaushik: Department of Environmental Sciences, Central University of Rajasthan, NH-8, Bandarsindri, Kishangarh, Ajmer, Rajasthan – 305817. garimakaushik@curaj.ac.in.

Gaurav Yadav: Protein Research Laboratory, Department of Bioengineering, Integral University, Lucknow, Uttar Pradesh, India. yadav.gauravbiotech@gmail.com.

Giuseppe Graziano: Department of Science and Technology, University of Sannio 82100, Benevento, Italy. graziano@unisannio.it

J.P. Mathi Guru: Division of Basic Science and Humanities, SKUAST-Kashmir, India. mathithemicrobiologist@gmail.com

Leila Khalili: Department of Health, Nutrition, and Food Sciences, Florida State University, Tallahassee, FL 32306, USA. lk21i@fsu.edu

Liliana Aguilar Marcelino: Cenid-Salud Animal e Inocuidad, Inifap., Carretera Federal Cuernavaca-Cuautla No. 8534, Col. Progreso, Jiutepec, Morelos, México. liliana@colpos.mx.

Madeha A. Alonazi: Department of Plant and Microbiology, College of Sciences, King Saud University, Riyadh, KSA. malonze@ksu.edu.sa.

Maimona Saeed: Department of Botany, Rawalpindi Women University, 6th Road, Satellite Town, Rawalpindi, 46300, Pakistan. maimonasaeed6@gmail.com.

Malik A. Aziz: Division of Basic Sciences, FOA, Wadura – 193201, SKUAST-Kashmir, India. asifmalik@skuastkashmir.ac.in.

Manuraj P.: Department of Environmental Sciences, Central University of Rajasthan, NH-8, Bandarsindri, Kishangarh, Ajmer, Rajasthan – 305817. manurajmkl@gmail.com

Marika Pellegrini: Department of Life, Health and Environmental Sciences, University of L'Aquila, 67100 L'Aquila, Italy. marika.pellegrini@univaq.it.

Mawish Mahroof: Centre of Research for Development, University of Kashmir, Srinagar – 190006, J&K, India. mawishmahroof@gmail.com

Md. Niamat Ali: Centre of Research for Development, University of Kashmir, Srinagar – 190006, J&K, India. mdniamat@hotmail.com

Mehnaz Shakeel: Division of Plant Pathology, FOA, Wadura, SKUAST-Kashmir, India. mehnasm31@skuastkashmir.ac.in.

Mei-Ling Tsai: Department of Seafood Science, National Kaohsiung University of Science and Technology, Kaohsiung City, Taiwan. mltsai@nkust.edu.tw.

Mina Najafi: Department of Soil Science and Plant Nutrition, Faculty of Agriculture, Ege University, Izmir – Türkiye. minanajafi.agri@gmail.com.

Mohd Sajjad Ahmad Khan: Department of Basic Sciences, Deanship of Preparatory Year and Supporting Studies, Imam Abdulrahman Bin Faisal University, P.O. Box 1982, Dammam 34212, Saudi Arabia. mohdsajjadahmadkhan@gmail.com.

Monika Chaudhary: School of Biotechnology and Life Sciences, Shobhit Institute of Engineering and Technology, Meerut – 250110, India. monika.lm1020@gmail.com.

Muhammad Izhar Shafi: Department of Soil and Environmental Sciences, Faculty of Crop Production Sciences, University of Agriculture, Peshawar Pakistan. mirzaizhar2008@gmail.com.

Nayereh Akbari: Department of Plant Protection, Faculty of Agriculture, Urmia University, Urmia Iran. Nayereh.akbari@gmail.com.

Neesa Majeed: Government Medical College, Baramulla – 193101, J&K, India. loneneesa@gmail.com.

Neha Bharti: Department of Environmental Sciences, Central University of Rajasthan, NH-8, Bandarsindri, Kishangarh, Ajmer, Rajasthan – 305817. nehabharti2500@gmail.com.

Noshin Ilyas: Department of Botany, PMAS Arid Agriculture University, Rawalpindi 46300 Punjab, Pakistan. noshinilyas@yahoo.com.

Nowsheen N.: Division of Fruit Science, FOH, Shalimar, SKUAST-Kashmir, India._nowsheennazir@skuastkashmir.ac.in._

Parushi Nargotra: Department of Seafood Science, National Kaohsiung University of Science and Technology, Kaohsiung City, Taiwan. parushi11nargotra@gmail.com.

Peter Poczai: Finnish Museum of Natural History, University of Helsinki, FL10044 Helsinki, Finland. peter.poczai@helsinki.fi.

Priti Raj: Department of Environmental Sciences, Central University of Rajasthan, NH-8, Bandarsindri, Kishangarh, Ajmer, Rajasthan – 305817. pritiraj231645@gmail.com.

Rihab Djebaili: Department of Life, Health and Environmental Sciences, University of L'Aquila, Via Vetoio, Coppito, 67100 L'Aquila, Italy. rihab.djebaili@.univaq.it.

Roohi: Protein Research Laboratory, Department of Bioengineering, Integral University, Lucknow, Uttar Pradesh, India. roohi0607@gmail.com.

Ruqeya Nazir: Centre of Research for Development, University of Kashmir, Srinagar – 190006, J&K, India. ruqeya@uok.edu.in.

S.A. Mir: Division of Agriculture Statistics, Faculty of Horticulture, Shalimar, SKUAST-Kashmir. Mir_98@msn.com.

Saadat A. Kanth: Bioresource Processing Institute of Australia (BioPRIA), Department of Chemical & Biological Engineering, Monash University, Clayton VIC 3800, Australia. SadaatAltaf@monash.edu.

Sania Tabassum: Genetic and Molecular Epidemiology Research Group, Department of Bioinformatics and Biosciences, Capital University of Science and Technology, Islamabad, Pakistan. sania.tabassum@cust.edu.pk.

Shayesta Islam: Department of Environmental Sciences, Cluster University of Kashmir, India. bhatshayesta@gmail.com.

Shivani D. Raval: Department of Microbiology and Biotechnology Centre, The Maharaja Sayajirao University of Baroda, Vadodara, Gujarat – 390002, India. sd.raval96@gmail.com.

Shrushti S. Patil: Research Sat, Adelaide, Australia. patilshrushti24@gmail.com.

Sohail: Department of Agriculture, Forestry and Bioresources, Forest Ecophysiology Lab. Seoul National University, Seoul, South Korea. sohail.botanist@hotmail.com.

Srishti Sharma: School of Bioengineering and Biosciences, Lovely Professional University (LPU), Jalandhar, Punjab, India and Blue Marble Space Institute of Science, Seattle, Washington, USA. srishtijalaj@gmail.com.

Suchitra Daslaniya: Department of Microbiology and Biotechnology Centre, The Maharaja Sayajirao University of Baroda, Vadodara, Gujarat 390002, India. patelsuchi2434@gmail.com.

Sumera Shabir: Department of Botany, PMAS Arid Agriculture University, Rawalpindi 46300 Punjab, Pakistan. sumerashabir6@gmail.com.

Surati Kumari: National Institute of Biomedical Genomics, Kalyani, West Bengal, India. surati.kr.51b@gmail.com.

Syeda Marriam Bakhtiar: Genetic and Molecular Epidemiology Research Group, Department of Bioinformatics and Biosciences, Capital University of Science and Technology, Islamabad, Pakistan. marriam@cust.edu.pk.

Taliha Ahmed: Genetic and Molecular Epidemiology Research Group, Department of Bioinformatics and Biosciences, Capital University of Science and Technology, Islamabad, Pakistan. taliha.ahmed@cust.edu.pk.

Vishal Sharma: Department of Seafood Science, National Kaohsiung University of Science and Technology, Kaohsiung City, Taiwan; School of Biotechnology, University of Jammu, Jammu, India and Department of Marine Environmental Engineering, National Kaohsiung University of Science and Technology, Kaohsiung City, Taiwan. sudans123@gmail.com.

Yilmaz Kaya: Department of Agricultural Biotechnology, Faculty of Agriculture, Ondokuz Mayis University, Samsun, Turkey. yilmaz.kaya@omu.edu.tr.

Younes Rezaee Danesh: Department of Plant Protection, Faculty of Agriculture, Van Yuzuncu Yil University, Van-Türkiye. y.rdanesh@yahoo.com.

Zaffar Bashir: Centre of Research for Development, University of Kashmir, Srinagar – 190006, J&K, India. zaffarsahib@gmail.com.

Zakir Amin: ICAR-Central Institute of Temperate Horticulture, Srinagar – 191132, India. zakiraminmlk@gmail.com.

Zakira Nourin: Department of Biological Sciences and Chemistry, College of Arts and Science, University of Nizwa, Sultanate of Oman. zakira@unizwa.edu.com.

Contents

CHAPTER

1

Psychrophiles: Diversity, Recent Discoveries, and Future Prospective

Mawish Mahroof[1*], Neesa Majeed[2], Burhan Hamid[1], Zaffar Bashir[1] and Leila Khalili[3]

[1] Centre of Research for Development, University of Kashmir, Srinagar – 190006, India
[2] Government Medical College, Baramulla – 193101, J&K, India
[3] Department of Health, Nutrition, and Food Sciences, Florida State University, Tallahassee, FL 32306, USA

1. Introduction

Extreme environments refer to the locations or conditions, where life on Earth is characterized by extreme temperatures, acidity, pressures, radiation levels, salinity, or scarcity of vital resources like water or oxygen (Merino et al., 2019). While these conditions are typically hostile to most life forms, certain extremophilic organisms uniquely suited to these environments have evolved to not just survive but flourish in such challenging circumstances (Basu et al., 2022). Psychrophiles are the organisms that inhabit extremely cold environments, with temperatures ranging from –20°C to 10°C. These microorganisms adapt to severe conditions by adjusting their genetic makeup which enable them to flourish in freezing temperatures (Tendulkar et al., 2021). Psychrophiles possess a protective layer similar to mucus called exopolymers, enabling them to thrive in harsh temperatures. This substance aids in regulating the internal temperature of psychrophiles, preventing ice formation or freezing (Radchenkova et al., 2024). In certain instances, microbial existence in cold environments is confined to unfrozen water within solid ice and brine pathways. These microorganisms harbor significant amounts of salts and exopolymers, with temperature regulation facilitated by fluid movement (Genilloud, 2018). The ability of psychrophilic

*Corresponding author: mawishmahroof@gmail.com

DOI: 10.1201/9781003717003-1

microorganisms to thrive and multiply in cryo conditions suggests that they have successfully tackled the challenging and harsh conditions of their environments. These challenges include, instability of enzyme systems, diminished flexibility of membranes, and reduced physiological processes such as transcription, translation, cell division, protein activity, and overall cellular growth slowdowns. These microorganisms that have adapted to subzero environments have developed genetic and physical traits which aid their survival in such cryo conditions. These microorganisms have established a permanent presence in cryo conditions ranging from the deep sea and Arctic to the Himalayan mountains and polar regions (Morita, 1975; Tendulkar et al., 2021). Psychrophiles release enzymes known as psychrozymes or cold-adapted enzymes, which have a substantial impact in biotechnology and various commercial applications (Hamid et al., 2012; Debroy & George, 2023). Psychrozymes, like cellulases, lipases, proteases, invertases, and amylases, find application in various processes across the food, biofuel, beverage, and detergent industries (Figure 2) (Hamid et al., 2024; Kuddus et al., 2024; Hamid et al., 2022). These enzymes are identified by their optimal catalytic performance at lower temperatures, although they are also sensitive to heat and quickly lose activity at mild temperatures. But at the same time psychrozymes demonstrate notable biocatalytic performance in low and moderate temperatures in comparison to enzymes that thrive in moderate and high temperatures (Hamid et al., 2022; Gerday et al., 2000; Chattopadhyay, 2006; Sahay et al., 2013). Psychrophiles are very important to the scientific world as well because they contribute in various scientific fields such as in medical research; psychrozymes derived from psychrophiles find application in activities such as protein engineering, DNA manipulation, and the development of pharmaceuticals (Annapure & Pratisha, 2022). Psychrophiles in bioremediation are utilized to decontaminate cold environments polluted with various substances (de Melo Carlos et al., 2023). Psychrozymes derived from psychrophiles are utilized in various industrial sectors like food processing, enhancing the effectiveness of tasks such as refrigeration and cold storage processes (Gupta et al., 2023). In order to flourish in cold environments, psychrophiles need to exhibit a variety of distinctive traits such as the structure of the cell envelope in psychrophiles undergoing alterations, while their primary metabolic activities typically decrease when subjected to cold temperatures (Collins & Margesin, 2019). Moreover, psychrophiles generate proteins and small molecules that serve to prevent freezing and maintaining osmotic balance (Collins & Margesin, 2019). The primary adaptation to cold conditions primarily revolves around the functionality of proteins, acting as essential regulators of metabolism and the cell cycle (Ghosh & Pulicherla, 2021). Enzymes derived from psychrophiles exhibit greater flexibility and a reduction in intramolecular interactions within proteins (Parvizpour et al., 2021). Consequently, the molecular mechanisms behind the cold adaptation of these enzymes, along with their considerable potential for biotechnology, have garnered significant interest from contemporary researchers. However, it is equally important to explore all facets of the physiology, ecology, and molecular adaptation of psychrophiles, as well as their contribution in biogeochemical cycles (Ghosh & Pulicherla, 2021). The exceptional ability of psychrophiles to withstand and thrive in extremely cold environments emphasizes their vital importance across a range of disciplines, spanning from biotechnology to environmental remediation. Their capacity to flourish in subzero conditions, driven by distinct genetic and physiological characteristics, has opened up avenues for groundbreaking applications in both industrial and scientific realms. Researchers are studying them more to figure out exactly how they do it and how they can help us in different areas. It's exciting because this could help us in making big advancements in many fields in the future.

2. Psychrophiles: Habitat and Diversity

The term 'psychrophile' originates from Greek, with 'Psychros' meaning cold and 'Philos' meaning loving, thus meaning 'cold-loving' organisms. These microorganisms thrive in low temperatures that have been documented since ancient times (Farrell & Rose, 1967). Psychrophiles are classified into two categories psychrophilic (cold-loving) and psychrotolerant (cold-adapted), both

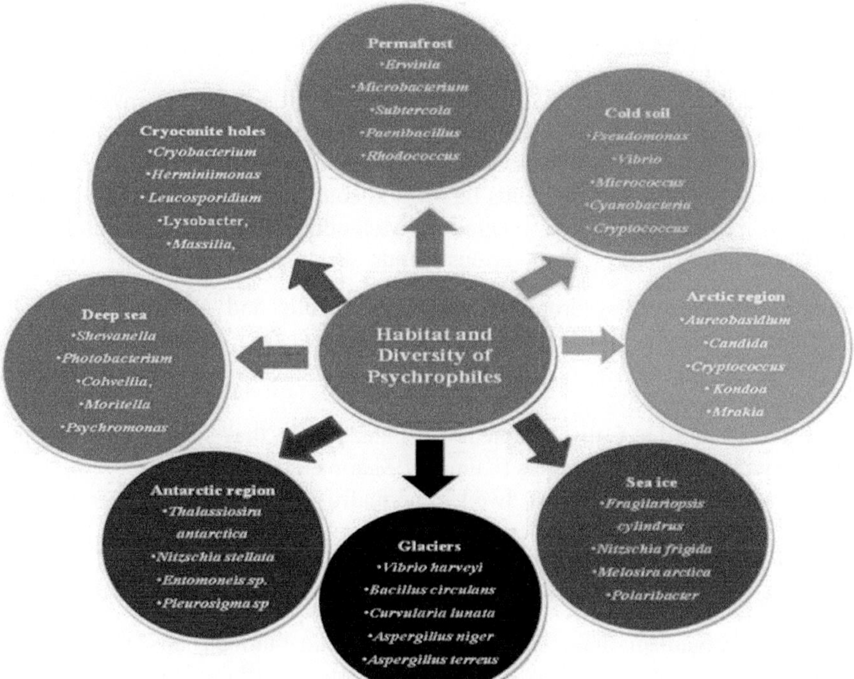

Figure 1: Habitat and diversity of psychrophiles.

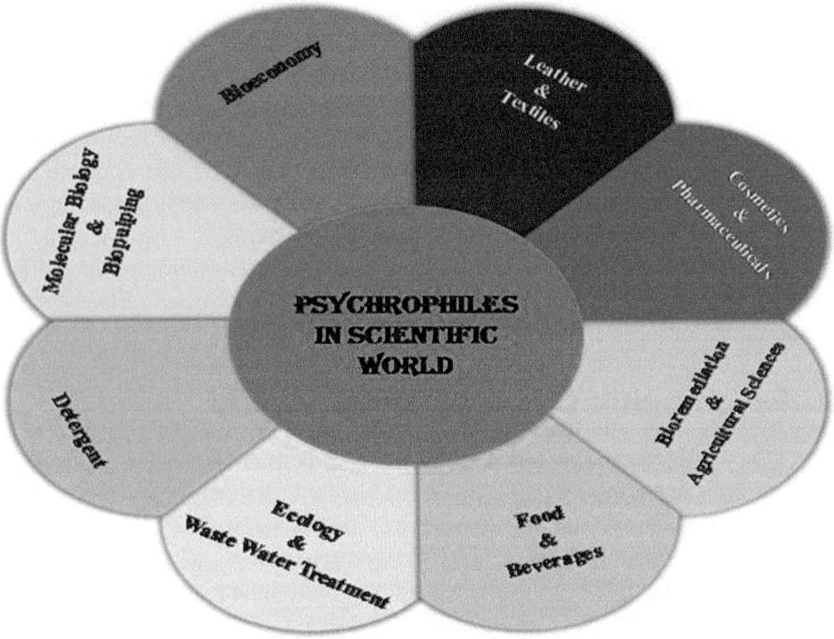

Figure 2: Application of psychrophiles in the scientific world.

categories inhabit various cold environments on the Earth. These environments encompass glaciers, high mountains, shallow subterranean areas, and even refrigeration appliances. Additionally, these microorganisms can be located on the surfaces of animals and plants thriving in cold habitats, where temperatures typically remain below 5°C (Hamid & Mohiddin, 2018). These cold loving microorganisms also reside in various habitat such as:

- **Polar Regions:** Psychrophiles inhabit the Arctic and Antarctic, enduring temperatures plunging below freezing (Anisimov et al., 2007).
- **Deep Ocean:** Psychrophiles thrive in the perpetual cold of deep-sea realms (Taguchi et al., 2022).
- **Glaciers and Ice Caps:** Adapted to icy landscapes, they survive and flourish in environments like glaciers (Kumar & Sharma, 2021).
- **Permafrost:** Psychrophiles dwell in permafrost, characterized by prolonged freezing of soil or rock (Finore et al., 2020).
- **Cryoconite Holes:** Psychrophilic organisms have also evolved to survive in cryoconite holes (Pittino et al., 2023).

These organisms have evolved specialized adaptations, this involves the creation of proteins that act as antifreeze and prevent cellular ice formation and maintain flexible cell membranes that function effectively in low temperatures (Pathania et al., 2022). Figure 1 shows the habitat and diversity of psychrophiles.

3. Psychrophile in Arctic and Antarctic Environment

In polar regions like the Arctic and Antarctic, psychrophiles inhabit diverse environments like ice, snow, permafrost, and even within certain organisms. They fulfill crucial ecological functions in these areas by engaging in nutrient cycling and serving as a vital food source for higher organisms (Dasauni et al., 2022). Psychrobacter, a bacterial genus, is frequently found in cold habitats such as glaciers, snow, and sea ice. Species belonging to Psychrobacter have been identified in the Antarctic region (Heuchert et al., 2006). Polaromonas are bacteria specialized for cold environments commonly found in polar regions. They possess the ability to decompose organic material even in low temperatures, playing a significant role in nutrient recycling within cold ecosystems. Species belonging to Polaromonas have been identified in both Antarctic and Arctic areas (Gawor et al., 2016). Colwellia, another group of bacteria adapted to cold conditions, are frequently observed in marine habitats, particularly in polar oceans. Polaromonas species have been detected in both the Antarctic and Arctic areas (Bowman et al., 1997; Yu et al., 2011). Antarctic yeasts refer to specific species of yeast that have been separated from soils and ice in Antarctica. These yeasts possess adaptations enabling them to endure and propagate in exceedingly frigid environments (Nikitin et al., 2023).

4. Psychrophile in Cryoconite Holes

Cryoconite holes refer to submerged indentations found in areas of glacier melt known as ablation zones, commonly observed in polar and mountainous regions (Gribbon, 1979; Hodson et al., 2008). These holes are extensively studied as key habitats within glacial ecosystems. The holes emerge as a result of decreased ice reflectivity, attributed to the accumulation of dark inorganic and organic matter on the glacier's surface, leading to localized melting (Hodson et al., 2008). Another significant factor for the creation of cryoconite holes are the microbial communities inhabiting these distinctive environments. The collaborative actions of primary producers like bacteria and algal mats within cryoconite holes lead to the accumulation of dark organic material and decrease the reflectivity of ice surfaces (Anesio et al., 2009). Furthermore, cryoconite holes are produced by cyanobacteria (Sheikh et al., 2022; Hodson et al., 2010) participating in the organic carbon cycling processes within glaciers (Anesio et al., 2009).

5. Psychrophile in Deep Sea Environments

The deep sea is recognized as an extreme habitat characterized by intense hydrostatic pressures, primarily cold temperatures ranging from 2 to 4°C, darkness, and limited nutrient availability, although there is typically enough dissolved oxygen present (Nogi, 2017; Alam & Tiwary, 2023). In the deep-sea realm, psychrophiles are commonly known as psychropiezophiles. These microorganisms uniquely adapt to survive in the depths, thriving best under high pressure conditions. While some can tolerate regular atmospheric pressure, those unable to do so are termed obligatory piezophiles. Interestingly, change in temperature affects them more as compared to pressure change. Although over more than half of the Earth's surface (60%) is submerged beneath oceans that reach depths surpassing 1,000 meters, our understanding of microbes inhabiting the deep sea remains largely incomplete. Initially, deep-sea psychropiezophilic bacteria cultures consisted solely of species linked to one of five genera within the Gammaproteobacteria subgroup: *Photobacterium*, *Shewanella*, *Moritella*, *Psychromonas*, and *Colwellia*. However, in recent times, species belonging to Firmicutes and Alphaproteobacteria have also been discovered. The genome analysis of several of these bacteria has unveiled distinctive traits inherent to these microorganisms (Nogi, 2017).

6. Psychrophiles Associated with Plants

Abiotic and biotic stresses, particularly when unexpected, are significant causes for crop loss (Tapia-Vázquez et al., 2020). Interaction with Plant Growth Promoting Microbes (PGPM) have been shown to shield plants from abiotic stresses like high salinity (Sridhar et al., 2025a, b; Jabborova et al., 2025; Lata et al., 2018; Paul et al., 2017) or drought (Vafa et al., 2024; Ferioun et al., 2025; Eswaran et al., 2024). The mechanism by which microorganisms facilitate plant growth promotion are intricate, involving various factors such as humidity, temperature, salt levels, soil pH, nutrient availability, and interactions with other microbes, including potential plant pathogens (Gowtham et al., 2024; Souza et al., 2015). Extreme environments offer promise as a reservoir of PGPM and numerous studies have investigated the interactions between plants and extremophilic microbes (Mawar et al., 2025; Yadav et al., 2015, 2016). Bacteria that have adjusted to cold extreme environments like the extreme soils of the Indian Himalayan region show significant promise in hosting PGPM (Balcazaret al., 2015). A recent study by Rondón et al. (2019) revealed that cold-loving *Pseudomonas* sp, found in Venezuelan tropical glaciers, exhibit traits that promote plant growth. This suggests that psychrophilic microorganisms offer a diverse array of potent biofertilizers, particularly in harsh environments. While the majority of research focuses on bacteria, there are documented instances of plant growth-promoting (PGP) fungi in extreme environments as well (Acuña-Rodríguez et al., 2019; Yadav et al., 2018).

7. Psychrophiles Associated with Animals

Psychrophiles have attracted considerable interest due to their distinctive adaptations and possible utility across diverse domains (Xu et al., 2024). Despite extensive investigations into psychrophiles within the terrestrial and aquatic ecosystem, their association with animals has been relatively neglected in research. Despite being neglected, 431 psychrophiles or psychrotolerant microorganisms were isolated from the operational areas of Irish beef slaughterhouses and spoiled vacuum-packed beef, employing PCR and 16S rRNA sequencing techniques (Moschonas et al., 2011). Foodborne illnesses have been linked to anaerobic clostridia that thrive in cold temperatures and low-oxygen environments (Gupta et al., 2005), additionally, the deterioration of vacuum-sealed chilled meats (Broda et al., 2009). These bacteria have been identified in beef slaughterhouses and surrounding areas, which includes animal waste, hides, gastrointestinal tract, soil, sewage, and water sources (Boerema et al., 2003; Broda et al., 2002; Broda et al., 2009). Nevertheless, the occurrence of other pathogenic or spoilage-causing anaerobic bacteria that thrive in cold conditions in beef slaughterhouses is yet to be determined (Moschonas et al., 2011).

8. Psychrophiles in Artificial Environments

In context to artificial environments, psychrophiles have garnered attention for their possible utility across diverse sectors. Their applications range from environmental bioremediation and preservation of food to serving as cell factories for manufacturing different enzymes. Additionally, they have been explored for their role in mitigating oil spills in marine ecosystems (Sharma et al., 2022). Cold-adapted fungi, both psychrophilic and psychrotrophic, play a crucial role in biotechnology and pharmaceutical sectors due to their varied traits developed through adaptation to harsh conditions. These traits include the synthesis of enzymes that function effectively in cold temperatures, production of pharmaceutical or bioactive compounds, as well as exo-polysaccharides. Additionally, they hold promise for applications in bioremediation and can serve as biofertilizers (Hassan et al., 2016).

9. Recent Discoveries on Psychrophiles

Psychrophiles, particularly those in the ocean, dominate a significant portion of the ecosystem, comprising roughly 70% of the oceans below 1,000 meters at temperatures around 4–5 °C. Additionally, they thrive in extreme environments like the Arctic permafrost at –10 °C, porous dry rocks in valleys of Antarctica at –60°C, and the exteriors of glaciers located within cryoconite pits (Son et al., 2023). Numerous recently discovered varieties of psychrotolerant bacteria have been extracted from temperate and cold regions (Baba et al., 2021) and soil contaminated with oil in Antarctica and extensively analyzed (Parvizpour et al., 2021). The advancement of Next-Generation Sequencing (NGS) technologies has led to a surge in new datasets of psychrophile genome sequences. This has significantly improved our comprehension of these intriguing organisms and their survival strategies (Bowman, 2017).

10. Reports on Psychrophilic Bacteria

Recent research has shed light on the physical and functional adjustments of cold-loving bacteria when faced with temperature challenges. For example, the psychrophilic bacterium *Pseudarthrobacter psychrotolerans* YJ56, found in Antarctic soil, displayed robust growth at 13°C but failed to thrive at 30°C. Interestingly, when exposed to 25°C, strain YJ56 transformed its morphology into a filamentous bacterium with multiple branches (Son et al., 2023). *Cryobacterium* strains, comprising 14 variants, were extracted from three Tibetan ice cores and compared to their mesophilic counterparts within the Microbacteriaceae family of Actinobacteria. The findings indicate that the cold-adapted *Cryobacterium* strains experienced more pronounced alterations in their genetic makeup compared to their counterparts thriving in moderate temperatures. Notably, the genomes of these psychrophilic *Cryobacterium* strains contain a notably greater abundance of genes associated with stress response, mobility, and chemotaxis compared to their mesophilic counterparts, with statistical significance (Liu et al., 2020). *Pseudomonas sivasensis* W-6, a bacterium thriving in cold conditions, was discovered in the Napahai plateau wetland. Its cold adaptation exhibits psychrophilic traits instead of psychrotrophic traits. This cold-adapted bacterium, *Pseudomonas sivasensis* W-6, has the capability to utilize glycogen and trehalose as energy sources, facilitated by carbohydrate-active enzymes, allowing it to survive in low-temperature environments. Furthermore, the mechanisms of cold adaptation in *Pseudomonas sivasensis* W-6 include adjustments in membrane fluidity through alterations in the profile of unsaturated fatty acids, the involvement of two-component regulatory systems, anti-sense transcription, and the function of rpsU genes in the translation process, among others (Xiong et al., 2023). Table 1 contains various recently discovered bacteria along with their respective applications.

11. Reports on Psychrophilic Fungi

Recent findings suggest that fungal communities in glacial ice from the Arctic and Patagonia, as well as soil from Antarctica, are primarily composed of Basidiomycota and Ascomycota taxa. Additionally, there has been a notable increase in the discovery of secondary metabolites originating from cold-adapted microbes, especially psychrophilic fungi (Hassan et al., 2016). Five psychrotrophic cellulose-degrading fungi were isolated, out of which one isolate of cellulose-degrading fungi adapted to cold environments was identified as a new fungal species named *Aureobasidium paleasum* sp. nov., displaying significant potential for breaking down straw. Through enzyme activity assays and FITR spectroscopy, it was observed that this psychrotrophic fungus exhibited robust cellulolytic capabilities at lower temperatures, with a broad thermal tolerance range spanning from 5°C to 50°C (with an optimal activity observed at 10°C). *A. paleasum* effectively decomposed rice straws and cellulose at 10°C,

Table 1: Diverse array of recently discovered psychrophilic microorganisms, accompanied by their respective applications

Microorganism	Microorganism Type	Discovered From	Applications	References
Psychrobacter sp. NJ228	Bacteria	Antarctica	It is used as a source of cold adaptive enzyme used in biotechnology	Zhang et al., 2023
Subtercola endophyticus sp. AK-R2A1-2^T	Bacteria	Abies Koreana	It greatly enhances the growth of rice seedlings in cold conditions	Jiang et al., 2022
Exiguobacterium sibiricum K1	Bacteria	Agro-residual waste	Enhance agricultural productivity in an environmentally sustainable manner within cold environments	Kumari et al., 2023; Kumari et al., 2024
Myxobacterial Strain	Bacteria	Mirgund Wetland from the North-Western Himalayas	It is capable of generating secondary compounds that exhibit broad spectrum antibiotics or antifungal properties	Dhanjal et al. 2023; Bhat et al., 2021
Arthrobacter oryzae BIM B-1663	Bacteria	Antarctic green snow	It efficiently uses both organic and inorganic substrates as metabolic substrates, thereby serving as a valuable asset for agricultural bioremediation efforts	Smirnova et al., 2023; Roy and Kumar, 2020
Pseudomonas spp.	Bacteria	Cold regions of Leh Ladakh India	*Pseudomonas* spp. has valuable applications in the bioremediation, biotechnology, plant growth promotion, and biological control	Kasana and Singh et al., 2024; Mehmood et al., 2023
Pseudomonas S2TR-14	Bacteria	Contaminated cold-climate site	It is capable of producing cold active enzyme in molecular biology	Miri et al., 2021

(*Contd.*)

(*Contd.*)

Microorganism	Microorganism Type	Discovered From	Applications	References
Bacillus pumilus	Bacteria	Himalayan glacier soil, India	It serves as a key component in fungicides for agricultural purposes	Farooq et al., 2021
Rahnella inusitata	Bacteria	Antarctica	It produces β-Galactosidase which helps in eliminating lactose from dairy items to cater to those with lactose intolerance	Núñez-Montero et al., 2021; Husain et al., 2010
Trichoderma gamsii R1	Fungi	Shrimp shell waste	It is used as biocontrol agent	Wang et al., 2023; Risoli et al., 2023; Afrouz et al., 2023
Penicillium chrysogenum	Fungi	Antarctica	It is used as a potent antifungal agent	Glodowsky et al., 2020; Pacios-Michelena et al., 2024
Glaciozyma antarctica	Fungi	Soya coast, East Antarctica	It releases various cold active enzymes which are potentially used in biotechnology	Yusof et al., 2021
Cladosporium parasphaerospermum	Fungi	Sohag city, Egypt	Producing bioactive molecules and their capacity to stimulate plant growth in action	Moharram et al., 2022; Gautam et al., 2019
Rhodotorula mucilaginosa BPT1	Fungi	Baramullah (J&K) India	It is identified as an ideal candidate for biorefining carotenoids, lipids, enzymes, and other functional bioproducts, leveraging low-cost agricultural waste substrates	Parmar et al., 2023; Li et al., 2022

Halorubrum lacusprofundi	Archaea	Deep Lake, Antarctica	It is used in several biotechnological processes	Gebhard et al., 2023
Monoraphidium sp. (strain ZBD-06)	Algae	Almaty Lake	Antibacterial activity against various pathogens like *Bacillus subtilis, Enterococcus faecalis, Staphylococcus aureus, Staphylococcus epidermidis, Escherichia coli* , etc.	Balouch et al., 2020
Chlorella vulgar KNUA0 07	Algae	King George island, Antarctica	It stands out as one of the most frequently utilized alga in aquaculture	Jo et al., 2020
Chlamydomonas sp. KNM0029	Algae	Arctic sea ice	It serves as a model organism in the field of molecular biology	Kim et al., 2020

outperforming the common cellulose-degrading fungus *Penicillium oxalicum*. Furthermore, compared to *P. oxalicum*, *A. paleasum* accelerated the thermophilic stage, leading to improved compost maturity and overall compost quality (Jiang et al., 2021). Table 1 displays a variety of recently identified fungi along with their various applications.

12. Reports on Psychrophilic Archaea

Cold-loving archaea are plentiful and play essential roles across the vast cold regions of the Earth's biosphere (Allen et al., 2009). The majority of isolated cold-adapted archaea are methanogens, while others consist of a halophile, a euryarchaeon found in association with bacteria, and acrenarchaeon residing as a symbiont in marine sponges (Cavicchioli, 2006). Two psychrophilic methanotrophic species *Methylobacter psychrophilus* Z-0021 T (Omel'Chenko et al., 1996) and *Methylosphaera hansonii* (Bowman et al., 1997), both belonging to the Methylomonadaceae family. These species were identified and separated from soil in the Russian Arctic tundra and from surface sediments of an Antarctic lake with distinct layers, respectively. The complete genome sequencing of a cold-adapted methane-consuming bacterium belonging to the Methylobacter genus, which is obligated to thrive in low temperatures, was conducted. This organism was obtained from a boreal lake situated in Finland (Khanongnuch et al., 2022, 2023). Table 1 shows recently discovered archaea/algae along with their diverse applications.

13. Future Prospective of Psychrophiles

Psychrophiles are known for being a potential source of enzymes that can remain highly active even when temperatures are low. They achieve this by minimizing the reaction's reliance on temperature (Yayanos, 2009; Cavicchioli et al., 2011; Irwin, 2020; Rai & Kumar, 2015). Psychrozymes have been found to facilitate biotechnological processes that occur at low temperatures (Goodchild et al., 2004). Genetic modification provides benefits by altering enzyme properties to boost their performance, as demonstrated in the cold-tolerant Lipase B from *Candida antarctica*. A mutant version of this enzyme showed increased activity and stability (Zhang et al., 2003). Researchers have isolated and examined a heat-sensitive alkaline phosphatase sourced from a bacterium found in Antarctica (Kobori et al., 1984). In molecular biology and recombinant technology, these enzymes are utilized to dephosphorylate DNA vectors before cloning to prevent re-circularization. They are also employed to remove phosphate groups from the 5′-termini of nucleic acids prior to 5′-end labeling by polynucleotide kinase, as well as to eliminate dNTPs and pyrophosphate from PCR reactions. A heat-sensitive uracil-DNA N-glycosylase from Atlantic cod (*Gadus morhua),* known for its cold adaptation properties (Leiros et al., 2003), is also used to eliminate DNA contaminants in sequential PCR reactions. Psychrozymes have captured the interest of researchers across various fields, including pharmaceuticals, for their potential in improving product quality through enzyme-based enhancements. This has led to the development of numerous anti-cancer, anti-tumor, and anti-fungal agents derived from bacterial, *Streptomyces*, and fungal strains (Banerjee et al., 2016). The inherent heat sensitivity of the majority of enzymes adapted to cold environments allows them to be deactivated under moderate heat, rendering them resistant to inactivation by chemical agents. This characteristic feature of psychrophilic enzymes makes them promising candidates for the development of live vaccines. Additionally, through the engineering of mesophilic pathogens, thermolabile counterparts of crucial enzymes are produced, rendering them sensitive to temperature. These modified strains are designed to lose their pathogenicity at mammalian body temperatures while preserving their entire antigenic profile (Pinto & Nano, 2015; Parvizpour et al., 2021). Psychrophiles, organisms adapted to cold environments, offer numerous applications in present and future research endeavors. Their specialized enzymes, referred to as psychrozymes, demonstrate heightened effectiveness at low temperatures, rendering them indispensable in biotechnological applications like food preservation, bioremediation, and pharmaceutical manufacturing. Additionally, the study of psychrophiles yields valuable insights into their adaptation mechanisms to extreme conditions, thereby contributing to fields such as climate change research and astrobiology. Moreover, their involvement in bioprospecting presents opportunities for the discovery of novel enzymes and bioactive compounds with potential applications in industry and medicine. In summary, psychrophiles stand as a promising avenue for tackling a wide array of challenges across diverse disciplines, ranging from biotechnology to environmental science and beyond.

14. Conclusion

Psychrophiles, microorganisms that thrive in extremely cold environments, exhibit notable diversity and distinctive adaptations that enable them to survive in freezing conditions. This research explores their habitat diversity in Antarctic, deep-sea, artificial, and icy ecosystems, emphasizing their ecological roles. Psychrophiles play a significant role in maintaining ecological balance in polar regions and deep marine environments, and they also show resilience in artificial environments such as cryogenic laboratories. Recent advancements in research have provided insights into their genetic and metabolic adaptations, opening up possibilities for their use in biotechnological applications like medicine, bioremediation, and food preservation. Understanding their role in mitigating climate change and enhancing ecosystem resilience is crucial. Overall, psychrophiles offer promising opportunities in scientific and industrial fields, contributing to our comprehension of extreme cold environments and offering potential solutions to diverse challenges.

References

Acuña-Rodríguez, I.S., Hansen, H., Gallardo-Cerda, J., Atala, C., & Molina-Montenegro, M.A. 2019. Antarctic extremophiles: Biotechnological alternative to crop productivity in saline soils. *Frontiers in Bioengineering and Biotechnology*, 7, 22.

Afrouz, M., Sayyed, R.Z., Fazeli-Nasab, B., Piri, R., Almalki, W.H., & Fitriatin, B.N. 2023. Seed bio-priming with beneficial *Trichoderma harzianum* alleviates cold stress in maize. *PeerJ.*, 11, e15644. http://doi.org/10.7717/peerj.15644.

Alam, M., & Tiwary, B.K. (Eds.). 2023. *Extremophiles: Diversity, Adaptation, and Applications*. Bentham Science Publishers.

Allen, M.A., Lauro, F.M., Williams, T.J., Burg, D., Siddiqui, K.S., De Francisci, D. et al. 2009. The genome sequence of the psychrophilic archaeon, *Methanococcoides burtonii:* The role of genome evolution in cold adaptation. *The ISME Journal*, 3(9), 1012–1035.

Anesio, A.M., Hodson, A.J., Fritz, A., Psenner, R., & Sattler, B. 2009. High microbial activity on glaciers: Importance to the global carbon cycle. *Global Change Biology*, 15(4), 955–960.

Anisimov, O.A., Vaughan, D.G., Callaghan, T.V., Furgal, C., Marchant, H., Prowse, T.D., & Walsh, J.E. 2007. Polar regions (arctic and antarctic). *Climate Change*, 15, 653–685.

Annapure, U.S., &Pratisha, N. 2022. Psychrozymes: A novel and promising resource for industrial applications. In: *Microbial Extremozymes*, pp. 185–195. Academic Press.

Baba, Z.A., Hamid, B., Sheikh, T.A., Alotaibi, S., Enshasy, H.E., Ansari, M.J. et al. 2021. Psychrotolerant *Mesorhizobium* sp. isolated from temperate and cold desert regions solubilize potassium and produces multiple plant growth promoting metabolites, *Molecules*, 26, 5758. https://www.mdpi.com/1420-3049/26/19/5758/htm.

Balcazar, W., Rondón, J., Rengifo, M., Ball, M.M., Melfo, A., Gómez, W. et al. 2015. Bioprospecting glacial ice for plant growth promoting bacteria. *Microbiological Research*, 177, 1–7.

Balouch, H., Demirbag, Z., Zayadan, B.K., Sadvakasaova, A.K., Bolatkhan, K., Gencer, D. et al. 2020. Isolation, identification, and antimicrobial activity of psychrophilic freshwater microalgae Monoraphidium sp. from Almaty region. *International Journal of Biology and Chemistry*, 13(1), 14–23.

Banerjee, R., Halder, A., & Natta, A. 2016. Psychrophilic microorganisms: Habitats and exploitation potentials. *Eur. J. Biotechnol. Biosci.*, 4(3), 16–24.

Basu, B., Roy, C., Saha, S., Tirkey, A.S., Biswas, R., Dey, A. , & Mitra, A.K. 2022. Extremophiles to polyextremophiles: Survival of the fittest. *Extremophiles: A Paradox of Nature with Biotechnological Implications*, 1, 221.

Bhat, M.A., Mishra, A.K., Bhat, M.A., Banday, M.I., Bashir, O., Rather, I.A., & Jan, A.T. 2021. Myxobacteria as a source of new bioactive compounds: A perspective study. *Pharmaceutics*, 13(8), 1265.

Boerema, J.A., Broda, D.M., & Bell, R.G. 2003. Abattoir sources of psychrophilic clostridia causing blown pack spoilage of vacuum-packed chilled meats determined by culture-based and molecular detection procedures. *Letters in Applied Microbiology*, 36(6), 406–411.

Bowman, J.P., McCammon, S.A., & Skerrat, J.H. 1997. *Methylosphaera hansonii* gen. nov., sp. nov., a psychrophilic group I methanotroph from Antarctic marine-salinity, meromictic lakes. *Microbiology*, 143(4), 1451–1459.

Bowman, J.P., Gosink, J.J., McCammon, S.A., Lewis, T.E., Nichols, D.S., Nichols, P.D., & McMeekin, T.A. 1998. *Colwellia demingiae* sp. nov., *Colwellia hornerae* sp. nov.,

Colwellia rossensis sp. nov. and *Colwellia psychrotropica* sp. nov.: Psychrophilic Antarctic species with the ability to synthesize docosahexaenoic acid (22: 6 ω3). *International Journal of Systematic and Evolutionary Microbiology*, 48(4), 1171–1180.

Bowman, J.P. 2017. Genomics of psychrophilic bacteria and archaea. *Psychrophiles: From Biodiversity to Biotechnology*, 345–387.

Broda, D.M., Bell, R.G., Boerema, J.A., & Musgrave, D.R. 2002. The abattoir source of culturable psychrophilic Clostridium spp. causing 'blown pack' spoilage of vacuum-packed chilled venison. *Journal of Applied Microbiology*, 93(5), 817–824.

Broda, D.M., Boerema, J.A., & Brightwell, G. 2009. Sources of psychrophilic and psychrotolerant clostridia causing spoilage of vacuum-packed chilled meats, as determined by PCR amplification procedure. *Journal of Applied Microbiology*, 107(1), 178–186.

Cavicchioli, R. 2006. Cold-adapted archaea. *Nature Reviews Microbiology*, 4(5), 331–343.

Cavicchioli, R., Charlton, T., Ertan, H., Omar, S.M., Siddiqui, K.S., & Williams, T. 2011. Biotechnological uses of enzymes from psychrophiles. *Microbial Biotechnology*, 4(4), 449–460.

Chattopadhyay, M.K. 2006. Mechanism of bacterial adaptation to low temperature. *Journal of Biosciences*, 31, 157–165.

Collins, T., & Margesin, R. 2019. Psychrophilic lifestyles: Mechanisms of adaptation and biotechnological tools. *Applied Microbiology and Biotechnology*, 103, 2857–2871.

Dasauni, K., Divya, & Nailwal, T.K. 2022. Cyanobacteria in cold ecosystem: Tolerance and adaptation. *Survival Strategies in Cold-adapted Microorganisms*, pp. 1–29. Springer, Singapore.

de Melo Carlos, L., Camacho, K.F., Duarte, A.W., de Oliveira, V.M., Boroski, M., Rosa, L.H., & Passarini, M.R. 2023. Bioprospecting the potential of the microbial community associated to Antarctic marine sediments for hydrocarbon bioremediation. *Brazilian Journal of Microbiology*, 1–15.

Debroy, A., & George, N. 2023, February. Psychrophiles and psychrozymes: Structural adaptation and applications. In: *AIP Conference Proceedings* (Vol. 2558, No. 1). AIP Publishing.

Dhanjal, D.S., Singh, S., Kumar, V., Ramamurthy, P.C., Chopra, C., Wani, A.K., ... & Singh, J. 2023. Isolation and characterization of cellulase-producing myxobacterial strain from the unique niche of Mirgund Wetland from the North-Western Himalayas. *Journal of Applied Biology and Biotechnology*, 11(5), 119–125.

Environmental Protection Agency. 2003. *Bacillus pumilus* strain GB 34 (006493) Fact Sheet.

Eswaran, S.U.D., Sundaram, L., Perveen, K., Bukhari, N.A & Sayyed R.Z. 2024. Osmolyte-producing microbial biostimulants regulate the growth of *Arachis hypogaea* L. under drought stress. *BMC Microbiology*, 4, 165. https://doi.org/10.1186/s12866-024-03320-6.

Falciatore, A., Merendino, L., Barneche, F., Ceol, M., Meskauskiene, R., Apel, K., & Rochaix, J.D. 2005. The FLP proteins act as regulators of chlorophyll synthesis in response to light and plastid signals in Chlamydomonas. *Genes and Development*, 19(1), 176–187.

Farooq Saleem, Ruqeya Nazir, Shabir Ahmad Ganai, & Bashir Ahmad Ganai. 2021. Isolation and characterization of a new cold-active protease from psychrotrophic bacteria of Western Himalayan glacial soil. *Scientific Reports*, 11(1), 12768.

Farrell, J., & Rose, A. 1967. Temperature effects on microorganisms. *Annual Reviews in Microbiology*, 21(1), 101–120.

Ferioun, M., Zouitane, I., Bouhraoua, S., Elouattassi, Y., Belahcen, D., Errabbani, A. et al. 2025. Applying microbial biostimulants and drought-tolerant genotypes to enhance barley growth and yield under drought stress. *Frontiers in Plant Science*s, 15, 1494987. https://doi.org/10.3389/fpls.2024.1494987._

Finore, I., Vigneron, A., Vincent, W.F., Leone, L., Di Donato, P., Schiano Moriello, A., ... & Poli, A. 2020. Novel psychrophiles and exopolymers from permafrost thaw lake sediments. *Microorganisms*, 8(9), 1282.

Gautam, A.K., & Avasthi, S. 2019. Fungal endophytes: Potential biocontrol agents in agriculture. In: *Role of Plant Growth Promoting Microorganisms in Sustainable Agriculture and Nanotechnology*, pp. 241–283. Woodhead Publishing.

Gawor, J., Grzesiak, J., Sasin-Kurowska, J., Borsuk, P., Gromadka, R., Górniak, D., ... & Zdanowski, M.K. 2016. Evidence of adaptation, niche separation, and microevolution within the genus Polaromonas on Arctic and Antarctic glacial surfaces. *Extremophiles*, 20, 403–413.

Gebhard, L.J., Duggin, I.G., & Erdmann, S. 2023. Improving the genetic system for *Halorubrum lacusprofundi* to allow in-frame deletions. *Frontiers in Microbiology*, 14, 1095621.

Genilloud, O. 2018. Mining actinomycetes for novel antibiotics in the omics era: Wre we ready to exploit this new paradigm. *Antibiotics*, 7(4), 85.

Gerday, C., Aittaleb, M., Bentahir, M., Chessa, J.P., Claverie, P., Collins, T., ... & Feller, G. 2000. Cold-adapted enzymes: From fundamentals to biotechnology. *Trends in Biotechnology*, 18(3), 103–107.

Ghosh, M., & Pulicherla, K.K. 2021. Psychrophiles as the source for potential industrial psychrozymes. *Recent Developments in Microbial Technologies*, 355–366.

Godowsky, A.P., Ruberto, L.A., Martorell, M.M., Mac Cormack, W.P., & Levin, G.J. 2020. Cold active transglutaminase from antarctic Penicillium chrysogenum: Partial purification, characterization, and potential application in food technology. *Biocatalysis and Agricultural Biotechnology*, 29, 101807.

Goodchild, A., Saunders, N.F., Ertan, H., Raftery, M., Guilhaus, M., Curmi, P.M., & Cavicchioli, R. 2004. A proteomic determination of cold adaptation in the Antarctic archaeon, Methanococcoides burtonii. *Molecular Microbiology*, 53(1), 309–321.

Gowtham, H.G., Murali, M., Shilpa, N., Amruthesh, K.N., Gafur, A., Antonius, S., & Sayyed, R.Z. 2024. Harnessing Abiotic Elicitors to Bolster Plant's Resistance against Bacterial Pathogens. *Plant Stress*, 11, 100371. https://doi.org/10.1016/j.stress.2024.100371.

Gribbon, P.W.F. 1979. Cryoconite holes on Sermikavsak, west Greenland. *Journal of Glaciology*, 22(86), 177–181.

Gupta, A., Sumner, C.J., Castor, M., Maslanka, S., & Sobel, J. 2005. Adult botulism type F in the United States, 1981–2002. *Neurology*, 65(11), 1694–1700.

Gupta, V., Bhaskar, P., Thoudam, J., Bisht, S., Sharma, A., & Tripathi, R. 2023. Psychrophiles as a novel and promising source of cold-adapted industrial enzymes. *The Applied Biology and Chemistry Journal* (*TABCJ*), 4(2), 54–68.

Hamid, B., Singh, P., Rana, R.S., & Sahay, S. 2012. Isolation and identification of psychrophilic yeast from the soil of northern region of India. Biotechnology, 18, 103–107.

Hamid, B., & Mohiddin, F.A 2018. Cold-active enzymes in food processing. In: *Enzymes in Food Technology: Improvements and Innovations*, pp. 383–400. Springer Nature, Singapore Pte. Ltd.

Hamid, B., Bashir, Z., Yatoo, A.M., Mohiddin, F., Majeed, N., Bansal, M., & Alfaifi, M.Y. 2022. Cold-active enzymes and their potential industrial applications: A review. *Molecules*, 27(18), 5885. https://doi.org/10.3390/molecules27185885.

Hamid, B., Baba, Z.A., Sheikh, T.A., Perveen, K., & Sayyed, R.Z. 2024. Investigating potential protease activity of psychrotrophic bacteria from a municipal landfill for solid waste management. *Biomass Conversion and Biorefinery*. https://doi.org/10.1007/s13399-024-05621-2.

Hassan, N., Rafiq, M., Hayat, M., Shah, A.A., & Hasan, F. 2016. Psychrophilic and psychrotrophic fungi: A comprehensive review. *Reviews in Environmental Science and Bio/Technology*, 15, 147–172.

Heuchert, A., Glöckner, F.O., Amann, R., & Fischer, U. 2004. Psychrobacter nivimaris sp. nov., a heterotrophic bacterium attached to organic particles isolated from the South Atlantic (Antarctica). *Systematic and Applied Microbiology*, 27(4), 399–406.

Hodson, A., Anesio, A.M., Tranter, M., Fountain, A., Osborn, M., Priscu, J., ... & Sattler, B. 2008. Glacial ecosystems. *Ecological Monographs*, 78(1), 41–67.

Hodson, A., Cameron, K., Bøggild, C., Irvine-Fynn, T., Langford, H., Pearce, D., & Banwart, S. 2010. The structure, biological activity, and biogeochemistry of cryoconite aggregates upon an Arctic valley glacier: Long year breen, Svalbard. *Journal of Glaciology*, 56(196), 349–362.

Husain, Q. 2010. β Galactosidases and their potential applications: A review. *Critical Reviews in Biotechnology*, 30(1), 41–62.

Irwin, J.A. 2020. Physiological and biotechnological aspects of extremophiles. *Overview of Extremophiles and Their Food and Medical Applications*, pp. 65–87. Academic, New York.

Jabborova, D., Jabbarov, Z., Abdrakhmanov, T., Fayzullaev, O., Saharan, B.S., Perveen, K et al. 2025. Assessing the synergistic effects of biochar, hydrogel, and biofertilizer on growth and physiological traits of wheat in saline environments. *Functional Plant Biology*, 52, FP24277. https://doi.org/10.1071/FP24277.

Jiang, G., Chen, P., Bao, Y., Wang, X., Yang, T., Mei, X., ... & Shen, Q. 2021. Isolation of a novel psychrotrophic fungus for efficient low-temperature composting. *Bioresource Technology*, 331, 125049.

Jiang, L., Peng, Y., Seo, J., Jeon, D., Jo, M.G., Lee, J.H., ... & Lee, J. 2022. Subtercola endophyticus sp. nov., a cold-adapted bacterium isolated from Abies koreana. *Scientific Reports*, 12(1), 12114.

Jo, S.W., Do, J.M., Kang, N.S., Park, J.M., Lee, J.H., Kim, H.S., ... & Yoon, H.S. 2020. Isolation, identification, and biochemical characteristics of a cold-tolerant Chlorella vulgaris KNUA007 isolated from King George Island, Antarctica. *Journal of Marine Science and Engineering*, 8(11), 935.

Kasana, R.C., & Singh, S.K. 2024. Isolation and identification of a siderophore producing and antagonistic cold tolerant (psychrotrophic) Pseudomonas sp. LHSP1: Identification of cold tolerant pseudomonas sp. *Annals of Arid Zone*, 63(1), 23–31.

Khanongnuch, R., Mangayil, R., Svenning, M.M., & Rissanen, A.J. 2022. Characterization and genome analysis of a psychrophilic methanotroph representing a ubiquitous Methylobacter spp. cluster in boreal lake ecosystems. *ISME Communications*, 2(1), 85.

Khanongnuch, R., Mangayil, R., & Rissanen, A.J. 2023. Conversion of methane to organic acids is a widely found trait among gamma proteobacterial methanotrophs of freshwater lake and pond ecosystems. *Microbiology Spectrum*, 11(6), e01742–23.

Kim, E.J., Kim, S., Choi, H.G., & Han, S.J. 2020. Co-production of biodiesel and bioethanol using psychrophilic microalga Chlamydomonas sp. KNM0029C isolated from Arctic sea ice. *Biotechnology for Biofuels*, 13, 1–13.

Kobori, H., Sullivan, C.W., & Shizuya, H. 1984. Heat-labile alkaline phosphatase from Antarctic bacteria: Rapid 5′ end-labeling of nucleic acids. *Proceedings of the National Academy of Sciences*, 81(21), 6691–6695.

Kuddus, M., Roohi, Bano, N., Sheik, G.B., Joseph, B., Hamid, B., ... & Madhavan, A. 2024. Cold-active microbial enzymes and their biotechnological applications. *Microbial Biotechnology*, 17(4), e14467.

Kumar, R., & Sharma, R.C. 2021. Psychrophilic microbial diversity and physicochemical characteristics of glaciers in the Garhwal Himalaya, India. *Journal of Microbiology, Biotechnology, and Food Sciences*, 10(5), e2096–e2096.

Kumari, S., Kumar, A., & Kumar, R. 2023. A cold-active cellulase produced from Exiguobacterium sibiricum K1 for the valorization of agro-residual resources. *Biomass Conversion and Biorefinery*, 13(16), 14777–14787.

Kumari, S., Kumar, A., Lepcha, A., & Kumar, R. 2024. Cold-adapted Exiguobacterium sibiricum K1 as a potential bioinoculant in cold regions: Physiological and genomic elucidation of biocontrol and plant growth promotion. *Gene*, 148439.

Lata, R., Chowdhury, S., Gond, S.K., & White Jr, J.F. 2018. Induction of abiotic stress tolerance in plants by endophytic microbes. *Letters in Applied Microbiology*, 66(4), 268–276.

Leiros, I., Moe, E., Lanes, O., Smalås, A.O., & Willassen, N.P. 2003. The structure of uracil-DNA glycosylase from Atlantic cod (*Gadus morhua*) reveals cold-adaptation features. *Acta Crystallographica Section D: Biological Crystallography*, 59(8), 1357–1365.

Li, Z., Li, C., Cheng, P., & Yu, G. 2022. Rhodotorula mucilaginosa: Alternative sources of natural carotenoids, lipids, and enzymes for industrial use. *Heliyon*, 8(11).

Liu, Y., Shen, L., Zeng, Y., Xing, T., Xu, B., & Wang, N. 2020. Genomic insights of *Cryobacterium* isolated from ice core reveal genome dynamics for adaptation in glacier. *Frontiers in Microbiology*, 11, 1530.

Mawar, R., Ranawat, M.·, Ram, L., & Sayyed, R.Z. 2023. Harnessing drought tolerant PGPM in arid agro ecosystem for plant disease management and soil amelioration. In: Ritu Mawar, R.Z. Sayyed, S.K. Sharma, & K.S. Sattiraju (Eds.), *Plant Growth Promoting Microorganisms of Arid Region: Status and Prospects*, pp. 27–43. Springer, Singapore. https://doi.org/10.1007/978-981-19-4124-5_2.

Mehmood, N., Saeed, M., Zafarullah, S., Hyder, S., Rizvi, Z.F., Gondal, A.S., & Kupe, M. 2023. Multifaceted impacts of plant-beneficial pseudomonas spp. in managing various plant diseases and crop yield improvement. *ACS Omega*, 8(25), 22296–22315.

Merino, N., Aronson, H.S., Bojanova, D.P., Feyhl-Buska, J., Wong, M.L., & Giovannelli, D. 2019. Living at the extremes: Extremophiles and the limits of life in a planetary context. *Frontiers in Microbiology*, 10, 447668.

Miri, S., Perez, J.A.E., Brar, S.K., Rouissi, T., & Martel, R. 2021. Sustainable production and co-immobilization of cold-active enzymes from Pseudomonas sp. for BTEX biodegradation. *Environmental Pollution*, 285, 117678.

Moharram, A.M., Zohri, A.N.A., Hesham, A.E.L., Abdel-Raheam, H.E., Al-Ameen Maher, M., and Al-Bedak, O.A.H. 2022. Production of cold-active pectinases by three novel Cladosporium species isolated from Egypt and application of the most active enzyme. *Scientific Reports*, 12(1), 15599.

Morita, R.Y. 1975. Psychrophilic bacteria. *Bacteriol Rev.*, 39, 144–167.

Moschonas, G., Bolton, D.J., McDowell, D.A., & Sheridan, J.J. 2011. Diversity of culturable psychrophilic and psychrotrophic anaerobic bacteria isolated from beef abattoirs and their environments. *Applied and Environmental Microbiology*, 77(13), 4280–4284.

Nikitin, D.A. 2023, February. Ecological characteristics of Antarctic fungi. In: *Doklady Biological Sciences* (Vol. 508, No. 1, pp. 32–54). Moscow: Pleiades Publishing.

Nogi, Y. 2017. Microbial life in the deep sea: Psychropiezophiles. *Psychrophiles: From Biodiversity to Biotechnology* (2nd Edn.), pp. 133–152. Springer.

Núñez-Montero, K., Salazar, R., Santos, A., Gómez-Espinoza, O., Farah, S., Troncoso, C., ... & Barrientos Diaz, L. 2021. Antarctic Rahnella inusitata: A producer of cold-stable β-galactosidase enzymes. *International Journal of Molecular Sciences*, 22(8), 4144.

Omel'Chenko, M.V., Vasil'Eva, L.V., Zavarzin, G.A., Savel'Eva, N.D., Lysenko, A.M., Mityushina, L.L., ... & Trotsenko, Y.A. 1996. A novel psychrophilic methanotroph of the genus. *Methylobacter*, 65(3), 384–389.

Pacios-Michelena, S., Valdés, R.A., Saldaña-Mendoza, S.A., Tucuch-Pérez, M.A., González, C.N.A., Alvarez-Perez, O.B., & Ilina, A. 2024. Encapsulation and application of cell-free extracts produced by *Penicillium chrysogenum* R1 in tomato plants for the control of Fusarium oxysporum. *Environmental Quality Management*, 34, 1–7.

Parmar, T.S., Ahirwar, R., & Sahay, S. 2023. Isolation, purification and characterization of carboxymethyl cellulase (CMCase) from psychrotolerant yeast *Rhodotorula mucilaginosa* BPT1. *Materials Today: Proceedings*, 72, 2768–2772.

Parvizpour, S., Hussin, N., Shamsir, M.S., & Razmara, J. 2021. Psychrophilic enzymes: Structural adaptation, pharmaceutical and industrial applications. *Applied Microbiology and* ***B****iotechnology*, 105, 899–907.

Pathania, S., Solanki, P., Putatunda, C., Bhatia, R.K., & Walia, A. 2022. Adaptation to cold environment: The survival strategy of Psychrophiles. In: *Survival Strategies in Cold-adapted Microorganisms*, pp. 87–111. Springer, Singapore.

Paul, S., Dukare, A.S., Bandeppa, Manjunatha, B.S., & Annapurna, K. 2017. Plant growth promoting rhizobacteria for abiotic stress alleviation in crops. In: *Advances in Soil Microbiology: Recent Trends and Future Prospects: Volume 2: Soil-Microbe-Plant Interaction*, pp. 57–79. Springer.

Pinto, C.T., & Nano, F.E. 2015. Stable, temperature-sensitive recombinant strain of Mycobacterium smegmatis generated through the substitution of a psychrophilic ligA gene. *FEMS Microbiology Letters*, 362(18), fnv152.

Pittino, F., Ambrosini, R., Seeger, M., Azzoni, R.S., Diolaiuti, G., Alviz Gazitua, P., & Franzetti, A. 2023. Geographical variability of bacterial communities of cryoconite holes of Andean glaciers. *Scientific Reports*, 13(1), 2633.

Radchenkova, N., & Rusinova-Videva, S. 2024. Extremophiles as a "green source" of new exopolysaccharides with ecological importance and multifunctional applications. *Polymer-Plastics Technology and Materials*, 63(3), 247–269.

Rai, A.K., & Kumar, R. 2015. Potential of microbial bio-resources of Sikkim Himalayan region. *ENVIS Bull. Himalayan Ecol.*, 23, 99–105.

Risoli, S., Petrucci, A., Vicente, I., & Sarrocco, S. 2023. Trichoderma gamsii T6085, a biocontrol agent of Fusarium head blight, modulates biocontrol-relevant defence genes expression in wheat. *Plant Pathology*, 72(8), 1442–1452.

Rondón, J.J., Ball, M.M., Castro, L.T., & Yarzábal, L.A. 2019. Eurypsychrophilic Pseudomonas spp. isolated from Venezuelan tropical glaciers as promoters of wheat growth and biocontrol agents of plant pathogens at low temperatures. *Environmental Sustainability*, 2, 265–275. https://doi.org/10.1007/s42398-019-00072-2

Roy, P., & Kumar, A. 2020. Arthrobacter. In: *Beneficial Microbes in Agro-Ecology*, pp. 3–11. Academic Press.

Sahay, S., Hamid, B., Singh, P., Ranjan, K., Chauhan, D., Rana, R.S., & Chaurse, V.K. 2013. Evaluation of pectinolytic activities for oenological uses from psychrotrophic yeasts. *Letters in Applied Microbiology*, 57(2), 115–121.

Sharma, S., Chaturvedi, U., Sharma, K., Vaishnav, A., & Singh, H.B. 2022. An overview of survival strategies of psychrophiles and their applications. In: *Survival Strategies in Cold-adapted Microorganisms*, pp. 133–151. Springer, Singapore.

Sheikh, T., Baba, Z., Hamid, B., Iqbal, S., Yatoo, A., Fatima, S., Nabi, A. et al. 2022. Extracellular polymeric substances in psychrophilic cyanobacteria: A potential bioflocculant and carbon sink to mitigate cold stress. *Biocatalysis and Agricultural Biotechnology*, 42, 102375. https://doi.org/10.1016/j.bcab.2022.102375.

Smirnova, M., Losada, C.B., Akulava, V., Zimmermann, B., Kohler, A., Miamin, U., ... & Shapaval, V. 2023. New cold-adapted bacteria for efficient hydrolysis of feather waste at low temperature. *Bioresource Technology Reports*, 23, 101530.

Son, Y., Min, J., Shin, Y., & Park, W. 2023. Morphological and physiological adaptations of psychrophilic *Pseudarthrobacter psychrotolerans* YJ56 under temperature stress. *Scientific Reports*, 13(1), 14970.

Souza, R.D., Ambrosini, A., & Passaglia, L.M. 2015. Plant growth-promoting bacteria as inoculants in agricultural soils. *Genetics and Molecular Biology*, 38, 401–419.

Sridhar, D., Alherwairini, S.S., Eswaran, S.U., Barasarathi, J., Lalitha, S., & Sayyed, R. 2025a. The soil microbiome enhances sesame growth and oil composition, and soil nutrients under saline conditions. *Scientific Reports*, 15(1), 1–17. https://doi.org/10.1038/s41598-025-15589-2.

Sridhar, D., Alheswairini, S.S., Barasarathi, J., Enshasy, H.A.E., Lalitha, S., Mir, S.H., Nithyapriya, S. et al. 2025b. Halophilic rhizobacteria promote growth, physiology, and salinity tolerance in *Sesamum indicum* L. grown under salt stress. *Frontiers in Microbiology*, 16, 1590854. https://doi.org/10.3389/fmicb.2025.1590854.

Taguchi, M., Guo, Y., Nishizawa, T., Chohnan, S., & Kurusu, Y. 2022. Complete genome sequence of a psychrophilic bacterium, Pseudoalteromonas sp. strain APM04, isolated from the seafloor of the South Mariana Trough, Pacific Ocean. *Microbiology Resource Announcements*, 11(8), e00374–22.

Tapia-Vázquez, I., Sánchez-Cruz, R., Arroyo-Domínguez, M., Lira-Ruan, V., Sánchez-Reyes, A., del Rayo Sánchez-Carbente, M., ... & Folch-Mallol, J.L. 2020. Isolation and characterization of psychrophilic and psychrotolerant plant-growth promoting microorganisms from a high-altitude volcano crater in Mexico. *Microbiological Research*, 232, 126394.

Tendulkar, S., Hattiholi, A., Chavadar, M., & Dodamani, S. 2021. Psychrophiles: A journey of hope. *Journal of Biosciences*, 46(3), 64.

Vafa, Z.N., Sohrabi, Y., Heidari, G., Rizwan, M., & Sayyed R.Z. 2024. Wheat growth and yield response are regulated by mycorrhizae application and supplemental irrigation. *Chemosphere*, 364, 143068 https://doi.org/10.1016/j.chemosphere.2024.143068.

Wang, J., Zhu, M., Wang, P., & Chen, W. 2023. Biochemical properties of a cold-active chitinase from marine Trichoderma gamsii R1 and its application to preparation of chitin oligosaccharides. *Marine Drugs*, 21(6), 332.

Xiong, L., Li, Y., Yu, H., Wei, Y., Li, H., & Ji, X. 2023. Whole genome analysis and cold adaptation strategies of *Pseudomonas sivasensis* W-6 isolated from the Napahai plateau wetland. *Scientific Reports*, 13(1), 14190.

Xu, H., Xu, D., & Liu, Y. 2024. Molecular biology applications of psychrophilic enzymes: Adaptations, advantages, expression, and prospective. *Applied Biochemistry and Biotechnology*, 1–25.

Yadav, A.N., Sachan, S.G., Verma, P., & Saxena, A.K. 2015. Prospecting cold deserts of north western Himalayas for microbial diversity and plant growth promoting attributes. *Journal of Bioscience and Bioengineering*, 119(6), 683–693.

Yadav, A.N., Sachan, S.G., Verma, P., & Saxena, A.K. 2016. Bioprospecting of plant growth promoting psychrotrophic Bacilli from cold desert of north western Indian Himalayas. *Indian J. Exp. Biol.*, 54(2), 142–150.

Yadav, A.N., Verma, P., Sachan, S.G., Kaushik, R., & Saxena, A.K. 2018. Psychrotrophic microbiomes: Molecular diversity and beneficial role in plant growth promotion and soil health. In: *Microorganisms for Green Revolution: Volume 2: Microbes for Sustainable Agroecosystem*, pp. 197–240. Springer.

Yayanos, A.A. (2009). High-pressure habitats. In: *Encyclopedia of Microbiology* (3rd Edn.), pp. 228–239 https://doi.org/10.1016/B978-012373944-5.00283-2.

Yu, Y., Li, H.R., & Zeng, Y.X. 2011. *Colwellia chukchiensis* sp. nov., a psychrotolerant bacterium isolated from the Arctic Ocean. *International Journal of Systematic and Evolutionary Microbiology*, 61(4), 850–853.

Yusof, N.A., Hashim, N.H.F., & Bharudin, I. 2021. Cold adaptation strategies and the potential of psychrophilic enzymes from the antarctic yeast, Glaciozyma antarctica PI12. *Journal of Fungi*, 7(7), 528.

Zhang, A., Hou, Y., Wang, Y., Wang, Q., Shan, X., & Liu, J. 2023. Highly efficient low-temperature biodegradation of polyethylene microplastics by using cold-active laccase cell-surface displaysystem. *Bioresource Technology*, 382, 129164.

Zhang, N., Suen, W.C., Windsor, W., Xiao, L., Madison, V., & Zaks, A. 2003. Improving tolerance of *Candida antarctica* lipase B towards irreversible thermal inactivation through directed evolution. *Protein Engineering*, 16(8), 599–605.

CHAPTER
2

Polyextremophiles as Model Microorganisms to Search for Life on Mars

Gaurav Yadav[1], Shrushti S. Patil[2], Aamina Shahab[1], Surati Kumari[3], Srishti Sharma[4,5] and Roohi[1*]

[1] Protein Research Laboratory, Department of Bioengineering, Integral University, Uttar Pradesh, India
[2] Research Sat, Adelaide, Australia
[3] National Institute of Biomedical Genomics, Kalyani, West Bengal, India
[4] School of Bioengineering & Biosciences, Lovely Professional University (LPU), Jalandhar, Punjab, India
[5] Blue Marble Space Institute of Science, Seattle, Washington, USA

1. Polyextremophiles: An Introduction

There are many environments on Earth that could be classified as extreme. Environments that constitute a large portion of the biosphere, such as cold environments (e.g., >80% of the Earth's biosphere is permanently below 5°C), are considered extreme compared to the standard conditions of temperature, pressure, humidity, air quality, light, nutrient supply, etc. This is not just because the extreme cold greatly restricts the existence of humans there, but rather because it is extreme relative to the typical standard conditions (Hamid et al., 2022). Thomas D. Brock, a microbiologist, defined extreme environments as "they are environments with 'restricted species diversity and the absence of some taxonomic groups'". Extremes in one or more physical or chemical conditions can be found in a variety of environments on the surface of our planet, particularly in subsurface regions. Like the Earth, other planetary bodies might have various environments with a wide range of possibilities for each condition. Life can be found in a variety of extreme environments, including

*Corresponding author: roohi0607@gmail.com

DOI: 10.1201/9781003717003-2

volcanic and sulfuric hot springs, dry, hot deserts, deep-sea hydrothermal vents, acid-mine drainages, extremely pressured deep seas, cold, high UV radiation polar environments, subsurface caves, and super-saturated salt lakes (Schultz et al., 2023).

Over the past few decades, we have significantly improved our understanding of life in these extreme environments. Prokaryotic life has dominated a large portion of our planet's evolutionary history and has evolved to occupy nearly all environmental niches. On this planet Earth, there are three main lineages of organisms: the eukaryotes (or Eukarya), which are comprised of all animals, plants, fungi, and numerous unicellular microorganisms; and the prokaryotes, which are further divided into the two lineages of bacteria (or Eubacteria) and the Archaea (or Archaebacteria). Archaea are divided into three main groups: extreme halophiles, extreme thermophiles, and methanogens. So, many archaea are considered extremophiles (Seckbach et al., 2013). Most extremophiles are included in the Archaea and Bacteria domains, whereas very few extremophiles belong to the Eukarya domain.

Microorganisms that not only tolerate but also thrive in these conditions that were previously thought to destroy biomolecules and organisms, such as extremes in temperature, pH, pressure, salinity, ionizing radiation, and a variety of planetary environments within the solar system as well as the galaxy, are called extremophiles (Figure 1). Was it possible to imagine that life could thrive beyond

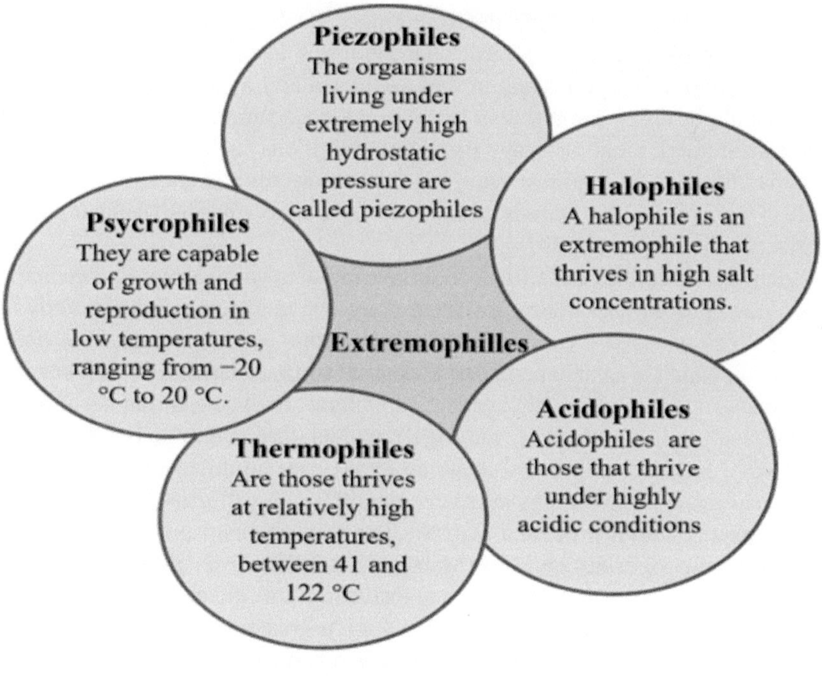

Figure 1: Different types of extremophiles.

the boiling point of water in the 1970s? But now, we have discovered a large number of microorganisms that could flourish at temperatures up to 113°C and even at 121°C (Stan-Lotter, 2008). Over the past 50 years, an ever-growing variety of extremophiles have been discovered, which has increased our understanding of life's physiological and biochemical capacities and shed light on the long history of evolution on our planet (DasSarma et al., 2020). Extremophiles, which are classified according to the physicochemical conditions in which they grow, continue to be a novel group of microorganisms found in various biosphere-wide environments. Microorganisms found in hot, acidic environments necessarily require a low pH and high temperature to thrive, and these microorganisms are called thermoacidophiles, where 'phile' refers to 'love', or more accurately, "a requirement". Halophiles are microorganisms that require high levels of salt in their growth medium. *Deinococcus radiodurans, Methanosarcina* sp. (archaea), and *Pyrococcus furiosus* (archaea) from hydrothermal vents are all highly resistant to radiation and are therefore classified as radiation-resistant extremophiles or radioresistant microbes (Hamid et al., 2022). Psychrophiles are the extremophiles that are able to survive in cold environments, such as polar regions, deep seas, and high mountain altitudes; thermophiles and hyperthermophiles can thrive under high temperature ranges, including in volcanoes, desert hot springs, and hydrothermal vents. The extremophiles that survive at pH less than 5 and are localized in acid-mine drainage sites and acidic lakes are called acidophiles. Alkaliphiles, on the other hand, survive at high pH ranges, such as in sodic lakes. Piezophiles prefer high pressure ranges, such as deep inside the ocean; halophiles prosper under high salt concentrations in seas, salars, saline lakes, and brine pools; xerophiles require low water availability so they can survive in deserts; oligotrophes can thrive in low nutrient concentrations; and metalophiles can tolerate high metal or heavy metal concentrations (Schultz et al., 2023). To survive, life usually requires water, but lithotrophs are microorganisms that are able to survive for long periods with low water activity on or just below the rock surfaces (Bjelland et al., 2011).

The extremophiles that are able to survive in a number of extreme environments are termed polyextremophiles, and these represent an important area of study for astrobiology and space exploration. Such an example is *Brevibacterium linens* AE038-8, which is an arsenic-tolerant bacterial strain isolated from well water in Tacuman, Argentina. It has the tendency to tolerate high concentrations of heavy metals such as Cd(II), Cr(VI), and Cu(II); temperature ranges of 10–30°C; a pH range of 3 to 1; the ability to tolerate up to 5M concentrations of NaCl; and the ability to grow under combined stress conditions. It can grow in particular in environments with low pH and high NaCl concentrations, which are relevant to the habitability of brines on Mars (Maizel et al., 2019). Each of the complex sets of conditions that an organism needs to thrive in a specific environment must be kept within its tolerance range; if any condition is greater than either its minimum or maximum tolerance, the organism won't survive (Stan-Lotter, 2008). High radiation doses, extreme temperatures, different gravity, pressure, pH, salinity, energy sources, and nutrient scarcity are just a few of the harsh environmental factors that make outer space uninhabitable and harmful for the growth of life. Microbial life may be able to endure the harsh conditions of space because it can endure inhospitable habitats and a wide range of physicochemical conditions on Earth. Environmental factors that can restrict microbial life include extreme temperatures, pH, salinity, water availability, pressure, radiation, and nutrients. Microorganisms, or extremophiles, thrive in these harsh conditions, whereas polyextremophiles are able to grow at their best under several extreme stresses in the same environment (Schultz et al., 2023). Polyextremophile research has influenced our understanding of the origins, evolution (Bertrand et al., 2015), and potential for life on other planets (Schulze-Makuch et al., 2018). It has also resulted in numerous discoveries in molecular biology and medicine (Coker, 2016; Durvasula & Rao, 2018).

Extremophiles can survive in any extreme environment, but there are some polyextremophilic combinations that have not yet been studied on Earth. These gaps may be filled by synthetic biology. It is possible to create novel polyextremophiles by engineering different biochemical properties and survival mechanisms. An organism that can already survive at low temperatures might acquire radio-resistance, for instance, if extreme genome repair pathways are introduced into a psychrophile (Galhardo & Rosenberg, 2009). This transformation would result in an organism that could survive in environments that were extremely cold and radiation-exposed. It's entirely possible that the first organism to live in an extraterrestrial environment will be created, not found (Seckbach et al., 2013).

The extraterrestrial environment's conditions are mostly considered extreme with respect to the typical standard conditions of life on Earth. Such as the presence of salt hydrates and brines on the surface of Mars (Maizel et al., 2019; Ojha et al., 2015). It has also been found that these brines could provide a habitable environment (Leuko et al., 2014). The only living organisms that could survive in brine are the halophiles, but along with surviving in extremely salty concentrations, the other challenges that could be faced by life could be extremely low temperatures and high radiation. Therefore, finding a suitable living form that could be able to survive in these extreme conditions is done by looking for polyextremophiles (Maizel et al., 2019). The advancement in methodology in the development of extremophile studies includes: utilizing proteomics for studying extremophiles (Gao et al., 2020), using RNA stable isotope probing for studying methanotrophic communities (Wigley et al., 2022), measuring the abundance of syntrophic acetate-oxidizing microorganisms in ammonia-stressed reactors using quantitative PCR (Yan et al., 2021), utilizing metagenomics for studying viruses from hypersaline lakes (Le Lay et al., 2023), using a multiplex approach to quantify fine-scale temperature fluctuations in the proteome (Gao et al., 2020), developing a gene transfer system for manipulating halophiles (Galiana et al., 2022), thermophiles (Chung et al., 2020), and psychrophiles (Hamid et al., 2022).

The primitive Earth's environment was thought to have been warmer than today, with a high concentration of carbon dioxide gas and anaerobic conditions. The main gases in the early atmosphere were carbon dioxide, water vapour, hydrogen sulfide, nitrogen, and methane, with only trace amounts of oxygen allegedly present (Kasting, 1993). And it is assumed that the first prokaryotic form of life was a thermophilic microorganism, or at least a thermotolerant one, or one that survived the late heavy bombardment. According to Martin et al. (2008), several scientists think that the hydrothermal vents at high temperatures are where life first began. The most deeply rooted microorganisms (found in hydrothermal vents at the bottom of the ocean) are thermophiles, which is an intriguing finding and raises the possibility that the earliest common ancestor was a thermophilic cell (Martin et al., 2008). It is believed that the atmosphere was not oxidized until the appearance of cyanobacteria, which 'filled' the atmosphere with their oxygen release. Finally, the atmosphere's current level of >20% oxygen was reached. As a result, the earliest living organisms existed under the above-mentioned conditions as well as post-biotic ones (pressure, high temperature, anaerobic atmosphere) (Seckbach et al., 2013).

The presence of liquid water, sources of energy, and an abundance of organic molecules are among the requirements and priorities for the origins of life beyond Earth. These elements play a crucial role in cellular metabolism. There are good chances of finding life anywhere there is liquid water (Chela-Flores & Seckbach, 2011; Seckbach et al., 2013). One of the main areas of focus in the search for extraterrestrial life today is the investigation of the existence (or lack thereof) of life on Mars. The cold, dry world of the present-day Martian surface (2°C to –65°C, and 20 °C to –126°C) is similar to the frozen desert of Antarctica (McKay et al., 2011). Photos taken over Mars by a variety of rovers rambling across the Martian surface demonstrate that this planet was once warm and wet with an abundance of water (about 3.5 billion years ago). Also found on Mars is the unstable chemical methane, which may support some bio-sources from methanogenic microbial activities (Seckbach et al., 2013). Extremely northern regions of Canada, which resemble those on Mars, were able to support methane-eating bacteria. So, based on all this evidence, it seems possible to think that life might be present or was present on Mars, but it is uncertain that there is life on Mars until samples from Mars are brought back to Earth and researched.

2. Classification of Polyextremophiles

Researchers have long been investigating microbes living in extreme conditions based on their adaptation to usually a single physiochemical condition such as microbes adapted to hot temperature (hyperthermophiles and thermophiles), cold temperature (psychrophiles), salinity (halophile), low and high pH (acidophile and alkaliphile respectively), high doses of radiation (radioresistant), desiccation (xerophile), pressure (piezophile), high concentration of metals and nutrients and low availability (Merino et al., 2019).

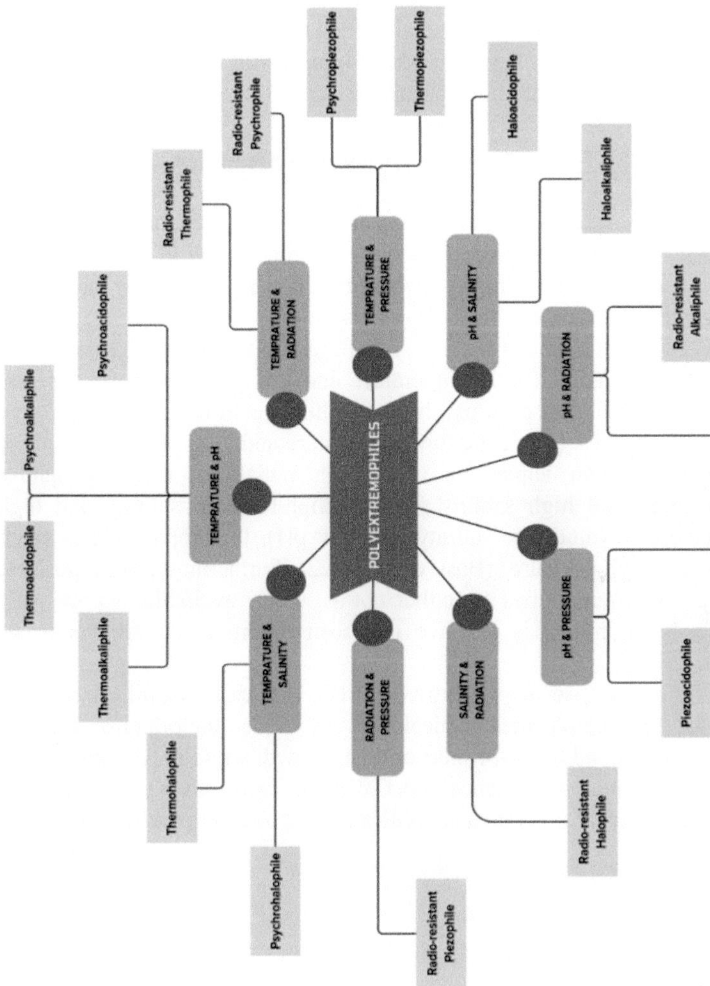

Figure 2: Different forms of polyextremophiles based on the interaction of extremes.

However, as depicted in Figure 2, many environments exhibit more than one extreme condition (poly-extreme regions), resulting in 'polyextremophilic' microbes with a tendency to tolerate or adapt to multiple stressors (Chela-Flores, 2013). These microbes are exceptionally relevant for future interplanetary missions and to delve into the search for "limits of habitability" boundary conditions that constrain life, the origin of life and signs of life beyond Earth (Maizel et al., 2019). In further sub-sections, polyextremophiles are grouped based on advancements made in their exploration: thoroughly explored and rarely explored polyextremophiles. Few polyextremophiles with three or more extreme physiochemical conditions are also discussed.

3. Thoroughly Explored Polyextremophiles

In the absence of a systematic classification, an attempt has been made to delineate various forms of polyextremophiles based on the intersection of extremes, such as hypersaline alkaline lakes (high salt concentration and high pH), acidic hot springs (high temperature and low pH), black smokers (high temperature and pressure), and so on (Capece et al., 2013). Different forms of polyextremophiles include thermoacidophiles (high temperature and low pH), thermoalkaliphiles (high temperature and high pH), psychroacidophiles (cold temperature and low pH), psychroalkaliphiles (moderate low temperature and high pH), thermophilic radiation-resistant (high temperature and high radiation), thermohalophiles (high temperature and high salinity), haloalkaliphile (high salinity and high pH), haloacidophile (moderate salinity and low pH), thermopiezophiles (high temperature and high pressure) (Basu et al., 2022). Furthermore, other possible combinations of extremophiles are either not as widely available as previously mentioned polyextremophiles or have not been reported and are yet to be sought after.

Typically, in the case of polyextremophiles, true extremophilic behaviour is not seen for multiple physicochemical stressors, rather, the organism requires one of the extreme conditions for their optimal growth while only being able to tolerate other extremes, for instance, most of the microbes living in cold acidic regions are either psychrotolerant acidophiles or acid-tolerant psychrophile (Basu et al., 2022).

Temperature and pH

Environments combining these extremes are seen across various geographic locations worldwide. While theoretically, extremophiles adapted to low temperature and high pH (psychroalkaliphiles), and low temperature and low pH (psychroacidophiles), high temperature and high pH (thermoalkaliphiles), high temperature and low pH (thermoacidophiles) can exist, it is quite unusual to encounter psychroalkaliphiles and psychroacidophiles in nature (Capece et al., 2013). Low temperature and acidic conditions rarely overlap, thereby providing limited opportunities for the development of true psychrophilic acidophiles (Capece et al., 2013). However, certain extremotolerant species still have been found in such habitats. According to Dedysh et al. (2004), they isolated three psychrophilic but moderately acidophilic strains of *Methylocella* growing optimally at pH 5.5–6.0 and 15°C temperature. Another exploration into the microbial life of the Arctic zone marked the first reported case of true psychrophilic and moderate acidophilic bacteria, unveiling a new genus, *Lichenicoccus*, comprising three novel strains exhibiting optimal growth at the temperature range of 10–15°C and pH 4.5–5.5 (Pankratov et al., 2020). Similarly, many extremotolerant species from cold alkaline niches have been described but none showed true extremophile, until the discovery of 'Champion extremophiles', *Arsukibacterium ikkense* and *Rhodonellum psychrophilum* (Hoover & Pikuta, 2009). These psychroalkaliphiles show growth within the same optimal pH range of 9.2–10.0 but require different optimum temperatures, *A. ikkense* favouring 15°C (Schmidt et al., 2007), while *R. psychrophilum* prefers 5°C (Schmidt et al., 2006). However, unlike true psychroalkaliphiles, true psychroacidophiles are yet to be discovered.

Thermoacidophiles and thermoalkaliphiles are among the most frequently encountered and studied polyextremophiles, especially thermoacidophiles (due to widely distributed acidic hot springs). Thermoacidophilic microbes are distributed across various genera, though not limited to, *Acidianus*, *Caldivirga*, *Caldococcus*, *Ignicoccus*, *Acidilobus*, *Alicyclobacillus*, *Caldariella*, *Caldococcud*, *Cyanidium*, *Ferroplasma*, *Palaeococcus*, *Picrophilus*, *Pyrobaculum*, *Pyrococcus*, *Pyrodictium*, *Pyrolobus*, *Sulfolobus*, *Staphylothermus*, *Sulfurisphaera*, *Thermodiscus*, *Thermoproteus*, and *Vulcanisaeta*; whereas thermoalkaliphilic strains have been reported to belong to genera *Aeribacillus*, *Aeropyrum*, *Caloramator*, *Chloroflexus*, *Clostridium*, *Isosphaera*, *Meiothermus*, *Methanohalophilus*, *Thermobranchium*, *Thermomicrobium*, and *Truepera* (Capece et al., 2013). However, some genera such as *Thermococcus* include both thermoacidophilic and thermoalkaliphilic microbes, for instance, *Thermococcus celer* grows optimally at 88°C at pH 5.8 (Zillig et al., 1983) while *Thermococcus alcaliphilus* show optimal growth at 85°C at pH 9.0 (Keller et al., 1995). Interestingly, optimal growth conditions of thermoalkaliphiles found so far indicate a potential antagonistic relationship between temperature and pH (Capece et al., 2013). It was observed that as the temperature optimum reaches towards extreme value, the pH optimal tends to decrease and vice versa (Kumar & Satyanarayana, 2013).

Temperature and Salinity

Cold saline springs of Antarctic and Arctic regions harbour microbes capable of withstanding both cold temperature and high salinity (Hoover & Pikuta, 2009). For instance, Lost Hammer Spring maintains a sub-zero temperature throughout the year with ~24% salinity and is home to various psychrotolerant and halotolerant microorganisms of genera *Bacteroidetes, Proteobacteria*, and *Actinobacteria* (Goordial et al., 2013). Two novel species, *Psychrobacter submarinus* and *Psychrobacter marincola*, both psychrophilic and moderate halophilic, were isolated and found to survive temperature ranges of 4 to 35°C and 7 to 35°C, respectively (Goordial et al., 2013). Similarly, different water bodies of McMurdo Dry Valley were reported to have 12 strains of psychrophilic halotolerant microbes with two of them proposed as type strains of novel species *Psychrobacter vallis* CMS 39^T and *Psychrobacter aquaticus* CMS 56^T (Shivaji et al., 2005). It has been noted that many psychrophiles, mainly those that demonstrate 'salt-in' adaptive measures, typically show some extent of adaptation for saline conditions (Capece et al., 2013). Interestingly, it's the mesophilic species, rather than psychrophilic or psychrotolerant ones, that display the highest tolerance for salinity (Hoover & Pikuta, 2009).

On the other end of the spectrum are the microbes inhabiting hot and saline environments, referred to as thermohalophiles, some of which are also adapted to alkaline pH (haloalkalithermophiles). Majority of thermohalophiles identified thus far, are either thermotolerant or moderate thermophiles, for example, *Haloterrigena thermotolerans* is a moderate thermophilic halophile growing optimally at 50°C and 3–3.5M NaCl (Montalvo-Rodríguez et al., 2000). However, *Halothermothrix orenii* and *Thermohalobacter berrensis* can survive lower salt ranges compared to other halophilic archaea, while demonstrating a higher degree of adaptation to high temperatures. This trade-off between extreme tolerance to temperature and salt is likely due to the high energy expenditure required to maintain cellular integrity, as both stressors result in a significant increase in cellular permeability (Mesbah & Wiegel, 2005). Other examples of thermohalophiles include *Halobacterium salinarum, Haloferax denitrificans, Halorubrum saccharovorum, Halorubrum coriense, Halobacterium marismortui*, and *Methanohaobium evestigatum* (Mesbah & Wiegel, 2005).

Temperature and Radiation

In the quest to search for life in extreme environments, researchers came across a unique form of extremophilic microbe with the ability to withstand higher doses of radiation. These radioresistant microbes survive damages induced by UV and ionizing radiation (alpha, beta, and gamma rays) by employing an extensive DNA-damage repair system, maintaining polyploidy, and reducing free-radical levels in cells (Ranawat & Rawat, 2017). There have been contradictory opinions on whether radiation resistance is the 'byproduct' of an adaptive strategy against other stressors, e.g., desiccation or not (Beblo-Vranesevic et al., 2018; Gabani & Singh, 2013). They can be found in various habitats (e.g., deserts, arid and semi-arid river basins, hydrothermal vents, hot springs, and artificial environments) and mostly overlap with other extreme conditions, such as temperature, salinity, pH, or desiccation (Gabani & Singh 2013). Thermophilic radio-resistant microbes are suitable candidates for future astrobiological missions since they inhabit similar analogues of other planetary bodies. Thermophilic microbes, *Thermococcus gammatolerans* and *Pyrococcus abyssi* can withstand extremely high doses of radiation up to 6 kGy and 2.5 kGy, respectively (Capece et al., 2013). Some researchers even suggested that adaptation towards extremely high temperatures is the reason behind tolerance seen against radiation and desiccation (Ranawat & Rawat, 2017).

At the other end, extremophiles inhabiting cold regions are less studied for their potential to resist radiation in comparison to thermophilic radio-resistant counterparts but are equally fascinating. Due to ozone layer depletion over colder regions of the Antarctic and Arctic regions, microbial species tend to show tolerance towards UV radiation as well. However, in our limited knowledge, it is typically psychrotolerants rather than true psychrophiles that have been found to resist solar radiation. In another study, four novel psychrophilic strains of ionizing-sensitive *Deinococcus* strains were isolated, although they exhibited poor radiation resistance in comparison to *D. radiodurans* (Callegan et al., 2008). Further investigation would be needed to explore the interactions between radiation and temperature extremes.

Temperature and Pressure

Piezophilic organisms flourish in high-pressure environments (normal pressure for humans is 0.1 MPa) such as deep sea and deep subsurface (Scheffer & Gieg, 2023). High-pressure environments usually coincide with temperature extremes; thus, apart from pressure, piezophiles are also adapted to cold and hot environments. Deep sea environments (low temperature and high pressure) harbour the majority of known psychropiezophiles while thermopiezophiles are typically seen in deep subsurface and hydrothermal vents (high temperature and high pressure). Examples include *Colwellia marinimaniae* MT-41, a psychropiezophile from the bottoms of Mariana trench with optimum growth at 69 MPa pressure at 2°C (Kusube et al., 2017), and *Pyrococcus yayanosii* CH1 a thermopiezophile found in a deep-sea hydrothermal vent of Ashadze site, with optimum growth at 52 MPa at 98°C (Birrien et al., 2011).

Despite living in different temperature extremes, psychropiezophiles and thermopiezophiles share a common adaptive measure in response to high pressure, i.e., the induction of both heat-shock and cold-shock proteins (Vanlint et al., 2011). Other adaptive measures seen in the case of psychropiezophiles are: (i) incorporation of monosaturated and polyunsaturated fatty acids, and (ii) production of extracellular polymeric substance; while in the case of thermopiezophiles, the measures are: (i) neutralization of opposite effects of temperature and pressure maintains cellular integrity, and (ii) proteins with strong hydrophobic cores with little surface charge (Capece et al., 2013). However, further exploration of piezophiles is necessary to understand the co-production of heat-shock and cold-shock proteins in both forms of piezophiles and the neutralization effect observed in thermopiezophiles.

pH and Salinity

Hot springs, hydrothermal vents, deep sea brines, and some of the anthropogenic sites have been reported to harbour haloacidophiles adapted to low pH and high salinity, and haloalkaliphiles adapted to high pH and high salinity (Minegishi, 2013). In the past several years, due attention has been given to haloalkaliphiles; however, not much is known about haloacidophiles. This knowledge gap may stem from the rarity of acidic and saline environments compared to alkaline and saline systems (Belilla, 2020). In a study, it was implied that the effect caused by the adaptive strategies of halophiles was both synergistic as well as antagonistic to that of the acidophiles, moreover, the antagonistic effect was more prominent than the benefits provided by synergy (Capece et al., 2013). This might be the reason behind the rare occurrence of true haloacidophiles.

Haloalkaliphiles inhabit highly saline and alkaline environments, for example, salt salterns, soda lakes, and saline-alkaline deserts. Based on the level of salt and pH requirement for optimum growth, haloalkaliphiles can be categorized into extreme haloalkaliphiles including *Thioalkalivibrio versutus*, moderate haloalkaliphiles such as *Ectothiorhodospira haloalkaliphile*, and low-salt tolerant natronophile such as *Thioalkalimicrobium aerophilum* AL3^T (Banciu & Sorokin, 2013). Haloalkaphiles demonstrate microbes with a diverse range of pH and salt extremophile, e.g., *Ornithinibacillus salinisoli* is slightly alkaliphilic and moderately halophilic having optimum growth at pH 8.0 and 10–15% NaCl (Gan et al., 2018), while *Anaerobacillus isosaccharinicus* is an obligate alkaliphile and halotolerant growing optimally at pH 9.8–10 and able to tolerate up to 6.0% NaCl (Bassil & Lloyd, 2019). Like halophiles, haloalkaliphiles adapt to saline-alkaline conditions by employing the 'salt-in' or 'salt-out' strategy (Banciu & Sorokin 2013). They maintain cellular homeostasis through a combination of compatible solute accumulation, modification of cell membrane composition, antiporters to maintain pH balance and enzyme adaptation (Xue et al., 2021).

4. Rarely Explored Polyextremophiles

Other forms of polyextremophilic microbes have also been reported, though they have not been explored to the same extent as the aforementioned polyextremophiles. Several factors contribute to this disparity. Firstly, these microbes may inhabit regions where the intersection of multiple stressors is rare, making them less accessible for study. Secondly, the limited accessibility of such extreme environments poses challenges for sample collection. Additionally, maintaining the extreme physiochemical conditions required for studying these microbes presents technical challenges. Moreover, these microbes may receive less attention compared to other extremophiles that are perceived to have more immediate biotechnological relevance. Finally, financial constraints are one of the major reasons that can hamper studies needed for conducting in-depth experiments. Despite numerous novel extremophilic strains being discovered each year, there is a lack of extensive studies done to discern their potential adaptation to stressors found beyond their natural habitat. Furthermore, minimal focus is directed towards modification in their adaptive mechanism, when faced with multiple extreme conditions simultaneously. Majority of research on extremophiles tends to focus on their ability to tolerate individual extreme conditions, such as resistance to radiation, alkalinity, or pressure. However, there is a lack of emphasis on understanding how these organisms survive at the intersection of multiple extremes. This holistic view, which considers the interplay between different environmental stressors, is crucial for gaining insights into polyextremophile and its relevance to future astrobiological studies. Therefore, there is a need for a more systematic research design aimed at elucidating the adaptations of such extraordinary microbes in polyextreme environments. Prime examples of less-explored polyextremophiles include microbes adapted to pH and pressure (piezoacidophile and piezoalkaliphile), pH and radiation (acidic radio-resistant and alkaliphilic radio-resistant), salinity and radiation (radiation-resistant halophiles), salinity and pressure (halopiezophiles).

High-pressure environments with extreme pH such as acidic and alkaline deep-sea hydrothermal vents allow the microbial communities to adapt to these polyextreme conditions. Metabolic and pressure-related studies on piezophiles have revealed a significant effect of high pressure on reducing the pH, suggesting the presence of acidophilic properties in piezophiles while limiting the possibility of alkalipiezophiles. Although ongoing research explores the interaction between temperature and pressure extremes, there is a critical need to delve deeper into understanding the interplay between pressure and pH extremes. In a survey study, acidophilic yeast *Rhodotorula taiwanenis* MD1149, extracted from acid mine drainage displayed resistance to gamma radiation with a reported D10 value of 2.5 kGy at pH optima of 2.3 (Tkavc et al., 2018). In a comparative analysis of acidophiles, *Leptospirillum ferriphilum* and *Acidithiobacillus ferrooxidans*, *L. ferriphilum* reported better tolerance to oxidative stress and UV-C radiation, illustrating another example of a radiation-resistant acidophile (Farías et al., 2021). Kottemann et al. (2005) reported radiation resistance in halophilic archaea *Halobacterium* sp. NRC-1, although they suggested that this is a result of adaptation against desiccation (Kottemann et al., 2005). In similar studies, it was found that saline conditions provide a kind of cellular ‘protection’ and adaptation to desiccation has been seen to give rise to resistance against other environmental stressors such as vacuum and radiation (Kottemann et al., 2005). Halopiezophiles are usually distributed across deep hypersaline anoxic basins (DHABs) such as Urania, Discovery, Kyros, and Thetis basins among a few others (Bartlett, 2009). DHABs comprise diverse microbial communities usually with sulphur oxidizing properties, for example, *Chromohalobacter israelensis, Marinobacter flavimaris*, and *Halomonas meridiana* (Varrella et al., 2020).

5. Adaptation Against Two or More Extremes

Polyextreme environments represent the intersection of extreme conditions. In the previous section, to simplify the classification, polyextremophiles were discussed keeping the focus on only two extremes; however, such regions can exhibit more than two extreme conditions. Albeit the extremities, polyextreme environments are reported to have diverse living organisms. One such polyextreme environment is the High-altitude Andean Lakes (HAALs), comprising salt flats and shallow lakes at an altitude higher than 3000 m in the Puna-Andes region with multiple environmental stressors such as high concentration of Arsenic (As) content, low precipitation, wide range of daily temperature variation, high solar radiation, high salinity and alkaline pH (Zannier et al., 2019). Runzheimer et al. (2024) explored the diversity of microbes inhabiting five HAALs and reported two novel isolates via proteotyping technique (Runzheimer et al., 2024). Another exploration resulted in the discovery of three novel As-resistant strains belonging to the family of *Vibrionaceae* and novel extremozymes from *Acinetobacter* capable of functioning in these pristine conditions (Albarracín et al., 2015).

Dallol, one of the most extreme regions of Earth, is a unique system with the combination of hostile factors such as extreme temperature, high metal concentration, salinity, and acidic conditions; comprising various lakes with hyper-salinity and hyper-acidic conditions, salt canyons, cave reservoirs, and hydrothermal salt plains (Belilla, 2020). Owing to its unique characteristic features similar to Nili Patera Caldera and Gusev Crater on Mars, Dallol is considered as an excellent analogue site for the Martian hydrothermal region (Cavalazzi et al., 2019). In 2019, Gómez et al. (2019) reported the evidence of thermohaloacidophilic ultra-small microorganisms in the samples taken from a furamole at the Dallol hydrothermal region in Danakil depression showing close relatedness with members of Order *Nanohaloarchaea* (Gómez et al., 2019). This study expanded our knowledge of boundary conditions of life on Earth and holds the potential to shape future astrobiology missions to detect signs of life on other planetary bodies.

Salar de Huasco (SH), another polyextreme environment, is a high-altitude wetland with a wide range of temperature differences daily, extreme salinity and arid conditions, high doses of radiation and arsenic content (Cortés-Albayay et al., 2019). Castro-Severyn et al. (2021) examined different sites of Salar de Huasco and found diverse microbial communities mainly dominated by *Proteobacteria, Bacteroidetes*, and *Cyanobacteria* (Castro-Severyn et al., 2021). In another study, 12 isolates of *Streptomyces*, found from different sites of Salar de Huasco, were proposed as novel species with the majority of them having antifungal, antibacterial and cytotoxic activity (Cortés-Albayay et al., 2019). However, where success in finding microbial life in several polyextreme environments is considered a remarkable event, a 'failure' to detect microbial life in a polyextreme environment can be equally interesting as it provides a different direction for scientists to work upon. Recently, a team of researchers led by Nicholas Dragone tested 204 ice-free samples of Antarctic soil from the Shackleton Glacier region for the presence of microbial life and reported their findings stating "...microbial habitability is constrained by near-continuous exposure to cold, dry, and salty conditions, establishing the environmental conditions that limit microbial life in terrestrial surface soils" (Dragone et al., 2021). Their inability to detect microbial life indicates that either we are close to finding unknown "limits of life" or there exists a limit in our present handling and culturing techniques. It would be an interesting expedition leading research either towards 'lifeless' soil or potentially 'unknown' highly extreme microorganisms.

6. Polyextreme Environment and Mars: Analogue Environment on Earth

Polyextreme environments exist in various locations across the globe, including deep-sea hydrothermal vents, hypersaline lakes, acidic hot springs, polar regions, and desert ecosystems. Occupying the outer limits of habitability, these environments push organisms to their physiological thresholds, compelling them to adapt to the harshest of conditions. Researchers are intensely drawn to the study of polyextreme environments, as they offer valuable insights into the limits of life on Earth and the potential existence of life in extraterrestrial realms. By unravelling the survival strategies employed by organisms (Polyextremophiles) in polyextreme environments, scientists aim to deepen their understanding of the fundamental principles that govern life and the possibilities for life to thrive under extreme conditions beyond our planet. Recent discoveries have revealed the presence of sulphur-oxidizing bacteria in high-pressure, acidic hydrothermal vents in the Guaymas Basin and psychrophilic, halotolerant bacteria in the Siberian permafrost (Lonsdale, 1977; Shcherbakova et al., 2009). The presence of polyextremophiles in the cold, highly pressurized environments of subglacial lakes on Earth strengthens the likelihood of extraterrestrial life residing within subsurface lakes on moons like Europa and Enceladus (Marion et al., 2003; Parrilli et al., 2011). While extremophiles as a group can survive in individual extreme conditions, there are still certain combinations of polyextremophiles that remain uncharacterized on Earth. The exploration of these remarkable microorganisms broadens our knowledge of life's resilience and further supports the potential for life to exist in the extreme environments found beyond our planet. Recent advancements in technology have revolutionized the exploration of polyextreme environments. High-throughput sequencing, metagenomics, and molecular tools allow researchers to identify diverse microbial communities thriving in extreme habitats. Understanding the genetic and physiological adaptations of organisms in these environments not only benefits astrobiology but also has practical applications in environmental conservation, biotechnology, and developing innovative strategies for coping with extreme conditions on Earth. By studying the resilience of organisms in polyextreme environments, researchers gain knowledge applicable to diverse fields, from biotechnological processes to environmental challenges.

7. Significance of Studying Polyextreme Environments for Mars Exploration

One of the primary scientific motivations for studying polyextreme environments lies in their analogical nature to Mars. These environments serve as terrestrial testbeds, providing crucial data for assessing the habitability of Mars' surface and subsurface regions. By understanding how life copes with extreme conditions on Earth, scientists can evaluate the potential for Martian habitats to support microbial life and guide the selection of landing sites for future Mars missions. By leveraging our understanding of polyextreme environments, scientists can enhance our chances of successfully exploring Mars and unraveling the mysteries of potential life beyond Earth. Following are the key scientific motivations for studying polyextreme environments include:

- **Habitability Assessments:** Polyextreme environments help determine the habitability potential of different regions on Mars by identifying the thresholds beyond which life becomes unsustainable.
- **Astrobiological Framework:** Polyextreme environments provide insights into the fundamental requirements and adaptations of life in extreme conditions, guiding strategies to search for and characterize potential Martian life.
- **Resource Utilization:** Understanding polyextreme environments aids in identifying potential resources and sustainable strategies for resource utilisation on Mars, utilizing insights from extremophiles thriving on Earth.
- **Mission Planning and Risk Mitigation:** Studying polyextreme environments informs mission planning, technology development, and risk mitigation strategies to ensure the safety and well-being of future Mars astronauts.
- **Biological Contamination Control:** Research on polyextreme environments contributes to effective measures, protocols, and techniques for preventing contamination of Mars with terrestrial microorganisms during spacecraft missions.
- **Environmental Monitoring:** Polyextreme environments serve as benchmarks for monitoring and assessing long-term environmental changes on Mars, including temperature, radiation levels, atmospheric composition, and habitability potential.

8. Overview of Martian Environment

Mars, the fourth planet from the Sun, presents a diverse and intriguing environment, attracting intense scientific study and exploration. Its surface conditions feature a wide temperature range, varying from an average of –80°C (–112°F) to 20°C (–4°F) near the equator during daytime. The thin atmosphere results in an average surface pressure of 6.36 millibars, composed mainly of carbon dioxide, with trace amounts of other gases. Moreover, Mars' thin atmosphere and lack of a strong magnetic field expose its surface to higher levels of cosmic and solar radiation. Water resources on Mars hold great importance due to their potential for life and future human missions. Polar ice caps contain vast amounts of frozen water, while evidence suggests the presence of subsurface water ice across significant areas. Seasonal changes create transient features, indicating the flow of salty water during warmer periods. Despite its extreme conditions, Mars' subsurface environments offer the potential for microbial life and protective habitats for humans. Geologically, Mars showcases diverse landscapes, such as the massive volcano Olympus Mons and a network of valleys and channels carved by past flowing water. The planet's surface also boasts impact craters, plains, highlands, canyons, and polar ice caps, contributing to a captivating and varied terrain. Overall, the study of Mars continues to deepen our understanding of planetary processes and provides insights into the potential for future exploration and habitation by humans.

9. Earth's Polyextreme Environments

Earth's polyextreme environment provides a vast array of extreme conditions that are home to remarkable microorganisms known as polyextremophiles. These unique and resilient extremophiles thrive in multiple harsh habitats, defying conventional notions of the limits of life. Polyextremophiles have been discovered in environments such as hydrothermal vents in the deep ocean, where they endure high temperatures, crushing pressures, and chemically-rich surroundings. Furthermore, they have been found in hypersaline lakes, brine pools, and arid deserts, where they withstand extreme salinity, heat, and aridity. Some polyextremophiles inhabit the frigid temperatures and high radiation levels of polar regions and high-altitude environments, showcasing their adaptability to extreme cold and intense UV radiation. These polyextremophiles have evolved specialized mechanisms and biochemistry that enable them to thrive in these seemingly inhospitable conditions, expanding our understanding of life's diversity and its potential to survive in some of the harshest environments on Earth and beyond. Studying polyextremophiles offers valuable insights into the resilience and adaptability of life, and it may have implications for astrobiology and the search for life on other planets. One notable example of polyextremophiles can be found in hydrothermal vents, deep within the ocean floor (Table 1). These extremophiles, such as thermophilic archaea and bacteria, endure scalding temperatures of up to 400°C (752°F) and pressures exceeding 300 times that of Earth's surface, making them some of the most heat-tolerant organisms known.

In hypersaline environments, like the iconic Dead Sea or salt flats, halophilic archaea flourish, coping with salt concentrations up to ten times that of seawater. These organisms have adapted to the osmotic stress by accumulating compatible solutes, enabling them to maintain cell integrity and function in high saline conditions. Additionally, they possess specialized cell membrane proteins that counteract the deleterious effects of salt on their cellular structures. In arid deserts,

Table 1: Polyextreme environments and the potential martian analog sites on the Earth

Polyextreme Environment	**Sub-Category**	**Polyextremophiles Model Microorganisms**	**Potential Martian Analog Sites**	**References**
Deep-sea hydrothermal vents	High temperature	*Thermococcus gammatolerans,*	Mid-Atlantic Ridge	Blöchl et al., 1997; Reysenbach and Cady, 2001; Jolivet et al., 2003; Kelley et al., 2005
	High pressure	*Pyrolobus fumarii,*	Lost City Hydrothermal Field	
	High radiation	*Thermococcus onnurineus*	Guaymas Basin	
Acidic hot springs	High temperature	*Acidithiobacillus ferrooxidans,* *Acidianus brierleyi*	Yellowstone National Park Rio Tinto, Spain	Valdés et al., 2008
Hypersaline environments	High salinity	*Salinibacter ruber,*	Great Salt Lake, Utah	Antón et al., 2002
	High UV radiation	*Halobacterium salinarum*	Atacama Desert, Chile	
Polar regions	Low temperature	*Psychrobacter cryohalolentis, Chlamydomonas*	McMurdo Dry Valleys, Antarctica	Misra et al., 2006
Radioactive sites	High radiation	*Deinococcus radiodurans,*	Chernobyl Exclusion Zone	Daly, 2009; Takacs-Vesbach et al., 2010
	Low temperature	*Candidatus; Desulforudis audaxviator*	Mount Erebus, Antarctica	

where water is scarce and temperatures can soar, xerophilic fungi and bacteria have evolved to withstand extreme desiccation and heat. They form protective spores or other survival structures, preventing cellular damage and ensuring their survival during extended dry periods. Moreover, some xerophiles produce enzymes and proteins that remain active at high temperatures, allowing them to thrive in hot environments. Polyextremophiles also inhabit the frigid and UV-radiation-rich environments of polar regions. For instance, psychrophilic bacteria are capable of growth and reproduction at near-freezing temperatures. They have adapted to cope with low temperatures by modifying their cell membranes to remain flexible and functional in the cold.

At high altitudes, where oxygen levels are significantly reduced, organisms like hypoliths, which are photosynthetic cyanobacteria, thrive beneath translucent rocks in extreme cold and low atmospheric pressure. These microorganisms receive just enough light for photosynthesis, and they have adapted their cellular processes to function efficiently under these challenging conditions. The study of polyextremophiles not only enhances our understanding of life's adaptability on Earth but also has important astrobiological implications. These resilient microorganisms provide insights into the potential for life to exist in extraterrestrial environments with extreme conditions, expanding the frontiers of our search for life beyond Earth. By unravelling the unique strategies and mechanisms of polyextremophiles, scientists are paving the way for advancements in biotechnology, pharmaceuticals, and agriculture, as well as gaining valuable insights into the origin and evolution of life on our planet.

10. Mars Analogue Polyextreme Environments on Earth

Early in its history, Mars underwent a polyextreme environmental evolution with diverse extreme conditions. The preservation of subaerial habitats, ecotones, and microbial dispersal pathways during the Noachian to Amazonian transition depended on fluctuating interactions between various environmental extremes and their dominance over time, influenced by factors like the passage of time and changes in Mars' axial tilt. To understand the potential habitability of Mars and the possibility of life existing there, scientists have turned to studying extreme environments on Earth known as "Martian analog sites". By studying Martian analog sites on Earth, we can access essential data using advanced technologies and instruments, including mineralogical, geochemical, geochronological, and microbiological information. This research sheds light on water cycles, habitability in extreme environments, the limits of life, and the detection of biosignatures. Furthermore, it offers valuable insights into the search for life beyond our planet and contributes to refining planetary protection measures.

Following are the various analog sites on Earth that offer valuable opportunities to study polyextremophiles and gain insights into their adaptability and potential for life in extreme environments, providing essential knowledge for astrobiological research and the search for life on Mars:

- **Rio Tinto Mars Analogue Site:** Located in the Iberian Pyritic Belt in Southwest Spain, the Rio Tinto Mars Analogue site is characterized by the highly coloured, deep red Rio Tinto, also known as the "Red River". This river supports microbial life under extremely acidic conditions with a pH of around 2 or even lower. The area is rich in iron (~25,000 ppm) and sulphur (~110,000 ppm), contributing to the blood-red pigmentation of the river. Microbial communities have adapted to this extreme environment, forming biofilms and mats in the water's edge and isolated pools in the Jaroso Ravine. The mineral-rich environment also hosts significant quantities of sulfates like jarosite, halotrichite, copiapite, and gypsum. The presence of similar geological deposits on Mars, identified by the Mossbauer spectrometer on board the MER 'Opportunity' rover, makes this site particularly relevant for studying Martian environments.
- **Atacama Desert:** The Atacama Desert in North Chile is the driest place on Earth, maintaining extreme aridity and high UV radiation, making it a compelling analog for Martian conditions. In this hyperarid desert, researchers have discovered fascinating endolithic habitats within rocks, which provide shelter to phototrophic and heterotrophic microorganisms. These endolithic communities serve as a stress avoidance strategy, as the mineral substrate protects them from lethal UV and PAR radiation, provides physical stability, and enhances moisture availability. The Atacama Desert's microbial communities thrive in various rock substrates, such as halites, Ca-sulfate crusts, ignimbrites, calcites, and rhyolites. These communities include prokaryotes like Cyanobacteria and diverse eukaryotic microorganisms, such as non-lichenized and lichenized fungi and algae. Metagenomic analyses have revealed autotrophic CO_2 fixation and photorespiration carried out primarily by cyanobacteria, indicating their functional significance in this extreme environment.
- **Lava Caves:** Lava caves on Earth offer a unique environment for studying microbial mats with diverse colours ranging from white to yellow to pink. These microbial communities vary in size from small scattered colonies to extensive areas covering cave walls or ceilings. The microbial mats in moist Hawaiian and Azorean caves tend to be more extensive compared to the more arid caves in Hawaii and New Mexico. Cyanobacteria, though rarely found in caves beyond the Twilight Zone, are present in some caves. The dominant bacterial communities in the surface volcanic terrains in Hawaii comprise Acidobacteria, Alpha- and Gammaproteobacteria, Actinobacteria, and Cyanobacteria. These subsurface microbial mats in lava caves have significant potential as analogs for studying microbial survival strategies in light-deprived environments.
- **Lonar Lake:** Located in the Buldhana district, Maharashtra, India, Lonar Lake is a shallow lake occupying a young impact crater formed between 15–570 thousand years ago. The lake is saline (8–40 ppt) and hyperalkaline

(pH 9.5–10.5) due to evaporative concentration, making it an extreme environment for life. Despite these harsh conditions, microbial life thrives in Lonar Lake, including cyanobacteria, planctomycetes, spirochaetes, and alkalophilic methanogens. The lake sediments preserve a variety of biomarkers, providing insights into past environmental conditions and carbon cycling rates. The microbial communities in Lonar Lake offer valuable information on how life adapts and survives in challenging environments, with potential implications for understanding Martian habitats.

11. Similarities and Differences between Mars and Earth's Polyextreme Environments

The study of polyextreme environments on Mars and Earth not only deepens our understanding of the limits of life but also sheds light on the broader field of astrobiology. The following will be a comparative study between Mars and Earth's polyextreme environments to understand life's adaptability; it also holds promise for advancements in biotechnology, pharmaceuticals, and our broader quest to unlock the mysteries of life's origins and existence in the cosmos.

Temperature and Pressure Ranges

Mars: The temperature on Mars varies significantly due to its thin atmosphere, ranging from an average of –80°C (–112°F) to a maximum of about 20°C (–4°F) near the equator during daytime. The key factor here is the thin atmosphere, which cannot effectively retain heat, leading to extreme temperature fluctuations. Additionally, the average surface pressure on Mars is a mere 6.36 millibars, which is less than 1% of Earth's atmospheric pressure. This low pressure affects the boiling and freezing points of water on Mars, making it challenging for liquid water to exist on the surface.

Earth's Polyextreme Environments: Earth's polyextreme environments encompass a wide range of temperature and pressure conditions. For instance, hydrothermal vents in the deep ocean can reach temperatures as high as 400°C (752°F) and experience extreme pressure due to the depth of the ocean. In contrast, polar regions can reach extremely low temperatures, sometimes dropping below –80°C (–112°F). Earth's diverse environments offer an array of temperature and pressure extremes, exceeding the ranges observed on Mars, thanks to Earth's thicker atmosphere and varying geophysical characteristics.

Availability and State of Water

Mars: Mars has polar ice caps consisting of water and dry ice (frozen carbon dioxide). Subsurface water ice is also believed to exist across significant areas of the planet. Seasonal changes have led to the formation of recurring slope lineae (RSL), suggesting the transient flow of salty liquid water during warmer periods. However, the overall availability of liquid water on the surface is limited, primarily due to low temperatures and low atmospheric pressure.

Earth: Earth boasts a wealth of water sources, from vast oceans to freshwater lakes, rivers, underground aquifers, and even extreme environments like hydrothermal vents. In hydrothermal vent systems, water is abundant, albeit at extremely high temperatures and pressures, creating unique ecosystems. Conversely, arid deserts can be extremely dry, but other environments offer a rich variety of liquid water sources. This diversity of water availability on Earth contrasts with Mars, where liquid water is a precious resource.

Radiative and UV Conditions

Mars: Mars' thin atmosphere and lack of a strong magnetic field expose its surface to elevated levels of cosmic and solar radiation, including harmful UV radiation. The limited atmospheric shielding on Mars makes it a hostile environment for potential life forms, as exposure to high levels of radiation can be detrimental to cellular structures and genetic material.

Earth: Earth's robust atmosphere and magnetic field provide a protective shield against cosmic and solar radiation, including harmful UV rays. While some extreme environments on Earth, such as high-altitude regions and deserts, can still experience elevated UV radiation levels, the overall protection afforded by Earth's atmosphere and magnetic field allows for a broader range of habitats where life can thrive.

Soil and Mineral Compositions

Mars: Mars' surface is composed of various minerals, including iron-rich minerals like hematite and magnetite. Additionally, sulfates such as jarosite, halotrichite, copiapite, and gypsum have been identified in Martian soil and rocks, indicating past interactions with water.

Earth: Earth's extreme environments exhibit a diverse range of soil and mineral compositions specific to each location. For example, hydrothermal vents are characterized by minerals rich in sulphides due to their association with high-temperature, chemically-rich hydrothermal fluids. In contrast, desert soils may contain high levels of salts and sand. The range of geological compositions on Earth is extensive and influenced by local conditions, creating unique habitats for extremophiles.

Other Factors

Mars: Mars experiences significant dust storms, which can affect surface conditions and visibility. The planet also boasts remarkable geological features, including Olympus Mons (the largest volcano in the solar system), valleys, channels carved by past water flows, and numerous impact craters.

Earth: Earth's polyextreme environments encompass a wide array of geological features, from underwater mountain ranges and deep-sea trenches to glaciers, deserts, and high-altitude regions. Each of these environments contributes to the planet's rich diversity of ecosystems and geological formations.

12. Polyextremophiles from Mars Analogue Site as Analogue Model Microorganisms for Life on Mars

Polyextremophiles of the Grit Crust of Atacama Desert as a Model Microorganisms

The microbial population exhibits remarkable adaptation to the challenging environmental circumstances found in the Atacama Desert, including the greatest levels of solar radiation, significant temperature fluctuations, and pronounced wet-dry cycles. In addition, this biocrust possesses certain distinctive characteristics that set it apart from biocrusts seen in other desert biomes on Earth. Jung et al. (2022) emphasize the capacity of the grit crust to serve as a prototype for astrobiology, specifically in relation to the establishment of extraterrestrial microbial communities and the development of biotechnological solutions that facilitate human colonization of celestial bodies (Jung et al., 2022). The presence of terrestrial extremophile microbes has always served as a source of inspiration for ideas on life strategies in extra-terrestrial settings, including Mars, as well as their potential applicability in space activities driven by biotechnology. Ultratrophic organisms typically encompass a variety of hetero- and chemoautotrophic bacteria, archaea, fungus, photoautotrophic cyanobacteria, and eukaryotic algae (Coleine & Delgado-Baquerizo, 2022; Sundarasami et al., 2019). Biocrust-forming bacteria, which are the earliest terrestrial microbial associations on Earth, can provide insights on ancient human inquiries, such as the existence of life alone and the origins and development of life. This is related to their very tolerant nature, as well as their early origins and their capacity to interact with one another. Cyanobacteria, a group of microorganisms that underwent evolutionary changes in the late Archaean period (Sánchez-Baracaldo & Cardona, 2020), have a wide distribution across many environments on Earth and engage in symbiotic relationships with fungi (Kaasalainen et al., 2017) and plants (Chang et al., 2019). Cyanobacteria not only experienced extreme UV conditions but also played a significant role in the Great Oxygenation Event, which altered Earth's atmosphere and served as the precursor to all other eukaryotic photosynthetic species, including algae and plants. This suggests that cyanobacteria (and other microbes) have resemblance to the notion of basic life forms that possess a significant capacity to adapt to harsh environments. Therefore, similar life forms might potentially exist in other planets, even if the abiotic circumstances vary. Regardless of their current status as living or extinct, these organisms have left their imprints on Earth. By analyzing the patterns they have left behind, scientists can uncover biosignatures that can aid in the exploration for life on other celestial bodies, particularly within the Solar System, and specifically on worlds like Mars.

The Atacama Desert is widely recognized as a region with elevated levels of radiation, with the highest recorded irradiation on Earth (Rondanelli et al., 2015). Constituents of the microbial community are diverse micro-lichens (*Buellia* sp., *Pleopsidium chlorophanum*, *Tetramelas pulverulentus*, *Aspicilia* sp., *Diploschistes* sp.), with green algal photobionts (*Trebouxia* sp., *Myrmecia* sp.), free-living eukaryotic green algae (*Stichococcus* sp., *Pseudostichococcus* sp., *Chlorella* sp., *Klebsormidium* sp.) and cyanobacteria (*Nostoc* sp., *Microcoleus vaginatus*, *Kastovskya adunca*, *Myxacorys chilensis*, *Pleurocapsa* sp.), as well as melanized microcolonial fungi (*Constantinomyces* sp., *Lichenothelia* sp.) (Jung et al., 2020). The challenging environmental circumstances in which they flourish suggest that the microbial community responsible for constructing the grit crust, or at least the specific strains present within it, might potentially serve as viable candidates for assessing their viability in outer space or under Martian conditions (Sánchez-Baracaldo & Cardona, 2020). It is important to note that unicellular living forms emerged throughout the early stages of Earth's history. Conversely, the climate of the Atacama Desert is deemed to be consistent for a minimum of 150 million years (Hartley et al., 2005), suggesting that the microscopic organisms responsible for constructing the grit crust of the Atacama Desert are highly suited to harsh conditions. The microbial community of the grit crust is well-suited for biotechnological purposes in non-natural environments, such as the deliberate colonization of other planets by humans. In these cases, cyclic artificial wetting is crucial, but unlike other biocrusts, the grit crust is not lethal due to its specialized ability to tolerate desiccation. Although the grit crust bacteria colonize the top of the quartz pebbles, they also reside in the undersurface and naturally existing fissures. These areas are shielded from intense UV light, yet they are still able to carry out photosynthesis since the stones are clear to translucent. Furthermore, previous studies have demonstrated that these microorganisms play an active role in facilitating bioweathering processes related to the absorption of nutrients. This, in turn, might result in the development of a terrestrial protopedon as an early step in pedogenesis (Jung, Baumann, Emrich et al., 2020).

Microalgae, Lichens, and Fungus found in the Grit Crust as a Model Microorganism

Phototrophic extremophiles have been observed in challenging habitats on Earth, including arid deserts, hot springs with varying acidity levels, caves, and Antarctic sub-glacial lakes. These settings are regarded analogous to Mars and the ice moons of Jupiter and Saturn (Sánchez-Baracaldo & Cardona, 2020). The prokaryotic cyanobacterial order *Chroococcidiopsidales* has garnered significant attention in astrobiology research due to its intriguing presence in many biological niches within extreme settings. The tolerance of some strains belonging to the genus *Chroococcidiopsis* sensu lato has been demonstrated to withstand a minimum of 4 years of air-drying, as well as exposure to UVC radiation up to 13 kJ m^{-2}, X-rays up to 15 kGy, and γ-radiation up to 12 kGy (Jung et al., 2022). These strains were able to endure 2.5 years of exposure to Mars and Space-like environments in their desiccated form (Jung et al., 2022). Furthermore, a thorough examination of the entire genome sequences revealed that the space-derived strain did not exhibit any higher number of genetic variations compared to triplicates of the reference strain kept on the ground. This is because *Aliterella chasmolithica* (Jung et al., 2022) and *Gloeocapsopsis diffluens* (Jung et al., 2021) possess robust mechanisms to safeguard themselves against harm during extended periods of dryness, which is particularly advantageous for exposure to space-like conditions during travel or terraformation endeavours. Lichens, which are a constituent of the grit crust found in the Coastal Range of the Atacama Desert, constitute a micro-ecosystem including a fungus and an algae photobiont partner. This micro-ecosystem encompasses a diverse microbiome (Jung et al., 2022). The grit crust has been shown to include many lichen species, namely, those belonging to the genera *Aspicilia*, *Buellia*, and *Pleopsidium* (Jung et al., 2020). Additionally, some micro-colonial fungus from the genera *Lichenothelia* and *Constantinomyces*, which exhibit polyextremotolerance (Grube et al., 2013), have also been observed. *Pleopsidium chlorophanum*, a lichen commonly found in the grit crust, was subjected to continuous exposure to simulated circumstances indicative of a protected Martian surface niche (269 kJ m^{-2}) for a duration of 34 days. The lichen not only endured and maintained its photosynthetic activity, but also underwent physiological adaptation by augmenting its photosynthetic activity over a period of 34 days (de Vera et al., 2014). Moreover, black micro-colonial fungi (MCF), commonly known as rock-inhabiting fungi (RIF), exhibit remarkable adaptability to a wide range of environmental conditions, including extreme temperatures, acidity, osmotic stress, salinity, dehydration, solar and UV radiation, and even radioactivity (Onofri et al., 2007). Notably, certain MCF strains have demonstrated the ability to endure both actual space missions and simulated Martian environments (Onofri et al., 2007). This finding suggests that lichens, together with their photobionts and RIFs as extremophile organisms, have the potential to serve as valuable markers for assessing and enhancing the habitability of a planet. Additionally, they can aid in the exploration of potential life-sustaining environments on celestial bodies like Mars (Joseph et al., 2020).

13. Endolithic Polyextremophiles as Model Microorganisms

The study conducted by Wierzchos et al. (2015) highlights the remarkable adaptation mechanisms employed by chlorophototrophic and eukaryotic algae, as well as chlorophototrophic and prokaryotic cyanobacteria, in response to the dry and very solar radiation-rich environment prevalent in this desert region (Wierzchos et al., 2015). It was observed that the primary adaptation strategy employed by organisms was the avoidance of harsh environmental circumstances through the establishment of colonies in cryptoendolithic and hypoendolithic environments within gypsum deposits. The occurrence of cryptoendolithic colonization took place at a depth of a few millimetres under the surface of gypsum. This colonization exhibited a series of well-structured layers consisting of algae and cyanobacteria, a phenomenon that has not been previously documented in the context of endolithic microbial communities. The researchers also detected a significant abundance of carotenoids in the upper cryptoendolithic algal habitat, as well as scytonemin in the hypoendolithic habitat of cyanobacteria. The endolithic habitat is a microhabitat characterized by a network of holes and fissures that are interconnected with the surface of transparent rock. Endolithic colonization can be seen as a strategy to avoid stress, where the mineral substrate above provides effective protection against harmful UV radiation and excessive photosynthetic radiation. It also serves as a thermal buffer, shields against freeze-thaw events, ensures physical stability, and increases moisture availability (De Los et al., 2014; Wierzchos et al., 2012). The hyperarid zone of the Atacama Desert in northern Chile is home to notable instances of lithic ecosystems.

An Anaerobic Polyextremophiles as Model Microorganisms

Anoxic settings pose the greatest challenges in terms of sample collection, isolation, and culturing of microorganisms among polyextreme environments. The objective of Cockell et al. (2018) is to present a comprehensive framework for conducting future field work and sample operations (Cockell et al., 2018). Several anoxic settings were chosen based on their features that resemble historical and contemporary regions on Mars. The samples collected encompassed anoxic springs abundant in sulphur from Germany, the Boulby Mine in the UK, a lake situated in a basaltic environment in Iceland, acidic sediments found in the Rio Tinto in Spain, glacier samples from Austria, and permafrost samples (Cockell et al., 2018). The samples were utilized to employ a culturing methodology aimed at enhancing the presence of anaerobic organisms. This was achieved by employing a specific medium that would enable the cultivation of organisms under consistent circumstances for further physiological comparisons. The MASE study focuses on the examination of chosen organisms in relation to their survival and development in the presence of Mars-related stressors. Artificial fossilisation is employed to generate biosignatures, which are subsequently examined and utilized to assess the effectiveness of life detection instruments in the context of planetary missions (Cockell et al., 2018). Anoxic analogue habitats are a category of terrestrial ecosystems that have received very little research attention in terms of elucidating the boundaries and spatial distribution of liveable extraterrestrial conditions. However, the majority of known extraterrestrial habitats are either devoid of oxygen or have extremely low levels of this gas. As an illustration, the Martian atmosphere is characterized by a low oxygen content of 0.14%. Consequently, it is anticipated that the surface and subsurface of the planet are exclusively conducive to the growth of microbes that possess the ability to thrive in anaerobic environments. Efforts to streamline the process of enriching and isolating anaerobic bacteria have been made for a considerable duration (La Scola et al., 2014). However, these methods continue to pose difficulties because of the meticulous nature of anaerobic organisms, which often exhibit slower growth rates compared to aerobic organisms. This phenomenon can be attributed to the fact that while numerous anaerobic isolates have been acquired from various habitats over time, the isolation of these microbes from analogous conditions is less frequently pursued compared to aerobic species. Ideally, the process of acquiring anaerobic organisms from analogous settings should adhere to a series of procedures. To begin with, it is imperative to acquire samples using rigorous anoxic sampling techniques that effectively reduce the risk of oxygen exposure. Additionally, it is imperative to employ a well-defined medium for the isolation of organisms, ensuring their reproducibility. The core chemical formula can be consistent across all enrichments, enabling the subsequent cultivation of diverse species under comparable medium conditions for the purpose of comparing their physiology and biochemistry. Furthermore, it is advisable to store the acquired isolates in a global culture repository to facilitate their accessibility to fellow researchers, thus circumventing the arduous and often disorganized procedure of directly acquiring them from various researchers. In their study, Cockell et al. (2018) provide a comprehensive account of the systematic methodologies employed by our consortium in the European Commission's Mars Analogues for Space Exploration (MASE) project, along with the outcomes derived from this approach. The project involves several steps, including identifying suitable anoxic field sites, collecting anoxic samples for sequencing to characterize the entire anaerobic community, enriching and isolating anaerobic organisms, and storing them in a culture collection for long-term storage (Cockell et al., 2018).

Polyextremophiles of Caves as Model Microorganisms

In their study, Northup and Lavoie (2001) present a comprehensive analysis of geomicrobiological interactions within caves, which are characterized by restricted nutrition availability and the presence of diverse redox interfaces (Northup & Lavoie, 2001). The breakdown or precipitation of host rock and speleothems (secondary mineral formations) occurs as a result of the interactions between cave microorganisms and mineral environments. The metabolic activities of bacteria that oxidize sulphur, iron, and manganese can produce significant acidity, leading to the dissolution of cave walls and formations. Instances of corrosion that may have been influenced by microorganisms include the presence of corrosion residues (such as those found in the Lechuguilla and Spider caves in New Mexico, USA), moon milk from various caves (such as

Spider Cave in New Mexico and caves in the Italian Alps), and the formation of sulphuric acid speleogenesis and the expansion of caves (such as Movile Cave in Romania and Cueva de Villa Luz in Mexico). The development of carbonates, moon milk, silicates, clays, iron and manganese oxides, sulphur, and saltpetre in caves has been seen to be triggered by microorganisms. This mineralization process occurs at many sizes, ranging from microscopic to landscape biokarst. A significant number of microorganisms found in deep caverns exhibit similar characteristics to those found on the surface, being opportunistic and only active under favourable growth circumstances (E. Northup & Lavoie, 2001). The majority of them are non-residents who are carried by water, air, sediment, and animals. Nevertheless, the enrichment-based and cultural investigations have mostly concentrated on conventional heterotrophic microorganisms identified through surface research. These methodologies have demonstrated the ability to cultivate fewer than 1% of the microorganisms found in a given environment (E. Northup & Lavoie, 2001).

Dark oligotrophic volcanic ecosystems (DOVEs) are ideal habitats for studying life in the absence of sunlight due to their low inorganic content, abundance of chemical reactants, and well-documented chemical interactions with Earth's surface processes. The polar alpine environment on the summit of Mt. Erebus (Antarctica) provides ice caves that serve as habitats for DOVEs. These caves are characterized by a scarcity of organic matter and the presence of oxygenated hydrothermal activity inside highly reducing host rock. In a study, Tebo et al. (2015) conducted a survey on microbial communities in sediment samples collected from three distinct caves (Tebo et al., 2015). These caves were identified using various techniques including polymerase chain reaction (PCR), cloning, sequencing, and analysis of the *small subunit* (*16S*) *ribosomal* and *Ribulose-1,5-bisphosphate Carboxylase/Oxygenase* (*RubisCO*) genes. Notably, two caves were completely dark, while the third cave received snow-filtered sunlight on a seasonal basis. The predominant microbial communities seen in all three caves consist of Bacteria and fungus, with no presence of Archaea. The bacterial communities found in these ice caves exhibit limited phylogenetic variety. However, there is a significant diversity of *RubisCO* genes, including newly formed clades that branch deeply. This suggests that the Calvin-Benson-Bassham (CBB) cycle is involved in the process of CO_2 fixation. An in-depth analysis was conducted on the microbial communities in Warren Cave, a dark cave known for its low phylogenetic diversity. The purpose was to acquire insights on the energetic foundation of the microbial ecology in the cave. The presence of atmospheric carbon, namely, carbon dioxide (CO_2) and carbon monoxide (CO), which may originate from volcanic eruptions, is believed to provide carbon and/or contribute to the energy needs of chemoautotrophic microbial populations in Warren Cave and maybe other caves associated with Mount Erebusice. Our research provides a first insight into the potential of Mt. Erebusice caves as natural experimentation sites for investigating carbon, energy, and nutrient origins in the underground biosphere, as well as the nutritional constraints on life.

Polyextremophiles of Antarctic Dry Valleys as Model Microorganisms

Microorganisms that exhibit extreme radiation resistance have the ability to endure levels of ionizing radiation that surpass those found in the natural environment. The resistance to radiation is thought to be an unintended adaptation to the resistance to desiccation, as both dangers result in comparable cellular harm. Hence, desert soils present a great opportunity for the exploration of novel radiation-resistant breeds. Musilova et al. (2015) was the pioneer in isolating radiation-resistant microorganisms from the Antarctic Dry Valleys, a very cold desert on Mars, utilizing gamma-ray exposure preselection (Musilova et al., 2015). The halomonads, which were discovered by the sequencing of the 16S rRNA gene, had the largest number of survivals among the various irradiation treatments. This study represents the initial investigation of the desiccation and irradiation survival characteristics of the subjects under examination. Furthermore, there has been a lack of quantitative investigation into the correlation between desiccation and radiation resistance across a wide range of microorganisms. A significant association was seen between desiccation and radiation resistance, suggesting that a 5-day increase in desiccation resistance is associated with a corresponding increase in the survival of room-temperature irradiation by 1 kGy. The microbial radiation resistance is enhanced by a factor of 9 when exposed to irradiation at a temperature of –79°C, which is indicative of the usual surface temperatures on Mars. Hence, the capacity of the organisms found here to withstand freezing temperatures, desiccation, and radiation has significant implications for the possibility of sustaining life in a latent or cryopreserved state on Mars.

Other Polyextremophiles as Model Microorganisms

The Rio Tinto, located in Southwestern Spain inside the Iberian Pyritic Belt, is characterized by its very acidic conditions. This acidity is mostly attributed to the chemolithotrophic behaviour of microorganisms that flourish in the extensive deposits of pyrite found in the Iberian Pyritic Belt (Gómez & Parro, 2012). A variety of minerals, including gypsum, jarosite, goethite, and hematites, have been identified in various locations of Mars due to the presence of elevated levels of ferric iron and sulphates resulting from the metabolic processes of pyrite (Fernández-Remolar et al., 2004). Microorganisms residing in protected habitats, such as permafrost, arctic regions beneath ice shield layers, hot springs and hot vent geothermal zones, and acidic conditions, are of significant astrobiological significance. Psychrophiles, hyperthermophiles, acidophiles, halophiles, and bacteria that can withstand high osmotic pressures, as well as alkaliphiles and microorganisms that have adapted to high radiation doses, are particularly intriguing for astrobiology (Rothschild & Mancinelli, 2001). Permafrost settings are mostly inhabited by anaerobic bacteria, including methanogens, and effectively retain organic materials. This characteristic makes them particularly valuable for the identification of biomarkers and the development of automated techniques for life detection. Cryptoendolithic microenvironments refer to uncomplicated ecological niches that are situated within the pores and voids of rocks, at close proximity to their surface. Typically, these safeguarded microenvironments serve as habitats for photosynthetic organisms, which use the sunlight that permeates the rock. This specific scenario underscores the significant significance of accurately determining the appropriate locations to search for indications of life on celestial worlds other than the Earth, in the absence of human intervention. The preservation of organic matter in sedimentary deposits from acidic settings is a subject of particular interest, primarily due to the well-documented challenges associated with oxidative stress. There is ongoing debate on whether bacteria have the ability to attain the most favourable 'design' of this environment through the creation of habitability conditions or homeostatic processes (Fernandez-Remolar et al., 2008). The adaptation of severe environmental circumstances to 'optimal' ones may be facilitated by homeostatic processes and the investment made by microorganisms. The astrobiological significance lies in the outcomes of adaptation to acidic severe conditions and the formation of subterranean ecosystems. Subsurface ecosystems provide optimal protection (Fernández-Remolar et al., 2008) for microbial populations from external radiation and severe circumstances. Additionally, some physico-chemical components such as minerals, precipitates, and iron solutions can serve as shielding materials (Gómez et al., 2007). The survival of life on Earth relies heavily on the presence of redox couples, which are essential for extracting sufficient redox potential to generate proton gradients. The observed metabolic diversity on Earth serves as evidence for the extensive range of redox pairs that can be utilized as a viable energy source. Once again, extremophiles offer valuable insights into their capacity to get energy from redox couples with little disparities in redox potentials. The investigation of metabolic activity by means of redox chemistry has been previously examined during the Viking missions. However, forthcoming mission designs will also use this methodology. The inquiry regarding the presence of extinct life beyond Earth may alone be resolved by missions that assess metabolic activity. These procedures would ensure the attainment of conclusive positive outcomes regarding the existence of viable microorganisms.

Proteins undergo denaturation and unfolding when subjected to temperatures beyond their typical range, leading to the disruption of intracellular linkages, which poses a threat to the organism in question. In order to mitigate the breakdown of proteins caused by elevated temperatures, thermophilic microorganisms employ the production of chaperones or thermosomes. These structures enable the microorganisms to restore their protein structure and functionality, even in challenging environmental conditions (Ali et al., 2023). Bacteria that exhibit robust growth in elevated temperatures have developed hydrogen bonding mechanisms that engage with hydrophobicity, therefore impeding the process of protein unfolding. In contrast, the structural stability of enzymes in thermophilic bacteria can be attributed to the presence of a significant number of salt and disulfide bridges. Furthermore, the thermo-resistance of a system is influenced by several factors such as structural compactness, oligomerization, glycosylation, and hydrophobic interactions among its subunits (Chakraborty et al., 2022). These factors play a crucial role in maintaining stability. The involvement of heat shock proteins (*HSPs*) in protein folding in thermophiles is facilitated by the chaperones *DnaK*, *GroEL*, and *GroES*. The DNA repair mechanism also exhibits a response to DNA damage. In order to stabilize their membranes, thermophiles employ branched chain fatty acids and polyamines, such as spermidine (Orellana et al., 2018). Ultraviolet-resistant extremophiles have evolved a diverse range of mechanisms to counteract the detrimental effects of UV exposure. These methods encompass DNA repair, chaperone activation, and active defence against UV-induced oxidative stress, such as the build-up of glutathione (Pérez et al., 2017). The microorganisms' radio tolerance is linked to their capacity to repair DNA damage due to their accumulation of a significant amount of intracellular manganese and a little amount of iron, which grants them resistance to UV radiation (Paulino-Lima et al., 2016).

The survival of psychrophiles in extremely cold conditions is facilitated by their cellular cold-adaptability systems (Jin et al., 2022). Membrane fluidity loss is prevented by the presence of unsaturated fatty acids, cyclopropane-containing fatty acids, and short chain fatty acids (Bradburn et al., 2003). Another process involves the enhanced production of cold-shock proteins (CSPs) and chaperones, which serve to safeguard the synthesis of RNA and proteins (Godin-Roulling et al., 2015). In the third process, antifreeze proteins (AFPs) are produced and attach to ice crystals, resulting in thermal hysteresis (Muñoz et al., 2017). Mannitol accumulates as a cryo-protectant, constituting a fourth pathway (Baraúna et al. 2017). Acidophiles maintain a cytoplasmic pH close to neutrality to protect acid-sensitive cellular components. The process of adapting to an acidic environment encompasses several methods. Proton flux systems actively transport protons to maintain a pH of 1 in the first mechanism. The outflow of protons through transport pumps in the electron transport chain, as well as the inflow of protons through *F0F1-type ATP synthases*, has been shown in several reports (Michels & Bakker, 1985). Primary proton pumps, also known as symporters, and secondary proton pumps, such as cation/H^+ antiporters, are two types of proton flow systems. One further method inhibits the influx of protons into the cytoplasm through the reduction of cell membrane permeability. The formation of a positive potential by K^+ ions within the membrane serves to impede the entry of protons (Christel et al., 2018). A further illustration of a microbial adaptation may be observed in the upregulation of a protein complex implicated in DNA repair, replication, and recombination in two *Halobacterium* sp. NRC-1 mutants, which have the ability to endure exceedingly high amounts of radiation (LD50 > 11 kGy). Alkaliphile bacteria employ both symporters and antiporters to cope with elevated pH levels. Symporters facilitate the entry of Na^+ and other solutes into cells, while electrogenic antiporters generate a gradient of Na^+ and H^+ concentrations (Krulwich et al., 1998).

If a biosphere had previously existed on Mars, resembling the one that emerged on early Earth, it is hypothesized that it remains in an abiotic condition within the planet's interior, akin to microbial communities residing in the ancient permafrost of the Arctic and Antarctic regions. In the present-day Martian regolith, this ancient life form appears to lack the ability to repair damage (or these processes happen over a long period of time). Ionizing radiation is the primary factor that hinders the existence of such non-living life. In their study, Cheptsov et al. (2018) examined soil samples obtained from two distinct extreme habitats on Earth. These habitats included ancient permafrost found in the Dry Valleys of Antarctica and Xerosol soil from the mountain desert in Morocco. The soil samples were subjected to gamma irradiation with a dose of 40 kGy under conditions of low pressure (1 Torr) and low temperature (≤50°C) (Cheptsov et al., 2018). The samples exhibited microbial communities that demonstrated significant resistance to the applied effects, maintained a substantial number of live cells, exhibited metabolic activity, and displayed a high level of diversification. The findings suggest that the potential biosphere has the capacity to remain in a latent condition for a minimum of 500 thousand years and 8 million years in the surface layer of Mars regolith and at a depth of 5 m, respectively, under the prevailing ionizing radiation intensity.

On June 6, 2017, the Geological Survey of India documented the occurrence of a carbonaceous chondrite meteorite fall near Mukundpura, India. Research was undertaken to examine the microbial community that endured the impact of the meteorite. Meteorite impact soil was found to include *Actinobacteria*, *Proteobacteria*, and *Acidobacteria*, as determined by 16S rRNA metagenomic sequencing. The comparative phylogenetic analysis showed a remarkable prevalence of the class Bacilli in the soil affected by the effect. From a rock that had been struck by the Mukundpura meteorite, *Bacillus thermocopriae* IR-1, a semi-thermotolerant bacterium, was identified. The robustness of *B. thermocopriae* IR-1 to environmental stressors and impact shock in a Reddy shock tube was examined by Thombre et al. (2019). At a peak shock pressure of 300 kPa, temperature of 400 K, and Mach number of 1.47 (Thombre & Parag, 2019), *B. thermocopriae* IR-1 exhibited a survival rate of 28.82% when subjected to the impact of shock waves. This study provides the initial documentation on the impact shock's influence on *B. thermocopriae* IR-1. The present work represents the inaugural investigation into the examination of microbial diversity and the isolation of bacteria from soil within impact craters in the early aftermath of a meteorite impact event.

14. Conclusion

The research on polyextreme environments and Mars-analogue environments for polyextremophiles has the potential to revolutionize our understanding of life and its potential to exist beyond Earth. By studying these extremophilic organisms and the environments in which they thrive, we can gain insights into the limits of life and develop new technologies to explore and colonize extreme environments, such as those on Mars or other moons and planets. One of the foremost implications is its relevance to astrobiology. Studying polyextremophiles in Earth's extreme environments helps us understand the adaptability and resilience of life in harsh conditions. Identifying and characterizing extremophiles on Earth informs our strategies for searching for life on Mars and beyond, guiding the design of future missions and experiments.

Research in Mars analogue environments helps scientists refine their understanding of what constitutes a habitable environment on Mars. The discovery of polyextremophiles on Earth informs the criteria for potential habitability on Mars, including the search for biosignatures, the indicators of past or present life. By studying how extremophiles thrive in analogous settings on Earth, we gain insights into what types of organisms and biosignatures to look for on Mars.

Also, understanding polyextremophiles and their adaptability is essential for planetary protection measures. Preventing contamination of other planets and celestial bodies with Earth organisms is crucial to avoid interfering with potential native ecosystems or masking the detection of indigenous life. Knowledge gained from studying Earth's extremophiles helps establish protocols and guidelines for spacecraft sterilization and mission planning. Research on polyextremophiles also contributes to our understanding of the origin and evolution of life. By studying life forms that thrive in extreme environments, scientists gain insights into the early Earth conditions and the potential habitats where life could have originated. This knowledge aids in the development of hypotheses and experiments related to abiogenesis (the origin of life from non-living matter).

References

Albarracín, Virginia H., Daniel Kurth, Omar F. Ordoñez, Carolina Belfiore, Eduardo Luccini, Graciela M. Salum, et al. 2015. High-up: A remote reservoir of microbial extremophiles in Central Andean Wetlands. *Front. Microbiol.*, 6, 167950.

Ali, Nawab, Muhammad Nughman, & Syed Majid Shah. 2023. *Extremophiles and Limits of Life in a Cosmic Perspective. in Life in Extreme Environments: Diversity, Adaptability, and Valuable Resources of Bioactive Molecules*. IntechOpen.

Antón, Josefa, Aharon Oren, Susana Benlloch, Francisco Rodríguez-Valera, Rudolf Amann, & Ramón Rosselló-Mora. 2002. Salinibacter ruber gen. nov., sp. nov., a novel, extremely halophilic member of the bacteria from Saltern Crystallizer Ponds. *Int. J. Syst. Evol. Microbiol.*, 52(2), 485–491.

Banciu, Horia L., & Dimitry Y. Sorokin. 2013. Adaptation in haloalkaliphiles and natronophilic bacteria. In: *Polyextremophiles: Life Under Multiple Forms of Stress*, pp. 121–178. Springer.

Baraúna, Rafael A., Dhara Y. Freitas, Juliana C. Pinheiro, Adriana R.C. Folador, & Artur Silva. 2017. A proteomic perspective on the bacterial adaptation to cold: Integrating OMICs data of the psychrotrophic bacterium Exiguobacterium Antarcticum B7. *Proteomes*, 5(1), 9.

Bartlett, Douglas H. 2009. Microbial life in the trenches. *Mar. Technol. Soc. J.*, 43(5),128–131.

Bassil, Naji M., & Jonathan R. Lloyd. 2019. Anaerobacillus isosaccharinicus sp. nov., an alkaliphilic bacterium which degrades isosaccharinic acid. *Int. J. Syst, Evol. Microbiol.*, 69(12), 3666–3671.

Basu, Bashita, Camellia Roy, Samayita Saha, Annie Selma Tirkey, Rupanjana Biswas, Asmita Dey et al. 2022. Extremophiles to polyextremophiles: Survival of the fittest. In: *Extremophiles: A Paradox of Nature with Biotechnological Implications*, 1, 221–244. De Gruyter.

Beblo-Vranesevic, Kristina, Maria Bohmeier, Alexandra K. Perras, Petra Schwendner, Elke Rabbow, Christine Moissl-Eichinger et al. 2018. Lack of correlation of desiccation and radiation tolerance in microorganisms from diverse extreme environments tested under anoxic conditions. *FEMS Microbiol. Lett.*, 365(6), fny044.

Belilla, Jodie. 2020. *Unveiling Microbial Communities along the Polyextreme Physicochemical Gradients of Dallol and Its Surroundings* (Danakil Depression, Afar Region, Ethiopia). PhD Thesis. Universite Paris-Saclay.

Bertrand, Jean-Claude, Céline Brochier-Armanet, Manolo Gouy, & Frances Westall. 2015. For three billion years, microorganisms were the only inhabitants of the Earth. In: *Environmental Microbiology: Fundamentals and Applications: Microbial Ecology*, 75–106. Springer Science & Business Media Dordrecht.

Birrien, Jean-Louis, Xiang Zeng, Mohamed Jebbar, Marie-Anne Cambon-Bonavita, Joël Quérellou, Philippe Oger, et al. 2011. Pyrococcus yayanosii sp. nov., an obligate piezophilic hyperthermophilic archaeon isolated from a deep-sea hydrothermal vent. *Int. J. Syst. Evol. Microbiol.*, 61(12), 2827–2881.

Bjelland, Torbjørg, Martin Grube, Solveig Hoem, Steffen L. Jorgensen, Frida Lise Daae, Ingunn H. Thorseth et al. 2011. Microbial metacommunities in the lichen–rock habitat. *Environ. Microbiol. Rep.*, 3(4), 434–442.

Blöchl, Elisabeth, Reinhard Rachel, Siegfried Burggraf, Doris Hafenbradl, Holger W. Jannasch, & Karl O. Stetter. 1997. Pyrolobus fumarii, gen. and sp. nov., represents a novel group of archaea, extending the upper temperature limit for life to 113°C. *Extremophiles*, 1, 14–21. Springer Nature.

Bradburn, Michael J., Taane G. Clark, Sharon B. Love, & Douglas G. Altman. 2003. Survival Analysis Part III: Multivariate data analysis – Choosing a model and assessing its adequacy and fit. *Br. J. Cancer.*, 89(4), 605–611.

Callegan, Ryan P., M. Fernanda Nobre, Patrick M. McTernan, John R. Battista, Rafael Navarro-Gonzalez, Christopher P. McKay et al. 2008. Description of four novel psychrophilic, ionizing radiation-sensitive deinococcus species from alpine environments. *Int. J. Syst. Evol. Microbiol.*, 58(5), 1252–1258.

Capece, Mark C., Evan Clark, Jamal K. Saleh, Daniel Halford, Nicole Heinl, Samuel Hoskins et al. 2013. Polyextremophiles and the constraints for terrestrial habitability. *Polyextremophiles: Life under Multiple Forms of Stress*. pp. 3–59. Springer.

Castro-Severyn, Juan, Coral Pardo-Esté, Katterinne N. Mendez, Jonathan Fortt, Sebastian Marquez, Franck Molina et al. 2021. Living to the high extreme: Unraveling the composition, structure, and functional insights of bacterial communities thriving in the arsenic-rich Salar de Huasco Altiplanic ecosystem. *Microbiol. Spectr.*, 9(1), 10–1128.

Cavalazzi, B., Barbieri, R., Gómez, F., Capaccioni, B., Olsson-Francis, K., Pondrelli, M. et al. 2019. The Dallol Geothermal Area, Northern Afar (Ethiopia): An exceptional planetary field analog on Earth. *Astrobiology*, 19(4), 553–578.

Chakraborty, Anirban, Jayne E. Rattray, Sienna S. Drake, Stuart Matthews, Carmen Li, Bo Barker Jørgensen et al. 2022. Metabolic responses of thermophilic endospores to sudden heat-induced perturbation in marine sediment samples. *Front. Microbiol.*, 13, 958417.

Chang, Aimee Caye G., Tao Chen, Nan Li, & Jun Duan. 2019. Perspectives on endosymbiosis in coralloid roots: Association of cycads and cyanobacteria. *Front. Microbiol.*, 10, 461756.

Chela-Flores, Julian, & Joseph Seckbach. 2011. The Dry Valley Lakes, Antarctica: From sulfur stains on Earth to sulfur stains in the Jovian system. *Instruments, Methods, and Missions for Astrobiology XIV*. Vol. 8152, pp. 207–214. SPIE.

Chela-Flores, Julian. 2013. Polyextremophiles: Summary and conclusions. In: *Polyextremophiles: Life Under Multiple Forms of Stress*, pp. 611–615. Springer.

Cheptsov, Vladimir S., Elena A. Vorobyova, George A. Osipov, Natalia A. Manucharova, Polyanskaya Lubov'M, Mikhail V. Gorlenko et al. 2018. Microbial activity in Martian analog soils after ionizing radiation: Implications for the preservation of subsurface life on Mars. *AIMS Microbiol.*, 4(3), 541.

Christel, Stephan, Malte Herold, Sören Bellenberg, Mohamed El Hajjami, Antoine Buetti-Dinh, Igor V. Pivkin et al. 2018. Multi-omics reveals the lifestyle of the acidophilic, mineral-oxidizing model species Leptospirillum ferriphilum T. *Appl. Environ. Microbiol.*, 84(3), e02091-17.

Chung, Daehwan, Nicholas S. Sarai, Michael E. Himmel, & Yannick J. Bomble. 2020. Genetics of unstudied thermophiles for industry. In: *Metabolic Pathway Engineering*, pp. 5–19. SpringerLink.

Cockell, C.S., Schwendner, P., Perras, A., Petra Rettberg, Kristina Beblo-Vranesevic, Bohmeier, M. et al. 2018. Anaerobic microorganisms in astrobiological analogue environments: From field site to culture collection. *Int. J. Astrobiol.*, 17(4), 314–328.

Coker, James A. 2016. *Extremophiles and Biotechnology: Current Uses and Prospects*. F1000Research 5.

Coleine, Claudia, & Manuel Delgado-Baquerizo. 2022. Unearthing terrestrial extreme microbiomes for searching terrestrial-like life in the solar system. *Trends Microbiol.*, 30(11), 1101–1115.

Cortés-Albayay, Carlos, Johanna Silber, Johannes F. Imhoff, Juan A. Asenjo, Barbara Andrews, Imen Nouioui et al. 2019. The polyextreme ecosystem, Salar de Huasco at the Chilean Altiplano of the Atacama Desert Houses diverse Streptomyces spp. with promising pharmaceutical potentials. *Diversity*, 11(5), 69.

Daly, Michael J. 2009. A new perspective on radiation resistance based on Deinococcus radiodurans. *Nat. Rev. Microbiol.*, 7(3), 237–245.

DasSarma, Priya, André Antunes, Marta Filipa Simões, & Shiladitya DasSarma. 2020. Earth's stratosphere and microbial life. *Curr. Issues Mol. Biol.*, 38(1), 197–244.

De Los Ríos, Asunción, Jacek Wierzchos, & Carmen Ascaso. 2014. The lithic microbial ecosystems of Antarctica's McMurdo Dry Valleys. *Antarct. Sci.*, 26(5), 459–477.

de Vera, Jean-Pierre, Dirk Schulze-Makuch, Afshin Khan, Andreas Lorek, Alexander Koncz, Diedrich Möhlmann et al. 2014. Adaptation of an Antarctic lichen to Martian Niche conditions can occur within 34 days. *Planet. Space Sci.*, 98, 182–190.

Dedysh, Svetlana N., Yulia Y. Berestovskaya, Lina V. Vasylieva, Svetlana E. Belova, Valentina N. Khmelenina, Natalia E. Suzina et al. 2004. Methylocella tundrae sp. nov., a novel methanotrophic bacterium from acidic Tundra Peatlands. *Int. J. Syst. Evol. Microbiol.*, 54(1), 151–156.

Dragone, Nicholas B., Melisa A. Diaz, Ian D. Hogg, W. Berry Lyons, W. Andrew Jackson, Diana H. Wall et al. 2021. Exploring the boundaries of microbial habitability in soil. *J. Geophys. Res. Biogeosciences*, 126(6), e2020JG006052.

Durvasula, Ravi, & Subba Rao, D.V. 2018. Extremophiles: Nature's amazing adapters. In: *Extremophiles*, pp. 1–18. CRC Press.

Farías, Ricardo, Javiera Norambuena, Alonso Ferrer, Pamela Camejo, Claudia Zapata, Renato Chávez et al. 2021. Redox stress response and UV tolerance in the acidophilic iron-oxidizing bacteria Leptospirillum ferriphilum and Acidithiobacillus ferrooxidans. *Res. Microbiol.*, 172(3), 103833.

Fernández-Remolar, D., Gómez-Elvira, J., Gómez, F., Sebastian, E., Martıin, J., Manfredi, J.A. et al. 2004. The Tinto River, an extreme acidic environment under control of iron, as an analog of the Terra Meridiani Hematite site of Mars. Planet. *Space Sci.*, 52(1–3), 239–248.

Fernandez-Remolar, David C., Felipe Gomez, Olga Prieto-Ballesteros, Rachel T. Schelble, Nuria Rodriguez, & Ricardo Amils. 2008. Some ecological mechanisms to generate habitability in planetary subsurface areas by chemolithotrophic communities: The Río Tinto Subsurface Ecosystem as a Model System. *Astrobiology*, 8(1), 157–173.

Fernández-Remolar, David C., Olga Prieto-Ballesteros, Nuria Rodríguez, Felipe Gómez, Ricardo Amils, Javier Gómez-Elvira, & Carol R. Stoker. 2008. Underground habitats in the Río Tinto Basin: A model for subsurface life habitats on Mars. *Astrobiology*, 8(5), 1023–1047.

Gabani, Prashant, & Om V. Singh. 2013. Radiation-resistant extremophiles and their potential in biotechnology and therapeutics. *Appl. Microbiol. Biotechnol.*, 97(3), 993–1004.

Galhardo, Rodrigo S., & Susan M. Rosenberg. 2009. Extreme genome repair. *Cell*, 136(6), 998–1000.

Galiana, Carmen Pire, José María Miralles-Robledillo, Eric Bernabeu, Nadia Harfi, & Rosa María Martínez-Espinosa. 2022. Genetic and protein engineering of halophilic enzymes. In: *Extremozymes and Their Industrial Applications*. pp. 249–278. Elsevier.

Gan, Longzhan, Heming Zhang, Xiufeng Long, Jiewei Tian, Zhikuan Wang, Yuqin Zhang et al. 2018. Ornithinibacillus salinisoli sp. nov., a moderately halophilic bacterium isolated from a saline-alkali soil. *Int. J. Syst. Evol. Microbiol.*, 68(3), 769–775.

Gao, Lihua, Zhengfu Zhou, Xiaonan Chen, Wei Zhang, Min Lin, & Ming Chen. 2020. Comparative proteomics analysis reveals new features of the oxidative stress response in the polyextremophilic bacterium Deinococcus radiodurans. *Microorganisms*, 8(3), 451.

Godin-Roulling, Amandine, Philipp A.M. Schmidpeter, Franz X. Schmid, & Georges Feller. 2015. Functional adaptations of the bacterial chaperone trigger factor to extreme environmental temperatures. *Environ. Microbiol.*, 17(7), 2407–2420.

Gómez, Felipe, Angeles Aguilera, & Ricardo Amils. 2007. Soluble ferric iron as an effective protective agent against UV radiation: Implications for early life. *Icarus.*, 191(1), 352–359.

Gómez, Felipe, & Víctor Parro. 2012. Applications of extremophiles in astrobiology: habitability and life detection strategies. In: *Adaption of Microbial Life to Environmental Extremes: Novel Research Results and Application*, pp. 199–229. Springer.

Gómez, Felipe, Barbara Cavalazzi, Nuria Rodríguez, Ricardo Amils, Gian Gabriele Ori, Karen Olsson-Francis et al. 2019. Ultra-small microorganisms in the polyextreme conditions of the Dallol Volcano, Northern Afar, Ethiopia. *Sci. Rep.*, 9(1), 1–9.

Goordial, Jacqueline, Guillaume Lamarche-Gagnon, Chih-Ying Lay, & Lyle Whyte. 2013. Left out in the cold: Life in cryoenvironments. In: *Polyextremophiles: Life Under Multiple Forms of Stress*, pp. 335–363. Springer.

Grube, Martin, Lucia Muggia, & Cene Gostinčar. 2013. Niches and adaptations of polyextremotolerant black fungi. In: *Polyextremophiles: Life Under Multiple Forms of Stress*, pp. 551–566. Springer.

Hamid, Burhan, Zaffar Bashir, Ali Mohd Yatoo, Fayaz Mohiddin, Neesa Majeed, Monika Bansal et al. 2022. Cold-active enzymes and their potential industrial applications: A review. *Molecules*, 27(18), 5885.

Hartley, Adrian J., Guillermo Chong, John Houston, & Anne E. Mather. 2005. 150 million years of climatic stability: Evidence from the Atacama Desert, Northern Chile. *J. Geol. Soc.*, 162(3), 421–424.

Hoover, Richard B., & Elena V. Pikuta. 2009. Life in ice: Implications to astrobiology. In: *Instruments and Methods for Astrobiology and Planetary Missions XII*. Vol. 7441, pp. 245–258. SPIE.

Jin, Shanshan, Yizhe Wang, & Xihong Zhao. 2022. Cold-adaptive mechanism of psychrophilic bacteria in food and its application. *Microb. Pathog.*, 169, 105652.

Jolivet, Edmond, Stéphane L'Haridon, Erwan Corre, Patrick Forterre, & Daniel Prieur. 2003. Thermococcus gammatolerans sp. nov., a hyperthermophilic archaeon from a deep-sea hydrothermal vent that resists ionizing radiation. *Int. J. Syst. Evol. Microbiol.*, 53(3), 847–851.

Joseph, R.J., Graham, L., Burkhard Büdel, Patrick Jung, Kidron, G.J., Latif, K. et al. 2020. Mars: Algae, lichens, fossils, minerals, microbial mats, and stromatolites in Gale Crater. *Journal of Astrobiology and Space Science Reviews*, 3(1), 40–111.

Jung, Patrick, Karen Baumann, Dina Emrich, Armin Springer, Vincent JMNL Felde, Stefan Dultz et al. 2020. Lichens bite the dust: A Bioweathering scenario in the Atacama Desert. *Science*, 23(11).

Jung, Patrick, Karen Baumann, Lukas W. Lehnert, Elena Samolov, Sebastian Achilles, Michael Schermer et al. 2020. Desert breath: How fog promotes a novel type of soil biocenosis, forming the coastal Atacama Desert's living skin. *Geobiology*, 18(1), 113–124.

Jung, Patrick, Armando Azua-Bustos, Carlos Gonzalez-Silva, Tatiana Mikhailyuk, Daniel Zabicki, Andreas Holzinger, & Michael Lakatos. 2021. Emendation of the Coccoid cyanobacterial genus Gloeocapsopsis and description of the new species Gloeocapsopsis diffluens sp. nov. and Gloeocapsopsis dulcis sp. nov. isolated from the coastal range of the Atacama Desert (Chile). *Front. Microbiol.*, 12, 671742.

Jung, Patrick, Lukas W. Lehnert, Jörg Bendix, Guillaume Lentendu, Martin Grube, Fernando D. Alfaro et al. 2022. The grit crust: A poly-extremotolerant microbial community from the Atacama Desert as model for astrobiology. *Front. Astron. Space Sci.*, 342.

Kaasalainen, Ulla, Alexander R. Schmidt, & Jouko Rikkinen. 2017. Diversity and ecological adaptations in Palaeogene lichens. *Nat. Plants*, 3(5), 1–8.

Kasting, James F. 1993. Earth's early atmosphere. *Science*, 259(5097), 920–926.

Keller, Martin, Franz-Josef Braun, Reinhard Dirmeier, Doris Hafenbradl, Siegfried Burggraf, Reinhard Rachel, & Karl O. Stetter. 1995. Thermococcus Alcaliphilus sp. nov., a new hyperthermophilic Archaeum growing on polysulfide at alkaline pH. *Arch. Microbiol.*, 164, 390–395.

Kelley, Deborah S., Jeffrey A. Karson, Gretchen L. Fruh-Green, Dana R. Yoerger, Timothy M. Shank, David A. Butterfield et al. 2005. A serpentinite-hosted ecosystem: The lost city hydrothermal field. *Science*, 307(5714), 1428–1434.

Kottemann, Molly, Adrienne Kish, Chika Iloanusi, Sarah Bjork, & Jocelyne DiRuggiero. 2005. Physiological responses of the halophilic archaeon Halobacterium sp. strain NRC1 to desiccation and gamma irradiation. *Extremophiles*, 9, 219–227.

Krulwich, Terry Ann, Masahiro Ito, David B. Hicks, Raymond Gilmour, & Arthur A. Guffanti. 1998. pH homeostasis and ATP synthesis: Studies of two processes that necessitate inward proton translocation in extremely alkaliphilic Bacillus species. *Extremophiles*, 2, 217–222.

Kumar, Vikash, & Tulasi Satyanarayana. 2013. Thermoalkaliphilic microbes. In: *Polyextremophiles: Life under Multiple Forms of Stress*, pp. 271–296. Springer.

Kusube, Masataka, Than S. Kyaw, Kumiko Tanikawa, Roger A. Chastain, Kevin M. Hardy, James Cameron, & Douglas H. Bartlett. 2017. Colwellia marinimaniae sp. nov., a Hyperpiezophilic species isolated from an amphipod within the challenger deep, Mariana Trench. *Int. J. Syst. Evol. Microbiol.*, 67(4), 824–831.

La Scola, B., Khelaifia, S., Lagier, J.C. & Didier Raoult. 2014. Aerobic culture of anaerobic bacteria using antioxidants: A preliminary report. *Eur. J. Clin. Microbiol. Infect. Dis.*, 33, 1781–1783.

Le Lay, Callum, Joshua N. Hamm, Timothy J. Williams, Mang Shi, Ricardo Cavicchioli, & Edward C. Holmes. 2023. Viral community composition of hypersaline lakes. *Virus Evol.*, 9(2), vead057.

Leuko, Stefan, Petra Rettberg, Ashleigh L. Pontifex, & Brendan P. Burns. 2014. On the response of Halophilic archaea to space conditions. *Life*, 4(1), 66–76.

Lonsdale, Peter. 1977. Clustering of suspension-feeding macrobenthos near abyssal hydrothermal vents at oceanic spreading centers. *Deep-Sea Res.*, 24(9), 857863.

Maizel, Daniela, Laura Edith Alche, & Pablo Jacobo David Mauas. 2019. Polyextremophiles: Exploring the limits of habitability. *Memorie della Societa Astronomica Italiana*, 90, 658.

Marion, Giles M., Christian H. Fritsen, Hajo Eicken, & Meredith C. Payne. 2003. The search for life on Europa: Limiting environmental factors, potential habitats, and earth analogues. *Astrobiology*, 3(4), 785–811.

Martin, William, John Baross, Deborah Kelley, & Michael J. Russell. 2008. Hydrothermal vents and the origin of life. *Nat. Rev. Microbiol.*, 6(11), 805–814.

McKay, Christopher P., Nadia C.S. Mykytczuk, & Lyle G. Whyte. 2011. Life in ice on other worlds. In: *Polar Microbiology: Life in a Deep Freeze*, pp. 290–304. ASM Press.

Merino, Nancy, Heidi S. Aronson, Diana P. Bojanova, Jayme Feyhl-Buska, Michael L. Wong, Shu Zhang, & Donato Giovannelli. 2019. Living at the extremes: Extremophiles and the limits of life in a planetary context. *Front. Microbiol.*, 10, 780.

Mesbah, Noha M., & Juergen Wiegel. 2005. Halophilic thermophiles: A novel group of extremophiles. *Microbial Diversity: Current Perspective and Potential Applications*, pp. 91–118. I.K. International Pvt. Ltd.

Michels, Monika, & Evert P. Bakker. 1985. Generation of a large, protonophore-sensitive proton motive force and pH difference in the acidophilic bacteria Thermoplasma acidophilum and Bacillus acidocaldarius. *J. Bacteriol.*, 161(1), 231–237.

Minegishi, Hiroaki. 2013. Halophilic, acidophilic, and haloacidophilic prokaryotes. In: *Polyextremophiles: Life Under Multiple Forms of Stress*, pp. 201–213. Springer.

Misra, Hari S., Nivedita P. Khairnar, Swathi Kota, Smriti Shrivastava, Vasudha P. Joshi, & Shree K. Apte. 2006. An exonuclease I-sensitive DNA repair pathway in Deinococcus radiodurans: A major determinant of radiation resistance. *Mol. Microbiol.*, 59(4), 1308–1316.

Montalvo-Rodríguez, R.J., López-Garriga, R.H. Vreeland, Oren, A., Ventosa, A., & Kamekura, M. 2000. Haloterrigena thermotolerans sp. nov., a halophilic archaeon from Puerto Rico. *Int. J. Syst. Evol. Microbiol.*, 50(3), 1065–1071.

Muñoz, Patricio A., Sebastián L. Márquez, Fernando D. González-Nilo, Valeria Márquez-Miranda, & Jenny M. Blamey. 2017. Structure and application of antifreeze proteins from Antarctic bacteria. *Microb. Cell Fact.*, 16, 1–13.

Musilova, Michaela, Gary Wright, John M. Ward, & Lewis R. Dartnell. 2015. Isolation of radiation-resistant bacteria from Mars analog Antarctic Dry Valleys by preselection, and the correlation between radiation and desiccation resistance. *Astrobiology*, 15(12), 1076–1090.

Northup Diana, E. & Kathleen H. Lavoie. 2001. Geomicrobiology of caves: A review. *Geomicrobiol. J.*, 18(3), 199–222.

Ojha, Lujendra, Mary Beth Wilhelm, Scott L. Murchie, Alfred S. McEwen, James J. Wray, Jennifer Hanley et al. 2015. Spectral evidence for hydrated salts in recurring slope lineae on Mars. *Nat. Geosci.*, 8(11), 829–832.

Onofri, Silvano, Laura Selbmann, Gerrit S. de Hoog, Martin Grube, Donatella Barreca, Serena Ruisi, & Laura Zucconi. 2007. Evolution and adaptation of fungi at boundaries of life. *Adv. Space Res.*, 40(11), 1657–1664.

Orellana, Roberto, Constanza Macaya, Guillermo Bravo, Flavia Dorochesi, Andrés Cumsille, Ricardo Valencia, & Michael Seeger. 2018. Living at the frontiers of life: Extremophiles in Chile and their potential for bioremediation. *Front. Microbiol.*, 9, 392041.

Pankratov, Timofey A., Denis S. Grouzdev, Ekaterina O. Patutina, Tatiana V. Kolganova, Julia J. Berestovskaya, & Aleksandr A. Ashikhmin. 2020. Lichenicoccus roseus gen. nov., sp. nov., the first bacteriochlorophyll a-containing, psychrophilic and acidophilic Acetobacteraceae bacteriobiont of lichen Cladonia species. *Int. J. Syst. Evol. Microbiol.*, 70(8), 4591–4601.

Parrilli, Ermenegilda, Filomena Sannino, Gennaro Marino, & Maria Luisa Tutino. 2011. Life in icy habitats: New insights supporting panspermia theory. *Rend. Lincei.*, 22, 375–383.

Paulino-Lima, Ivan Glaucio, Kosuke Fujishima, Jesica Urbina Navarrete, Douglas Galante, Fabio Rodrigues et al. 2016. Extremely high UV-C radiation resistant microorganisms from desert environments with different manganese concentrations. *Journal of Photochemistry and Photobiology B: Biology*, 163, 327–336.

Pérez, Vilma, Martha Hengst, Lenka Kurte, Cristina Dorador, Wade H. Jeffrey, Ruddy Wattiez, Veronica Molina, & Sabine Matallana-Surget. 2017. Bacterial survival under extreme UV radiation: A comparative proteomics study of Rhodobacter sp., isolated from high altitude wetlands in Chile. *Front. Microbiol.*, 8, 268467.

Ranawat, Preeti, & Seema Rawat. 2017. Radiation resistance in thermophiles: Mechanisms and applications. *World J. Microbiol. Biotechnol.*, 33, 1–22.

Reysenbach, Anna-Louise, & Sherry L. Cady. 2001. Microbiology of ancient and modern hydrothermal systems. *Trends Microbiol.*, 9(2), 79–86.

Rondanelli, Roberto, Alejandra Molina, & Mark Falvey. 2015. The Atacama surface solar maximum. *Bull. Am. Meteorol. Soc.*, 96(3), 405–418.

Rothschild, Lynn J., & Rocco L. Mancinelli. 2001. Life in extreme environments. *Nature*, 409(6823), 1092–1101.

Runzheimer, Katharina, Clément Lozano, Diana Boy, Jens Boy, Roberto Godoy, Francisco J. Matus et al. 2024. Exploring Andean High-altitude lake extremophiles through advanced proteotyping. *J. Proteome Res.*, 23(3), 891–904.

Sánchez-Baracaldo, Patricia, & Tanai Cardona. 2020. On the origin of oxygenic photosynthesis and cyanobacteria. *New Phytol.*, 225(4), 1440–1446.

Scheffer, Gabrielle, & Lisa M. Gieg. 2023. The mystery of Piezophiles: Understudied microorganisms from the deep, dark subsurface. *Microorganisms*, 11(7), 1629.

Schmidt, Mariane, Anders Prieme, & Peter Stougaard. 2006. Rhodonellum psychrophilum gen. nov., sp. nov., a novel psychrophilic and Alkaliphilic bacterium of the Phylum bacteroidetes isolated from greenland. *Int. J. Syst. Evol. Microbiol.*, 56(12), 2887–2892.

Schmidt, Mariane, Anders Priemé, & Peter Stougaard. 2007. Arsukibacterium ikkense gen. nov., sp. nov, a novel Alkaliphilic, enzyme-producing γ-proteobacterium isolated from a cold and alkaline environment in greenland. *Syst. Appl. Microbiol.*, 30(3), 197–201.

Schultz, Júnia, Alef dos Santos, Niketan Patel, & Alexandre Soares Rosado. 2023. Life on the edge: Bioprospecting extremophiles for astrobiology. *J. Indian Inst.*, 103(3), 721–737.

Schulze-Makuch, Dirk, Dirk Wagner, Samuel P. Kounaves, Kai Mangelsdorf, Kevin G. Devine, Jean-Pierre de Vera et al. 2018. Transitory microbial habitat in the hyperarid Atacama Desert. *PNAS*, 115(11), 2670–2675.

Seckbach, Joseph, Aharon Oren, & Helga Stan-Lotter. 2013. *Polyextremophiles: Life Under Multiple Forms of Stress*. Vol. 27. Springer Science & Business Media.

Shcherbakova, V.A., Chuvil'Skaya, N.A., Rivkina, E.M., Pecheritsyna, S.A., Suetin, S.V., Laurinavichius, K.S. et al. 2009. Novel Halotolerant bacterium from cryopeg in permafrost: Description of Psychrobacter muriicola sp. nov. *Microbiology*, 78, 84–91.

Shivaji, Sisinthy, Gundlapalli S.N. Reddy, Korpole Suresh, Pratima Gupta, Suresh Chintalapati, Peter Schumann et al. 2005. Psychrobacter vallis sp. nov. and Psychrobacter aquaticus sp. nov., from Antarctica. *Int. J. Syst. Evol. Microbiol.*, 55(2), 757–762.

Stan-Lotter, Helga. 2008. Extremophiles, the physicochemical limits of life (growth and survival). In: *Complete Course in Astrobiology*, pp. 121–150. doi: 10.1002/9783527618996.ch5.

Sundarasami, Abhilash, Akshaya Sridhar, & Kabilan Mani. 2019. Halophilic archaea as beacon for exobiology. In: *Recent Advances and Future Challenges*, pp. 197–214. ABSR.

Takacs-Vesbach, Cristina, Lydia Zeglin, J.E. Barrett, Michael N. Gooseff, John C. Priscu, Doran, P. & Lyons, W.B. 2010. Factors promoting microbial diversity in the McMurdo Dry Valleys, Antarctica. In: *Life in Antarctic Deserts and Other Cold Dry Environments: Astrobiological Analogs*, 5, 221. ResearchGate.

Tebo, Bradley M., Richard E. Davis, Roberto P. Anitori, Laurie B. Connell, Peter Schiffman, & Hubert Staudigel. 2015. Microbial communities in dark oligotrophic volcanic ice cave ecosystems of Mt. Erebus, Antarctica. *Front. Microbiol.*, 6, 109324.

Thombre Rebecca, S., & A. Vaishampayan Parag. 2019. Survival of extremotolerant bacteria from the Mukundpura meteorite impact crater. *Astrobiology*, 2019.

Tkavc, Rok, Vera Y. Matrosova, Olga E. Grichenko, Cene Gostinčar, Robert P. Volpe, Polina Klimenkova et al. 2018. Prospects for fungal bioremediation of acidic radioactive waste sites: Characterization and genome sequence of Rhodotorula taiwanensis MD1149. *Front. Microbiol.*, 8, 305317.

Valdés, Jorge, Inti Pedroso, Raquel Quatrini, Robert J. Dodson, Herve Tettelin, Robert Blake et al. 2008. Acidithiobacillus ferrooxidans metabolism: From genome sequence to industrial applications. *BMC Genomics*, 9, 1–24.

Vanlint, Dietrich, Chris W. Michiels, & Abram Aertsen. 2011. Piezophysiology of the model bacterium Escherichia coli. In: *Extremophiles Handbook*. Springer Japan.
Varrella, Stefano, Michael Tangherlini, & Cinzia Corinaldesi. 2020. Deep hypersaline anoxic basins as untapped reservoir of polyextremophilic prokaryotes of biotechnological interest. *Mar. Drugs*, 18(2), 91.
Wierzchos, Jacek, Asunción de los Ríos, & Carmen Ascaso. 2012. *Microorganisms in Desert Rocks: The Edge of Life on Earth*. International Microbiology.
Wierzchos, Jacek, Jocelyne DiRuggiero, Petr Vítek, Octavio Artieda, Virginia Souza-Egipsy, Pavel Skaloud et al. 2015. Adaptation strategies of endolithic chlorophototrophs to survive the hyperarid and extreme solar radiation environment of the Atacama Desert. *Front. Microbiol.*, 6, 154656.
Wigley, Kathryn, Emmanuel Egbadon, Carlo R. Carere, Louise Weaver, Kim Baronian, Lee Burbery et al. 2022. RNA stable isotope probing and high-throughput sequencing to identify active microbial community members in a methane-driven denitrifying biofilm. *Journal of Applied Microbiology*, 132(2), 1526–1542.
Xue, Qiong, Dahe Zhao, Shengjie Zhang, Heng Zhou, Zhenqiang Zuo, Jian Zhou et al. 2021. Highly integrated adaptive mechanisms in Spiribacter halalkaliphilus, a bacterium abundant in Chinese Soda-Saline Lakes. *Environ. Microbiol.*, 23(11), 6463–6482.
Yan, Miao, Xinyu Zhu, Laura Treu, Giulia Ravenni, Stefano Campanaro, Estelle Maria Goonesekera et al. 2021. Comprehensive evaluation of different strategies to recover methanogenic performance in ammonia-stressed reactors. *Bioresour. Technol.*, 336, 125329.
Zannier, Federico, Luciano Raúl Portero, Omar Federico Ordoñez, Luciano José Martinez, María Eugenia Farías, & Virginia Helena Albarracin. 2019. Polyextremophilic bacteria from high altitude Andean Lakes: Arsenic resistance profiles and biofilm production. *Biomed Res. Int.*, 2019.
Zillig, W., Holz, I., Janekovic, D., Schäfer, W., & Reiter, W.D. 1983. The Archaebacterium thermococcus celer represents, a novel genus within the thermophilic branch of the archaebacteria. *Syst. Appl. Microbiol.*, 4(1), 88–94.

CHAPTER

3

Climate Change and Its Impact on Psychrophilic Microbes in Polar Regions

Mehnaz Shakeel[1], Asma Shakeel[2], Malik Asif Aziz[3*], Shayesta Islam[4], Nowsheen Nazir[5], Madeha Alonazi[6], Zakira Nourin[7] and Jegatheesan Pillathil Mathi Guru[3]

[1] Division of Plant Pathology, FOA, Wadura, SKUAST-Kashmir, India
[2] Division of Soil Science, FOA, Wadura, SKUAST-Kashmir, India
[3] Division of Basic Science and Humanities, FOA, Wadura – 193201, SKUAST-Kashmir, India
[4] Department of Environmental Sciences, Cluster University of Srinagar, J&K, India
[5] Division of Fruit Science, FOH, Shalimar, SKUAST-Kashmir, India
[6] Department of Plant and Microbiology, College of Sciences, King Saud University, Riyadh, KSA
[7] Department of Biological Sciences and Chemistry, College of Arts and Science, University of Nizwa, Sultanate of Oman

1. Introduction

Climate change has become an escalating global concern, capturing heightened attention over recent decades (Misra, 2014). While Earth's history is punctuated by natural climate variations, the present warming trend is of unparalleled significance, primarily attributed to human activities and advancing at an unprecedented rate (Folke et al., 2021). This phenomenon poses formidable challenges to polar regions, impacting both the Arctic and Antarctic ecosystems through warming temperatures, diminishing sea ice cover, permafrost thawing, and shifts in precipitation patterns (Figure 1) (Bertler & Barrett, 2010). These environmental shifts have far-reaching consequences, particularly for psychrophilic bacteria, specially adapted to thrive in frigid environments (Lyon & Mock, 2014). Despite being located at opposite ends of the Earth, the Arctic

*Corresponding author: asifmalik@skuastkashmir.ac.in

DOI: 10.1201/9781003717003-3

Figure 1: Impact of climate change on polar regions.

and Antarctic regions share striking similarities, experiencing extreme cold with unique ecosystems. Both regions, characterized by their harsh conditions, host extremophilic organisms known as psychrophiles or cryophiles (Edwards et al., 2020). These remarkable entities, spanning bacteria, archaea, and certain eukaryotes, exhibit a remarkable ability to thrive and reproduce in extremely low temperatures ranging from –20°C to 20°C. The Earth's past has witnessed seven cycles of glacial advance and retreat over the last 650,000 years, indicating the dynamic nature of its climate (Bosson et al., 2023). However, the current surge in global temperatures is undeniably linked to human-induced factors, primarily the combustion of fossil fuels and widespread deforestation. The resultant increase in atmospheric carbon dioxide (CO_2), a primary greenhouse gas, has surged by over a third since the Industrial Revolution, intensifying the greenhouse effect and culminating in global warming (Hay, 2014). Polar regions, characterized by their extreme cold and unique ecosystems, are undergoing rapid climate change, posing significant challenges to microbial communities crucial for these environments. The warming of both the Arctic and Antarctic regions, coupled with increasing extreme heat events, is altering the habitat of psychrophilic bacteria, impacting their distribution and survival (Collins & Margesin, 2019). The reduction in sea ice cover, particularly in the Arctic, directly affects the specialized bacteria adapted to cold environments, potentially leading to disruptions and contractions in their ecosystems (Khan et al., 2020). As the permafrost thaws in the Arctic, releasing trapped organic matter, it serves as a substrate for microbial activity, including psychrophilic bacteria (Nikrad et al., 2016). However, this release also contributes to the emission of greenhouse gases, further exacerbating climate change. Additionally, the intrusion of non-polar species poses a threat to indigenous flora and fauna, potentially altering key features of polar ecosystems and biodiversity. The escalating temperatures in polar regions could lead to expanded anaerobic and methane-based ecosystems, further challenging polar ecosystems. The Intergovernmental Panel on Climate Change (IPCC) underscores the imminent threat posed to polar regions, with more pronounced effects observed in the Arctic compared to the Antarctic (Ottersen et al., 2022). The urgency to address human-induced climate change is paramount, with the IPCC emphasizing that measures taken by 2030 will be crucial. Changes in precipitation patterns, driven by global warming, have the potential to disrupt not only the compositions but also the functionality of plant and soil microbial communities, impacting essential processes in polar ecosystems.

In conclusion, the trajectory of climate change jeopardizes the delicate balance of polar ecosystems. Human-induced alterations in global climate patterns are disrupting not only the physical environment but also the intricate web of relationships among organisms, including psychrophilic microbes. Urgent and concerted efforts are required to mitigate the impacts of climate change, safeguard polar ecosystems, and preserve the biodiversity essential for the well-being of our planet. The call for action is pressing, and the choices made in the next decade will undoubtedly shape the future of polar regions and, by extension, the health of our entire planet. Table 1 represents statistical data revealing the effect of climate change on polar regions.

Table 1: Overview of the recent statistical data related to the melting of polar regions due to climate change

Year Range	Statistic	Detail
1990s–2017	Ice Melt Increase	From 0.8 trillion tonnes/year to 1.3 trillion tonnes/year; a 65% increase in the rate of ice loss
2000–2019	Glacier Melt Acceleration	Meltwater accounted for 21% of sea level rise; annual glacier thinning rates nearly doubled
Early 21st Century	Glacier Ice Loss	Nearly 270 billion tonnes of ice lost per year
Recent Years	Polar Ice Effects	Melting polar ice is warping Earth's crust, changing sea levels, and the planet itself
Oct 2020–Sep 2021	Arctic Temperature	Seventh warmest on record; lowest post-winter sea ice volume in April 2021; third-lowest Eurasian Arctic snow cover in June

2. Impact of Climate Change on Polar Microbial Communities: An In-depth Analysis

Temperature Changes and Microbial Dynamics

Temperature is a pivotal factor intricately linked to the dynamics of microbial communities in polar regions (Verde et al., 2016). The accelerated warming witnessed in these areas, surpassing the global average, is anticipated to endure, fundamentally altering the parameters governing soil microbial activity. This transformative shift has profound implications, potentially reshaping polar carbon cycles and amplifying the release of carbon dioxide (CO_2) into the atmosphere (Bates & Mathis, 2009). Seasonal cycles within marginal ice zones in polar regions undergo significant temperature fluctuations, exerting a pronounced influence on microbial abundance and physiology (Anisimov et al., 2007). The anticipated repercussions of climate change extend to the modulation of distribution and activity among key microbial players, such as cyanobacteria and algal species. These microorganisms, instrumental in global carbon fixation and oxygen production, face disruptions in their ecological niches (Ibáñez et al., 2023). The reawakening of metabolic activity in microorganisms that have remained cryogenically preserved for millennia introduces a dynamic facet to microbial responses in the wake of climate change. A nuanced understanding of the intricate relationship between temperature changes and microbial dynamics is imperative for anticipating consequences on polar carbon cycles, greenhouse gas emissions, and the overall resilience of microbial communities (Deal et al., 2019).

Altered Precipitation Patterns

Another formidable force shaping the fate of microbial communities in polar regions is the changing paradigm of precipitation patterns. This encompasses alterations in the quantity, intensity, and temporal distribution of rainfall, manifesting as a significant influencer on plant and soil microbial ecosystems. The global trend portraying shifts in precipitation patterns, a phenomenon observed over recent decades and anticipated to persist, carries far-reaching consequences. Experimental endeavors, exemplified by precipitation manipulation studies in subtropical forests, provide valuable insights into the intricate dynamics of altered precipitation (Dolos, 2013). The intentional reduction of dry-season rainfall coupled with augmented wet-season rainfall yields tangible shifts in soil water content (SWC), exerting discernible impacts on microbial community composition and enzyme activities. Such empirical investigations illuminate the sensitivity of microbial communities to precipitation variations, underscoring potential ramifications for critical biogeochemical processes, including litter decomposition, and soil carbon and nitrogen cycling. Unraveling the interconnections between precipitation patterns and microbial community dynamics furnishes indispensable information for forecasting the resilience and adaptability of polar ecosystems amidst the ongoing climate change.

Ice Movements and Microbial Habitats

The intricate ballet of ice movements in polar regions introduces an unparalleled dimension to the exploration of microbial communities. Habitats once perceived as inhospitable, namely glacial ice habitats, now emerge as havens for microbial life, exhibiting active metabolism and growth (Microbes, 2021). Recent strides in microbiological analysis techniques have unveiled a tapestry of diverse microbial strains thriving in these glacial ice bastions. Far from being barren, these habitats provide not only a glimpse into contemporary biological diversity but also act as repositories of invaluable information regarding molecular systematics, metabolic profiles, and evolutionary adaptations. Microorganisms entombed in glacial ice for millennia serve as living archives, unraveling insights into past climates and ecosystems. Understanding the repercussions of ice movements on microbial habitats proves pivotal in deciphering the tenacity of these communities in the face of climate change. The microbial tapestry within glacial ice habitats not only augments the existing biodiversity of polar regions but also functions as a unique portal into Earth's climatic history, offering profound insights that stretch beyond the purview of contemporary microbial ecology.

3. Dynamics in Distribution and Diversity of Psychrophilic Microbes Due to Climate Change

Changes in Microbial Composition

Climate change, particularly the increase in temperature observed in polar regions, has significant repercussions on the composition of microbial communities (Kirchman et al., 2009). The intricate web of interactions within these ecosystems is delicate, and alterations in environmental conditions can trigger substantial shifts in microbial populations. Understanding the changes in microbial composition is crucial for comprehending the broader implications for polar ecosystems. As global temperatures rise due to climate change, the growth of soil microorganisms, including psychrophilic microbes, is impacted. The composition of microbial communities is intricately linked to soil properties, and alterations in temperature can disrupt this balance (De Vries & Shade, 2013). Studies indicate that global warming can lead to a reduction in soil labile carbon, favoring the proliferation of Oligotrophic bacteria while suppressing fungi and Actinobacteria (Wan et al., 2023). Oligotrophic bacteria, known for their ability to thrive in nutrient-poor environments, may experience increased dominance, altering the nutrient dynamics of polar soils. Meanwhile, the decline in fungi and Actinobacteria, which play crucial roles in organic matter decomposition, can have cascading effects on nutrient cycling processes (Sharma et al., 2022). These changes in microbial composition highlight the vulnerability of polar ecosystems to the impacts of climate change, emphasizing the need for in-depth research to grasp the intricate relationships within these microbial communities. Furthermore, climate change-induced alterations extend beyond local effects, influencing the geographical range of microbial species (Hutchins et al., 2019). This shift in distribution can impact the diversity and abundance of microbial communities, fostering changes in microorganism-microorganism and plant-microorganism interactions. For example, psychrotolerant microbes, adapted to grow near sub-zero temperatures, may face challenges as optimum growth conditions shift with rising temperatures (Baba et al., 2021; Kumari et al., 2021). Understanding these shifts is crucial for predicting the future dynamics of polar microbial communities and their roles in ecosystem processes.

Dynamics of Microbial Communities

The dynamics of microbial communities in polar regions are governed by a myriad of factors, each playing a unique role in shaping the intricate tapestry of life in these extreme environments. Temperature, precipitation, and ice movements are among the key influencers that orchestrate the dynamics of microbial populations in polar ecosystems. Life in polar environments unveils a paradox, beyond the harshness of ice and the relentless force of winds, microbial life thrives, organizing itself into complex community structures. The delicate balance of interactions and unique metabolic pathways allow microorganisms to flourish even in seemingly inhospitable conditions. Understanding the dynamics of microbial communities is essential for unraveling the mysteries of life in polar regions and predicting their responses to ongoing climate change. Temperature stands out as a pivotal factor influencing microbial dynamics (Kirchman et al., 2009). Polar microorganisms, including psychrophilic and psychrotolerant microbes, are finely tuned to temperature variations. As temperatures rise, the dynamics of microbial communities undergo intricate changes. The climate regime, characterized by extreme cold, plays a vital role in determining the diversity and dynamics of these communities. Niche characteristics further shape microbial community dynamics in polar regions. The availability of nutrients, moisture levels, and the unique features of polar soils create niches that different microbial taxa exploit. The interactions within these niches contribute to the overall biodiversity and resilience of microbial communities in the face of environmental challenges. Ice movements, another critical factor, introduce additional layers of complexity to microbial dynamics. Glacial ice habitats, once considered devoid of life, reveal microbial communities with active metabolism and growth (Kirchman et al., 2009). Recent advancements in microbiological analysis techniques have unveiled the unexpected diversity within these icy environments. The microbial life within glacial ice not only contributes to current biodiversity but also acts as a biological archive, offering insights into Earth's climatic history (Edwards & Cameron, 2017).

Effects of Biodiversity

Climate change exerts profound effects on the biodiversity of psychrophilic microbes in polar regions, amplifying the urgency of understanding these impacts to safeguard the delicate balance of these unique ecosystems (Khairnar et al., 2022, 2025; Bandh et al., 2022). As global temperatures rise, bacteria, including psychrophiles, undergo adaptations to hotter conditions. However, this adaptation comes at a cost, as the increased respiration rate releases more carbon into the atmosphere. This feedback loop has the potential to accelerate climate change beyond current models' predictions. Bacteria and archaea, related organisms, play pivotal roles in carbon cycling, and alterations in their abundance and activity can have far-reaching consequences (Priyanka et al., 2024). The release of more carbon into the atmosphere contributes to climate warming, creating a feedback loop that intensifies the challenges posed by climate change. This feedback mechanism emphasizes the interconnectedness of microbial communities and global climate dynamics, underscoring the intricate relationship between microorganisms and the broader Earth system (Bardgett et al., 2008). Moreover, changes in microbial biodiversity can affect the functioning of polar ecosystems. Bacteria, fungi, and other microorganisms play essential roles in nutrient cycling, organic matter decomposition, and overall ecosystem health (Sahu et al., 2017). The decline or proliferation of specific microbial taxa can disrupt these vital processes, potentially leading to cascading effects on higher trophic levels and ecosystem stability. Understanding the effects of climate change on the biodiversity of psychrophilic microbes involves unraveling the mechanisms behind microbial adaptations, exploring the resilience of these communities to environmental stressors, and predicting the potential consequences for polar ecosystems (Table 2). As these microbial communities are integral components of Earth's biodiversity, their fate in the face of climate change is intertwined with the broader challenges and opportunities for global conservation efforts.

Table 2: Impact of climate change on psychrophilic microbes in polar regions

Impact	Description	Reference
Community Content Alteration	Large-scale climate changes could lead to the substitution of true psychrophilic species by psychrotolerant ones, potentially due to gene transfer and other processes.	NASA, 2010
Genomic Traits	Psychrophilic Arthrobacter showed distinct genomic and physiological characteristics linked to their global environmental distribution, indicating niche adaptation.	Shen et al., 2021
Soil Microbial Response	Simulated climate change across polar ecosystems indicated changes in soil microbial responses.	Khan and Ball, 2024
Metabolomic Responses	Seasonal cycles in polar regions, including shifts in temperature and salinity, influence microbial abundance and physiology.	Dawson et al., 2023

4. Ecological Functions of Psychrophilic Microbes in Polar Regions

Nutrient Cycling in Changing Environments

Nutrient cycling is a fundamental ecological process that sustains life by ensuring the flow of essential elements through ecosystems. In polar regions, where extreme conditions prevail, this process is finely tuned and largely orchestrated by microorganisms, including psychrophilic microbes. Understanding nutrient cycling is not only vital for ecological balance but also holds implications for sustainable agriculture and environmental management. Microbes, as primary actors in nutrient cycling, play a pivotal role in determining how much carbon is stored in polar soils (Nikrad et al., 2016). The intricate dance of carbon through various pathways, from microbial biomass to atmospheric release, is a critical aspect of nutrient cycling. Considering nutrient cycling as a consumer-driven process provides insights into the mechanisms underlying this ecological phenomenon. It allows us to delve into the complexities of microbial interactions with their environment, unraveling the factors that govern nutrient cycling pathways. In the context of psychrophilic microbes, the changing environmental conditions in polar regions due to climate change have tangible effects on their survival strategies and diversity (King-Miaow et al., 2019). Short-term exposure of Antarctic soils to low temperatures, for instance, alters microbial dynamics by increasing biomass with lower bacterial diversity. This shift in microbial composition, particularly with a prominence of the phyla proteobacteria, firmicute, and actinobacteria, highlights the sensitivity of nutrient cycling pathways to temperature variations (Sharma et al., 2020). Understanding these nuances is crucial for predicting and mitigating the consequences of climate change on polar nutrient cycling dynamics.

Carbon Storage and Sequestration

The role of psychrophilic microbes in carbon storage and sequestration emerges as a linchpin in the global fight against climate change. Carbon, a cornerstone element in microbial growth, undergoes a transformative journey within microbial cells. When utilized for growth, carbon becomes incorporated into microbial biomass, fostering its storage in polar soils (Sheikh et al., 2022). This process assumes paramount importance as it directly impacts the atmospheric carbon dioxide levels, a potent greenhouse gas. Conversely, when carbon is employed for metabolic activities, it is released into the air as carbon dioxide. Psychrophilic microbes, with their unique adaptations to cold environments, wield a significant influence on the delicate balance between carbon storage and atmospheric release (Panikov, 2014). Their role in carbon sequestration is magnified by the fact that microbial processes in the soil surpass other mechanisms, including biomatter decomposition, by at least fourfold. The significance of microbial contributions to carbon storage cannot be overstated. Earth's soils, holding three times more carbon than the atmosphere, become crucial carbon sinks. Psychrophilic microbes, by mediating carbon dynamics in polar soils, play a pivotal role in mitigating climate change (Dey, 2024). Understanding the intricacies of this process, including the specific mechanisms employed by psychrophilic microbes, is essential for devising effective climate change mitigation strategies.

Maintaining Ecosystem Balance

The role of psychrophilic microbes extends beyond nutrient cycling and carbon storage; these microorganisms significantly contribute to maintaining the delicate balance of polar ecosystems. Their impact reverberates through various facets of ecosystem health, ranging from nutrient balance in the soil to biodiversity preservation. Nutrient balance in the soil is a key ecological function facilitated by psychrophilic microbes. As primary drivers of nutrient cycling, these microorganisms ensure that essential elements are available to support plant growth and sustain the broader ecosystem (Cary et al., 2010). The intricate dance of nutrient availability and microbial activity influences the biodiversity of polar regions, shaping the unique composition of flora and fauna adapted to extreme cold conditions (Carroll & Carroll, 2020). The applications of psychrophilic microbes extend beyond their ecological roles. Their significance in the food industry, waste processing, agriculture, medicines, and environmental bioremediation underscores their versatility. Harnessing the capabilities of these microorganisms for practical applications is not only a testament to their adaptability but also presents opportunities for sustainable practices in diverse sectors (Hamid et al., 2022; Kuddus et al., 2024). However, the impact of climate change on psychrophilic microbes poses a significant challenge to maintaining the ecosystem balance. The changing environmental conditions, including temperature fluctuations, altered precipitation patterns, and shifts in microbial composition, can disrupt the intricate web of interactions within polar ecosystems (Ernakovich et al., 2014). Understanding these impacts is crucial for predicting future climate scenarios and developing strategies to mitigate the effects of climate change on psychrophilic microbes.

5. Adaptations and Responses of Psychrophilic Microbes

Coping Mechanisms to Temperature Shifts

Psychrophiles, with their remarkable ability to thrive in extremely cold conditions, have evolved an array of sophisticated coping mechanisms to navigate temperature shifts (Teufel & Morgan-Kiss, 2018). These adaptations are crucial for their survival in environments characterized by sub-zero temperatures, where the challenges of maintaining cellular integrity and functionality are immense (Collins & Margesin, 2019). One key adaptation involves the accumulation of low molecular mass carbohydrate cryoprotectants. These cryoprotectants act as shields, safeguarding psychrophile cells from the detrimental effects of ice formation (Collins & Margesin, 2019). By preventing the formation of ice crystals within the cellular structure, these molecules play a pivotal role in maintaining cellular integrity during temperature shifts. Another crucial adaptation lies in the adjustment of membrane lipid composition. Psychrophiles modify the structure of their cell membranes to enhance flexibility and functionality in cold environments (Parvizpour et al., 2021). This adjustment is vital for ensuring membrane fluidity, a property that becomes challenging to maintain at low temperatures. Cold-active small molecules and antifreeze proteins are additional tools in the psychrophile's arsenal. These molecules actively inhibit the recrystallization of ice, preventing the formation of large ice crystals that could damage cellular structures (Abeysinghe et al., 2022). The synthesis of cold-adapted enzymes further exemplifies the coping mechanisms employed by psychrophiles (Figure 2). These enzymes are tailored to function optimally at low temperatures, ensuring the continuation of essential biochemical processes even in cold environments (Tattersall et al., 2012). The abnormal thermo-sensitivity of psychrophiles highlights their adaptation to the specific challenges posed by extreme cold conditions. The enhanced regulation of genes in response to freezing temperatures is a testament to the dynamic nature of their adaptive strategies. The upregulation of cold-active enzymes in the face of hostile temperatures underscores the importance of these molecular tools in maintaining cellular functionality in freezing conditions. Understanding these coping mechanisms provides valuable insights into the resilience of psychrophiles and their ability to thrive in some of the harshest environments on Earth.

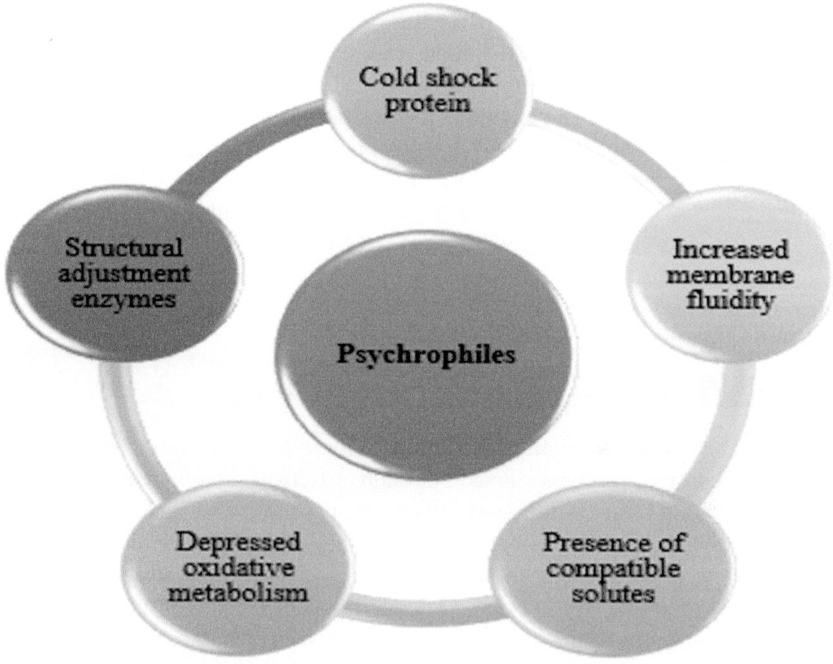

Figure 2: Adaption mechanisms of psychrophiles.

Strategies for Changing Habitat Conditions

Psychrophilic microorganisms, faced with the challenges of surviving in cold conditions, have evolved an arsenal of strategies to adapt to changing habitat conditions. These strategies, ranging from molecular adaptations to complex cellular mechanisms, highlight the versatility of psychrophiles in responding to their dynamic environments (Abeysinghe et al., 2022). Antifreeze proteins and cold shock proteins stand out as key players in the adaptation of psychrophilic microorganisms. These proteins actively counter the formation of ice crystals, preventing cellular damage during temperature fluctuations (Rizzo & Lo Giudice, 2022). Cold-active enzymes further contribute to their adaptive toolbox, ensuring that essential biochemical processes continue to operate at optimal levels in cold conditions. Increased membrane flexibility is another critical strategy employed by psychrophiles. The ability to adjust the flexibility of their cell membranes allows these microorganisms to maintain membrane integrity and functionality in the face of extreme cold (Collins & Margesin, 2019). This adaptation is particularly essential for sustaining cellular processes and preventing damage caused by freezing temperatures. Beyond molecular adaptations, psychrophilic microbes utilize a range of biomolecules and cellular structures to thrive in cold environments. Antibiofilm molecules, surfactants, glycoproteins, and polysaccharides contribute to their resilience against additional stress factors associated with cold habitats (Angelin & Kavitha, 2022). These include desiccation, variations in pH, radiation, excessive UV exposure, high osmotic pressure, and low nutrient availability. Production of outer membrane vesicles and modifications to the cell envelope further exemplify the strategies employed by psychrophiles (Zlatkov, 2021). These adaptations play crucial roles in cellular communication, nutrient acquisition, and overall survival in dynamic and challenging polar environments. Understanding the breadth of these strategies provides insights into the adaptability of psychrophilic microorganisms. Their ability to respond to changing habitat conditions showcases the complexity of their evolutionary journey and the dynamic nature of their interactions with extreme cold environments.

Microbial Resilience to Climate-induced Stress

The resilience of psychrophilic microbes to climate-induced stress is a testament to their adaptive prowess honed through extensive biological evolution. Throughout long-term evolutionary processes, psychrophiles have developed a suite of low-temperature adaptation mechanisms at the cellular and molecular levels, effectively countering the adverse effects of low temperatures on growth and survival (Morgan et al., 2006). One key arena of adaptation lies in the fine composition and structural changes at the level of the cell membrane, protein, and enzyme molecules (Janmohammadi et al., 2015). Psychrophiles have evolved to modify the composition and structure of their cell membranes, ensuring optimal functionality in cold environments. These changes are pivotal for maintaining membrane fluidity, a critical factor in sustaining cellular processes in sub-zero temperatures (De Maayer et al., 2014). Proteins and enzymes, essential players in biochemical reactions, undergo structural alterations to operate optimally in low-temperature conditions. Psychrophiles showcase a remarkable ability to fine-tune these molecular structures, allowing for efficient functioning despite the challenges posed by cold environments. This adaptive flexibility in the molecular realm is a cornerstone of their resilience to climate-induced stress. The long-term biological evolution of psychrophilic microbes has equipped them with the ability to overcome the hurdles presented by low temperatures. This resilience extends beyond molecular adaptations to encompass broader cellular mechanisms that support growth and survival in polar habitats.

6. Permafrost Melting: Implications for Psychrophilic Microbes

Opening New Habitats for Microbial Colonization

Permafrost, characterized by its frozen state, conceals a myriad of geomorphological features beneath its icy surface. These features, including ice wedges, taliks, cryopegs, massive ground ice, frost boils, thermokarst lakes, organic accumulations, and broken soil horizons, serve as unique habitats for microbial life (Altshuler et al., 2017). As permafrost begins to thaw due to global climate change, the repercussions for psychrophilic microbes are profound (Yarzábal et al., 2021). The microbial communities residing within permafrost are intricately connected to the organic matter trapped in its frozen embrace. As thawing progresses, these microbes become active, consuming the once-frozen organic material. This microbial activity releases carbon dioxide and methane into the atmosphere, contributing to the greenhouse gas inventory and exacerbating climate change. Simultaneously, the thawing of permafrost unravels new habitats for microbial colonization. Thermokarst lakes emerge as hotspots for microbial activity in the wake of permafrost thaw (Magnússon, 2022). These lakes, born from the melting of once-frozen ground, offer a unique environment for psychrophilic microbes. The newfound liquid water and nutrient availability create favorable conditions for microbial growth and biogeochemical cycling. This microbial colonization in response to permafrost thaw not only impacts local ecosystems but also introduces an additional dynamic to global biogeochemical cycles (Ernakovich et al., 2022). Furthermore, the thawing process exposes ancient organic matter previously locked away in permafrost. This influx of organic material serves as a nutrient source for microbial communities, stimulating their growth and metabolic activities. The opening of these new habitats, coupled with the release of greenhouse gases, reshapes the ecological landscape and underscores the interconnectedness of permafrost dynamics, microbial life, and climate change (Vincent, 2020).

Release of Ancient Organic Carbon

The vast expanses of permafrost hold a staggering 1,500 billion tons of carbon, nearly double the current atmospheric carbon content. The thawing of permafrost sets in motion a complex interplay between microbial communities and the ancient organic matter they consume. As permafrost melts, the microbial decomposition of frozen organic material becomes a significant player in the global carbon cycle (Ottoni et al., 2022). The microbial communities within permafrost, adapted to cold conditions, spring to life as temperatures rise. Their metabolic activities involve consuming the once-frozen organic matter and releasing carbon dioxide and methane into the atmosphere (Gilman, 2015). This microbial respiration, once initiated by thawing, accelerates the release of carbon compounds that have been sequestered for thousands of years. The implications of this release of ancient organic carbon extend beyond local ecosystems (Chaopricha & Marín-Spiotta, 2014). The carbon stored in permafrost serves as a historical archive of past climates, and its sudden release injects a substantial amount of greenhouse gases into the contemporary atmosphere. The deepening of the permafrost's active layer, facilitated by thawing, allows water to flow more easily, washing away carbon from long-dead plants and animals and contributing to the rising concentrations of greenhouse gases. Understanding the magnitude of this release is crucial for accurately assessing the impact on the global carbon cycle. The sudden infusion of carbon, with its origins dating back thousands of years, introduces a dynamic element to climate change dynamics, adding complexity to the intricate web of interactions between permafrost, microbes, and the Earth's climate system.

Impact of Metabolic Activities and Greenhouse Gas Emissions

Permafrost thawing triggers a cascade of events that significantly impact microbial activities and greenhouse gas emissions. In its frozen state, permafrost acts as a vast reservoir of carbon, largely inert due to the immobilization of microbial activity. However, as permafrost thaws, microbial decomposition of organic matter intensifies, leading to the release of substantial amounts of greenhouse gases. The intricate dance between permafrost, microbial communities, and greenhouse gas emissions hinges on the metabolic activities of psychrophilic microbes. The deep-frozen soil keeps carbon inert, but the thawing process unleashes microbial activity, transforming once-stable carbon into volatile carbon dioxide and methane to a lesser extent. The consequences of this process have far-reaching implications for climate change, potentially accelerating the ongoing global warming trend. As permafrost thaws, the changes in temperature and nutrient availability can impact the metabolic activities of psychrophilic microbes (Vrionis et al., 2011). The intricate balance of nutrient cycling and microbial growth is disrupted, influencing the rates of greenhouse gas emissions. The release of ancient carbon, combined with alterations in microbial metabolic pathways, contributes to the complex dynamics observed in thawing permafrost environments (Lavelle et al., 2005). Moreover, the impact of permafrost thaw extends beyond immediate greenhouse gas emissions. Changes in nutrient availability, soil structure, and microbial composition can have lasting effects on ecosystem dynamics (Vonk et al., 2015). The interconnectedness of these factors underscores the need for comprehensive research to elucidate the nuanced interactions between permafrost thaw, microbial responses, and greenhouse gas fluxes.

7. Conclusion

The rising concentration of atmospheric CO_2 is a pressing global issue that is rapidly transforming ecosystems throughout the world. Among the most affected areas are the polar regions, including the Arctic and Antarctic, where significant changes in temperature, precipitation, and ice movement are occurring. These environmental shifts are having a profound impact on the delicate microbial communities that have adapted to the harsh cold conditions, known as psychrophilic microbes. These specialized microorganisms thrive in extremely cold environments and play a crucial role in important ecological processes such as nutrient cycling, carbon storage, and overall ecosystem health. However, the escalating effects of climate change are fueling a cascade of transformations in the polar regions, further endangering these unique and important microbial communities. The distribution, diversity, and ecological functions of psychrophilic microbes in polar regions are heavily impacted by climate change. As the polar environments warm, microbial communities face changes in their composition and dynamics, which can disrupt the delicate balance of these ecosystems. The melting of ice and permafrost also opens up previously inaccessible habitats for microbial colonization, further altering their distribution. Additionally, the release of ancient organic carbon from melting permafrost acts as a source of nutrients for psychrophilic microbes, potentially increasing their metabolic activities and contributing to greenhouse gas emissions. To accurately predict the feedback loops between climate change and these microbes, it is crucial to understand and study these complex interactions. Urgent action is needed to curb the rising levels of atmospheric CO_2 to mitigate the adverse effects of climate change and protect the delicate ecosystems of the polar regions.

References

Abeysinghe, G., De Zoysa, H.K.S., Bamunuarachchige, T.C., & Zakeel, M.C.M. 2022. Cold-tolerant and cold-loving microorganisms and their applications. In: *Trends of Applied Microbiology for Sustainable Economy*, pp. 185–205. Academic Press.

Altshuler, I., Goordial, J., & Whyte, L.G. 2017. Microbial life in permafrost. In: *Psychrophiles: From Biodiversity to Biotechnology*, pp. 153–179. Springer.

Angelin, J., & Kavitha, M. 2022. Molecular mechanisms behind the cold and hot adaptation in extremozymes. In: *Extremozymes and Their Industrial Applications*, pp. 141–176. Academic Press.

Anisimov, O.A., Vaughan, D.G., Callaghan, T.V., Furgal, C., Marchant, H., Prowse, T.D. et al. 2007. *Polar Regions (Arctic and Antarctic). Climate Change. Impacts, Adaptation, and Vulnerability*. pp. 653–685. In: M.L. Parry, O.F. Canziani, J.P. Palutikof, P.J. van der Linden & C.E. Hanson (Eds.). Contribution of Working Group II to the Fourth Assessment Report of the Intergovernmental Panel on Climate Change. Cambridge University Press, Cambridge.

Baba, Z.A., Hamid, B., Sheikh, T.A., Alotaibi, S., Enshasy, H.E., Ansari, M.J. et al. 2021. Psychrotolerant *Mesorhizobium* sp. isolated from temperate and cold desert regions solubilize potassium and produces multiple plant growth promoting metabolites. *Molecules*, 26, 5758. https://www.mdpi.com/1420-3049/26/19/5758/htm.

Bandh, S.A., Parray, J.A., & Shameem, N. 2022. *Climate Change and Microbial Diversity: Advances and Challenges*, 296 pp. Apple Academic Press, New York,.

Bardgett, R.D., Freeman, C., & Ostle, N.J. 2008. Microbial contributions to climate change through carbon cycle feedbacks. *The ISME Journal*, 2(8), 805–814.

Bates, N.R., & Mathis, J.T. 2009. The Arctic Ocean marine carbon cycle: Evaluation of air-sea CO_2 exchanges, ocean acidification impacts, and potential feedbacks. *Biogeosciences*, 6(11), 2433–2459.

Bertler, N.A., & Barrett, P.J. 2010. Vanishing polar ice sheets. In: *Changing Climates, Earth Systems, and Society*, pp. 49–83. Springer Netherlands.

Bosson, J.B., Huss, M., Cauvy-Fraunié, S., Clément, J.C., Costes, G., Fischer, M. et al. 2023. Future emergence of new ecosystems caused by glacial retreat. *Nature*, 620, 562–569.

Carroll, M.L., & Carroll, J. 2020. The arctic seas. In: *Biogeochemistry of Marine Systems*, pp. 127–156. Blackwell Publishing.

Cary, S.C., McDonald, I.R., Barrett, J.E., & Cowan, D.A. 2010. On the rocks: The microbiology of Antarctic Dry Valley soils. *Nature Reviews Microbiology*, 8(2), 129–138.

Chaopricha, N.T., & Marín-Spiotta, E. 2014. Soil burial contributes to deep soil organic carbon storage. *Soil Biology and Biochemistry*, 69, 251–264.

Collins, T., & Margesin, R. 2019. Psychrophilic lifestyles: Mechanisms of adaptation and biotechnological tools. Applied *Microbiology and Biotechnology*, 103, 2857–2871.

Dawson, H.M., Connors, E., Erazo, N.G., Sacks, J.S., Mierzejewski, V., Rundell, S.M. et al. 2023. Microbial metabolomic responses to changes in temperature and salinity along the western Antarctic Peninsula. *The ISME Journal*, 17(11), 2035–2046.

De Maayer, P., Anderson, D., Cary, C. & Cowan, D.A. 2014. Some like it cold: Understanding the survival strategies of psychrophiles. *EMBO Reports*, 15(5), 508–517.

De Vries, F.T., & Shade, A. 2013. Controls on soil microbial community stability under climate change. *Frontiers in Microbiology*, 4, 265.

Deal, C., Gibson, G., & Weijer, W. 2019. *Marine Biogeochemistry in the Coastal Arctic: Towards Improved Quantitative Understanding of the Controls on Marine Biogeochemical Processes in the Arctic Coastal Zone, and Their Impacts on Climate and the Food Web: A White Paper for DOE's Regional and Global Model Analysis (RGMA) Program* (No. LA-UR-19-28702). Los Alamos National Lab. (LANL), Los Alamos, NM (United States).

Dey, S. 2024. Role of microbes in climate change and its impact: A critical review. In: *Functional Metagenomics*, pp. 169–190. ResearchGate.

Dolos, K. 2013. *Drivers of Forest Dynamics: Joint Effects of Climate and Competition* (Doctoral dissertation). ResearchGate.

Edwards, A., & Cameron, K.A. 2017. Microbial life in supraglacial environments. In: *Psychrophiles: From Biodiversity to Biotechnology*, pp. 57–81. Springer Nature.

Edwards, A., Cameron, K.A., Cook, J.M., Debbonaire, A.R., Furness, E., Hay, M.C., & Rassner, S.M.E. 2020. Microbial genomics amidst the Arctic crisis. *Microbial Genomics*, 6(5), e000375.

Ernakovich, J.G., Hopping, K.A., Berdanier, A.B., Simpson, R.T., Kachergis, E.J., Steltzer, H. et al. 2014. Predicted responses of arctic and alpine ecosystems to altered seasonality under climate change. *Global Change Biology*, 20(10), 3256–3269.

Ernakovich, J.G., Barbato, R.A., Hewitt, R.E., Doherty, S.J., Whalen, E.D., Abbott, B.W. et al. 2022. Microbiome assembly in thawing permafrost and its feedbacks to climate. *Glob. Chang. Biol.*, 28(17), 5007–5026.

Folke, C., Polasky, S., Rockstrom, J., Galaz, V., Westley, F., Lamont, M. et al. 2021. Our future in the Anthropocene biosphere. *Ambio*, 50, 834–869.

Gilman, F.R. 2015. *Microbial Responses and Contributions to Climate Change in Greenland*. University of Montana.

Hamid, B., Bashir, Z., Yatoo, A.M., Mohiddin, F., Majeed, N., Bansal, M. et al. 2022. Cold-active enzymes and their potential industrial applications: A review. *Molecules*, 27(18), 5885.

Hay, W.W. 2014. The accelerating rate of global change. *Rendiconti Lincei*, 25, 29–48.

Hutchins, D.A., Jansson, J.K., Remais, J.V., Rich, V.I., Singh, B.K. & Trivedi, P. 2019. Climate change microbiology: Problems and perspectives. *Nature Reviews Microbiology*, 17(6), 391–396.

Ibáñez, A., Garrido-Chamorro, S., & Barreiro, C. 2023. Microorganisms and climate change: A not so invisible effect. *Microbiology Research*, 14(3), 918–947.

Janmohammadi, M., Zolla, L., & Rinalducci, S. 2015. Low temperature tolerance in plants: Changes at the protein level. *Phytochemistry*, 117, 76–89.

Khairnar, M., Alhewairini, S.S., Kajale, S.C., Dwibedi, V., Barasarathi, J., Mastinu, A., & Sayyed, R. 2025. Comprehensive genomic analysis of *Sinorhizobium meliloti* isolates associated with Fenugreek (*Trigonella foenum-graecum* Linn.) from diverse agroclimatic regions of India. *Chemical and Biological Technologies in Agriculture*, 12, 135. https://doi.org/10.1186/s40538-025-00811-0.

Khairnar, M., Hagir, A., Parmar, K., Sayyed, R., James, E., & Rahi, P. 2022. Phylogenetic diversity and plant growth-promoting activities of rhizobia nodulating fenugreek (*Trigonella foenumgraecum* Linn.) cultivated in different agroclimatic regions of India. *FEMS Microbiology Ecology*, 98, 2, fiac014. https://doi.org/10.1093/femsec/fiac014.

Khan, A., & Ball, B.A. 2024. Soil microbial responses to simulated climate change across polar ecosystems. *Science of The Total Environment*, 909, 168556.

Khan, I. 2020. Microbes and environment: Global warming reverting the frozen zombies. In: *Environment, Climate, Plant, and Vegetation Growth*, pp. 607–633. Springer Nature Switzerland AG.

King-Miaow, K., Lee, K., Maki, T., Lacap Bugler, D.C. & Archer, S. 2019. Airborne microorganisms in Antarctica: Transport, survival, and establishment. In: *The Ecological Role of Micro-Organisms in the Antarctic Environment*, pp. 163–196. Springer International Publishing.

Kirchman, D.L., Morán, X.A.G. & Ducklow, H. 2009. Microbial growth in the polar oceans: Role of temperature and potential impact of climate change. *Nature Reviews Microbiology*, 7(6), 451–459.

Kuddus, M., Bano, N.R., Sheik, G.B., Joseph, B., Hamid, B., Sindhu, R. et al. 2024. Cold-active microbial enzymes and their biotechnological applications. *Microbial Biotechnology*, 17, e14467.

Kumari, A., Upadhyay, J., Joshi, R. 2021. Psychrotrophic microbes: Biodiversity, adaptation, and implications. In: *Microbial Metatranscriptomics Belowground*, pp. 273–293. Springer Nature Singapore Pte.

Lavelle, P., Dugdale, R., & Scholes, R. 2005. Nutrient cycling. In: *Ecosystems and Human Well-Being: Current State and Trends: Findings of the Condition and Trends Working Group.* Island Press, Washington, Covelo, London.

Lyon, B.R., & Mock, T. 2014. Polar microalgae: New approaches towards understanding adaptations to an extreme and changing environment. *Biology*, 3(1), 56–80.

Magnússon, R.I. 2022. *Greening, Browning, or Drowning? Permafrost–Vegetation–Climate Interactions in the Siberian Lowland Tundra* (Doctoral dissertation). Wageningen University and Research.

Microbes UEA. 2021. *Extreme Environments*. CRC Press.

Misra, A.K. 2014. Climate change and challenges of water and food security. *International Journal of Sustainable Built Environment*, 3(1), 153–165.

Morgan-Kiss, R.M., Priscu, J.C., Pocock, T., Gudynaite-Savitch, L. & Huner, N.P.A. 2006. Adaptation and acclimation of photosynthetic microorganisms to permanently cold environments. *Microbiology and Molecular Biology Reviews*, 70(1), 222–252.

NASA. 2010. https://ntrs.nasa.gov/citations/20100002095.

Nikrad, M.P., Kerkhof, L.J., & Häggblom, M.M. 2016. The subzero microbiome: Microbial activity in frozen and thawing soils. *FEMS Microbiology Ecology*, 92(6), fiw081.

Ottersen, G., Constable, A.J., & Hollowed, A.B. 2022. Climate change impacts on polar marine ecosystems: Toward robust approaches for managing risks and uncertainties. *Frontiers in Climate*, 3.

Ottoni, J.R., de Oliveira, V.M., & Passarini, M.R.Z. 2022. Microbes in thawing permafrost: Contributions to climate change. In: *Microbiome Under Changing Climate*, pp. 1–28. Woodhead Publishing.

Panikov, N.S. 2014. Subzero activity of cold-adapted yeasts. In: *Cold-adapted Yeasts: Biodiversity, Adaptation Strategies, and Biotechnological Significance*, pp. 295–323. Springer Nature Link.

Parvizpour, S., Hussin, N., Shamsir, M.S., & Razmara, J. 2021. Psychrophilic enzymes: Structural adaptation, pharmaceutical and industrial applications. *Applied Microbiology and Biotechnology*, 105, 899–907.

Priyanka, P., Chattopadhyay, J., & Srivastava, N. 2024. Response of microbial communities to environmental changes. In: *Functional Metagenomic*s, pp. 143–153. Academic Press.

Rizzo, C., & Lo Giudice, A. 2022. Life from a snowflake: Diversity and adaptation of cold-loving bacteria among ice crystals. *Crystals*, 12(3), 312.

Sahu, N., Vasu, D., Sahu, A., Lal, N., & Singh, S.K. 2017. Strength of microbes in nutrient cycling: A key to soil health. In: *Agriculturally Important Microbes for Sustainable Agriculture: Volume I: Plant-soil-Microbe Nexus*, pp. 69–86. Springer Nature.

Sharma, B.P., Adhikari, S., & Paudel, G. 2020. Microbial diversity in the glacial ecosystem of Antarctic, Arctic, and Tibetan Plateau: Properties and response to the environmental condition. *Janapriya Journal of Interdisciplinary Studies*, 9(1), 231–250.

Sharma, P., Sangwan, S., & Kumari, A. 2022. Impact of climate change on soil microorganisms regulating nutrient transformation. In: *Plant Stress Mitigators: Action and Application*, pp. 145–172. Springer Nature Singapore.

Sheikh, T., Baba, Z., Hamid, B., Iqbal, S., Yatoo, A., Fatima, S. et al. 2022. Extracellular polymeric substances in psychrophilic cyanobacteria: A potential bioflocculant and carbon sink to mitigate cold stress. *Biocatalysis and Agricultural Biotechnology*, 42, 102375. https://doi.org/10.1016/j.bcab.2022.102375.

Shen, L., Liu, Y., Allen, M.A., Xu, B., Wang, N., Williams, T.J. et al. 2021. Linking genomic and physiological characteristics of psychrophilic Arthrobacter to metagenomic data to explain global environmental distribution. *Microbiome*, 9(1), 136.

Tattersall, G.J., Sinclair, B.J., Withers, P.C., Fields, P.A., Seebacher, F., Cooper, C.E. et al. 2012. Coping with thermal challenges: Physiological adaptations to environmental temperatures. *Comprehensive Physiology*, 2(3), 2151–2202.

Teufel, A.G., & Morgan-Kiss, R.M. 2018. Physiological and biochemical adaptations of Psychrophiles. In: *Extremophiles*, pp. 185–208. CRC Press.

Verde, C., Giordano, D., Bellas, C.M., di Prisco, G., & Anesio, A.M. 2016. Polar marine microorganisms and climate change. *Advances in Microbial Physiology*, 69, 187–215.

Vincent, W.F. 2020. Arctic climate change: Local impacts, global consequences, and policy implications. In: *The Palgrave Handbook of Arctic Policy and Politics*, pp. 507–526. Springer Link.

Vonk, J.E., Tank, S.E., Bowden, W.B., Laurion, I., Vincent, W.F., Alekseychik, P. et al. 2015. Reviews and syntheses: Effects of permafrost thaw on Arctic aquatic ecosystems. *Biogeosciences*, 12(23), 7129–7167.

Vrionis, H.A., Warner, K., Whyte, L.G., & Millar, R. 2011. Climate change, ozone depletion, and life at the poles. In: *Polar Microbiology: Life in a Deep Freeze*, pp. 265–289. Wiley.

Wan, P., Zhang, F., Zhang, K., Li, Y., Qin, R., Yang, J. et al. 2023. Soil warming decreases carbon availability and reduces metabolic functions of bacteria. *Catena*, 223, 106913.

Yarzábal, L.A., Salazar, L.M.B., & Batista-García, R.A. 2021. Climate change, melting cryosphere, and frozen pathogens: Should we worry? *Environmental Sustainability*, 4(3), 489–501.

Zlatkov, N., Nadeem, A., Uhlin, B.E. & Wai, S.N. 2021. Eco-evolutionary feedbacks mediated by bacterial membrane vesicles. *FEMS Microbiology Reviews*, 45(2), fuaa047.

CHAPTER
4

Cold Loving Archaea: Diversity, Adaptation, and Applications

Shivani D. Raval[1*], Suchitra Daslaniya[1] and Bilal Ahmed[2]

[1] Department of Microbiology and Biotechnology Centre, The Maharaja Sayajirao University of Baroda, Vadodara, Gujarat – 390002, India

[2] Department of Agricultural & Biological Engineering, Purdue University, West Lafayette, United States

1. Introduction

Temperature, pH, nutrients, humidity, and other discrete yet interrelated life-sustaining abiotic conditions are all necessary for the growth and sustenance of life on Earth. The most important one among them is the temperature. While there have been reports of life in a variety of environmental niches with temperatures ranging from –20°C to 110°C, surviving at temperature extremes is incredibly difficult. The most common temperature extremes on Earth are found in the cold ecological niches, which are defined by average annual temperatures of less than 5°C. Many habitats that are inhabited by diverse microorganisms emerged during their evolution and metamorphosis into their current form, while certain deep, dark habitats found in the ocean and colder parts of the planet are still inhabited by cold-loving microbes. For their growth, these extremophilic cold-loving bacteria or archaea typically prefer temperatures of 15 to 0°C or lower. Cold-loving archaea play an ecological and astrobiological role in cold habitats with their cellular metabolism for survival and rigid growth at sub-zero temperatures (Pathania et al., 2022; Vothknecht & Tumbula, 1999).

In 1977 Carl Woese and George E. Fox first classified archaea independently from bacteria as the third domain of life (Figure 1) based on ribosomal RNA (rRNA) genes. Recent surveys on genome diversity and cultured-based approaches

*Corresponding author: sd.raval96@gmail.com

DOI: 10.1201/9781003717003-4

used to understand the newly proposed 27 phyla of archaea and their evolution (Baker et al., 2020). In terms of genomic structure (such as introns, histones, and multiple replication origins), transcriptional and translational machinery, and the absence of peptidoglycan in cell walls, single-celled without nuclei with the incorporation of ether lipids, specifically archaeols, in the cell membranes of archaea is one of biochemistry's distinctive features (Eme & Doolittle, 2015). Archaea are found in cold terrestrial and aquatic environments. From cold environments different archaea isolated were Ace Lake Antarctica containing *Methanogenium frigidum* and *Methanococcoides burtonii*, Soppen Lake, Switzerland containing *Methanosarcina lacustris, Methanogenium marinum, Methanosarcina baltica, Methanococcoides alaskense* (Skan Bay, Alaska), *Halorubrum lacusprofundi* (Deep Lake Antarctica), *Halorubrum lacusprofundi* (Ocean) maintaining microbial symbiotic communities (Kodera et al., 2022). In cold conditions, the maximum growth temperature (T_{max}) and the temperature at which cold-loving archaea develop the fastest (T_{opt}) can both be significantly higher than the temperature of their natural habitat.

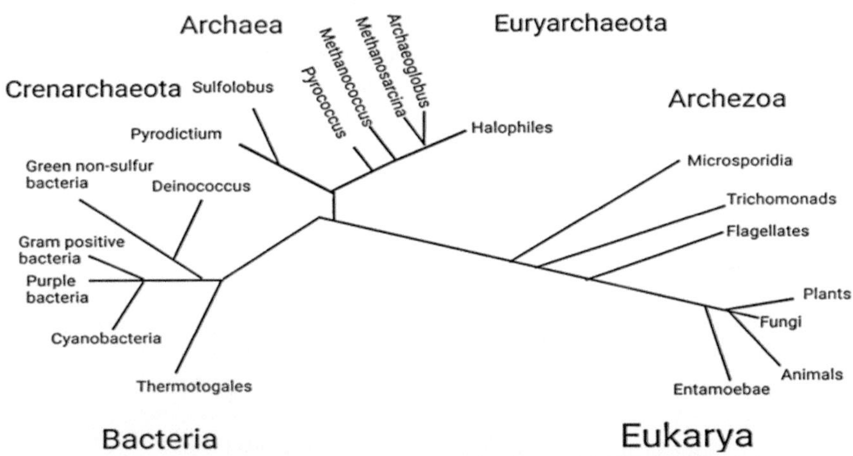

Figure 1: Phylogenetic tree of three domains of life by Carl Woese.

The first isolated archaeon *Methanogenium frigidum*, a stenopsychrophile, grows at a slower rate (<0.1 generations/day) at 4°C than the eurypsychrophilic Antarctic archaeon *Methanococcoides burtonii* (0.3 generations/day). This is true even if *M. frigidum*'s T_{max} (~18°C) is lower than *M. burtonii*'s (~28°C). Cold-loving archaea grow at 0–10°C, metabolize at –20°C to –40°C under ice and snow, and can endure at –45°C (Anani et al., 2023). Around 75% of earth's biosphere is surrounded by cold environments, and settings that are consistently cold (less than 5°C), such as alpine habitats, polar regions, the deep ocean, caverns, the subsurface of the land and ocean, and the high atmosphere, are home to cold-loving archaeon. They are found in large numbers at the chilly ocean depths (1–5°C; below ~1,000 m), which cover 70% of the lithosphere. Psychrophilic archaea serve a variety of purposes in a variety of cold climates, and their number in the ocean-estimated at approximately 10^{28} cells demonstrates their influence on the environment (Cavicchioli, 2006). Remarkably little is known about cold-loving archaea due to debate on psychrophilic terms. This chapter outlines cold-loving archaeal diversity, its adaptation to cold conditions, and applications in various sectors of industries, research, medical, agricultural, and environmental as a novel microbial strategy for the benefit of society.

2. Diversity of Cold Loving Archaea

The recycling of nitrogen and organic and inorganic carbon compounds by cold-loving archaea has been shown to lead to global energy cycles; this processing takes place at nearly all stages, from the C1 cycle to anaerobic ammonia oxidation to heterotrophic change (Moissl-Eichinger et al., 2018). Archaea's phylogenetic and functional diversity (Figure 2), quantity, and range of cold biotopes they inhabit all indicate how much they alter the cold biosphere. Even if there aren't many culturable isolates, research on the nutrient uptake of uncultivated archaea is fast expanding our knowledge of their functional implications (Cavicchioli, 2006). The variety of cold biotopes that archaea inhabit, their quantity, and their phylogenetic and functional diversity (Figure 2) all indicate how much they alter

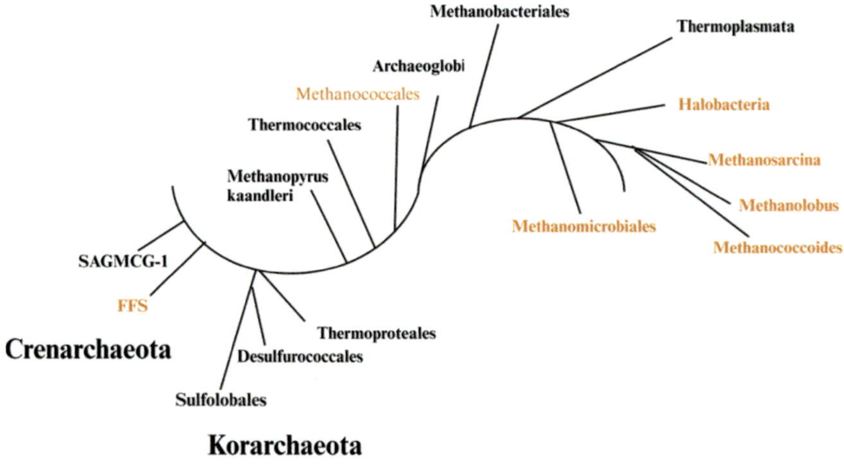

Figure 2: Diversity of cold-loving archaea from cold (Orange) environments and its comparison with other archaeal groups (Black).

the cold biosphere. Despite the paucity of culturable isolates, several studies, such as uptake of nutrient measurements, genomic studies use fluorescence *in situ* hybridization (FISH) probes to monitor active populations, studies that target 16Sr RNA genes in combination with activational genes (like nitrogenase and methyl coenzyme M reductase), and conclusions about biology, ecology, and evolution derived from molecular studies, are expanding quickly on our knowledge of the practical consequences of uncultivated archaea (Spang et al., 2015).

In coastal, open ocean, and deep-sea waters, archaea can make up a significant portion of the microbial population. They are found in cold marine waters and sea ice. In cold, marine-based lakes, marine silt, the deepest parts of the ocean, and the subsurface seafloor (100–200 m) are all home to archaea as shown in Table 1. Microbial methanogenesis in the subsurface of the seabed produces

Table 1: Cold-loving archaeal diversity in cold environments

Cold Environments	Cold-loving Archaeon Diversity	References
Arctic shelf ecosystem (Particle rich coastal water and riverine)	*Euryarchaeota*	Galand et al., 2008
Elevated arctic lakes	*Euryarchaeota* and *Crenarchaeota*	Galand et al., 2008; Pouliot et al., 2009
Cold Deep marine waters (Deep sea)	*Crenarchaeota* (Ammonia oxidizing archaea)	Galand et al., 2008
Alpine soils (High-altitude soils)	Ammonia oxidizing archaea *Thaumarchaeota*	Siles and Margesin, 2016
Antarctic soils	Majority of *Crenarchaeota*	Shravage et al., 2007
Arctic cold soils	Methanogenic archaea (*Euryarchaeota*)	Hoj et al., 2005
Cold sulphidic marsh water	*Crenarchaeum* (Cre1)	Koch et al., 2006
Polar/Alpine soils	Ammonia oxidizing archaea (*Thaumarchaeota*) *Candidatus Nitrosopolaris*	Pessi et al., 2022
Alpine Permafrost and Siberian	*Euryarchaeota (Methanosaetaceae Methanomicrobiaceae, Methanosarcinaceae*) and lower abundance of *Crenarchaeota*	Ganzert et al., 2007; Rivkina et al., 2007; Yang et al., 2008
Sea ice	*Thaumarchaeaota* (Marine group I)	Deming and Eric, 2017a
Cold Seeps	*Methanotrophic archaea* (ANME group 2 and 1)	Knittel et al., 2005

methane-hydrate deposits, which have accumulated to an estimated level of four to eight times the amount of carbon in all living species on Earth (Trouche et al., 2021). *Crenarchaeota* plays significant ecological functions in cold regions as nitrifiers in the rhizosphere. Alpine lakes, glacier ice, and streams are also home to archaea. *Cenarchaeum symbiosum* belonging to marine group I *Crenarchaeota* are found in cold Arctic Ocean environments in Californian maintaining carbon and nitrogen cycles (Galand et al., 2009).

The diversity of archaea in the arctic meltwater region was studied through sequencing and found Creanarchaeota (*Nitrososphaeraceae* unclassified and *Candidatus Nitrocosmicus*) and *Methanoregula* at the intertidal and subtidal zone maintain carbon and nitrogen cycle in ecology (Qin et al., 2023). The Arctic Ocean of the western region was dominated by some archaeal five phylum *Thermoplasmatota, Crenarchaeota, Euryarchaeota, Halobacterota, Nanoarchaeota* (*Nanoarchaeia*), and Marine Group II archaea dominated in deep water masses (Vipindas et al., 2023). *Methanosarcinales* is the order in which the acetoclastic archaea are described. So far, only two genera, i.e., *Methanosarcina* and *Methanosaeta*, have been found in freshwater silt and soils in cold climates (Dellagnezze et al., 2023). A diverse archaeal population was studied in the cold Antarctic brine marine habitat belonging to the genus of methanogens (*Methanoculleus*) (Guglielmin et al., 2023). There was very little archaeal abundance in Livingston Island Antarctica; *Woesearchaeales*, *Nitrosarchaeum*, *Candidatus Nitrosopumilus*, and Marine Group II were the most common groups (Doytchinov et al., 2024). In the coastal western Antarctic peninsula, the domain archaea were dominated by phylum *Thermoplasmatoata* and at genus level with *Nitrosopumilus* and *Thalassarchaeum* in depth drive distribution studies (Dutta et al., 2023).

3. Cold Loving Archaeal Adaptation

As we currently understand the microbial ecology and diversity of Earth's cryosphere, even the coldest and most frozen settings support a vast array of viable, metabolically active, and remarkably diverse microbial communities that represent important evolutionary families. Despite their varied metabolic capacities and other environmental factors, the primary trait shared by all cold-adapted microorganisms is their capacity to effectively withstand the adverse effects of low temperatures by developing a variety of structural and functional adaptations (Margesin & Miteva, 2011; Rabani et al., 2023).

The capacity to recognize temperature changes is necessary for adaptation to low temperatures. Low temperatures have an impact on the physical characteristics and functionality of membranes because they cause a decrease in membrane fluidity, the start of the gel phase transition, and cell function loss. To maintain the ideal level of membrane fluidity, lipid fatty acyl chains linked to the cell membrane can be modified. Saturated acyl fatty acid is changed into unsaturated acyl chains by an enzyme called desaturates, which is activated by cold shock.

As a result, there is an increase in methyl-branched fatty acid, which helps cold-adapted microbes survive in harsh conditions. Temperature changes can cause a downshift or upward shift in the fluidity of cell membranes, which can have an impact on active transport and protein release (Phadtare & Severinov, 2010).

The cell membrane, which serves as a barrier between the internal and exterior surroundings, is one of the main sensors for colds. The membrane stiffens up in colder weather, which triggers a membrane-associated sensor. The response regulator receives the signal from the sensor and upregulates the expression of several genes related to membrane fluidity modulation, which in turn causes more expression of several genes involved in the cold adaptation of archaea (Shivaji & Prakash, 2010). A review of proteome analyses and genome sequences of several psychrophilic archaea reveals that the primary upregulated processes for low-temperature growth are engaged in synthesis of proteins (transcription and translation), Protein and RNA folding (modulation of protein structure at low temperature), membrane fluidity preservation, cryoprotective substance synthesis and uptake (compatible solutes and exogenous polysaccharides), antioxidant activities, and metabolic pathways control (Figure 3) (Goyal et al., 2022; Shen et al., 2021).

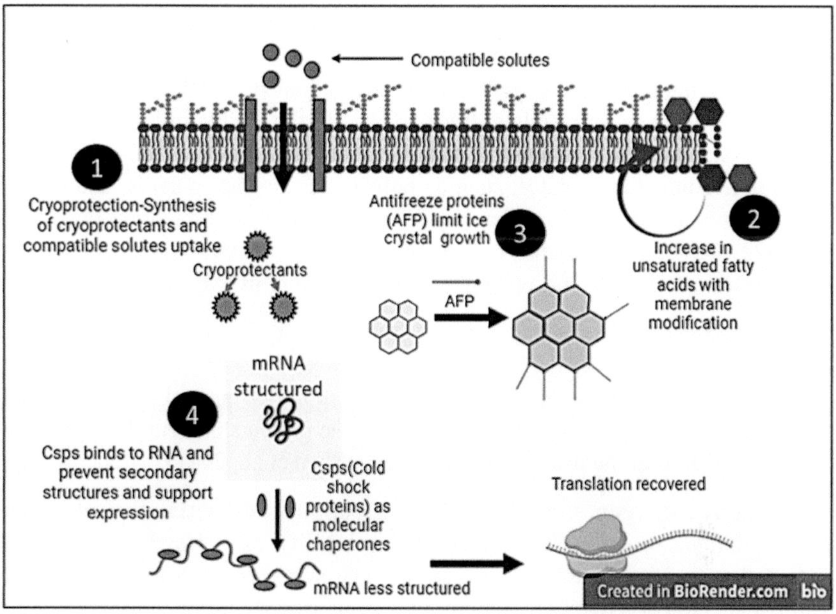

Figure 3: Cellular adaptation mechanism of cold-loving archaea in cold stress.

Various tactics are used by archaea to promote fluidity in their membranes. When exposed to low temperatures, the structure of the phospholipid and fatty acid cell walls changes, increasing the fluidity of the membrane (Al-Ghanayem

& Joseph, 2020). Modification in isomerized form of fatty acid, increment of methyl-branched fatty acids, polar carotenoids, and the ratio of anteiso/iso-branched fatty acids, along with a decrease in mean chain length of fatty acids and sterol/phospholipids ratio, are additional modes of modification of membrane fluidity (Singh, 2022; Shivaji & Prakash, 2010). The cold shock response exhibited by abrupt temperature changes by cold-loving archaea differs from that of thermophiles and mesophiles: (1) the first group does not suppress the production of housekeeping protein chains, and (2) the other type has a higher and increasing number of cold-shock proteins in response to cold shock. Furthermore, during low-temperature growth, cold-adapted archaea persistently create one set of proteins known as "cold-acclimation proteins", which then increase in steady-state levels when their temperature is dropped (Shen et al., 2021).

Cold-loving archaea have evolved to prevent cold-induced protein coagulation by maintaining optimal fluidity through the excretion of polymeric substances, such as EPS, ice-nucleating proteins, compatible solutes, and bio-surfactant. Microbes exposed to cold environments experience osmotic drainage and dehydration of the cell interior, which results in the activation of enzymes and adverse effects on cell function and survival. To shield the cell from cryoinjuries, organic osmolytes build up inside the cell and inhibit the freezing of solutes inside the cell membrane (Shen et al., 2021). Additionally, extracellular polymeric substances are secreted when the cell membrane is mechanically disrupted by ice. Cold-loving archaea or psychrophilic microorganisms synthesize a variety of substances to prevent intracellular freezing or to lessen the harmful consequences of ice crystal formation (Singh, 2022; Shivaji & Prakash, 2010). The formation of suitable compatible solutes, such as polyamines, sugars, polyols, and amino acids, allows bacterial and archaeal cells to achieve osmoprotection (Goyal et al., 2022). Ice-binding proteins called antifreeze proteins (AFPs) can change the structure of ice crystals and prevent ice from growing in two different ways.

- They cause water's freezing point to drop before it freezes without changing the melting point (thermal hysteresis activity).
- When frozen, AFPs exhibit ice re-crystallization detrimental activity, which means that at below-freezing temperatures, the proteins prevent the formation of big crystals at the price of small crystals (Shen et al., 2021).

Reactive oxygen species (ROS) protection is essential for tolerance at low temperatures, as it increases the solubility of gases. ROS have the potential to seriously harm cell structures. The synthesis of large quantities of antioxidant enzymes (catalase, superoxide dismutases, dioxygen-consuming lipid desaturases) or the lack of pathways for ROS production are two psychrophilic archaeon (*Methanolobus psychrophilus* R15) cold adaptive techniques for the detoxification of ROS (Dong & Chen, 2012; Chen et al., 2012). The processes of cold-living, and the ecological roles of cold-loving archaea will be better understood with the help of expanding the number of sequenced genomes of cold-loving microorganisms and the wider intervention of novel metagenomic and proteomic technologies. High concentrations of exopolymeric materials are produced by bacteria that live in cold aquatic conditions, such as sea ice in the Antarctic archaeon *Methanococcoides burtonii* (Doyle et al., 2011; Burg, 2009; Pathania et al., 2022). Exopolymeric materials have important roles in adherence to moist surfaces and the creation of biofilms, which trap nutrients, shield cells from adverse environmental conditions, and facilitate biochemical interactions (Deming & Young, 2017).

4. Application of Cold Loving Archaea

Cold-loving archaea have many different applications in medical, industrial, environmental bioremediation, and agricultural areas for plant growth promotion at low temperatures (Figure 4).

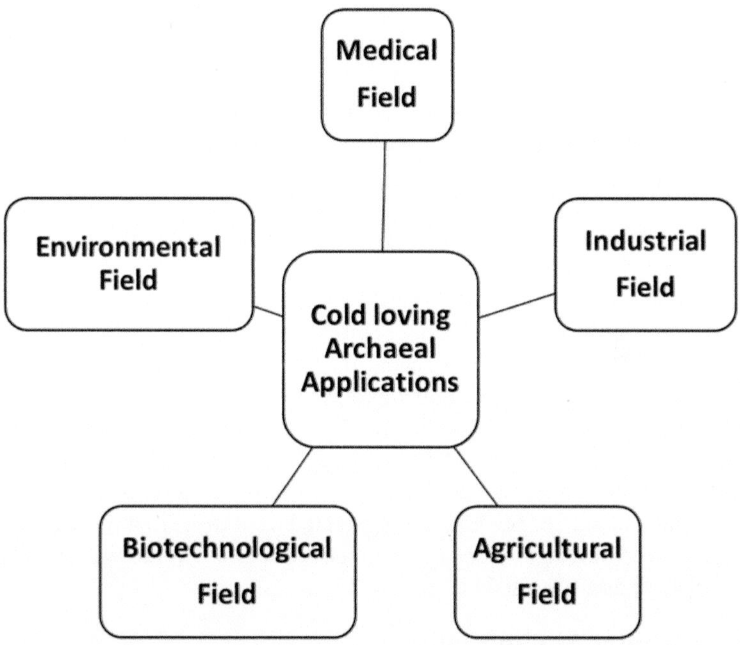

Figure 4: Applications of cold-loving archaea in different fields.

Medical and Industrial Applications

Cold-loving archaea in Cryosurgery—a minimally invasive technique in which tumours are frozen and killed using chilled probes—can be performed more successfully with elevated AFPs concentrations. Because ice crystals develop quickly and needle-like around AFPs at high concentrations (5–10 mg/mL), AFPs are more effective in killing tumour cells (Baskaran et al., 2021). Lastly, foods (including fruits, vegetables, and ice cream) may be thawed using AFPs without suffering from severe cellular damage or quality loss from ice recrystallization. AFPs are useful for food preservation (Bhatia et al., 2021).

Since many species are unable to produce the requisite PUFAs, the polyunsaturated fatty acids (PUFA) of psychrophilic archaea have prospective applications in human nutraceuticals and aquaculture as dietary supplements. Humans need PUFAs like eicosapentaenoic acid (EPA) and decosahexaenoic acid (DHA) because they are components of membrane acyl-lipids in the brain and retina and precursors of regulatory substances including prostaglandins, thromboxanes, and leukotrienes (Lauritano & Coppola, 2023). PUFAs may play a role in the transportation of cholesterol and triglycerides chosen for the nervous system and cardiovascular health (Hamdan, 2018). Much research is being done on the development of psychrophilic archaea as a substitute source of dietary PUFA, as they have substantial PUFA contents and can guarantee high production rates and reliable product results. Aquaculture, agriculture, cryopreservation, cryosurgery, and food storage are among the fields in which Archaeal homologues of AFPs have many applications. Fish raised with AFPs are resistant to freezing, and by using naturally cold waters as possible culture sites, aquaculture can reap significant financial rewards. AFPs can also be used to protect crops from ice-related cell and tissue damage. Similarly, by reducing the thawing temperature and rate (which is typically 600°C min^{-1}), low concentrations of AFPs can help cryopreserve cells, tissues, and organs while also lowering costs and potential heat-related damages (Lauritano & Coppola, 2023).

One of the numerous benefits is the increased use of cold-active enzymes in food preservation at low temperatures to prevent spoiling and preserve food's appealing qualities, such as flavour and nutritional content. β-galactosidases or lactases are used to hydrolyze and remove lactose from milk in cold storage. Additionally, this is used to make lactose-free milk and its byproducts for those who are sensitive to lactose (Al-Ghanayem & Joseph, 2020; Hamid et al., 2022). Another well-known application of cold-active pectinases is utilized in the making of wine (Malik et al., 2018). Proteases generated from psychrophilic archaea have been suggested for usage in detergents intended for cold water washing (Hamid et al., 2024; Aparici-Carratalá et al., 2023).

Environmental Applications

Bioremediation solutions remain a big problem in many places. Cold-adapted archaea are essential and cost-effective in bioremediation (or bioaugmentation) of various pollutants such as radionuclides (uranium, technetium, and cobalt), biopolymers (cellulase, chitin, proteins, lignin, and chlorinated compounds), phenols, nitriles, aromatic amines, dodecane, hexadecane, naphthalene, and toluene and triacylglycerols), as well as keratinous waste (originating from the poultry and leather industries), which includes additional archaeal enzymes in industrial waste and hazardous materials at low temperatures (Al-Ghanayem & Joseph, 2020; Dempsey, 2017; Gupta et al., 2020). The exploration of cold-active enzymes as a cleaning agent has expanded due to commercial considerations, rising regulatory expectations, and environmental sustainability since certain cold-loving archaea generate stable enzymes that can be used in bioremediation as economical and environmentally beneficial techniques (Hamid et al., 2012; Al-Ghanayem & Joseph, 2020). Commonly utilized cold-adapted enzymes such as lignin peroxidase, manganese peroxidase, tyrosinase, laccase, amidases, cellulases, amylases, glucoamylases, pectinases, proteases, lipases, and amyloglucosidases have been detected in garbage systems of management from various cold-adapted archaeal habitats and environments, such as deep seas, sediments from the Antarctic and Arctic Oceans, cold mountain soils, Antarctic soil. Petroleum exploration has been heavily drilled, which has resulted in oil spills that have contaminated several habitats cold-loving archaea that actively participate in the biodegradation of oils in marine settings (Bhatia et al., 2021; Gupta et al., 2020; Koul et al., 2021; Hualpa et al., 2023).

Agricultural Applications

The incapacity of existing biofertilizers to adhere plant roots owing to decreased metabolite activity at lower temperatures is one of the main obstacles to their use in high-altitude agriculture (Rawat et al., 2020; Afrouz et al., 2023; Baba et al., 2021). Therefore, the use of cold-loving archaea in agriculture plays an efficient role in promoting plant development, either as defence molecules or as cold-adapted plant growth-promoting rhizobacteria (PGPR), or as other microbes as bioinoculants that offer cold-stress protection and facilitate the assimilation of essential plant nutrients like phosphates (Sheikh et al., 2022). The new integrated omics technologies have resulted in the discovery of several candidates that can be employed as PGP bacteria in cold regions, whereas conventional methods have found cold-adapted microbes at a slower phase (Rai et al., 2023). Microbes known as plant growth-promoting (PGP) agents have a wide range of functions that either directly or indirectly aid in the growth and development of plants (Bright et al., 2025; Ferioun et al., 2025; Sridhar et al., 2025a, 2025b). However, the inhibitory impact of the low temperatures may limit the activity of PGP bacteria in cold climes. Although PGP may be genetically modified to have cold-loving genes for improved activity, this is not a practical solution due to the multigene structure of PGP and the environmental risks associated with releasing genetically modified organisms into the environment (Puranik et al., 2022).

Biotechnological Applications

Recently cold-loving archaea are widely used in biotechnological aspects due to their stable attributes which exhibit the ideal candidate for applications in various sectors (Table 2).

Table 2: Biotechnological applications of cold-loving archaea

Biotechnological Areas	Applications of Cold Loving Archaea
Biotechnology (Food processes)	Exopolysaccharides as emulsifying agents in the coating of foods and protease to maintain flavour and aroma during cold storage, Lactase in lactose-free milk processing, Ice-nucleating proteins for beverages and frozen food industries.
Environmental	Decontamination of ecosystems and elimination of xenobiotic compounds and industrial waste, Biopolymers and Biofuels from polyhydroxyalkanoates.
Bioinformatics and Nanotechnology	S-layer glycoproteins are located in the membrane for bioanalytical sensors and ultrafiltration membranes and archaeal proteins in optical switches. Microfluidic devices from ice nucleating Proteins.
Agricultural	Biofertilizers-Ammonia oxidizing archaea in cold farming, Stress tolerance enhancement in plants by cyanophycins.
Industrial processes	Archaeosomes in the cosmetics industry for skin care creams, Polyhydroxyalkanoates in bioplastics production, Biomining of metals, and proteases used in detergents for cold water washing, Recombinant Protein production at low temperature (Chaperones).
Veterinary	Safeguarding the dog's heart following cardiac revascularization.
Medical	Drug administration with archaeosome,Cold-loving Archaeal Glucose dehydrogenase in blood tests, methanogenic archaea *Methanomassiliicoccus luminyensis* for medical therapy for trimethylaminuria disease (TMAU), Membrane pigment (Carotenoids) in health and Nutrition.
Marine	Cold-loving archaea in the deep sea as archaeobiotics (Twenty Green)® in aquaculture.

5. Conclusion

Studies on the microbial ecology of cold habitats are becoming increasingly popular. Both culture-dependent and culture-independent molecular techniques have been used to comprehend the variety and microbial populations' ecological role in cold ecosystems for cold-loving archaeal diversity studies. More fresh information about cold-loving archaeal lifestyles will be revealed by the developing fields of proteome and genome analysis. Due to the enormous potential and exceptional ability to adapt to vastly different habitats, psychrophilic archaeal has piqued curiosity for possible applications in industry, and in agriculture as a plant growth booster for crops grown in different extreme habitats in cold climates, and in medicine for disease diagnosis and its treatment. Presumably, new cold-loving archaea with unexpected traits will be found that will expand their field of study as 'omics' and bioinformatics-based technologies advance, exploration strategies for cold hostile environments get better, and a new generation of genomes is sequenced.

References

Afrouz, M., Sayyed, R.Z., Fazeli-Nasab, B., Piri, R., Almalki, W.H., & Fitriatin, B.N. 2023. Seed bio-priming with beneficial *Trichoderma harzianum* alleviates cold stress in maize. *PeerJ.*, 11, e15644 http://doi.org/10.7717/peerj.15644.

Al-Ghanayem, A.A., & Joseph, B. (2020). Current prospective in using cold-active enzymes as eco-friendly detergent additive. *Applied Microbiology and Biotechnology*, 104(7), 2871–2882.

Anani, O.A., Shahnawaz, M., Gupta, P., Osarenotor, O., & Olisaka, F.N. 2023. 13 Ecological roles of psychrophiles. In: *Soil Microbiome of the Cold Habitats: Trends and Applications*, 179 pp. CRC Press.

Aparici-Carratalá, D., Esclapez, J., Bautista, V., Bonete, M.J., & Camacho, M. 2023. Archaea: Current and potential biotechnological applications. *Research in Microbiology*, 174(7), 104080.

Baba, Z.A., Hamid, B., Sheikh, T.A., Alotaibi, S., Enshasy, H.E., Ansari, M.J., ... & Sayyed, R.Z. 2021. Psychrotolerant *Mesorhizobium* sp. Isolated from temperate and cold desert regions solubilize potassium and produces multiple plant growth promoting metabolites. *Molecules*, 26(19), 5758.

Baker, B.J., De Anda, V., Seitz, K.W., Dombrowski, N., Santoro, A.E., & Lloyd, K.G. 2020. Diversity, ecology, and evolution of Archaea. *Nature Microbiology*, 5(7), 887–900.

Baskaran, A., Kaari, M., Venugopal, G., Manikkam, R., Joseph, J., & Bhaskar, P.V. 2021. Anti-freeze proteins (Afp): Properties, sources and applications: A review. *International Journal of Biological Macromolecules*, 189, 292–305.

Bhatia, R.K., Ullah, S., Hoque, M.Z., Ahmad, I., Yang, Y.H., Bhatt, A.K. et al. 2021. Psychrophiles: A source of cold-adapted enzymes for energy efficient biotechnological industrial processes. *Journal of Environmental Chemical Engineering*, 9(1), 104607.

Bright, J.P., Maheshwari, H.S., Thangappan, S., Perveeen, K., Bukhari, N.A., Mitra, D., ... & Mastinu, A. 2025. Biofilmed-PGPR: A next-generation bioinoculant for plant growth promotion in rice (*Oryza sativa* L.) under changing climate. *Rice Science*, 32(1), 94–106.

Burg, D.W. (2009). *Cold Adaptation in the Antarctic Archean Methanococcoides Burtonii: The Role of the Hydrophobic Proteome and Variations in Cellular Morphology* (Doctoral dissertation), UNSW Sydney.

Cavicchioli, R. (2006). Cold-adapted archaea. *Nature Reviews Microbiology*, 4(5), 331–343.

Chen, Z., Yu, H., Li, L., Hu, S., & Dong, X. 2012. The genome and transcriptome of a newly described psychrophilic archaeon, *Methanolobus psychrophilus* R 15, reveal its cold adaptive characteristics. *Environmental Microbiology Reports*, 4(6), 633–641.

Dellagnezze, B.M., Bovio-Winkler, P., Lavergne, C., Menoni, D.A., Mosquillo, F., Cabrol, L., ... & Etchebehere, C. 2023. Acetoclastic archaea adaptation under increasing temperature in lake sediments and wetland soils from Alaska. *Polar Biology*, 46(4), 259–275.

Deming, J.W., & Eric Collins, R. 2017a. Sea ice as a habitat for bacteria, archaea, and viruses. In: *Sea Ice*, pp. 326–351. Wiley-Blackwell.

Deming, J.W., & Young, J.N. 2017. The role of exopolysaccharides in microbial adaptation to cold habitats. *Psychrophiles: From Biodiversity to Biotechnology*, pp. 259–284. Springer.

Dempsey, M.J. 2017. Nitrification at low temperature for purification of used water. *Psychrophiles: From Biodiversity to Biotechnology*, pp. 601–612. Springer.

Dong, X., & Chen, Z. 2012. Psychrotolerant methanogenic archaea: Diversity and cold adaptation mechanisms. *Science China Life Sciences*, 55, 415–421.

Doyle, S., Dieser, M., Broemsen, E., & Christner, B. 2011. General characteristics of cold-adapted microorganisms. *Polar Microbiology: Life in a Deep Freeze*, pp. 101–125. ASM Press.

Doytchinov, V.V., Peykov, S., & Dimov, S.G. 2024. Study of the bacterial, fungal, and archaeal communities structures near the Bulgarian Antarctic Research Base "St. Kliment Ohridski" on Livingston Island, Antarctica. *Life*, 14(2), 278.

Dutta, A., Connors, E., Trinh, R., Erazo, N., Ducklow, H.W., Steinberg, D.K., & Bowman, J.S. 2023. Depth drives the distribution of microbial ecological functions in the coastal western Antarctic Peninsula. *Frontiers in Microbiology*, 14, 1168507.

Eme, L., & Doolittle, W.F. 2015. Archaea. *Current Biology*, 25(19), R851–R855.

Ferioun, M., Zouitane, I., Bouhraoua, S., Elouattassi, Y., Belahcen, D., Errabbani, A., ... & El Ghachtouli, N. 2025. Applying microbial biostimulants and drought-tolerant genotypes to enhance barley growth and yield under drought stress. *Frontiers in Plant Sciences*, 15, 1494987.

Galand, P.E., Lovejoy, C., Pouliot, J., & Vincent, W.F. 2008. Heterogeneous archaeal communities in the particle-rich environment of an arctic shelf ecosystem. *Journal of Marine Systems*, 74(3–4), 774–782.

Galand, P.E., Casamayor, E.O., Kirchman, D.L., Potvin, M., & Lovejoy, C. 2009. Unique archaeal assemblages in the Arctic Ocean unveiled by massively parallel tag sequencing. *The ISME Journal*, 3(7), 860–869.

Ganzert, L., Jurgens, G., Münster, U., & Wagner, D. 2007. Methanogenic communities in permafrost-affected soils of the Laptev Sea coast, Siberian Arctic, characterized by 16S rRNA gene fingerprints. *FEMS Microbiology Ecology*, 59(2), 476–488.

Goyal, D., Swaroop, S., Prakash, O., & Pandey, J. 2022. Survival strategies in cold-adapted microorganisms. *Survival Strategies in Cold-adapted Microorganisms*, pp. 173–186. Springer Nature Singapore.

Guglielmin, M., Azzaro, M., Buzzini, P., Battistel, D., Roman, M., Ponti, S., ... & Lo Giudice, A. 2023. A possible unique ecosystem in the endoglacial hypersaline brines in Antarctica. *Scientific Reports*, 13(1), 177.

Gupta, S.K., Kataki, S., Chatterjee, S., Prasad, R.K., Datta, S., Vairale, M.G., ... & Gupta, D.K. 2020. Cold adaptation in bacteria with special focus on cellulase production and its potential application. *Journal of Cleaner Production*, 258, 120351.

Hamdan, A. 2018. Psychrophiles: Ecological significance and potential industrial application. *South African Journal of Science*, 114(5–6), 1–6.

Hamid, B., Singh, P., Rana, R.S., & Sahay, S. 2012. Isolation and identification of psychrophilic yeast from the soil of northern region of India. Biotechnology, 18, 103–107.

Hamid, B., Bashir, Z., Yatoo, A.M., Mohiddin, F., Majeed, N., Bansal, M., ... & Alfaifi, M.Y. 2022. Cold-active enzymes and their potential industrial applications: A review. *Molecules*, 27(18), 5885.

Hamid, B., Baba, Z.A., Sheikh, T.A., Perveen, K., & Sayyed, R.Z. 2024. Investigating potential protease activity of psychrotrophic bacteria from a municipal landfill for solid waste management. *Biomass Conversion and Biorefinery*. https://doi.org/10.1007/s13399-024-05621-2.

Hoj, L., Olsen, R.A., & Torsvik, V.L. 2005. Archaeal communities in High Arctic wetlands at Spitsbergen, Norway as characterized by 16S rRNA gene fingerprinting. *FEMS Microbiol. Ecol.*, 53, 89e102.

Hualpa-Cutipa, E., Acosta, R.A.S., Cariga, O.J.M., Espinoza-Medina, M.A., Chavez-Rojas, D.C., Medina-Cerna, D., ... & Salazar-Quiñonez, I. 2023. Metagenomic approach role of psychrotrophic and psychrophilic microbes in bioremediation. In: *Metagenomics to Bioremediation*, pp. 513–536. Academic Press.

Knittel, K., Lösekann, T., Boetius, A., Kort, R., & Amann, R. 2005. Diversity and distribution of methanotrophic archaea at cold seeps. *Applied and Environmental Microbiology*, 71(1), 467–479.

Koch, M., Rudolph, C., Moissl, C., & Huber, R. 2006. A cold-loving crenarchaeon is a substantial part of a novel microbial community in cold sulphidic marsh water. *FEMS Microbiology Ecology*, 57(1), 55–66.

Kodera, S.M., Das, P., Gilbert, J.A., & Lutz, H.L. 2022. Conceptual strategies for characterizing interactions in microbial communities. *Science*, 25(2), 103775.

Koul, B., Chaudhary, R., & Taak, P. 2021. Extremophilic microbes and their application in bioremediation of environmental contaminants. In: *Microbe Mediated Remediation of Environmental Contaminants*, pp. 115–128. Woodhead Publishing.

Lauritano, C., & Coppola, D. 2023. Biotechnological applications of products released by marine microorganisms for cold adaptation strategies: Polyunsaturated fatty acids, antioxidants, and antifreeze proteins. *Journal of Marine Science and Engineering*, 11(7), 1399.

Malik, S.M., Ahanger, F.A., Sahay, S., & Jain, K. 2018. Cold-active pectinases and their commercial applications: A review. *Int. J. Sci. Res. in Biological Sciences*, 5, 4.

Margesin, R., & Miteva, V. 2011. Diversity and ecology of psychrophilic microorganisms. *Research in Microbiology*, 162(3), 346–361.

Moissl-Eichinger, C., Pausan, M., Taffner, J., Berg, G., Bang, C., & Schmitz, R.A. 2018. Archaea are interactive components of complex microbiomes. *Trends in Microbiology*, 26(1), 70–85.

Pathania, S., Solanki, P., Putatunda, C., Bhatia, R.K., & Walia, A. 2022. Adaptation to cold environment: The survival strategy of psychrophiles. *Survival Strategies in Cold-adapted Microorganisms*, pp. 87–111. Springer Nature Singapore.

Pessi, I.S., Rutanen, A., & Hultman, J. 2022. Candidatus nitrosopolaris, a genus of putative ammonia-oxidizing archaea with a polar/alpine distribution. *FEMS Microbes*, 3, xtac019.

Phadtare, S., & Severinov, K. 2010. RNA remodeling and gene regulation by cold shock proteins. *RNA Biology*, 7(6), 788–795.

Pouliot, J., Galand, P.E., Lovejoy, C., & Vincent, W.F. 2009. Vertical structure of archaeal communities and the distribution of Ammonia monooxygenase a gene variants in two meromictic high Arctic lakes. *Environmental Microbiology*, 11(3), 687–699.

Puranik, S., Singh, S.K., & Shukla, L. 2022. An insight to cold-adapted microorganisms and their importance in agriculture. *Survival Strategies in Cold-adapted Microorganisms*, pp. 379–411. Springer Nature Singapore.

Qin, Y., Wang, N., Zheng, L., Li, Q., Wang, L., Xu, X., & Yin, X. 2023. Study of archaeal diversity in the arctic Meltwater Lake Region. *Biology*, 12(7), 1023.

Rabbani, G., Ahmad, E., Ahmad, A., & Khan, R.H. 2023. Structural features, temperature adaptation and industrial applications of microbial lipases from psychrophilic, mesophilic and thermophilic origins. *International Journal of Biological Macromolecules*, 225, 822–839.

Rai, S., Omar, A.F., Rehan, M., Al-Turki, A., Sagar, A., Ilyas, N., ... & Hasanuzzaman, M. 2023. Crop microbiome: Their role and advances in molecular and omic techniques for the sustenance of agriculture. *Planta*, 257(2), 27. https://doi.org/10.1007/s00425-022-04052-5.

Rawat, J., Yadav, N., & Pande, V. 2020. Role of rhizospheric microbial diversity in plant growth promotion in maintaining the sustainable agrosystem at high altitude regions. In: *Recent Advancements in Microbial Diversity*, pp. 147–196. Academic Press.

Rivkina, E., Shcherbakova, V., Laurinavichius, K., Petrovskaya, L., Krivushin, K., Kraev, G., ... & Gilichinsky, D. (2007). Biogeochemistry of methane and methanogenic archaea in permafrost. *FEMS Microbiology Ecology*, 61(1), 1–15.

Sheikh, T., Baba, Z., Hamid, B., Iqbal, S., Yatoo, A., Fatima, S., ... & Sayyed, R.Z. 2022. Extracellular polymeric substances in psychrophilic cyanobacteria: A potential bioflocculant and carbon sink to mitigate cold stress. *Biocatalysis and Agricultural Biotechnology*, 42, 102375.

Shen, L., Zhang, S., & Chen, G. 2021. Regulated strategies of cold-adapted microorganisms in response to cold: A review. *Environmental Science and Pollution Research*, 1–19.

Shivaji, S., & Prakash, J.S. 2010. How do bacteria sense and respond to low temperature? *Archives of Microbiology*, 192, 85–95.

Shravage, B.V., Dayananda, K.M., Patole, M.S., & Shouche, Y.S. 2007. Molecular microbial diversity of a soil sample and detection of ammonia oxidizers from Cape Evans, Mcmurdo Dry Valley, Antarctica. *Microbiological Research*, 162(1), 15–25.

Siles, J.A., & Margesin, R. 2016. Abundance and diversity of bacterial, archaeal, and fungal communities along an altitudinal gradient in alpine forest soils: What are the driving factors? *Microbial Ecology*, 72, 207–220.

Singh, V. 2022. Microbial genes responsible for cold adaptation. In: *Survival Strategies in Cold-adapted Microorganisms*, pp. 153–171. Springer Nature Singapore.

Spang, A., Saw, J.H., Jørgensen, S.L., Zaremba-Niedzwiedzka, K., Martijn, J., Lind, A.E., ... & Ettema, T.J. 2015. Complex archaea that bridge the gap between prokaryotes and eukaryotes. *Nature*, 521(7551), 173–179.

Sridhar, D., Alherwairini, S.S., Eswaran, S.U., Barasarathi, J., Lalitha, S., & Sayyed, R. 2025a. The soil microbiome enhances sesame growth and oil composition, and soil nutrients under saline conditions. *Scientific Reports*, 15(1), 1–17. https://doi.org/10.1038/s41598-025-15589-2.

Sridhar, D., Alheswairini, S.S., Barasarathi, J., Enshasy, H.A.E., Lalitha, S., Mir, S.H., ... & Sayyed, R.Z. 2025b. Halophilic rhizobacteria promote growth, physiology, and salinity tolerance in *Sesamum indicum* L. grown under salt stress. *Frontiers in Microbiology*, 16, 1590854.

Trouche, B., Brandt, M.I., Belser, C., Orejas, C., Pesant, S., Poulain, J. , ... & Maignien, L. 2021. Diversity and biogeography of bathyal and abyssal seafloor bacteria and archaea along a Mediterranean—Atlantic Gradient. *Frontiers in Microbiology*, 12, 702016.

Vipindas, P.V., Jabir, T., Venkatachalam, S., Yang, E.J., Jain, A., & Krishnan, K.P. 2023. Vertical segregation and phylogenetic characterization of archaea and archaeal ammonia monooxygenase gene in the water column of the western Arctic Ocean. *Extremophiles*, 27(3), 24.

Vothknecht, U.C., & Tumbula, D.L. 1999. Archaea: From genomics to physiology and the origin of life. *Trends in Cell Biology*, 9(4), 159–161.

Yang, D., Wang, J., Bai, Y., Xu, S., & An, L. 2008. Diversity and distribution of the prokaryotic community in near-surface permafrost sediments in the Tianshan Mountains, China. *Canadian Journal of Microbiology*, 54(4), 270–280.

CHAPTER

5

Psychrophilic and Psychrotrophic Bacteria and Fungi: Comprehensive Insights

Asma Shakeel[1], Mehnaz Shakeel[2], Malik Asif Aziz[3*], Shakeel Ahmad Mir[4], Shayesta Islam[5], Jegatheesan Pillathil Mathi Guru[3], Nowsheen Nazir[5] and Madeha Alonazi[6]

[1] Division of Soil Science, FOA, Wadura, SKUAST-Kashmir, India
[2] Division of Pathology, FOA, Wadura, SKUAST-Kashmir, India
[3] Division of Basic Science and Humanities, FOA, Wadura – 193201, SKUAST-Kashmir, India
[4] Division of Agriculture Statistics, Faculty of Horticulture, Shalimar, SKUAST-Kashmir, India
[5] Department of Environmental Sciences, Cluster University of Srinagar, J&K, India
[6] Department of Plant and Microbiology, College of Sciences, King Saud University, Riyadh, KSA

1. Introduction

In the vast expanse of Earth's biosphere, organisms have evolved remarkable adaptations to thrive in diverse ecological niches. Among these resilient inhabitants are psychrophilic and psychrotrophic bacteria and fungi, which have mastered the art of survival and growth in cold environments. From the icy landscapes of polar regions to the chilly depths of freshwater bodies, these microorganisms exhibit a remarkable ability to thrive at low temperatures, presenting a captivating avenue for exploration into the adaptability of life (Furhan, 2020). Psychrophilic microorganisms, characterized by their preference for temperatures below 20°C, and psychrotrophic microorganisms, which thrive optimally between 0°C and 30°C but also tolerate higher temperatures, are integral components of cold ecosystems worldwide (Kanekar & Kanekar, 2022; Suyal et al., 2021a; Baba et al., 2021). Their presence and activities shape the dynamics of these environments, influencing nutrient cycling, organic matter degradation,

*Corresponding author: asifmalik@skuastkashmir.ac.in

DOI: 10.1201/9781003717003-5

and symbiotic relationships with other organisms (Suyal et al., 2021b). At the molecular level, both psychrophilic and psychrotrophic bacteria and fungi have evolved a suite of specialized adaptations to thrive in cold conditions (Pathania et al., 2022; Ramasamy et al., 2023). These include modifications in membrane composition to maintain fluidity, the production of cold-shock proteins to protect cellular structures, and adjustments in metabolic pathways to ensure enzymatic activity at low temperatures (Garcia-Lopez et al., 2021). Such adaptations not only enable their survival in cold environments but also confer a competitive advantage, allowing them to outcompete other microorganisms and exploit ecological niches that would otherwise remain unoccupied (Dasila et al., 2022). Moreover, the unique traits and metabolic capabilities of psychrophilic and psychrotrophic bacteria and fungi hold significant promise for biotechnological

Table 1: Difference between psychrophilic and psychrotrophic microorganisms

Feature	Psychrophilic Microorganisms	Psychrotrophic Microorganisms
Temperature Preference	Below 20°C	0°C to 30°C (with tolerance for higher temperatures)
Optimal Growth Temperature	Typically, 0°C to 15°C for bacteria; slightly higher for fungi	Generally, between 20°C and 30°C
Adaptations	Specialized adaptations for survival and growth in cold environments, including modifications in membrane composition, production of cold-shock proteins, and alterations in metabolic pathways	Adaptations to tolerate cold environments, but may not require extreme adaptations as psychrophiles; capable of thriving in a broader temperature range
Ecological Distribution	Found in polar regions, alpine environments, and deep-sea habitats; play crucial roles in nutrient cycling, organic matter degradation, and symbiotic relationships with other cold-adapted organisms	Widely distributed in various environments, including soil, water, and food; contribute to nutrient cycling and organic matter decomposition in both cold and moderate environments
Biotechnological Applications	Enzymes produced by psychrophilic micro-organisms, such as cold-active lipases, proteases, and amylases, have applications in food processing, detergent formulation, bioremediation, etc.	Psychrotrophic microorganisms also produce enzymes with biotechnological potential, although their optimal growth temperature may be higher; applications in similar industries but may be more suitable for processes requiring moderate temperatures

applications (Furhan, 2020; Kuddus et al., 2024). Enzymes produced by these microorganisms, such as cold-active lipases, proteases, and amylases, find applications in various industries, including food processing, detergent formulation, and bioremediation (Al-Maqtari et al., 2019; Vivek et al., 2022). Additionally, their potential in bioprospecting offers opportunities for the discovery of novel bioactive compounds with pharmaceutical and agricultural relevance, further underscoring their value beyond their natural habitats.

As we embark on a journey to explore the fascinating world of psychrophilic and psychrotrophic bacteria and fungi, we uncover not only the intricacies of their cold-adapted lifestyles but also profound insights into the resilience and adaptability of life on Earth. By unraveling their characteristics, ecological roles, and biotechnological potential, we gain a deeper appreciation for the diversity and complexity of microbial life and its remarkable ability to thrive in extreme environments (Table 1). This chapter aims to provide a comprehensive exploration of psychrophilic and psychrotrophic bacteria and fungi, shedding light on their intriguing biology, ecological significance, and potential applications in biotechnology.

2. Understanding Psychrophilic and Psychrotrophic Bacteria and Fungi

Psychrophilic and psychrotrophic microorganisms represent fascinating examples of life's adaptability to extremely cold environments. Psychrophiles are organisms that thrive optimally at temperatures below 20°C, while psychrotrophs exhibit optimal growth between 0°C and 30°C but can also tolerate higher temperatures (Kanekar & Kanekar 2022). Both groups have evolved specialized adaptations to not only survive but thrive in cold conditions, demonstrating remarkable resilience in the face of environmental challenges (Hassan et al., 2016). At the cellular level, psychrophilic and psychrotrophic microorganisms employ a variety of mechanisms to cope with the challenges of low temperatures (Jin et al., 2022; García-Descalzo et al., 2022). One key adaptation involves modifications in membrane composition to maintain fluidity in cold environments. Membrane fluidity is crucial for maintaining cell integrity and facilitating the transport of nutrients and metabolic intermediates across the cell membrane. Psychrophilic and psychrotrophic microorganisms adjust the composition of their lipid membranes, often increasing the proportion of unsaturated fatty acids, to prevent membrane solidification at low temperatures. Additionally, psychrophilic and psychrotrophic microorganisms produce cold-shock proteins, which are specialized molecular chaperones that help protect cellular structures from cold-induced damage (Kuhn, 2012; Bhanu & Chatterjee, 2024). These proteins are rapidly synthesized in response to sudden drops in temperature and aid in maintaining the stability and functionality of essential cellular components, such as proteins and nucleic acids. Cold-shock proteins play a crucial role in allowing microorganisms to quickly adapt to changes in environmental temperature and ensure cellular survival under cold stress conditions. Furthermore, psychrophilic and psychrotrophic microorganisms undergo alterations in their metabolic pathways to enhance enzymatic activity at low temperatures (Hamid et al., 2022). Enzymes are the catalysts of biochemical reactions within cells, and their activity is often influenced by temperature. In cold environments, where reaction rates are typically slower, psychrophilic and psychrotrophic microorganisms have evolved enzymes with enhanced cold tolerance and catalytic efficiency (Kuddus et al., 2024). These enzymes may exhibit structural adaptations, such as increased flexibility or enhanced active site accessibility, to function optimally at low temperatures.

Overall, the ability of psychrophilic and psychrotrophic microorganisms to thrive in cold environments is a testament to their remarkable adaptability and evolutionary resilience. By employing specialized adaptations in membrane composition, production of cold-shock proteins, and alterations in metabolic pathways, these microorganisms have successfully colonized some of the coldest habitats on Earth, from polar ice caps to deep-sea trenches. Further exploration of the molecular mechanisms underlying their cold adaptation will not only deepen our understanding of life's diversity and resilience but also hold potential for biotechnological applications in various industries.

3. Ecological Roles and Distribution of Psychrophilic and Psychrotrophic Bacteria and Fungi

Psychrophilic and psychrotrophic microorganisms are integral components of diverse ecosystems across the globe, where they play crucial roles in nutrient cycling, organic matter degradation, and symbiotic relationships with other organisms. Their ability to thrive in cold environments allows them to occupy unique ecological niches, influencing ecosystem dynamics in profound ways.

- **Polar Regions:** In polar regions, such as the Arctic and Antarctic, psychrophilic and psychrotrophic microorganisms are ubiquitous, thriving in extreme cold and often nutrient-poor environments (Doytchinov & Dimov, 2022). These microorganisms participate in vital processes such as nutrient cycling, where they decompose organic matter, releasing nutrients that are essential for the growth of other organisms. Additionally, psychrophilic and psychrotrophic microorganisms form symbiotic relationships with plants, such as mosses and lichens, aiding in their growth and survival in harsh polar conditions (Rai & Sharma, 2022).
- **Alpine Environments:** Alpine environments, characterized by high elevations and cold temperatures, are also home to psychrophilic and psychrotrophic microorganisms. In these habitats, they contribute to soil formation and

nutrient availability through the decomposition of organic matter (Shen et al., 2021). Psychrophilic and psychrotrophic microorganisms play a crucial role in the breakdown of plant material, releasing nutrients that are essential for the growth of alpine vegetation (Hassan et al., 2016). Additionally, they interact with alpine flora, forming symbiotic relationships that enhance plant health and resilience to environmental stressors.

- **Deep-Sea Sediments:** The deep-sea environment, with its cold temperatures and high pressures, harbors diverse communities of psychrophilic and psychrotrophic microorganisms (Kaur et al., 2022). These microorganisms are involved in the decomposition of organic matter that accumulates on the ocean floor, playing a vital role in nutrient cycling in deep-sea ecosystems. Additionally, psychrophilic and psychrotrophic microorganisms participate in symbiotic relationships with deep-sea fauna, providing essential nutrients and contributing to the overall biodiversity of these unique habitats (Anani et al., 2023).
- **Cold Freshwater Bodies:** Cold freshwater bodies, such as lakes, rivers, and streams in temperate and polar regions, are also rich in psychrophilic and psychrotrophic microorganisms (Zakhia et al., 2008). These microorganisms play a critical role in the degradation of organic matter, contributing to the recycling of nutrients and the maintenance of water quality. Additionally, psychrophilic and psychrotrophic microorganisms form complex food webs with other aquatic organisms, serving as primary producers and consumers in these ecosystems (Grant & Long, 2013).
- **Carbon Sequestration and Climate Dynamics:** Psychrophilic and psychrotrophic microorganisms also influence global carbon cycling and climate dynamics through their role in the preservation of organic matter in permafrost and glacial ice. In these cold environments, organic material is often preserved for thousands of years, acting as a significant reservoir of carbon. As psychrophilic and psychrotrophic microorganisms decompose organic matter in permafrost and glacial ice, they release greenhouse gases such as carbon dioxide and methane, contributing to climate change dynamics on a global scale (Ottoni et al., 2022).

4. Diversity and Taxonomy of Psychrophilic and Psychrotrophic Microorganisms

Psychrophilic and psychrotrophic microorganisms exhibit a remarkable diversity that spans across all domains of life, including bacteria, archaea, and fungi (Figure 1). This diversity reflects the adaptability and evolutionary strategies of these microorganisms to thrive in cold environments, where they play crucial ecological roles and contribute to ecosystem functioning. There are diverse cold environments in which different psychrophilic and psychotrophic microorganisms proliferate (Table 2).

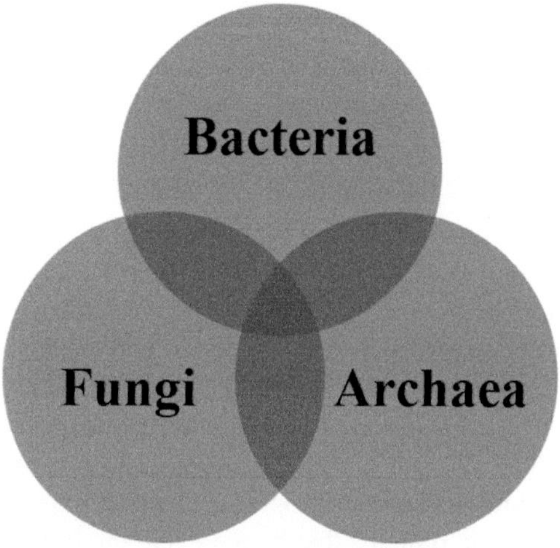

Figure 1: Diversity of microorganisms.

Table 2: Tabular representation of psychrophilic and psychrotrophic bacteria and fungi found in different cold environments

Environment	Psychrophilic Bacteria	Psychrotrophic Bacteria	Psychrophilic Fungi	Psychrotrophic Fungi
Polar Regions	*Psychrobacter* spp.	*Pseudomonas* spp.	*Penicillium* spp.	*Cladosporium* spp.
	Colwellia spp.	*Listeria* spp.	*Rhodotorula* spp.	*Geomyces* spp.
Alpine Environments	*Arthrobacter* spp.	*Bacillus* spp.	*Mrakia* spp.	*Pseudo-gymnoascus* spp.
	Janthino-bacterium spp.	*Micrococcus* spp.	*Oidiodendron maius*	*Mortierella* spp.
Deep-Sea Sediments	*Marinobacter* spp.	*Shewanella* spp.	*Aspergillus* spp.	*Mucor* spp.
	Psychromonas spp.	*Flavobacterium* spp.	*Leptolegnia* spp.	*Pseudeurotium* spp.
Glacial Environments	*Glaciimonas* spp.	*Acinetobacter* spp.	*Cryptococcus* spp.	*Fusarium* spp.
	Polaromonas spp.	*Carnobacterium* spp.	*Phoma* spp.	*Trichoderma* spp.
Cold Freshwater Bodies	*Sphingomonas* spp.	*Lactobacillus* spp.	*Leptolegnia* spp.	*Pseudo-gymnoascus* spp.
	Flavobacterium spp.	*Clostridium* spp.	*Metschnikowia* spp.	*Verticillium* spp.

Bacteria

Bacteria are single-celled microorganisms that are incredibly diverse and can be found in a wide range of environments, including cold habitats. Within the bacterial kingdom, there are two main groups of cold-adapted organisms: psychrophilic bacteria and psychrotrophic bacteria.

- **Psychrophilic Bacteria:** Psychrophilic bacteria are specialized to thrive in cold temperatures, typically below 20°C. These organisms have evolved unique physiological and biochemical adaptations to survive in such extreme conditions. One crucial adaptation is their cell membrane composition. Psychrophiles often have membrane lipids that are more fluid and flexible compared to those of mesophilic bacteria (organisms thriving at moderate temperatures). This allows their cell membranes to remain functional at low temperatures. Additionally, psychrophilic bacteria may produce antifreeze proteins or ice-binding proteins that help prevent the formation of ice crystals within their cells, which can be damaging (Białkowska et al., 2020). Enzymes produced by psychrophiles are optimized to function efficiently at cold temperatures. These enzymes have higher catalytic activity and flexibility, allowing them to carry out metabolic processes in cold environments.

  Psychrobacter includes species capable of thriving in various cold habitats, ranging from polar regions to cold marine environments (Pathania et al., 2022). *Psychrobacter* species have been found in sea ice, glaciers, and deep-sea sediments, among other places. *Colwellia* is another genus of psychrophilic bacteria commonly found in cold marine sediments (Zhang et al., 2017). *Colwellia* sp., play important roles in organic matter degradation and nutrient cycling in these environments.
- **Psychrotrophic Bacteria:** Psychrotrophic bacteria, although capable of growth at low temperatures, have broader temperature tolerances compared to psychrophiles (Bowman, 2017). They can grow over a wide temperature range, including both cold and moderate temperatures. These bacteria can also survive at refrigeration temperatures commonly used in food storage, posing challenges for food safety.

  While not as specialized as psychrophiles, psychrotrophic bacteria still possess adaptations that allow them to grow and metabolize efficiently at cold temperatures. They may produce enzymes with moderate cold tolerance, allowing them to function at lower temperatures than typical mesophilic bacteria. Some psychrotrophic bacteria, such as certain species of Listeria and Bacillus, can form biofilms in food processing environments, increasing their resistance to cleaning and sanitation procedures (Galié et al., 2018).

Archaea

Archaea are a group of single-celled microorganisms that constitute one of the three domains of life, alongside bacteria and eukaryotes. They are often found in extreme environments, such as hot springs, deep-sea hydrothermal vents, and highly saline environments, but they also inhabit more moderate environments, including soil and the human gut. In recent years, archaea have garnered increasing attention due to their unique biology and ecological roles.

- **Psychrophilic and Psychrotrophic Archaea:** Psychrophilic and psychrotrophic archaea have been discovered in cold environments such as polar ice caps, permafrost, and cold marine sediments (Pathania et al., 2022). These habitats are characterized by low temperatures, often below freezing, presenting extreme challenges for life. Psychrophilic and psychrotrophic archaea exhibit specialized adaptations to survive in these extreme cold conditions. One crucial adaptation is their membrane lipids. Unlike bacteria and eukaryotes, which typically use fatty acids in their membrane lipids, archaea use isoprenoid chains. These isoprenoid-based lipids provide greater stability and flexibility, allowing archaeal membranes to remain functional at low temperatures (Salvador et al., 2019). Additionally, psychrophilic and psychrotrophic archaea have metabolic pathways that are optimized for low temperatures. They produce enzymes with enhanced activity at cold temperatures, enabling them to carry out essential metabolic processes even in freezing environments. Some archaea may also produce antifreeze proteins or ice-binding proteins, similar to those found in psychrophilic bacteria, to prevent ice crystal formation within their cells (Białkowska et al., 2020).

  While the diversity and ecological roles of psychrophilic and psychrotrophic archaea are still being explored, they are recognized as important contributors to cold ecosystem dynamics. These archaea play roles in nutrient cycling, organic matter degradation, and other ecological processes in cold environments. In polar regions, where temperatures remain consistently low, psychrophilic and psychrotrophic archaea may represent significant components of microbial communities (Pearce, 2017). Their activities contribute to the functioning and resilience of cold ecosystems.

Fungi

Fungi are a diverse group of eukaryotic microorganisms that play crucial roles in various ecosystems, including cold environments. Psychrophilic fungi are specifically adapted to thrive in cold temperatures and are essential contributors to the functioning of cold ecosystems. Psychrophilic fungi have evolved specialized adaptations to survive and grow in cold environments, including polar regions, alpine environments, and cold freshwater bodies. One crucial adaptation is the production of antifreeze proteins and cryoprotectants (Teufel & Morgan-Kiss 2018; Furhan, 2020). These molecules help prevent ice crystal formation within fungal cells, protecting them from damage caused by freezing temperatures. Additionally, psychrophilic fungi have enzymes and metabolic pathways optimized for cold temperatures, allowing them to efficiently metabolize organic substrates even in chilly conditions.

Psychrophilic fungi play significant roles in cold ecosystems by contributing to organic matter decomposition and nutrient cycling (Hassan et al., 2016). They break down complex organic compounds, such as dead plant material and animal remains, releasing nutrients back into the environment (Joshi et al., 2021). By decomposing organic matter, psychrophilic fungi help to recycle carbon and other essential nutrients, supporting the growth of other organisms in cold environments. These fungi are also involved in symbiotic relationships with plants in cold habitats, such as mycorrhizal associations, where they facilitate nutrient uptake by the host plants (Oses-Pedraza et al., 2020).

Members of genera such as *Penicillium*, *Cladosporium*, and *Rhodotorula* are commonly found in cold environments and are considered psychrophilic fungi (Zalar et al., 2011). *Penicillium* spp, for example, are known to colonize a wide range of habitats, including polar regions, where they contribute to decomposition processes. *Cladosporium* spp. are often found in cold freshwater bodies, where they play roles in nutrient cycling and organic matter decomposition. *Rhodotorula* spp. are known for their ability to grow at low temperatures and are commonly found in cold environments worldwide.

The ability of psychrophilic fungi to metabolize organic substrates at cold temperatures makes them vital contributors to carbon and nutrient cycling in cold ecosystems. By breaking down organic matter, these fungi release nutrients that are essential for the growth and survival of other organisms in cold habitats, including plants, bacteria, and animals.

5. Adaptations to Cold Environments by Psychrophilic and Psychrotrophic Bacteria and Fungi

Surviving and thriving in cold environments present significant challenges for microorganisms, yet psychrophilic and psychrotrophic species have evolved a variety of adaptive strategies to overcome these challenges (Figure 2). These adaptations are crucial for maintaining cellular integrity, metabolic activity, and growth in subzero temperatures, allowing these microorganisms to colonize and flourish in cold habitats.

6. Production of Antifreeze Proteins (AFPs) and Ice-Binding Proteins (IBPs)

Antifreeze proteins (AFPs) and ice-binding proteins (IBPs) are crucial adaptations employed by psychrophilic and psychrotrophic microorganisms to thrive in cold environments (Suyal et al., 2021). These proteins play essential roles in preventing cellular damage caused by ice formation and allowing microorganisms to survive and function at freezing temperatures.

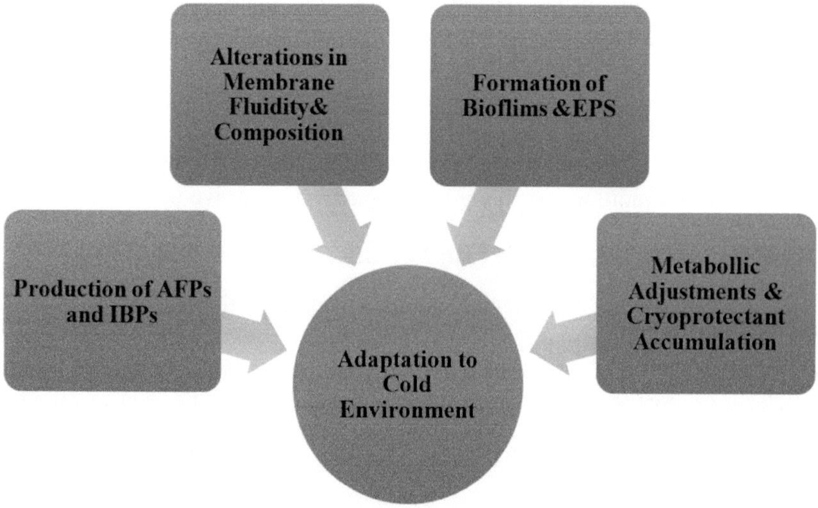

Figure 2: Adaptation mechanisms to cold environments.

Antifreeze Proteins (AFPs)

AFPs are small proteins produced by certain organisms, including bacteria, fungi, and fish, that inhibit the growth of ice crystals (Baskaran et al., 2021). These proteins bind to the surface of ice crystals, preventing their further growth. By binding to the ice crystal lattice, AFPs disrupt the formation of larger ice crystals, thereby lowering the freezing point of the solution. AFPs effectively inhibit ice recrystallization, which is the process by which small ice crystals grow at the expense of larger ones. This helps maintain the integrity of cells and tissues by preventing the formation of large, damaging ice crystals. AFPs typically have a structured, globular shape with ice-binding regions that interact with the surface of ice crystals. These proteins often contain hydrophobic residues that interact with the hydrophobic surfaces of ice crystals, facilitating binding.

Ice-Binding Proteins (IBPs)

IBPs are proteins produced by certain organisms, particularly cold-adapted microorganisms and organisms living in freezing environments like polar fish and insects (Oude et al., 2016). Unlike AFPs, which primarily inhibit ice crystal growth, IBPs bind to specific ice crystal faces and inhibit their growth, thereby modifying the morphology of ice crystals. By binding to specific ice crystal faces, IBPs can control the shape and size of ice crystals, preventing the formation of damaging ice structures within cells.

Some IBPs also function as ice nucleation inhibitors, preventing the spontaneous formation of ice crystals at temperatures below the freezing point. By inhibiting ice nucleation, IBPs help maintain the supercooled state of cellular fluids, further reducing the risk of ice formation and cellular damage. Both AFPs and IBPs are key adaptations that allow psychrophilic and psychrotrophic microorganisms to cope with cold temperatures. By minimizing ice crystal formation and growth, these proteins help prevent cellular damage and maintain cellular function in freezing environments (Karadayi et al., 2021). The production of AFPs and IBPs is a remarkable example of evolutionary adaptation to extreme environmental conditions, enabling microorganisms to thrive in cold habitats.

Alterations in Membrane Fluidity and Composition

In cold environments, maintaining optimal membrane fluidity is essential for the survival and functioning of microorganisms (Hassan et al., 2020). Low temperatures can lead to the solidification of membranes, which can impair cellular processes such as nutrient uptake, signal transduction, and membrane-bound enzyme activity. Psychrophilic and psychrotrophic microorganisms have evolved various adaptations to adjust membrane fluidity and composition in response to cold stress.

One key adaptation involves increasing the proportion of unsaturated fatty acids in membrane lipids. Unsaturated fatty acids have double bonds in their hydrocarbon chains, which introduce kinks and prevent close packing of lipid molecules. This increased unsaturation helps maintain membrane fluidity at low temperatures by preventing lipid molecules from tightly packing together, thus reducing the likelihood of membrane solidification. Psychrophilic and psychrotrophic microorganisms adjust the ratio of saturated to unsaturated fatty acids in their membrane lipids to optimize membrane fluidity for cold conditions. Some microorganisms produce cold-active enzymes involved in lipid metabolism, allowing them to regulate membrane lipid composition in response to changing environmental conditions. These enzymes function optimally at low temperatures, facilitating processes such as fatty acid desaturation and modification of membrane lipid composition. By producing cold-active enzymes, microorganisms can dynamically adjust membrane fluidity in cold environments, ensuring optimal cellular function.

Psychrophilic and psychrotrophic microorganisms may also modify the types of membrane lipids they produce in response to cold stress. For example, they may synthesize lipids with shorter hydrocarbon chains, which have greater fluidity and flexibility at low temperatures. Additionally, microorganisms may incorporate specialized lipids, such as glycolipids and betaine lipids, into their membranes to enhance cold tolerance.

Formation of Biofilms and Extracellular Polymeric Substances (EPS)

In cold environments, where survival can be challenging due to freezing temperatures and harsh conditions, psychrophilic and psychrotrophic microorganisms have evolved various strategies to enhance their cold tolerance and survival. One such strategy involves the formation of biofilms and the production of extracellular polymeric substances (EPS), which play crucial roles in protecting microbial communities from cold stress (Liu et al., 2023).

Biofilms are structured communities of microorganisms that adhere to surfaces and are embedded in a matrix of EPS. These biofilms provide protection and structural support to microbial communities, allowing them to withstand environmental challenges. In cold environments, biofilm formation allows microorganisms to adhere to surfaces, such as rocks, sediments, and ice, where they can access nutrients and shelter from freezing temperatures. By forming biofilms, microorganisms can create stable microenvironments within the matrix, which protect freezing and thawing cycles and other environmental stresses.

EPS are complex mixtures of polysaccharides, proteins, lipids, and nucleic acids produced by microorganisms within biofilms. In cold environments, EPS act as cryoprotectants, providing a protective barrier against cold stress and preventing cellular damage caused by ice crystal formation. EPS molecules surround microbial cells within the biofilm, forming a matrix that helps retain water and maintain cell hydration levels, even in freezing conditions. The hydrated matrix of EPS acts as a physical barrier, shielding microbial cells from freezing temperatures and reducing the risk of desiccation and cellular damage (Sheikh et al., 2022).

Biofilm formation and EPS production enhance the cold tolerance of microbial communities by providing protection against freezing temperatures and other environmental stresses (Bhagat et al., 2021). It allows microorganisms to access nutrients and resources in their environment, facilitating growth and survival in cold habitats. Biofilms also promote interactions between different microbial species within the community, enabling cooperation and mutualism that enhance overall community resilience. In environments characterized by freezing and thawing cycles, such as polar regions and alpine environments, biofilm formation and EPS production help microorganisms adapt to rapid changes in temperature and environmental conditions (Bright et al., 2025; Pathania et al., 2022). The structural stability provided by biofilms and the cryoprotective properties of EPS enable microbial communities to survive and persist despite fluctuations in temperature and other environmental factors

Metabolic Adjustments and Cryoprotectant Accumulation

In cold environments, psychrophilic and psychrotrophic microorganisms employ various metabolic adjustments to enhance their cold tolerance and maintain metabolic activity in subzero temperatures. These adjustments are essential for their survival and functioning in freezing conditions and involve the accumulation of cryoprotectants and the production of cold-active enzymes.

Cryoprotectants are small molecules that help microorganisms survive freezing temperatures by preventing cellular dehydration and stabilizing proteins and membranes (Wolkers & Oldenhof, 2021). Psychrophilic and psychrotrophic microorganisms accumulate cryoprotectants such as sugars (e.g., trehalose, glucose), polyols (e.g., glycerol, mannitol), and compatible solutes (e.g., proline, betaine) in their cells. Cryoprotectants lower the freezing point of intracellular fluids, reducing the risk of ice formation and cellular damage. They also protect cellular structures and macromolecules from the disruptive effects of freezing and thawing (Mishra et al., 2020).

In cold environments, metabolic processes can be significantly slowed down due to reduced enzyme activity. To counteract this, psychrophilic and psychrotrophic microorganisms produce cold-active enzymes that remain functional at low temperatures. These cold-active enzymes are adapted to function optimally in cold conditions, with enhanced catalytic activity and flexibility compared to enzymes from mesophilic organisms. Cold-active enzymes are involved in various metabolic pathways, including energy production, nutrient metabolism, and biosynthesis, allowing microorganisms to maintain essential cellular functions even in freezing temperatures.

Metabolic adjustments play a crucial role in enabling microorganisms to survive and thrive in cold environments by maintaining cellular homeostasis and metabolic activity (Garcia-Lopez et al., 2021). Cryoprotectant accumulation helps microorganisms cope with osmotic stress and dehydration caused by low temperatures, while also protecting cellular structures from damage (Elliott et al., 2017). Cold-active enzymes ensure the continuity of metabolic processes by facilitating the conversion of nutrients into energy and biomass, despite the constraints imposed by cold temperatures.

Psychrophilic and psychrotrophic microorganisms have evolved mechanisms to adapt to subzero temperatures and exploit cold environments as ecological niches. Their ability to undergo metabolic adjustments, accumulate cryoprotectants, and produce cold-active enzymes allows them to survive and function in freezing conditions, contributing to the ecological dynamics of cold habitats.

7. Biotechnological Applications of Psychrophilic and Psychrotrophic Bacteria and Fungi

The unique adaptations of psychrophilic and psychrotrophic microorganisms make them valuable resources for various biotechnological applications, offering advantages over enzymes and compounds derived from mesophilic organisms. These microorganisms produce a diverse array of enzymes and bioactive compounds that exhibit enhanced activity and stability at low temperatures, making them well-suited for use in a wide range of industrial processes.

- **Enzymes for Food Processing:** Psychrophilic and psychrotrophic microorganisms produce enzymes such as lipases, proteases, and amylases that are highly active at low temperatures, making them ideal candidates for food processing applications. Cold-active lipases, for example, are used in the dairy industry for the production of low-fat dairy products and in the baking industry for the modification of fats and oils (Rabbani et al., 2023). Similarly, cold-active proteases are employed in meat tenderization and cheese ripening processes, while cold-active amylases find applications in starch hydrolysis and beverage production (Kumari et al., 2021).

- **Detergent Formulation-Enzymes:** Enzymes produced by psychrophilic and psychrotrophic microorganisms are also used in detergent formulation for their ability to function effectively at low temperatures (Kumar et al., 2020). Cold-active enzymes such as proteases and lipases are incorporated into laundry detergents to enhance stain removal and fabric softening properties, particularly in cold water washing cycles (Hasan et al., 2010; Fernandes, 2023). By improving the efficiency of detergent formulations at lower temperatures, these enzymes contribute to energy savings and reduce environmental impact.
- **Bioremediation:** Psychrophilic and psychrotrophic microorganisms play a role in bioremediation processes, where they contribute to the degradation of pollutants in cold environments (Hamid et al., 2024; Hualpa et al., 2023). Cold-active enzymes produced by these microorganisms are utilized in bioremediation applications to accelerate the breakdown of contaminants such as hydrocarbons, pesticides, and heavy metals in cold soils, sediments, and water bodies (Hamid et al., 2022). By enhancing the efficiency of bioremediation processes in cold environments, these enzymes offer sustainable solutions for environmental clean-up and restoration.
- **Bioprospecting for Novel Compounds:** Beyond enzyme production, psychrophilic and psychrotrophic microorganisms are being explored for their potential in bioprospecting and the discovery of novel bioactive compounds with pharmaceutical and agricultural applications (Al-Maqtari et al., 2019). Cold environments are rich sources of biodiversity, and microorganisms adapted to these habitats produce bioactive compounds with unique chemical structures and biological activities. These compounds hold promise for the development of new drugs, antimicrobial agents, and agricultural bioproducts, addressing global health and agricultural challenges (Hamid et al., 2022).

8. Challenges and Future Directions

Psychrophilic and psychrotrophic microorganisms, though pivotal for ecological balance and biotechnological innovations, encounter significant hurdles hindering their comprehensive study and exploitation. Culturing these microorganisms in laboratory settings poses a formidable challenge due to their fastidious nature and specific growth requirements, including low temperatures and specialized nutrient compositions mimicking their natural habitats. Additionally, their slow growth rates make obtaining pure cultures and elucidating their physiological and biochemical characteristics laborious tasks. Furthermore, the limited accessibility of extreme environments, such as polar regions and deep-sea trenches where these microorganisms flourish, presents logistical and technical obstacles for scientific exploration. Harsh weather conditions, remote locations, and logistical constraints compound the difficulties in conducting fieldwork and environmental monitoring programs in these regions. Nevertheless, advancements in molecular biology, metagenomics, and bioinformatics offer promising solutions to these challenges. High-throughput sequencing technologies enable the characterization of microbial communities in extreme environments, unveiling their genetic diversity and metabolic potential. Metagenomic approaches facilitate the discovery of novel genes and pathways involved in cold adaptation, shedding light on the molecular mechanisms underlying microbial survival in cold environments. Moreover, bioinformatics tools aid in the analysis and interpretation of large-scale genomic and metagenomic datasets, accelerating our understanding of cold-adapted microbiomes. By integrating multi-omics approaches, including genomics, transcriptomics, proteomics, and metabolomics, researchers can gain comprehensive insights into the metabolic networks, regulatory mechanisms, and ecological roles of psychrophilic and psychrotrophic microorganisms. This holistic approach will not only advance our fundamental understanding of cold-adapted microbiomes but also pave the way for transformative discoveries and applications in biotechnology, environmental science, and beyond.

9. Conclusion

The exploration into psychrophilic and psychrotrophic microorganisms illuminates their extraordinary resilience in cold environments and underscores their pivotal ecological roles and vast biotechnological potential. Spanning across the domains of bacteria, archaea, and fungi, these microorganisms display a myriad of specialized adaptations, including the production of antifreeze proteins, adjustments in membrane fluidity, and the formation of protective biofilms, all tailored to thrive in the harshest cold conditions. Their contributions to nutrient cycling, organic matter decomposition, and symbiotic partnerships are instrumental in shaping the delicate balance of cold ecosystems, while their enzymatic arsenal finds invaluable applications across diverse industries, ranging from food production to environmental cleanup. Nonetheless, the study of psychrophilic and psychrotrophic microorganisms faces persistent challenges, notably in laboratory cultivation and accessing remote environments, necessitating continued advancements in molecular biology, metagenomics, and bioinformatics for a more comprehensive understanding. Looking ahead, harnessing these insights holds promise for unlocking new pathways towards sustainable development, environmental stewardship, and biomedical breakthroughs, thus charting an exciting course for the future of cold-adapted microbiology and its myriad applications.

References

Al-Maqtari, Q.A., Waleed, A.A., & Mahdi, A.A. 2019. Cold-active enzymes and their applications in industrial fields: A review. *Int. J. Res. Agric. Sci.*, 6(4), 2348–3997.

Anani, O.A., Shahnawaz, M., Gupta, P., Osarenotor, O., & Olisaka, F.N. 2023. 13 Ecological roles of psychrophiles. In: *Soil Microbiome of the Cold Habitats: Trends and Applications*. CRC Press, 179 pp.

Baba, Z.A., Hamid, B., Sheikh, T.A., Alotaibi, S., Enshasy, H.E., Ansari, M.J. et al. 2021. Psychrotolerant *Mesorhizobium* sp. Isolated from temperate and cold desert regions solubilize potassium and produces multiple plant growth promoting metabolites. *Molecules*, 26, 5758. https://www.mdpi.com/1420-3049/26/19/5758/htm.

Baskaran, A., Kaari, M., Venugopal, G., Manikkam, R., Joseph, J., & Bhaskar, P.V. 2021. Antifreeze proteins (Afp): Properties, sources and applications: A review. *International Journal of Biological Macromolecules*, 189, 292–305.

Bhagat, N., Raghav, M., Dubey, S., & Bedi, N. 2021. Bacterial exopolysaccharides: Insight into their role in plant abiotic stress tolerance. *Journal of Microbiology and Biotechnology*, 31(8), 1045.

Bhanu, L.M., & Chatterjee, S. 2024. Cellulases from psychrophilic and psychrotrophic microorganisms and their potential applications. In: *Harnessing Microbial Potential for Multifarious Applications*. Springer Nature Singapore, pp. 549–573.

Białkowska, A., Majewska, E., Olczak, A., & Twarda-Clapa, A. 2020. Ice binding proteins: Diverse biological roles and applications in different types of industry. *Biomolecules*, 10(2), 274.

Bowman, J.P. 2017. Genomics of psychrophilic bacteria and archaea. In: Psychrophiles: *From Biodiversity to Biotechnology*. Springer Link, pp. 345–387.

Bright, J.P., Maheshwari, H.S., Thangappan, S., Perveeen, K., Bukhari, N.A., Mitra, D. et al. 2025. Biofilmed-PGPR: A next-generation bioinoculant for plant growth promotion in rice (*Oryza sativa* L.) under changing climate. *Rice Science*, 2025(32). https://doi.org/10.1016/j.rsci.2024.08.008.

Dasila, H., Maithani, D., Suyal, D.C., & Debbarma, P. 2022. Cold-adapted microorganisms: survival strategies and biotechnological significance. In: *Survival Strategies in Cold-adapted Microorganisms*. Springer Nature, pp. 357–378.

Doytchinov, V.V., & Dimov, S.G. 2022. Microbial community composition of the Antarctic ecosystems: Review of the bacteria, fungi, and archaea identified through an NGS-based metagenomics approach. *Life*, 12(6), 916.

Elliott, G.D., Wang, S., & Fuller, B.J. 2017. Cryoprotectants: A review of the actions and applications of cryoprotective solutes that modulate cell recovery from ultra-low temperatures. *Cryobiology*, 76, 74–91.

Fernandes, P. 2023. 9 Role of microbial oxidative enzymes. In: *Microbial Oxidative Enzymes: Biotechnological Applications*. De Gruyter, 183 pp.

Furhan, J. 2020. Adaptation, production, and biotechnological potential of cold-adapted proteases from psychrophiles and psychrotrophs: Recent overview. *Journal of Genetic Engineering and Biotechnology*, 18(1), 36.

Galié, S., García-Gutiérrez, C., Miguélez, E.M., Villar, C.J., & Lombó, F. 2018. Biofilms in the food industry: Health aspects and control methods. *Frontiers in Microbiology*, 9, 315815.

García-Descalzo, L., García-López, E., & Cid, C. 2022. Comparative proteomic analysis of psychrophilic vs. mesophilic bacterial species reveals different strategies to achieve temperature adaptation. *Frontiers in Microbiology*, 13, 841359.

Garcia-Lopez, E., Alcazar, P. & Cid, C. 2021. Identification of biomolecules involved in the adaptation to the environment of cold-loving microorganisms and metabolic pathways for their production. *Biomolecules*, 11(8), 1155.

Grant, W.D., & Long, P.E. 2013. *Environmental Microbiology*. Springer Science & Business Media.

Hamid, B., Baba, Z.A., Sheikh, T.A., Perveen, K., & Sayyed, R.Z. 2024. Investigating potential protease activity of psychrotrophic bacteria from a municipal landfill for solid waste management. In: *Biomass Conversion and Biorefinery*. SpringerLink. https://doi.org/10.1007/s13399-024-05621-2.

Hamid, B., Bashir, Z., Yatoo, A.M., Mohiddin, F., Majeed, N., Bansal, M. et al. 2022. Cold-active enzymes and their potential industrial applications: A review. *Molecules*, 27(18), 5885.

Hasan, F., Shah, A.A., Javed, S., & Hameed, A. 2010. Enzymes used in detergents: Lipases. *African Journal of Biotechnology*, 9(31), 4836–4844.

Hassan, N., Rafiq, M., Hayat, M., Shah, A.A., & Hasan, F. 2016. Psychrophilic and psychrotrophic fungi: A comprehensive review. *Reviews in Environmental Science and Bio/Technology*, 15, 147–172.

Hassan, N., Anesio, A.M., Rafiq, M., Holtvoeth, J., Bull, I., Haleem, A. et al. 2020. Temperature driven membrane lipid adaptation in glacial psychrophilic bacteria. *Frontiers in Microbiology*, 11, 824.

Hualpa-Cutipa, E., Acosta, R.A.S., Cariga, O.J.M., Espinoza-Medina, M.A., Chavez-Rojas, D.C., Medina-Cerna, D. et al. 2023. Metagenomic approach role of psychrotrophic and psychrophilic microbes in bioremediation. In: *Metagenomics to Bioremediation*. Academic Press, pp. 513–536.

Jin, S., Wang, Y., & Zhao, X. 2022. Cold-adaptive mechanism of psychrophilic bacteria in food and its application. *Microbial Pathogenesis*, 169, 105652.

Joshi, S., Bajpai, A., & Johri, B.N. 2021. Extremophilic fungi at the interface of climate change. In: *Fungi Bio-Prospects in Sustainable Agriculture, Environment and Nano-Technology*. Academic Press, pp. 1–22.

Kanekar, P.P., & Kanekar, S.P. 2022. Psychrophilic, psychrotrophic, and psychrotolerant microorganisms. In: *Diversity and Biotechnology of Extremophilic Microorganisms from India*. Springer Nature Singapore, pp. 215–249.

Karadayi, G., Alaylar, B., Karadayi, M., Gürkök, S., Aksu, S., Egamberdieva, D. et al. 2021. Biodiversity, ecological, and commercial importance of psychrophilic microorganisms. In: *Microbial Communities and Their Interactions in the Extreme Environment*. Springer, pp. 225–247.

Kaur, J., Vishnu, A.L., Khipla, N., & Kaur, J. 2022. Microbial life in cold regions of the deep sea. In: *Survival Strategies in Cold-adapted Microorganisms*. Springer, pp. 63–86.

Kuddus, M., Bano, N., Sheik, G.B., Joseph, B., Hamid, B., Sindhu, R. et al. 2024. Cold-active microbial enzymes and their biotechnological applications. *Microbial Biotechnology*, 17(4), e14467.

Kuhn, E. 2012. Toward understanding life under subzero conditions: The significance of exploring psychrophilic "cold-shock" proteins. *Astrobiology*, 12(11), 1078–1086.

Kumar, A., Mukhia, S., Kumar, N., Acharya, V., Kumar, S., & Kumar, R. 2020. A broad temperature active lipase purified from a psychrotrophic bacterium of Sikkim Himalaya with potential application in detergent formulation. *Frontiers in Bioengineering and Biotechnology*, 8, 642.

Kumari, M., Padhi, S., Sharma, S., Phukon, L.C., Singh, S.P. & Rai, A.K. 2021. Biotechnological potential of psychrophilic microorganisms as the source of cold-active enzymes in food processing applications. *3 Biotech.*, 11(11), 479.

Liu, J., Wu, S., Feng, L., Wu, Y. & Zhu, J. 2023. Extracellular matrix affects mature biofilm and stress resistance of psychrotrophic spoilage Pseudomonas at cold temperature. *Food Microbiology*, 112, 104–214.

Mishra, P.K., Bisht, S.C., Pandey, B.M., Meena, V.S., Parihar, M., Mahanta, D. et al. 2020. Plant growth promoting rhizobacteria: Mechanisms and alleviation of cold stress in plants. In: *Plant Stress Biology: Strategies and Trends*. Springer Nature, pp. 185–226.

Oses-Pedraza, R., Torres-Díaz, C., Lavín, P., Retamales-Molina, P., Atala, C., Gallardo-Cerda, J. et al. 2020. Root endophytic Penicillium promotes growth of Antarctic vascular plants by enhancing nitrogen mineralization. *Extremophiles*, 24, 721–732.

Ottoni, J.R., de Oliveira, V.M., & Passarini, M.R.Z. 2022. Microbes in thawing permafrost: Contributions to climate change. In: *Microbiome Under Changing Climate*. Woodhead Publishing, pp. 1–28.

Oude Vrielink, A.S., Aloi, A., Olijve, L.L. & Voets, I.K. 2016. Interaction of ice binding proteins with ice, water and ions. *Biointerphases*, 11(1).

Pathania, S., Solanki, P., Putatunda, C., Bhatia, R.K. & Walia, A. 2022. Adaptation to cold environment: The survival strategy of Psychrophiles. In: *Survival Strategies in Cold-adapted Microorganisms*. Springer Nature, pp. 87–111.

Pearce, D.A. 2017. Extremophiles in Antarctica: Life at low temperatures. In: *Adaptation of Microbial Life to Environmental Extremes: Novel Research Results and Application*. Springer International Publishing, pp. 99–131.

Rabbani, G., Ahmad, E., Ahmad, A. & Khan, R.H. 2023. Structural features, temperature adaptation, and industrial applications of microbial lipases from psychrophilic, mesophilic, and thermophilic origins. *International Journal of Biological Macromolecules*, 225, 822–839.

Rai, A.K., & Sharma, H. 2022. Cold-adapted microorganisms and their potential role in plant growth. In: *Survival Strategies in Cold-adapted Microorganisms*. Springer Nature, pp. 321–342.

Ramasamy, K.P., Mahawar, L., Rajasabapathy, R., Rajeshwari, K., Miceli, C., & Pucciarelli, S. 2023. Comprehensive insights on environmental adaptation strategies in Antarctic bacteria and biotechnological applications of cold adapted molecules. *Frontiers in Microbiology*, 14, 1197797.

Salvador-Castell, M., Tourte, M., & Oger, P.M. 2019. In search for the membrane regulators of archaea. *International Journal of Molecular Sciences*, 20(18), 4434.

Sheikh, T., Baba, Z., Hamid, B., Iqbal, S., Yatoo, A., Fatima, S. et al. 2022. Extracellular polymeric substances in psychrophilic cyanobacteria: A potential bioflocculant and carbon sink to mitigate cold stress. *Biocatalysis and Agricultural Biotechnology*, 42, 102375. https://doi.org/10.1016/j.bcab.2022.102375.

Shen, L., Liu, Y., Allen, M.A., Xu, B., Wang, N., Williams, T.J. et al. 2021. Linking genomic and physiological characteristics of psychrophilic Arthrobacter to metagenomic data to explain global environmental distribution. *Microbiome*, 9(1), 136.

Suyal, D.C., Joshi, D., Kumar, S., Bhatt, P., Narayan, A., Giri, K. et al. 2021a. Himalayan microbiomes for agro-environmental sustainability: Current perspectives and future challenges. *Microbial Ecology*, 1–33.

Suyal, D.C., Soni, R., Yadav, A.N., & Goel, R. 2021b. Cold adapted microorganisms: Survival mechanisms and applications. In: *Microbiomes of Extreme Environments*. CRC Press, pp. 177–191.

Teufel, A.G., & Morgan-Kiss, R.M. 2018. Physiological and biochemical adaptations of psychrophiles. In: *Extremophiles*. CRC Press, pp. 185–208.

Vivek, K., Sandhia, G.S. & Subramaniyan, S.J.B.A. 2022. Extremophilic lipases for industrial applications: A general review. *Biotechnology Advances*, 60, 108002.

Wolkers, W.F., & Oldenhof, H. 2021. Principles underlying cryopreservation and freeze-drying of cells and tissues. In: *Cryopreservation and Freeze-drying Protocols*. Humana Press, pp. 3–25.

Zakhia, F., Jungblut, A.D., Taton, A., Vincent, W.F., & Wilmotte, A. 2008. Cyanobacteria in cold ecosystems. In: *Psychrophiles: From Biodiversity to Biotechnology*. SpringerLink, pp. 121–135.

Zalar, P., Sonjak, S., & Gunde-Cimerman, N. 2011. Fungi in polar environments. In: *Polar Microbiology: Life in a Deep Freeze*. ASM Press, pp. 79–99.

Zhang, C., Guo, W., Wang, Y., & Chen, X. 2017. Colwellia beringensis sp. nov., a psychrophilic bacterium isolated from the Bering Sea. *International Journal of Systematic and Evolutionary Microbiology*, 67(12), 5102–5107.

CHAPTER

6

Psychrophiles Challenging Life, Permafrost, and Deep Ocean Waters: An Overview

Vishal Sharma[1,3*], Parushi Nargotra[1], Bijender Kumar Bajaj[2], Mei-Ling Tsai[1] and Cheng-Di Dong[3]

[1] Department of Seafood Science, National Kaohsiung University of Science and Technology, Kaohsiung City, Taiwan

[2] School of Biotechnology, University of Jammu, Jammu, India

[3] Department of Marine Environmental Engineering, National Kaohsiung University of Science and Technology, Kaohsiung City, Taiwan

1. Introduction

In the vast tapestry of Earth's ecosystems, one such rare environment called the cryosphere pushes the boundaries of life's endurance to the extreme. The cryosphere includes ice sheets, icebergs, glaciers, and permafrost in polar regions and high mountainous areas like Hindu Kush, Karakorum, and the Himalayas, which make up around 13% of the Earth's surface (HKKH). Microorganisms are best preserved in these frozen habitats because of a few factors, including low oxygen, darkness, extreme cold, and minimal human activity. Researchers around the globe have always been interested in ecological biodiversity, and a wealth of data regarding the global distribution of microorganisms (Kaur et al., 2022). When considering extremophilic organisms, these life forms are highly specialized for extreme conditions, setting the standard for the metabolic and biochemical operations in which the organism can thrive (Sharma et al., 2020; Singh et al., 2014). Within this icy realm, a remarkable group of microorganisms known as psychrophiles thrive, defying conventional notions of habitability. Psychrophiles, derived from the Greek words 'psycho', meaning cold, and 'philo', meaning loving, are cold-adapted organisms that have adapted to flourish

*Corresponding author: sudans123@gmail.com

DOI: 10.1201/9781003717003-6

in the frozen biosphere's habitats, such as ice lakes, glaciers, and the deep sea (Bhatia et al., 2021). These microorganisms have evolved crucial mechanisms to address the challenges of low temperatures, achieving necessary adaptations in their cellular structures and functional organization.

Permanently frozen earth, or permafrost, makes up a significant portion of Earth's terrestrial cold environment and is thought to hold 50% of the soil organic carbon (SOC) found beneath (Hu et al., 2014). Permafrost is deemed an extreme environment due to the necessity for indigenous organisms to endure extended periods of sub-zero temperatures and background radiation over geological time spans (Hamid & Mohiddin, 2018). This occurs in a habitat characterized by low water activity and exceptionally slow rates of nutrient and metabolite transfer. The ability of psychrophiles for low-temperature survival and proliferation in permafrost suggests that they have surmounted significant challenges and obstacles such as reduced enzyme activity, protein cold-denaturation, low membrane fluidity, intracellular ice formation, altered transport of nutrients and waste products, inappropriate protein folding, decreased rates of cell division, transcription, and translation (D'Amico et al., 2006). They have developed special adaptations to endure and even flourish in frigid temperatures despite facing the obstacles they confront in their harsh environments. According to reports, some bacteria that live in permafrost soil and sea ice can withstand temperatures as low as –20°C, which appears to be the lowest temperature limit for life. At these temperatures, brine channels and the tiny amounts of unfrozen water within the ice or permafrost soil are the only places where microbiological activity occurs. These are characterized by high concentrations of exopolymeric materials, salts, and/or particulate matter. Temperature gradients and concentration gradients control fluid movement.

To thrive in harsh environments, these microbes require a wide range of changes on all fronts, ranging from cell structure to metabolites to specialized biocatalysts that catalyze different anabolic and catabolic processes (Dziurzynski et al., 2023). These extremophiles have produced specific stabilizing factors that interact with other structural proteins and help maintain the structural proteins' kinetic stability, compactness, and thermodynamic stability, which helps protect them against harsh conditions and chemical denaturation (Bhatia et al., 2021b). Although the majority of these related proteins from their mesophilic counterparts are generally stable, these factors which are often heat sensitive, cold-adapted proteins, and/or factors need to be structurally altered to improve their involvement in resistance to cold (Pathania et al., 2022). Strong evolutionary pressure can result in substitutions, insertions, or deletions in the nucleotide sequence, which can alter the amino acid residues in a polypeptide chain (Baraúna et al., 2017).

Many different proteins and enzymes are produced by the psychrophiles through their essential cold adaptation, and these biomolecules may find use in current and future biotechnological processes. Emerging biotechnological tools, including protein engineering, rDNA technology, and metagenomics, allow the direct separation of encoding enzymes, novel genes, and other biomolecules from microbial populations in addition to traditional culture methods so they can be effectively utilized for a variety of food applications (Corsaro et al., 2017; Kumari et al., 2021). Psychrophilic enzymes account for 30–40% of all enzymes used in industrial applications globally (Bhatia et al., 2021). Effective low-temperature recombinant protein secretion systems are essential for enabling the biotechnological use of psychrophiles and their derivatives. Wild-type strains typically produce too few cold-active (heat-labile) proteins to be produced on an industrial scale (Margesin & Feller, 2010). Numerous cold-active enzymes derived from psychrophiles, such as β-galactosidase, proteases, pectinase, amylases, lipases, xylanases, pullulanases, and β-mannanases, have established applications in the food industry (Kumari et al., 2021). The current chapter highlights the vast biodiversity of psychrophiles in permafrost and deep ocean waters, their role and adaptation strategies in these extreme regions. Moreover, the different biotechnological applications of psychrophiles isolated from these life-challenging environments has been discussed and analyzed.

2. Diversity of Psychrophiles in Permafrost and Deep Ocean Waters

With temperatures below 5°C, about 80% of the planet earth is permanently cold. Sea ice (13% of the coverage space) and permafrost (24% of the land surface), snow (35% of the land surface), glaciers (10% of the land surface), and the deep sea (which makes up nearly 71% of the Earth's oceans and 90% of their volume is below 5°C) make up the majority of this low-temperature environment (Pathania et al., 2022). The microorganisms found in the permafrost soil and sea ice have been found to have a minimum temperature threshold for life of about –20°C. Brine channels and the little amounts of unfrozen water within the ice or permafrost soil are the only places where microbiological activity can occur at these temperatures. Concentration and temperature gradients sustain fluid movement in these environments, which contain significant concentrations of salts, exopolymeric materials, and/or particulate matter (D'Amico et al., 2006). A wide variety of different microorganisms such as bacteria, archaea, fungi, microalgae, and others thrive in the permafrost and deep ocean waters.

Diversity of Psychrophilic Bacteria in Permafrost

Permafrost, characterized by a below 0°C temperature for at least two consecutive years, blankets the Northern Hemisphere (25%) and Earth's exposed land surface (17%) (Sannino et al., 2021). Permanently frozen permafrost is often covered by a seasonally thawed active layer, the depth of which is influenced by air temperatures, moisture levels, vegetation cover, and the presence of snow (da Silva et al., 2019). The polar regions provide a spectrum of environments conducive to microbial colonization and growth, including ice-covered oceans, ice shelves, glaciers, tundra, rivers, polar desert soils, and ice-influenced lakes, and wetlands. Notably, these extreme areas such as high arctic permafrost soils harbor distinctive psychrophilic fungi and bacteria (Finore et al., 2020). Permafrost has been described as an important microhabitat for extremophiles despite its extreme features, which include minimal water availability, prolonged exposure to low temperatures, and extremely low rates of nutrition and metabolite transport.

In these adverse environments, bacteria appear to be the predominant microbial species that has been studied by many researchers (da Silva et al., 2020). The phylogenetic groups of different bacterial isolates from diverse permafrost niches are majorly categorized as *Firmicutes*, *Actinobacteria*, *Spirochaetes*, *α-Proteobacteria*, β-Proteobacteria, γ-Proteobacteria, *Bacteroidetes*, *Actinobacteria*, *Deinococcus-Thermus*, and *Cytophaga-Flavobacterium-Bacteroides* (CFB) group (Hu et al., 2015; Reddy et al., 2016). Numerous bacterial isolates, both Gram-positive and Gram-negative, are found in permafrost samples. Spore-forming bacteria are also frequently recovered, even though their existence varies significantly in geographically dispersed areas.

One of the world's most severe habitats to survive is Antarctica, which has an average thickness of 1.9 km and is the coldest, driest, windiest, and iciest of all known habitats. In this region, bacteria are the predominant microbial species growing. The 16S rRNA approach is used to identify the dominant and non-dominant bacterial taxa from a variety of habitats, including permafrosts in an effort to fully understand the diversity of bacteria found in the continent (Reddy et al., 2016). Carr et al. (2013) noted that the *Chloroflexi* group was present in unusually high proportions in the Betaproteobacteria, and that the two genera *Thiobacillus* and *Teptidiphilus* made up the majority of the group (almost 45%).

In a study it was found that the genetic variety of microorganisms grew with depth, in contrast to the overall number of bacteria, which reduced approximately two orders of magnitude from the surface down to the mineral frozen layer of the peat core. A group of researchers discovered a preponderance of *Acidobacteria* in the genetic diversity of microbes in the peat core at the thawed peat–permafrost boundary (Aksenov et al., 2021). Besides Antarctica, the southern part of the Svalbard Archipelago is one of the High Arctic zones, which are said to be the coldest and most hostile places on Earth covered with snow and ice. The researchers looked into the microbial communities' cultivability and biodiversity in Arctic soils to obtain 14 bacterial strains that were both psychrotolerant and multi-metal resistant, and that also possessed biotechnologically significant traits, such as the ability to degrade petroleum hydrocarbons (Dziurzynski et al., 2023).

Diversity of Psychrophilic Fungi in Permafrost

Due to their exceptional ability to adapt to cold temperatures, psychrophilic fungi can be found in some of the world's coldest climates. However, fungal patterns have surprisingly received little attention, despite the fact that a great deal of research has been done on the prokaryotic variety found in permafrost, which is characterized by extremely low rates of nutrition and metabolite transfer, low water activity, and sub-zero temperatures. The majority of current research on permafrost mycology has come from permafrost in the Antarctic and Arctic, and it has shown that polar permafrost supports a variety of fungal communities, such as filamentous fungi, ectomycorrhizal (ECM) fungi, and basidiomycetous yeast (Hu et al., 2014). The research investigations have revealed that most common fungal species found in permafrost were those belonging to the genera *Penicillium*, *Mortierella*, Geomyces, *Cryptococcus*, *Phoma*, and *Rhodotorula*. Numerous mycological investigations conducted in Antarctica included a variety of fungus that can be found in lakes, soil, old forests, macroalgae, and plants. Antarctica is home to about 0.6% of fungi (water molds, Kingdom Chromista) and 99.4% of genuine fungi, which include unicellular yeasts and filamentous fungus (phylum *Ascomycota*, *Basidiomycota*, *Chytridiomycota*, and *Zygomycota*) (Hassan et al., 2016). The diversity and distribution of fungi in active layers and permafrost were investigated from three Maritime Antarctica islands where a total of 351 fungal isolates from the active layer and 213 from the permafrost were found (da Silva et al., 2020). Among them, *Pseudogymnoascus*, *Penicillium*, and *Oidiodendron* species were the most prevalent in permafrost. The most prevalent families in the active layer were *Helotiales*, *Pseudeurotium*, *Mortierellaceae*, and *Bionectriaceae*. According to da Silva et al. (2019), the phylum *Ascomycota* contains the majority of the fungal species found in Antarctic permafrost, followed by the phyla *Basidiomycota*, *Mortierellomycota*, and *Mucoromycota*. Fungi in the Antarctic permafrost have so far been found to belong to 32 families, 45 genera, and 70 species. *Cladosporium herbarum*, one of the 70 fungal species, is the most widely distributed and has been isolated from various Antarctic locations. Additional research on the expansion of fungal variety and its potential to spread from permafrost to other environments in Antarctica and beyond is warranted in light of the discovery of live fungi trapped in permafrost.

Recent studies have shown that permafrost has significantly degraded over the previous 10 years, according to monitoring data, and global permafrost temperatures have climbed by an average of 0.29°C (Jiang et al., 2023). The major signs of it include an increasing active layer thickness, a shrinking permafrost area, the island's gradual disappearance, and discontinuous permafrost, all of which pose a serious danger to the stability of permafrost ecosystems. Jiang et al. (2023) evaluated the effects of global warming on the Greater Xing'an Mountain's permafrost regions. Through fungal sequencing, a total of 11 phyla, 39 classes, 89 orders, 187 families, 361 genera, and 522 species were acquired which were separated into 1,463 amplicon sequence variations (ASVs). The leading phylum and genus were *Ascomycota* and *Dimorphospora*, respectively, and there were variations in the relative abundance responses of different groups to warming at the phylum and genus levels. Permafrost microbial communities may be more susceptible to climate change as a result of a loosening of microbial association networks brought on by warming, while the expansion of microbiota that uses a wider variety of substrates may change the C pools of the soil. Perez-Mon et al. (2022) explained that the climate warming in the mountains of the mid-latitudes will cause a shift in the balance of life within the permafrost, favoring the establishment of sapro-pathotrophic fungi replacing basidiomycetous yeasts.

Diversity of Psychrophilic Algae in Permafrost

Algae are typically associated with aquatic environments, but they can be found in a variety of habitats, including permafrost. The predominant photosynthetic primary producers in cold environments are cyanobacteria and eukaryotic algae, which may be found in a very wide range of niches, including sea ice and snowfields, as well as lakes that are permanently covered in ice (Cvetkovska et al., 2017). Algae have been found in certain polar and subpolar permafrost zones, albeit being less common in permafrost than in aquatic environments. The diversity of algae in permafrost is not as well-studied as in other environments, and research on this topic is relatively limited. Algae found in permafrost are likely to be adapted to extreme cold conditions and may include various species of green algae, diatoms, cyanobacteria, and other microorganisms (Karadayi et al., 2021). From 28 Siberian permafrost cores, non-axenic cyanobacterial strains (30) were identified. Cyanobacteria classified as non-heterocystous (*Oscillatoriales*) and filamentous heterocystous (*Nostocales*) were found (Vishnivetskaya, 2009). Although cyanobacteria and microalgae have been reported from Arctic surface soil and aquatic environments, little is known about their diversity or resilience to the extremes of permafrost. Microalgae and cyanobacteria are incredibly adaptable and can survive in harsh environments. However, when underlying permafrost is exposed to light following sediment collapse brought on by thawing, cyanobacteria and green algae will become a major component of the ecosystem. Even so, they are probably not the first organisms to begin growing after subsurface permafrost thaws (Vishnivetskaya et al., 2020).

Diversity of Psychrophilic Bacteria in Deep Ocean Waters

The ocean covers over 71% of the earth's surface, with over half of that region being over 3,000 meters deep. The lowest stratum of the ocean below the thermocline, at a depth of at least 1,828 meters, is known as the deep sea. The deep sea is the largest and most remote biome of the biosphere, with a vast range of ecosystems found there. This zone has decreasing light intensity as depth increases, and the lack of light stops photosynthetic yield. The two main elements of the deep water that regulate benthic production and biomass are the approaching freezing point and the dearth of organic matter at the ocean floor (Jain et al., 2023). The deep water has several difficult conditions for life at the same time, with its extreme cold being its most defining attribute. As a result, the microbes that live here have evolved a few unique traits that enable them to flourish in their surroundings (Karadayi et al., 2021). One of the most remarkable ecological phenomena of the last three million years has been the link between temperature and microbial diversity. Many speculative theories based on ecology and evolution have been proposed to explain the mechanisms underlying this link. Still, only a small number of taxa have access to accessible microbial diversity since it cannot reach the deep ocean region. The majority of taxonomic richness and biomass in deep ocean/sea sediments is made up of bacteria, archaea, and certain fungi (mainly found at deep sea hyperthermal vents) (Kaur et al., 2022). These organisms are crucial to the nutrition cycle and remineralization of organic matter.

Deep-sea bacteria are able to flourish in their harsh surroundings because of a number of peculiar traits. *Proteobacteria* are discovered to predominate in the sediments of the Atlantic Ocean and the deeper Arctic and Pacific Oceans. Seasonal differences have been seen in the Pacific Ocean, with *Bacteroidetes* being more common than *Proteobacteria* during the winter (Jain et al., 2023). Around 51 bacteria were isolated by Shao et al. (2015) from the deep seawater column above the South-West Indian Ridge. These bacteria belonged to the following genera: Gamma-proteobacteria: *Alkaligens*, *Alkanovorax*, *Idiomarina*, *Halomonas*, *Pseudoidiomarina*, *Muricauda*, *Marinobacter*, and *Pseudomonas*, and *Bacillus* (*Firmicutes*), *Salegentibacter* (*Bacteroides*), and *Microbacterium* (*Actinobacteria*). From the Central Indian Ocean Basin, Gawas et al. (2019) identified 43 heterotrophic bacteria, including members of the phyla *Firmicutes*, *Actinobacteria*, and *Proteobacteria*. This was among the first records of *Brachybacterium* (*Actinobacteria*) and *Oceanobacillus* (*Firmicutes*). A halotolerant bacterium *Halomonas sedimenti* was discovered in the deep marine sediment of the Southwest Indian Ocean (Qiu et al., 2021).

Diversity of Psychrophilic Fungi in Deep Ocean Waters

Fungi are ubiquitous microorganisms that belong to the microbiota and are present in all natural habitats, including fresh waters and surface to deep ocean waters, eutrophic to ultra-oligotrophic lakes, lagoons, rivers, ground waters, melting water, and glacier ice (Jain et al., 2023). These microbial species are a desirable source for getting new enzymes and secondary metabolites for use in biotechnology and medicines because of their ability to adapt to low temperatures. The first deep-sea fungal species were discovered in the Atlantic Ocean at a depth of about 4,450 meters and many of these fungal species have been discovered in sediments from around the world, even beyond the Atlantic ocean (Ogaki et al., 2020). Numerous physical and chemical characteristics of deep sub-seafloors, including high hydrostatic and lithostatic pressure, salinity, porosity, severe pH, low temperature, and low nutrition supply, limit the capacity of fungi to thrive. The majority of fungi that have been recovered from deep-sea sediments using both independent and culture-dependent methods show similarity with terrestrial species, suggesting that they may have entered the deep sea through wind or terrestrial runoff (Ogaki et al., 2020). Rosa et al. (2019) showed that in order for psychrophilic fungi to thrive in the harsh conditions of Antarctica, they may exhibit unique biochemical pathways that enable them to produce particular or novel compounds that could be used as prototypes to create new drugs against various biological targets. Furthermore, recent research indicates that fungi found in various Antarctic environments merit attention because they can be thought of as living factories that produce unknown bioactive secondary metabolites, and only a small number of species and strains have been studied chemically up to the present.

Among the different yeast species, *Rhodosporidium* spp., *Rhodotorula* spp., *Candida* spp., *Cryptococcus* spp., *Pichia* spp., *Sporobolomyces* spp., and *Trichosporon* spp. have been discovered in the deep sea/ocean (Hassan et al., 2016). Yeasts in deep-sea environments are still mostly unexplored compared to prokaryotic microbes, with little research done on their physiology. The majority of fungi found in deep ocean habitats are psychrotrophic, although occasionally isolates from these settings can also thrive at temperatures higher than 30°C (Singh et al., 2010). Owing to their survival in the stringent cold waters, psychrophilic and psychrotrophic fungi provide more potential applications in a range of biotechnological and medicinal domains (Hassan et al., 2016). While further research is needed to fully understand all physiological and ecological adaptive mechanisms of these fungi, some examples of these mechanisms include changes in the structure of membrane lipids or fatty acids, an increase in the synthesis of mycosporine and melanin, trehalose, compatible solutes, and cold-active enzymes.

Diversity of Psychrophilic Algae in Deep Ocean Waters

As sunlight is scarce in the deep water, psychrophilic algae have adapted to make use of whatever light that is available for photosynthesis. Certain algae can be found near hydrothermal vents or cold seeps, which are places with special geological and chemical properties that support ecosystems that sustain life even in the absence of sunshine. Researchers are constantly finding new species and learning how they have adapted to harsh conditions in the study of psychrophilic algae in deep-sea habitats. Our comprehension of the diversity and tenacity of marine life, even in some of the harshest settings on Earth, is aided by the study of these creatures. Psychrophilic algae possess distinct physical characteristics such as smaller cell sizes, higher surface area to volume ratios, and structural changes to the cell wall that enable them to endure in frigid climates. Additionally, certain species might have unique buoyancy-regulating features. However, there are many obstacles while studying psychrophilic algae in the deep ocean waters, such as the difficulty of conducting research in such an environment and the restricted accessibility. Subsequent investigations have to concentrate on enhancing the techniques of discovery, comprehending the molecular foundation of adaptations, and evaluating the possible consequences of alterations in the environment on these distinct creatures.

3. Role of Psychrophiles in Permafrost and Deep Ocean Waters

The role of psychrophiles in permafrost and deep ocean waters is quite significant, and they contribute to various ecological and biochemical processes (Margesin & Collins, 2019). It could seem senseless to think of bacteria as being in a condition of stasis and slumber in the icy permafrost habitats. Permafrost, however, contains biologically active microorganisms. Although the active layer of the permafrost has greater microbial variety and activity, active microorganisms can still be discovered in the permafrost despite its harsh environment (Finore et al., 2020). The ability of psychrophiles to grow and spread at low temperatures implies that the fundamental problems associated with persistently cold environments have been resolved. These difficulties include flexibility of the membrane, inactivation of enzymes, stopping the machinery involved in transcription and translation, etc. Research on the transcriptomes of permafrost and deep oceans has started to establish the foundation for our comprehension of the metabolic processes that are active in these ecosystems and how they might affect larger global processes like greenhouse gas emissions and biogeochemical cycles (Lee et al., 2023). Changes in any of these variables can have a substantial effect on the function of the microbial community (Raymond-Bouchard & Whyte, 2017). Microbial activity and metabolism are largely driven and regulated by community-specific microbial interactions, temperature variations, the degree of thawing, and the availability of nutrients, water, and organic matter.

The overall richness of deep ocean environments is enhanced by psychrophiles. For these species to live and prosper in the frigid and gloomy deep-water environments, they have developed specialized adaptations. They also take part in biogeochemical cycling, which involves the cycling of nitrogen, carbon, and other necessary components. Their involvement in several processes, including decomposition, nutrient recycling, and organic matter breakdown, is crucial for preserving the general equilibrium of the deep-sea environment. Since they are at the base of the food chain in deep sea waters, other creatures, such as bigger bacteria, invertebrates, and even some deep-sea fish, use them as food. To keep life in the deep ocean going, these microbial food web interactions are crucial. Psychrophiles and other deep-sea creatures can act as markers for changes in temperature and ocean currents, among other environmental variables. By keeping an eye on their quantity and existence, scientists can evaluate how climate change is affecting deep-sea ecosystems.

Adaptation Strategies of Psychrophiles in Permafrost and Deep Ocean Waters

Psychrophiles exhibit extraordinary potential for effectively colonizing or adapting to very cold environments, when survival appears nearly impossible and temperatures hover around –20°C. Lower temperature adversely affects the solute diffusion rates, macromolecular interactions, cell integrity, membrane fluidity, and enzyme kinetics, among other physiochemical factors on microbes' ability to operate under freezing conditions. Despite all these hurdles they face in these challenging environments, a notable diversity of microbes of fungi, bacteria, and microalgae exists. The selection and growth characteristics of individual microorganisms can be greatly influenced by a variety of biotic (e.g., cell-cell interactions, antibiotics, oxygen, pH, salinity, nutrient flow), abiotic (e.g., free life against added particles, sea ice versus sea water), and broader ecological factors (Karadayi et al., 2021). The different adaptation strategies possessed by psychrophiles include changes in membrane fluidity and structure, presence of cold-shock proteins, antifreeze proteins, cold adapted enzymes, etc. (Figure 1).

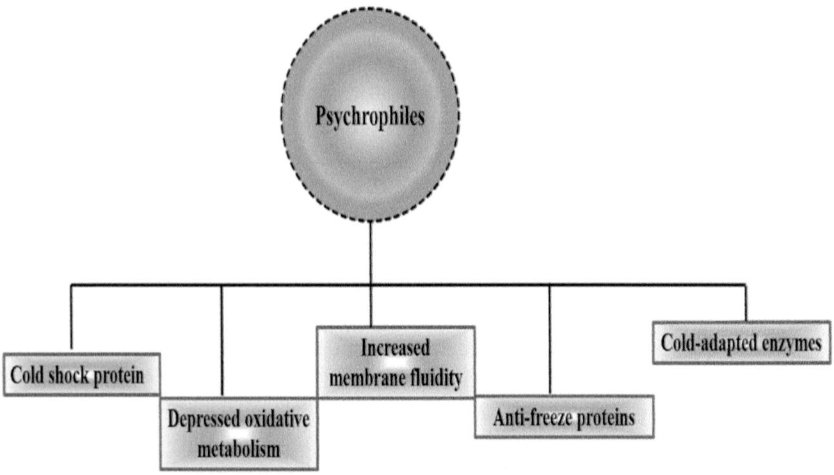

Figure 1: Different methods of adaptation exhibited by psychrophiles.

Changes in Membrane Structure

The adaptations of psychrophiles to cold environments lead to differing perspectives regarding basic characteristics of diverse biological processes, such as the genetic sequence that builds macromolecules that are stable even in harsh environments and biochemical constraints that may change macromolecule stability (Sharma et al., 2022). These microorganisms, which have evolved to harsh conditions, have unusual physiological potential in addition to a wide variety of metabolisms. The microbial cells are affected by low temperatures in a number of ways, including increased water viscosity, denaturation of proteins, slowed chemical processes, and decreased membrane stability (Hassan et al., 2016). In these conditions, cells suffer damage and therefore, cold-shock proteins expressed by the cell membranes mitigate such effects which are activated only upon receiving cold shock. Soon after it is synthesized, additional proteins are enlisted for growth synthesis, which causes the growth to occur under cold conditions, though slowly (Kaur et al., 2022).

The low temperature can cause metabolic imbalance and growth halt. Since cell membranes and envelopes are thought to be the primary conduits between a cell and its surroundings, structural changes to them are a key component of cold adaptation. The lipid makeup of the cell membrane can change at low temperatures, favoring shorter chains and reducing lipid saturation, among other commonly used strategies to control fluidity and prevent rigidity (Sharma et al., 2022). *Pseudomonas syringae* and *P. extremaustralis* rise strains from the Antarctic were found to have significant levels of hydroxy fatty acids, as well as changes to the fluidity and composition of LPS (Benforte et al., 2018). Cold adaptation is strongly associated with increased hydrophobicity and amounts of choline, peptidoglycan, and calcium carbonate along the cell membrane. In low-temperature environments, copies of the genes for peptidoglycan and carbonic anhydrase production increased, suggesting that bacteria were precipitating calcium carbonate. Lower temperatures cause components of cellular fluid to gel, which prevents proteins from working properly and causes leaking in the membrane of microorganisms. One of the main factors promoting cold tolerance is the polymers surrounding the microbial cells, called exopolysaccharides (Tribelli & López, 2018). A change in the membrane's composition, however, allows it to keep its fluidity, limiting gel formation and enhancing microbial survival at low temperatures.

To maintain the ideal level of membrane fluidity, lipid fatty acyl chains linked to the cell membrane can be modified. An increase in methyl branched fatty acids aids in the survival of cold microbial cells in difficult climatic conditions. Desaturases is an enzyme that is activated by cold shock and transforms saturated acyl fatty acids to unsaturated acyl chains. By creating steric limitations that decrease contact and alter the packing order in the membrane, composition changes play a critical role in improving membrane fluidity. *De novo* synthesis of less saturated fatty acids or decreasing the saturation of pre-existing fatty acids are two ways to increase the proportion of unsaturated fatty acids (Karadayi et al., 2021).

Presence of Cold-Shock Proteins

The microbial cells suffer damage when there is a sharp drop in temperature. Cold-shock proteins (CSPs) are proteins that mitigate such effects in these harsh conditions. Soon after it is synthesized, additional proteins are enlisted for growth synthesis, which causes the growth to occur under cold conditions, though slowly (Kaur et al., 2022). Numerous protein classes, including as hydrolases, glucanases, hydrogenases, isomerases, oxidoreductases, and nucleic acid-modifying enzymes have evolved over the time to be capable of functioning efficiently at temperatures that range from below freezing to well over 100°C (Karadayi et al., 2021). CSPs attach to the RNP1 and RNP2 binding motifs to preserve nucleic acids. In bacteria, CSPs are present in general to regulate various biological processes such as promotion of normal development and stress adaptation responses. They are also active during temperature shifts to withstand cold stress (Sharma et al., 2022). CSPs resemble nucleic acid chaperons to mitigate the effects of cold shock. The optimal activity of transcription and translation-related enzymes, such as RNA polymerase, peptidyl prolyl cis-trans isomerase, and elongation factor, is achieved at low temperatures by adaptation.

The extremely conserved nucleic acid-binding domain of cold shock proteins is called cold-shock domain (CSD) consisting of ribonucleoproteins 1 and 2 as the two important nucleic acid-binding motifs that assist the protein to bind to its target nucleic acid (Kaur et al., 2022). Many cold induced proteins (CIPs) have been identified in *E. coli*, including RNase R, RNA helicase CSD-A, CSPs, and exoribonuclease PNPase (Pathania et al., 2022). Because they interfere with RNA's secondary structure and facilitate smooth transcription and translation, CSPs are also referred to as molecular chaperones. The way that CSP attaches to RNA is crucial to this process. Their function as a molecular chaperon will be disrupted if CSP attaches itself firmly to RNA. Since CSPs prevent hairpin structures from forming, which stops transcription, they are often referred to as anti-terminators.

Anti-Freeze Proteins

Antifreeze proteins (AFPs) are non-colligative biological antifreezes that can bind to ice and prevent ice growth and recrystallization. They are sometimes referred to as ice structuring or thermal hysteresis proteins (Collins & Margesin, 2019). The ability of s AFPs to attach to ice crystals through a wide complementary surface results in the creation of thermal hysteresis, which lowers the temperature at which an organism can thrive (D'Amico et al., 2006). There are many different kinds of AFPs in nature that have a variety of structures in different psychrophiles. They are often lipidated and/or glycosylated and range in size from about 2–50 kDa. An Antarctic bacterium called *Marinomonas primoryensis* has been found to have a 1.5 MDa multidomain ice adhesion AFP (Bar Dolev et al., 2016). Other bacterial species known to exhibit AFP activity include *Enterobacter agglomerans*, *Marinomonas protea*, *Rhodococcus* sp., *Psychrobacter* sp., *Stenotrophomonas maltophilia* and *P. fluorescens* (Pathania et al., 2022). The mechanism by which AFPs work is by binding to particular ice crystal planes in an irreversible manner, which prevents secondary nucleation events and shapes a distinct ice morphology.

The ice surface expands between the adsorbed AFPs due to the Kelvin effect, locally curving the ice face and making it energetically unfavourable for further ice formation. This results in the freezing point being depressed due to thermal hysteresis (TH) (Collins & Margesin, 2019). AFPs are classified into two primary groups based on ice-binding patterns, i.e., hyperactive and active AFPs. The research is still going on what makes hyperactive AFPs different from active ones. Their selectivity for distinct ice faces may be the primary distinction; hyperactive AFPs attach to the ice's base and planes to impede expansion along the c-axis (Karadayi et al., 2021). AFPs are thought to help stabilize cell membranes, preserve structural integrity, and, in certain situations, use their adhesion domains and ice-binding function to help position cells on ice to improve access to oxygen and nutrients in the phototrophic zone.

Cold-Adapted Enzymes

Among diverse challenges that cold-adapted microorganisms face, the negative effect of low temperatures on reaction rates affect their metabolism. The Arrhenius Law states that temperature affects all reactions, including those that are catalyzed by enzymes, and that every drop in temperature causes an exponential drop in reaction rate. In low temperature conditions, the rate of chemical reactions is extremely slow because there is not enough kinetic energy to get past the barriers caused by enzyme activation. Enzyme activity is reduced 20–80 times when the temperature drops from 37°C to 0°C during a biological reaction that is taking place in a mesophyll at 37°C. However, creatures that are acclimated to cold temperatures have evolved a number of strategies to get around this restriction, such as the energy-intensive method of improved enzyme production.

The majority of the psychrophilic enzymes are environment-adapted and retain high specific activity that allow for proper metabolic rates in their frigid habitats. Enzymes preserve high activity at low temperatures and increase the flexibility of the active site needed for substrate binding by reducing the temperature dependency of the reaction (Km). The second goal is to preserve the catalytic center's specificity while enhancing catalysis at low temperatures through improved active site accessibility and advantageous electrostatic interactions with the substrate. Psychrophiles' stability curve demonstrates their stability at normal temperature, suggesting the role of hydrophobic forces in protein folding. Most of the cold-adapted enzymes that have been studied so far have decreased stability, higher catalytic activity at low to moderate temperatures, and a shift in the ideal temperature for activity towards lower temperatures when compared to mesophilic and thermophilic homologous enzymes. By reducing the activation enthalpy (ΔH^*) and temperature dependency of reactions, the "cold enzymes" effectively lower the activation energy (ΔG^*) (Collins & Margesin, 2019). This suggests a decrease in the quantity and/or intensity of enthalpic connections that break during the catalytic cycle. The genomes of psychrophilic archaea contained proteins that were identified by a higher concentration of uncharged polar amino acids (especially Gln and Thr), a lower concentration of hydrophobic amino acids (especially Leu), increased exposure of hydrophobic residues, and low charge linked to destabilizing the surface of psychrophilic proteins (Saunders et al., 2003).

The catalytic site of an enzyme is the only location where the increased structural flexibility of psychrophilic enzymes allows them to function in an unorganized manner. The degree of compatibility between the catalytic site and substrate is subsequently strengthened by increased flexibility. As a result, the rate at which substrates turnover increases and the activation energy decreases.

Psychrophilic enzymes have evolved a variety of strategies to increase their activity and flexibility while reducing their thermostability. One of these mechanisms is the reduction of proline and arginine, among other amino acids. These amino acids are recognized to reduce conformational flexibility by generating a large number of hydrogen bonds and salt bridges (Kaur et al., 2022). It was studied that some psychrophilic bacteria (*Shewanella* sp.) had lower alanine contents, whereas other psychrophilic bacteria (*Psychrobacter arcticus*) had lower proline/arginine contents (especially in those proteins involved in cell division and reproduction) (Zhao et al., 2010).

4. Application of Psychrophiles Isolated from Life Challenging Environments

Enzymes derived from microorganisms play a crucial role in various industrial, food, medical, and research applications due to their unique properties and versatility (Singh et al., 2014; Nargotra et al., 2018, 2023; Sharma et al., 2019). With advancements in research on cold-adaptive microbes, there is a notable and broad expansion in application domains associated with these psychrophilic microbes. Due to their cold adaptive mechanism, psychrophiles can be used for various biotechnological applications (Figure 2). Psychrophilic enzymes' natural qualities of exhibiting excellent activity at low to moderate temperature conditions and decreased stability provide numerous benefits in a range of applications (Hamid, 2021). They can be used to lower process temperatures,

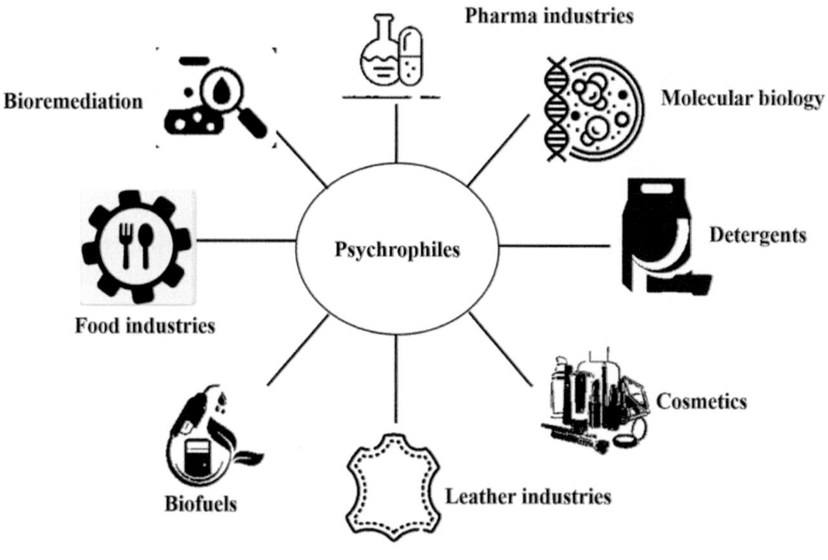

Figure 2: Different biotechnological applications of psychrophiles and their enzymes.

which improves process economics and environmental effect, and simplify enzyme inactivation (Parvizpour et al., 2021). They can also be utilized to increase the efficiency and economics of low- to moderate-temperature processes. In addition to being used in the food, feed, and technical sectors of the industrial enzymes market, cold enzymes are also finding new applications in the fine chemical and pharmaceutical industries (Hamid & Mohiddin, 2018; Kumari et al., 2021).

Enzymes and bioactive compounds produced by psychrophiles or whole cells hold promising applications in various sectors such as textile, detergents, food processing, bioremediation, tannery, cosmetics, moisturizers, sunscreens, food supplements, pharmaceuticals, wastewater treatment, etc. (Rafiq et al., 2019). Enzymes from psychrotolerant microorganisms are characterized by a common trend of decreased activation enthalpy (ΔH) and increased negative activation entropy (ΔS), which sets them apart from their thermophilic and mesophilic equivalents. This is attributed to their flexible structures, which likely serve as a compensatory mechanism for the limited kinetic energy in the cold environment (Cavicchioli et al., 2011). The use of psychrozymes can reduce unwanted chemical reactions that can happen at higher temperatures. The enzymes can be inactivated by heating and used for heat-sensitive reactions. These features are highly relevant in the food and feed industry, given the imperative to prevent spoilage and preserve the nutritional value and flavor of heat-sensitive substrates and products.

Molecular Biology Applications

Psychrozymes are also broadly applied in molecular biology applications owing to their high specificity and activity towards the substrate and thermolability (Parvizpour et al., 2021). Additionally, since heat-sensitive psychrozymes inactivate at temperatures below which double-stranded DNA melts, using them eliminates the need for chemical extraction procedures. Alkaline phosphatase (AP) is a psychrozyme extensively utilized in molecular biology. It is responsible for dephosphorylating DNA vectors to avoid self-ligation during cloning and for eliminating phosphates from DNA strands' 5′ termini prior to end-labelling by T4 polynucleotide kinase. Alkaline phosphatase by ArcticZymes produced from shrimp *Pandalus borealis* was the first psychrozyme AP to be commercially available in 1993. The most effective heat-inactivation properties to date are found in the shrimp alkaline phosphatase from ArcticZymes, which can fully inactivate at 65°C in 5 min or 75°C in 30 s (Kumari et al., 2021). Another known psychrophilic enzyme exploited in molecular biology experiments is Uracil-DNA glycosylase that repairs the DNA and causes the removal of uracil. Uracil-DNA glycosylase blocks the mutation in DNA by preventing the deamination of DNA cytosine to uracil. These enzymes have been extensively used since the 1990s in diagnostic labs to prevent the false-positive results in diagnostic PCRs (Mangiagalli et al., 2020). Commercially available uracil-DNA glycosylase from a marine bacterium *Gadus morhua* is produced by ArcticZymes, Merck (Rahway, NJ, USA) (Sarmiento et al., 2015).

Food Applications

Cold-adaptive enzymes are also used in food industries. Cold-active amylases present an intriguing option for baking, offering the potential to enhance bread softness and prevent stalling by being easily deactivated during the cooking phase. The patents for α-amylase produced from *Bacillus licheniformi* and heterologously expressed from *Clostridium* with activity at 10–60°C via protein engineering and lower than 10°C, respectively, are held by Novozymes (Bagsvaerd, Denmark) and ColdZYMES ApS, Greenland with industrial partners (Borchert et al., 2004; Stougaard et al., 2017). Proteases are one of the common classes of enzyme which is widely used in food processing such as bakery, brewing, dairy, meat tenderization, and cheese manufacture. The psychrophilic *Pseudoalteromonas* strain SM9913 protease is used to tenderize meat that is high in collagen (Mangiagalli et al., 2020). It has been suggested that the psychrophilic serine protease found in *Chryseobacterium* sp. could be useful in enhancing the quality of meat due to its capacity to withstand salt and function best at low temperatures (Mageswari et al., 2017). The cold-adapted β-galactosidase enzymes have found numerous applications through research, especially in the dairy and food processing industries at low temperature processing (Hamid et al., 2022).

Pharmaceutical Applications

The microbial enzymes exhibit high activity at low temperatures, as a result of which predicted efficiency for the goal activity is provided by their reduced concentration. Therefore, it is possible to arrange cost-effective processes by reducing the quantity of costly enzymes. Psychrophiles are also reported for their pharmaceutical applications with antimicrobial, anti-cancer, anti-tumor, anti-inflammatory, and antioxidants characteristics. The cold adapted enzymes from the psychrophiles are employed in the enzymatic modification or semi-synthesis of pharmaceuticals and fine compounds (Hamid et al., 2022). In the context of tissue regeneration and healing, as well as the prevention of periodontal disease and dental plaque removal, the proteases derived from psychrophilic *Bacillus subtilis* and *Euphausia superba* serve to eliminate barriers and remove dental plaque (Fornbacke & Clarsund, 2013). *Dichotomomyces cejpii* F31-1, a marine fungus generated the polyketide Scequinadoline A, which exhibited inhibitory effects against the development of the dengue causing virus serotype 2 (Wu et al., 2018). Furthermore, bioactive peptides that function as antioxidants and antihypertensives are synthesized using cold-tolerant protease enzymes. Cold-adapted proteases are utilized as therapeutic agents against bacterial and viral infections and to remove dead and dried skin cells from cosmetic products because of their antibacterial action (Fornbacke & Clarsund, 2013; Hamid et al., 2022, 2024). Overall, the unique properties of cold-adapted enzymes from psychrophiles make them promising candidates for various pharmaceutical applications, offering advantages such as increased efficiency, reduced energy consumption, and improved product quality. Ongoing research in this field continues to explore new ways to harness the potential of these enzymes in drug development and production processes.

Table 1: Application of psychrophiles in different biotechnological fields

Source	Enzymes	Applications	References
Bacillus subtilis, Rhodotorula mucilaginosa, Cystofilobasidium capitatum	Pectinase	Removal of pectin from coffee, fruit juice, tea, frozen plant fibers, softened vegetable and plant tissue, extraction of vegetable oil, and wine haze removal	Adapa et al., 2014; Martín and Ambrosini, 2013; Sahay et al., 2013
Genomic DNA of frozen soil	Xylanase	Dough processing, baking, juice clarification, textile treatment, brewing industry, animal feed preparation, and pulp bleaching	Qiu et al., 2017
Glaciozyma antarctica PI12	β-mannanases	Decrease viscosity of food materials	Parvizpour et al., 2018
Pseudoalteromonas arctica PAMC 21717	Protease	As laundry detergent enzymes	Park et al., 2018
Streptomyces coelicolor A3(2)	Esterase	Decompose fat stains into more hydrophilic compounds that are easily washable to remove them	Brault et al., 2012
Pseudoalteromonas sp. strain KMM 701	α-galactosidase	Conversion of B RBCs into blood type O cells	Santiago et al., 2016
Candida antarctica and Pseudomonas sp.	Lipase	Production of optically active amines	Kavitha, 2016
Glaciozyma antarctica PI12	α-amylase	Bioremediation and wastewater treatment	Ramli et al., 2013
Staphylococcus saprophyticus	Nitroreductase	Activation of cancer prodrugs	Çelik and Yetiş, 2012
Pseudoalteromonas haloplanktis	Xylanase	Breadmaking	Kumari et al., 2021
Rhizomucor miehei	Glycogen branching enzyme	Wheat breadmaking	Wu et al., 2014
Pedobacter	-	Production of pyrrhoxanthin, violaxanthin, fucoxanthin, and nostoxanthine 3-sulfate	Silva et al., 2021

Detergent and Cleaning Applications

Application of psychrophiles in the detergent and bioremediation sectors is another promising benefit (Sheikh et al., 2024; Baba et al., 2021). Lipoclean®, Lipex® and Lipolase® Ultra developed by Novozymes are the lipase enzymes which are active at low temperatures and effectively eliminate fatty stains (oil, butter, and sauces) from fabrics (Sarmiento et al., 2015). Celluzyme® from the fungus *Humicola insolens* developed by Novozymes exhibit low temperature (~15°C) activity, and Celluclean® a mix of cellulases for low temperature washing. Cellulase lessens the accumulation of fuzz and pills in knitted clothing, and therefore is employed in detergents for color and brightness care. Commercially available psychrozyme protease (Polarzyme®) and amylase (Stainzyme® Plus) developed by Genencor (Palo Alto, CA, USA) and Novozymes, respectively, are also used in the detergent industry for the removal of protein and starch stains. These enzymes are active at 5–60°C (protease) and ~20°C (amylase) (Sarmiento et al., 2015). Various industrial applications of psychrophilic enzymes are listed in Table 1.

In colder temperate climates, soil and wastewater can be bioremediated with the help of psychrophilic microorganisms. They have the ability to break down a range of organic pollutants that contaminate the environment, such as pesticides, phenolic compounds, and mineral oil hydrocarbons (Jin et al., 2022). Degradation of phenol by the production of hydroquinone hydroxylase, phenol hydroxylase, and catechol 1,2-dioxygenase from Antarctic Aspergillus fumigatus has been reported (Gerginova et al., 2013). Psychrophiles are also a source of unsaturated fatty acids, ice-binding protein, and natural pigments which can be used in various sectors such as health, food processing, textile, cosmetics, etc. (Białkowska et al., 2020; Silva et al., 2021). Additionally, with the progress made in the fields of proteomics, genomes, transcriptomics, and metabolomics, comprehensive research on psychrophiles can be conducted to better understand their potential advantages for human health.

5. Conclusion

Research on psychrophiles that live in deep ocean and permafrost has provided amazing insights into how adaptable life can be in harsh conditions. These microbes' ability to withstand low temperatures challenges accepted ideas about habitability and broadens our understanding of the possibility of life beyond traditional bounds. Psychrophiles flourish in the permafrost, an apparently inhospitable frozen zone that provides a unique home that shows how resilient life can be in the worst temperatures on Earth. Furthermore, the study of deep ocean environments has uncovered a multitude of psychrophilic variety, demonstrating the adaptability of these creatures to the extreme cold and pressure found there. Psychrophile adaptations, including the composition of their membranes and specialized enzymes, provide insight into the creative ways that life has evolved to deal with environmental obstacles. The adaptations seen in these resilient microbes open our eyes to fresh ideas about the limits of habitability and encourage more research into the planet's most extreme habitats and beyond. The most promising tools for isolating cold-tolerant enzyme systems are these distinct metabolic and enzymatic pathways. In terms of their diversity and prospective applications in industrial sectors, psychrophiles appear to have even more potential than other counterparts.

References

Adapa, V., Ramya, L.N., Pulicherla, K.K., & Rao, K.R.S.S. 2014. Cold active pectinases: Advancing the food industry to the next generation. *Appl. Biochem. Biotechnol.*, 172, 2324–2337. https://doi.org/10.1007/s12010-013-0685-1

Aksenov, A.S., Shirokova, L.S., Kisil, O.Y., Kolesova, S.N., Lim, A.G., Kuzmina, D. et al. 2021. Bacterial number and genetic diversity in a Permafrost Peatland (Western Siberia): Testing a link with organic matter quality and elementary composition of a peat soil profile. *Diversity*, 13, 328. https://doi.org/10.3390/d13070328.

Baba, Z.A., Hamid, B., Sheikh, T.A., Alotaibi, S., Enshasy, H.E., Ansari, M.J. et al. 2021. Psychrotolerant *Mesorhizobium* sp. isolated from temperate and cold desert regions solubilize potassium and produces multiple plant growth promoting metabolites. *Molecules*, 26, 5758. https://www.mdpi.com/1420-3049/26/19/5758/htm.

Bar Dolev, M., Bernheim, R., Guo, S., Davies, P.L., & Braslavsky, I. 2016. Putting life on ice: Bacteria that bind to frozen water. *Journal of The Royal Society Interface*, 13, 20160210. https://doi.org/10.1098/rsif.2016.0210.

Baraúna, R.A., Freitas, D.Y., Pinheiro, J.C., Folador, A.R.C., & Silva, A. 2017. A proteomic perspective on the bacterial adaptation to cold: Integrating OMICs data of the psychrotrophic bacterium Exiguobacterium antarcticum B7. *Proteomes*, 5, 9. https://doi.org/10.3390/proteomes5010009.

Benforte, F.C., Colonnella, M.A., Ricardi, M.M., Venero, E.C.S., Lizarraga, L., López, N.I., & Tribelli, P.M. 2018. Novel role of the LPS core glycosyltransferase WapH for cold adaptation in the Antarctic bacterium Pseudomonas extremaustralis. *PLOS One*, 13, e0192559. https://doi.org/10.1371/journal.pone.0192559.

Bhatia, R.K., Ullah, S., Hoque, M.Z., Ahmad, I., Yang, Y.-H., Bhatt, A.K., & Bhatia, S.K. 2021.. Psychrophiles: A source of cold-adapted enzymes for energy efficient biotechnological industrial processes. *Journal of Environmental Chemical Engineering*, 9, 104607. https://doi.org/10.1016/j.jece.2020.104607.

Białkowska, A., Majewska, E., Olczak, A., & Twarda-Clapa, A. 2020. Ice binding proteins: Diverse biological roles and applications in different types of industry. *Biomolecules*, 10, 274. https://doi.org/10.3390/biom10020274.

Borchert, T.V., Svendsen, A., Andersen, C., Nielsen, B., Nissen, T.L., & Kjærulff, S. 2004. α-amylase mutants. US6673589B2.

Brault, G., Shareck, F., Hurtubise, Y., Lépine, F., & Doucet, N. 2012. Isolation and characterization of EstC, a new cold-active esterase from Streptomyces coelicolor A3(2). *PLOS One*, 7, e32041. https://doi.org/10.1371/journal.pone.0032041.

Carr, S.A., Vogel, S.W., Dunbar, R.B., Brandes, J., Spear, J.R., Levy, R. et al. 2013. Bacterial abundance and composition in marine sediments beneath the Ross Ice Shelf, Antarctica. *Geobiology*, 11, 377–395. https://doi.org/10.1111/gbi.12042.

Cavicchioli, R., Charlton, T., Ertan, H., Omar, S.M., Siddiqui, K.S., & Williams, T.J. 2011. Biotechnological uses of enzymes from psychrophiles. *Microbial Biotechnology*, 4, 449–460. https://doi.org/10.1111/j.1751-7915.2011.00258.x.

Çelik, A., & Yetiş, G. 2012. An unusually cold active nitroreductase for prodrug activations. *Bioorganic and Medicinal Chemistry*, 20, 3540–3550. https://doi.org/10.1016/j.bmc.2012.04.004.

Collins, T., & Margesin, R. 2019. Psychrophilic lifestyles: Mechanisms of adaptation and biotechnological tools. *Appl. Microbiol. Biotechnol.*, 103, 2857–2871. https://doi.org/10.1007/s00253-019-09659-5.

Corsaro, M.M., Casillo, A., Parrilli, E., & Tutino, M.L. 2017. Molecular structure of lipopolysaccharides of cold-adapted bacteria. In: Margesin, R. (Ed.), *Psychrophiles: From Biodiversity to Biotechnology*. Springer International Publishing, Cham, pp. 285–303. https://doi.org/10.1007/978-3-319-57057-0_13.

Cvetkovska, M., Hüner, N.P.A., & Smith, D.R. 2017. Chilling out: The evolution and diversification of psychrophilic algae with a focus on Chlamydomonadales. *Polar Biol.*, 40, 1169–1184. https://doi.org/10.1007/s00300-016-2045-4.

da Silva, T.H., Silva, D.A.S., Thomazini, A., Schaefer, C.E.G.R., & Rosa, L.H. 2019. Antarctic permafrost: An unexplored fungal microhabitat at the edge of life. In: Rosa, L.H. (Ed.), *Fungi of Antarctica: Diversity, Ecology, and Biotechnological Applications*. Springer International Publishing, Cham, pp. 147–164. https://doi.org/10.1007/978-3-030-18367-7_7.

da Silva, T.H., Silva, D.A.S., de Oliveira, F.S., Schaefer, C.E.G.R., Rosa, C.A., & Rosa, L.H. 2020. Diversity, distribution, and ecology of viable fungi in permafrost and active layer of Maritime Antarctica. *Extremophiles*, 24, 565–576. https://doi.org/10.1007/s00792-020-01176-y.

D'Amico, S., Collins, T., Marx, J., Feller, G., Gerday, C., & Gerday, C. 2006. Psychrophilic microorganisms: Challenges for life. *EMBO Reports*, 7, 385–389. https://doi.org/10.1038/sj.embor.7400662.

Dziurzynski, M., Gorecki, A., Pawlowska, J., Istel, L., Decewicz, P., Golec, P. et al. 2023. Revealing the diversity of bacteria and fungi in the active layer of permafrost at Spitsbergen island (Arctic): Combining classical microbiology and metabarcoding for ecological and bioprospecting exploration. *Science of The Total Environment*, 856, 159072. https://doi.org/10.1016/j.scitotenv.2022.159072.

Finore, I., Vigneron, A., Vincent, W.F., Leone, L., Di Donato, P., Schiano Moriello, A. et al. 2020. Novel Psychrophiles and Exopolymers from Permafrost Thaw Lake Sediments. *Microorganisms*, 8, 1282. https://doi.org/10.3390/microorganisms8091282.

Fornbacke, M., & Clarsund, M. 2013. Cold-adapted proteases as an emerging class of therapeutics. *Infect. Dis. Ther.*, 2, 15–26. https://doi.org/10.1007/s40121-013-0002-x.

Gawas, V.S., Shivaramu, M.S., Damare, S.R., Pujitha, D., Meena, R.M., & Shenoy, B.D. 2019. Diversity and extracellular enzyme activities of heterotrophic bacteria from sediments of the Central Indian Ocean Basin. *Sci. Rep.*, 9, 9403. https://doi.org/10.1038/s41598-019-45792-x.

Gerginova, M., Manasiev, J., Yemendzhiev, H., Terziyska, A., Peneva, N., & Alexieva, Z. 2013. Biodegradation of phenol by antarctic strains of Aspergillus fumigatus. *Zeitschrift* für *Naturforschung*, C 68, 384–393. https://doi.org/10.1515/znc-2013-9-1006.

Hamid, B., & Mohiddin, F.A. 2018. Cold-active enzymes in food processing. In: Kuddus, M. (Ed.), *Enzymes in Food Technology: Improvements and Innovations*. Springer, Singapore, pp. 383–400. https://doi.org/10.1007/978-981-13-1933-4_19.

Hamid, B., Bashir, Z., Yatoo, A.M., Mohiddin, F., Majeed, N., Bansal, M. et al. 2022. Cold-active enzymes and their potential industrial applications: A review. *Molecules*, 27, 5885. https://doi.org/10.3390/molecules27185885.

Hamid, B., Baba, Z.A., Sheikh, T.A., Perveen, K., & Sayyed, R.Z. 2024. Investigating potential protease activity of psychrotrophic bacteria from a municipal landfill for solid waste management. *Biomass Conversion and Biorefinery*. https://doi.org/10.1007/s13399-024-05621-2.

Hassan, N., Rafiq, M., Hayat, M., Shah, A.A., & Hasan, F. 2016. Psychrophilic and psychrotrophic fungi: A comprehensive review. *Rev. Environ. Sci. Biotechnol.*, 15, 147–172. https://doi.org/10.1007/s11157-016-9395-9.

Hu, W., Zhang, Q., Li, D., Cheng, G., Mu, J., Wu, Q. et al. 2014. Diversity and community structure of fungi through a permafrost core profile from the Qinghai-Tibet Plateau of China. *Journal of Basic Microbiology*, 54, 1331–1341. https://doi.org/10.1002/jobm.201400232.

Hu, W., Zhang, Q., Tian, T., Cheng, G., An, L., & Feng, H. 2015. The microbial diversity, distribution, and ecology of permafrost in China: A review. *Extremophiles*, 19, 693–705. https://doi.org/10.1007/s00792-015-0749-y.

Jain, V., Arya, P., Yagnik, S.M., Raval, V.H., & Singh, N.A. 2023. Microbial diversity of cold-water reservoirs and their prospective applications. In: Soni, R., Suyal, D.C., Morales-Oyervides, L., Sungh Chauhan, J. (Eds.), *Current Status of Fresh Water Microbiology*. Springer Nature, Singapore, pp. 49–75. https://doi.org/10.1007/978-981-99-5018-8_3.

Jiang, Y., Yang, L., Wu, S., & Zhou, T. 2023. Warming changes the composition and diversity of fungal communities in permafrost. *Ann. Microbiol.*, 73, 7. https://doi.org/10.1186/s13213-022-01707-2.

Jin, S., Wang, Y., & Zhao, X. 2022. Cold-adaptive mechanism of psychrophilic bacteria in food and its application. *Microb. Pathog.*, 169, 105652. https://doi.org/10.1016/j.micpath.2022.105652.

Karadayi, G., Alaylar, B., Karadayi, M., Gürkök, S., Aksu, Ş., Egamberdieva, D., & Güllüce, M. 2021. Biodiversity, ecological, and commercial importance of psychrophilic microorganisms. In: Egamberdieva, D., Birkeland, N.-K., Li, W.-J., & Panosyan, H. (Eds.), *Microbial Communities and Their Interactions in the Extreme Environment, Microorganisms for Sustainability*. Springer, Singapore, pp. 225–247. https://doi.org/10.1007/978-981-16-3731-5_11.

Kaur, Jupinder, Vishnu, A.L., Khipla, N., & Kaur, Jaspreet. 2022. Microbial life in cold regions of the deep sea. In: Goel, R., Soni, R., Suyal, D.C., Khan, M. (Eds.), *Survival Strategies in Cold-Adapted Microorganisms*. Springer, Singapore, pp. 63–86. https://doi.org/10.1007/978-981-16-2625-8_3.

Kavitha, M. 2016. Cold active lipases: An update. *Frontiers in Life Science*, 9, 226–238. https://doi.org/10.1080/21553769.2016.1209134.

Kumari, M., Padhi, S., Sharma, S., Phukon, L.C., Singh, S.P., & Rai, A.K. 2021. Biotechnological potential of psychrophilic microorganisms as the source of cold-active enzymes in food processing applications. *3 Biotech*, 11, 479. https://doi.org/10.1007/s13205-021-03008-y.

Lee, H., Cho, Y.-J., Cho, A., & Kim, O.-S. 2023. Environmental adaptation of psychrophilic bacteria *Subtercola* spp. Isolated from various cryospheric habitats. *J. Microbiol.*, 61, 663–672. https://doi.org/10.1007/s12275-023-00068-y.

Mageswari, A., Subramanian, P., Chandrasekaran, S., Karthikeyan, S., & Gothandam, K.M. 2017. Systematic functional analysis and application of a cold-active serine protease from a novel *Chryseobacterium* sp. *Food Chem.*, 217, 18–27. https://doi.org/10.1016/j.foodchem.2016.08.064.

Mangiagalli, M., Brocca, S., Orlando, M., & Lotti, M. 2020. The "cold revolution": Present and future applications of cold-active enzymes and ice-binding proteins. *New Biotechnology*, 55, 5–11. https://doi.org/10.1016/j.nbt.2019.09.003.

Margesin, R., & Feller, G. 2010. Biotechnological applications of psychrophiles. *Environ. Technol.*, 31, 835–844. https://doi.org/10.1080/09593331003663328.

Margesin, R., & Collins, T. 2019. Microbial ecology of the cryosphere (glacial and permafrost habitats): Current knowledge. *Appl. Microbiol. Biotechnol.*, 103, 2537–2549. https://doi.org/10.1007/s00253-019-09631-3.

Martín, M.C., & Ambrosini, V.I.M. de. 2013. Cold-active acid pectinolytic system from psychrotolerant bacillus: Color extraction from red grape skin. *Am. J. Enol. Vitic.*, 64, 495–504. https://doi.org/10.5344/ajev.2013.13002.

Nargotra, P., Sharma, V., Gupta, M., Kour, S., & Bajaj, B.K. 2018. Application of ionic liquid and alkali pretreatment for enhancing saccharification of sunflower stalk biomass for potential biofuel-ethanol production. *Bioresour. Technol.*, 267, 560–568. https://doi.org/10.1016/j.biortech.2018.07.070.

Nargotra, P., Sharma, V., Lee, Y.-C., Tsai, Y.-H., Liu, Y.-C., Shieh, C.-J. et al. 2023. Microbial Lignocellulolytic enzymes for the effective valorization of lignocellulosic biomass: A review. *Catalysts*, 13, 83. https://doi.org/10.3390/catal13010083.

Ogaki, M.B., Coelho, L.C., Vieira, R., Neto, A.A., Zani, C.L., Alves, T.M.A. et al. 2020. Cultivable fungi present in deep-sea sediments of Antarctica: Taxonomy, diversity, and bioprospecting of bioactive compounds. *Extremophiles*, 24, 227–238. https://doi.org/10.1007/s00792-019-01148-x.

Park, H.J., Lee, C.W., Kim, D., Do, H., Han, S.J., Kim, J.E. et al. 2018. Crystal structure of a cold-active protease (Pro21717) from the psychrophilic bacterium, *Pseudoalteromonas arctica* PAMC 21717, at 1.4 Å resolution: Structural adaptations to cold and functional analysis of a laundry detergent enzyme. *PLOS One*, 13, e0191740. https://doi.org/10.1371/journal.pone.0191740.

Parvizpour, S., Razmara, J., & Shamsir, M.S. 2018. Temperature adaptation analysis of a psychrophilic mannanase through structural, functional and molecular dynamics simulation. *Mol. Simul.*, 44, 1270–1277. https://doi.org/10.1080/08927022.2018.1492721.

Parvizpour, S., Hussin, N., Shamsir, M.S., & Razmara, J. 2021. Psychrophilic enzymes: Structural adaptation, pharmaceutical and industrial applications. *Appl. Microbiol. Biotechnol.*, 105, 899–907. https://doi.org/10.1007/s00253-020-11074-0.

Pathania, S., Solanki, P., Putatunda, C., Bhatia, R.K., & Walia, A. 2022. Adaptation to Cold Environment: The survival strategy of psychrophiles. In: Goel, R., Soni, R., Suyal, D.C., & Khan, M. (Eds.), *Survival Strategies in Cold-Adapted Microorganisms*. Springer, Singapore, pp. 87–111. https://doi.org/10.1007/978-981-16-2625-8_4.

Perez-Mon, C., Stierli, B., Plötze, M., & Frey, B. 2022. Fast and persistent responses of alpine permafrost microbial communities to *in situ* warming. *Sci. Total Environ.*, 807, 150720. https://doi.org/10.1016/j.scitotenv.2021.150720.

Qiu, H., Li, Z., Wang, H., Zhang, H., Li, S., Luo, X. et al. 2017. Molecular and biochemical characterization of a novel cold-active and metal ion-tolerant GH10 xylanase from frozen soil. *Biotechnol. Biotechnol. Equip.*, 31, 955–963. https://doi.org/10.1080/13102818.2017.1359667.

Qiu, X., Yu, L., Cao, X., Wu, H., Xu, G., & Tang, X. 2021. Halomonas sedimenti sp. nov., a Halotolerant bacterium isolated from deep-sea sediment of the Southwest Indian Ocean. *Curr. Microbiol.*, 78, 1662–1669. https://doi.org/10.1007/s00284-021-02425-9.

Rafiq, M., Hassan, N., Rehman, M., & Hasan, F. 2019. Adaptation mechanisms and applications of psychrophilic fungi. In: Tiquia-Arashiro, S.M., & Grube, M. (Eds.), *Fungi in Extreme Environments: Ecological Role and Biotechnological Significance*. Springer International Publishing, Cham, pp. 157–174. https://doi.org/10.1007/978-3-030-19030-9_9.

Ramli, A.N.M., Azhar, M.A., Shamsir, M.S., Rabu, A., Murad, A.M.A., Mahadi, N.M. et al. 2013. Sequence and structural investigation of a novel psychrophilic α-amylase from Glaciozyma antarctica PI12 for cold-adaptation analysis. *J. Mol. Model.*, 19, 3369–3383. https://doi.org/10.1007/s00894-013-1861-5.

Raymond-Bouchard, I., & Whyte, L.G. 2017. From transcriptomes to metatranscriptomes: Cold adaptation and active metabolisms of psychrophiles from cold environments. In: Margesin, R. (Ed.), *Psychrophiles: From Biodiversity to Biotechnology*. Springer International Publishing, Cham, pp. 437–457. https://doi.org/10.1007/978-3-319-57057-0_18.

Rosa, L.H., Pellizzari, F.M., Ogaki, M.B., de Paula, M.T.R., Mansilla, A., Marambio, J. et al. 2019. Sub-antarctic and antarctic marine ecosystems: An unexplored ecosystem of fungal diversity. In: Rosa, L.H. (Ed.), *Fungi of Antarctica: Diversity, Ecology and Biotechnological Applications*. Springer International Publishing, Cham, pp. 221–242. https://doi.org/10.1007/978-3-030-18367-7_10.

Sahay, S., Hamid, B., Singh, P., Ranjan, K., Chauhan, D., Rana, R.S., & Chaurse, V.K. 2013. Evaluation of pectinolytic activities for oenological uses from psychrotrophic yeasts. *Letters in Applied Microbiology*, 57, 115–121. https://doi.org/10.1111/lam.12081.

Sannino, C., Borruso, L., Mezzasoma, A., Battistel, D., Ponti, S., Turchetti, B. et al. 2021. Abiotic factors affecting the bacterial and fungal diversity of permafrost in a rock glacier in the Stelvio Pass (Italian Central Alps). *Applied Soil Ecology*, 166, 104079. https://doi.org/10.1016/j.apsoil.2021.104079.

Santiago, M., Ramírez-Sarmiento, C.A., Zamora, R.A., & Parra, L.P. 2016. Discovery, molecular mechanisms, and industrial applications of cold-active enzymes. *Frontiers in Microbiology*, 7.

Sarmiento, F., Peralta, R., & Blamey, J.M. 2015. Cold and hot extremozymes: Industrial relevance and current trends. *Frontiers in Bioengineering and Biotechnology*, 3.

Sathyanarayana Reddy, G., Chattopadhyay, M.K., & Shivaji, S. 2016. Biodiversity, adaptation, and biotechnological importance of bacteria occurring in cold climates. In: Rampelotto, P.H. (Ed.), *Biotechnology of Extremophiles: Advances and Challenges, Grand Challenges in Biology and Biotechnology*. Springer International Publishing, Cham, pp. 47–81. https://doi.org/10.1007/978-3-319-13521-2_2.

Saunders, N.F.W., Thomas, T., Curmi, P.M.G., Mattick, J.S., Kuczek, E., Slade, R. et al. 2003. Mechanisms of thermal adaptation revealed from the genomes of the antarctic archaea Methanogenium frigidum and Methanococcoides burtonii. *Genome Res.*, 13, 1580–1588. https://doi.org/10.1101/gr.1180903.

Shao, Z., Yuan, J., Lai, Q., & Zheng, T. 2015. The diversity of PAH-degrading bacteria in a deep-sea water column above the Southwest Indian Ridge. *Frontiers in Microbiology*, 6.

Sheikh, T., Baba, Z., Hamid, B., Iqbal, S., Yatoo, A., Fatima, S. et al. 2022. Extracellular polymeric substances in psychrophilic cyanobacteria: A potential bioflocculant and carbon sink to mitigate cold stress. *Biocatalysis and Agricultural Biotechnology*, 42, 102375. https://doi.org/10.1016/j.bcab.2022.102375.

Sharma, S., Sharma, V., Nargotra, P., & Bajaj, B.K. 2020. Bioprocess development for production of a process-apt xylanase with multifaceted application potential for a range of industrial processes. *SN Appl. Sci.*, 2, 739. https://doi.org/10.1007/s42452-020-2541-6.

Sharma, S., Chaturvedi, U., Sharma, K., Vaishnav, A., & Singh, H.B. 2022. An overview of survival strategies of psychrophiles and their applications. In: Goel, R., Soni, R., Suyal, D.C., & Khan, M. (Eds.), *Survival Strategies in Cold-adapted Microorganisms*. Springer, Singapore, pp. 133–151. https://doi.org/10.1007/978-981-16-2625-8_6.

Sharma, V., Nargotra, P., & Bajaj, B.K. 2019. Ultrasound and surfactant assisted ionic liquid pretreatment of sugarcane bagasse for enhancing saccharification using enzymes from an ionic liquid tolerant *Aspergillus assiutensis* VS34. *Bioresour. Technol.*, 285, 121319. https://doi.org/10.1016/j.biortech.2019.121319.

Silva E, T.R., Silva Júnior, L.C.F., de Queiroz, A.C., Alexandre Moreira, M.S., de Carvalho Fraga, C.A., de Menezes, G.C.A. et al. 2021. Pigments from Antarctic bacteria and their biotechnological applications. *Crit. Rev. Biotechnol.*, 41, 809–826. https://doi.org/10.1080/07388551.2021.1888068.

Singh, P., Raghukumar, C., Verma, P., & Shouche, Y. 2010. Phylogenetic diversity of culturable fungi from the deep-sea sediments of the Central Indian Basin and their growth characteristics. *Fungal Diversity*, 40, 89–102. https://doi.org/10.1007/s13225-009-0009-5.

Singh, S., Gupta, P., Sharma, V., Koul, S., Kour, K., & Bajaj, B.K. 2014. Multifarious potential applications of keratinase of *Bacillus subtilis* K-5. *Biocatal. Biotransformation*, 32, 333–342. https://doi.org/10.3109/10242422.2014.978306.

Stougaard, P., Vester, J.K., & Glaring, M.A. 2017. Cold-active alpha-amylase. US20170044510A1.

Tribelli, P.M., & López, N.I. 2018. Reporting key features in cold-adapted bacteria. *Life*, 8, 8. https://doi.org/10.3390/life8010008.

Vishnivetskaya, T.A. 2009. Viable cyanobacteria and green algae from the permafrost darkness. In: Margesin, R. (Ed.), *Permafrost Soils, Soil Biology*. Springer, Berlin, Heidelberg, pp. 73–84. https://doi.org/10.1007/978-3-540-69371-0_6.

Vishnivetskaya, T.A., Almatari, A.L., Spirina, E.V., Wu, X., Williams, D.E., Pfiffner, S.M., & Rivkina, E.M. 2020. Insights into community of photosynthetic microorganisms from permafrost. *FEMS Microbiology Ecology*, 96, fiaa229. https://doi.org/10.1093/femsec/fiaa229.

Wu, D.-L., Li, H.-J., Smith, D.R., Jaratsittisin, J., Xia-Ke-Er, X.-F.-K.-T., Ma, W.-Z. et al. 2018. Polyketides and alkaloids from the marine-derived fungus *Dichotomomyces cejpii* F31-1 and the antiviral activity of Scequinadoline A against Dengue Virus. *Marine Drugs*, 16, 229. https://doi.org/10.3390/md16070229.

Wu, S., Liu, Y., Yan, Q., & Jiang, Z. 2014. Gene cloning, functional expression, and characterisation of a novel glycogen branching enzyme from *Rhizomucor miehei* and its application in wheat breadmaking. *Food Chem.*, 159, 85–94. https://doi.org/10.1016/j.foodchem.2014.02.161.

Zhao, J.-S., Deng, Y., Manno, D., & Hawari, J. 2010. Shewanella spp. genomic evolution for a cold marine lifestyle and *in situ* explosive biodegradation. *PLOS One*, 5, e9109. https://doi.org/10.1371/journal.pone.0009109.

CHAPTER
7

Microbiology of Antarctic Cryoconite Holes – Life in Cold

Saadat Altaf Kanth[1*], Peter Poczai[2], Burhan Hamid[3], Md. Niamat Ali[3] and Ruqeya Nazir[3]

[1] Bioresource Processing Institute of Australia (BioPRIA), Department of Chemical & Biological Engineering, Monash University, Clayton VIC 3800, Australia
[2] Finnish Museum of Natural History, University of Helsinki, FL10044 Helsinki, Finland
[3] Centre of Research for Development, University of Kashmir, Srinagar – 190006, India

1. Introduction

The challenging meteorological and geological dynamics characterize Antarctica as the southernmost continent. This extreme environment provides an exceptional natural laboratory for studying extremophiles and their adaptations, offering unparalleled opportunities for scientific exploration and discovery (Pearce, 2012). This frozen continent, renowned as the coldest, windiest, and driest place on Earth presents wide contrasts in climate across its vast territory of approximately 14 million square kilometres (Gonzalez & Vasallo, 2019). From the relatively milder temperatures around the Antarctic Peninsula to the bone-chilling cold of its inland plateau, Antarctica experiences mean annual temperatures ranging from –5°C to –50°C, interrupted by recurring freeze-thaw cycles (Convey & Biersma, 2022). Its atmospheric conditions, influenced by the surrounding Southern Ocean, maintain minimal precipitation, with air humidity of a staggering 99.6% of its surface covered in snow or ice. Life in Antarctica faces formidable challenges due to the scarcity of organic nutrients and water (Mallet et al., 2023). Intense solar radiation further limits the proliferation of life, creating a harsh landscape where only a handful of native vascular plant species cling to existence, primarily along the Antarctic Peninsula and maritime

*Corresponding author: SadaatAltaf@monash.edu

DOI: 10.1201/9781003717003-7

Antarctic archipelagos (Ramirez et al., 2024). Coastal regions attract marine vertebrates, while various invertebrate populations, including springtails, mites, tardigrades, rotifers, and nematodes, exhibit diverse distribution patterns across the continent (Varliero et al., 2024). Amidst this inhospitable terrain, microbial life dominates both in the number of species and their evolutionary history, across the continent, epitomizing its distinctive biodiversity. These microorganisms exhibit unique adaptations such as cold-active enzymes, antifreeze proteins, and specialized membrane lipids (Bendia et al., 2023; Jiang et al., 2022).

Microbial communities in Antarctica can be classified based on the ecological niches they occupy, including terrestrial and lithic communities, freshwater ecosystems from inland and glacial water bodies, communities in meltwater ponds, subglacial aquatic ecosystems, marine ecosystems, microbial mats and sedimentary ecosystems, cryoconite hole ecosystems, glacial ice and snow ecosystems, and airborne microbial ecosystems (Doytchinov & Dimov, 2022). Cryoconite holes are of particular interest for microbial studies due to their high and complex microbial diversity and productivity. Cryoconite holes present on the surfaces of glaciers are essentially meltwater ponds containing sediment at the bottom. They are highly susceptible to the impacts of climate change, highlighting the importance of studying their microbial communities (Pittino et al., 2023). Microbes in cryoconite holes play vital roles in nutrient cycling, carbon sequestration, and maintaining ecosystem stability; they also have the potential to produce bioactive compounds, offering prospects for industrial and pharmaceutical applications (Hassan et al., 2024). Hence, understanding the microbial diversity in these sensitive ecosystems is essential for developing effective conservation strategies, while also exploring areas for innovation.

As we delve deeper into the microbiology of Antarctic cryoconite holes, we unravel not only adaptation mechanisms and biotechnological potential but also fundamental insights into the origins of life on Earth. Safeguarding Antarctica's fragile ecosystems is a matter of preserving biodiversity and unlocking the secrets held within its microbial communities, paving the way for ground-breaking scientific discoveries.

2. Cryconite Holes

Cryoconite holes are small indentations found on glacier surfaces, which are filled with water and sediment, creating unique microenvironments for microbial colonization. The word 'cryoconite' is derived from a combination of two ancient Greek words, i.e., *kruos*, meaning ice, and *konis*, meaning dust, introduced by the explorer *Adolf Erik Nordenskiöld* (1832–1901) during his 1870 Greenland expedition (Rozwalak et al., 2022). Born in Finland and later working in Sweden, Nordenskiöld was a renowned Arctic explorer and scientist who made pioneering observations of particulate matter on glacier surfaces. With a background in mineralogy and geology, Nordenskiöld noted the accumulation of dark sediment on the ice, which absorbed sunlight and melted into shallow depressions, now known as cryoconite holes. His expedition combined scientific rigour with adventurous exploration, laying early foundations for the study of glacial microbiology. Nordenskiöld's interest in Arctic environments led to several expeditions to Greenland and the Spitsbergen archipelago, culminating in his historic Vega Expedition (1878–1880) (Figure 1), where he became the first to navigate the Northeast Passage. Nordenskiöld's meticulous recording of his observations and hypotheses during his Greenland journey laid the groundwork for modern interpretations of biogeochemical processes on glaciers. His work on glacial sedimentation was remarkably forward-thinking, as he hypothesized about the global transport of dust and its potential biological implications, topics that are central to current cryosphere research. These holes' form when wind-blown dust, soot, and other particulate matter accumulate on the glacier's surface, melt into the ice, and create water-filled indentations. Mineral particles covering the ice surface originate from various sources, including local moraines, surrounding valley walls, *in situ* melting of englacial material, long-range atmospheric transport, anthropogenic activity, and even extraterrestrial

Figure 1: Adolf Erik Nordenskiöld and his famous ship the SS Vega during the expedition in search of the Northeast Passage. Originally painted by Georg von Rosen in 1886. Image courtesy of the Historical Image Collection, Finnish Heritage Agency (Catalogue No. HK10002:252).

matter such as micro-meteorites (Kalińska et al., 2017; Kang et al., 2020; Lewandowski et al., 2020). These particles are essential for glacial ecosystems, providing nutrients and a substrate for microbial colonization (Zawierucha et al., 2019). Interaction between microorganisms and abiotic particles results in the formation of oval structures called cryoconite granules, driven by the production of organic compounds, primarily by cyanobacteria (Langford et al., 2014). The physical characteristics of cryoconite, including the size and shape of granules, play a crucial role in shaping the structure and dynamics of its microbial communities. Furthermore, the presence of dark sediments within the granules lowers the albedo of ice, altering its optical properties and increasing the energy absorbed by the glacier, which consequently impacts the glacier's radiative balance (Di Mauro et al., 2017). These granules, varying in size and shape, support diverse microbial communities in both oxygen-rich and oxygen-poor microenvironments (Segawa et al., 2020). As such, cryoconite holes are recognized as centres of high biodiversity on glaciers, hosting a wide range of organisms, including algae, bacteria, archaea, fungi, ciliates, and invertebrates, each contributing unique species and assemblages (Stibal et al., 2015).

Formation of Cryoconite Holes

Cryoconite holes are depressions found in the ablation zones of glaciers, formed through a combination of physical, chemical, and biological processes. The first stage is Deposition of Cryoconite, a dark, granular material composed of organic and inorganic particles like dust, soot, and microorganisms, that are transported by wind and then land on the surface of glaciers (Fountain et al., 2004). This sediment accumulates in surface depressions, often facilitated by the relatively low albedo of the weathering crust of superimposed ice (Lyons et al., 2016). The rate, frequency, and timing of sediment deposition are critical for hole formation, with maximum formation occurring just before the melt period starts (Wejnerowski et al., 2023). Cryoconite, being darker than the nearby ice, absorbs more sunlight, causing localized warming and accelerated melting of the underlying ice compared to the surrounding cleaner ice (Buda et al., 2020). Consequently, depressions or holes form in the ice, ranging from centimetres to metres in diameter. Meltwater accumulates in these depressions, causing them to deepen, which then creates unique microenvironments that support microbial life (MacDonell & Fitzsimons, 2008). The distribution of sediment and the pattern of shortwave radiation significantly influence hole formation. Thicker sediment layers generally produce deeper and wider holes due to increased heat absorption, but if the layer exceeds 3–5 mm in thickness or consists of coarse grains, it may insulate the surface, thereby reducing melting (Lokas et al., 2018). Over time, cryoconite holes can stabilize and grow deeper and wider. The maximum depth that a cryoconite hole can achieve is known as the equilibrium depth, which varies based on energy inputs and glacier characteristics (Fountain et al., 2004; Onuma et al., 2023).

Cryoconite holes host various microorganisms, creating a small ecosystem that can influence melting through biothermal energy from photosynthesis (Telling et al., 2012). The biomass in these holes depends on microbial pervasiveness and sediment grain size. For example, fine particles support higher biovolumes, thereby enhancing melt through increased photosynthesis and changes in solute chemistry (Poniecka et al., 2020). The morphology of cryoconite hole walls can be influenced by various factors such as the sun's path, effects of shade and shadows, and ice deformities. In high latitudes, cryoconite holes often have permanent ice lids due to the consistently low air temperatures, whereas in lower latitudes with higher year-round shortwave radiation and air temperatures, holes may remain liquid throughout the year, leading to more rapid development and higher evaporation rates (MacDonell & Fitzsimons, 2008). Cryoconite holes exhibit diverse physical and chemical properties influenced by their interaction with the atmosphere and surrounding environment. Depending on air temperature, cryoconite holes can either remain open, facilitating air and water exchange, or develop an ice lid, becoming isolated from atmospheric processes (Samui et al., 2018). This variation in openness affects the exchange of gases, nutrients, and microbes, consequently influencing the chemical and biological characteristics of the holes. Open cryoconite holes support diverse microbial communities due to the influx of fresh nutrients and microbes, thus impacting meltwater chemistry through metabolic activities. On the other hand, isolated holes may experience concentrated chemical constituents driven by ongoing biological and photochemical reactions (Foreman et al., 2007; Stibal et al., 2008; Webster et al., 2015).

Cryoconite holes can also be categorized as hydrologically connected or isolated. Hydrologically connected holes exchange water, microbes, and chemical components with nearby streams, sub-ice channels, or other cryoconite holes, maintaining dynamic chemical environments. In contrast, hydrologically isolated holes accumulate chemical constituents through the dissolution of debris and photochemically or biologically driven reactions, leading to unique chemical profiles (MacDonell et al., 2016). These holes can shrink due to the accumulation of ice on their walls or decay via water breaching through their walls, driven by expanding supraglacial drainage systems. These processes determine the lifespan of cryoconite holes and their role in storing and transferring chemical and microbial constituents on the glacier surface. As cryoconite holes decay, they contribute to the supraglacial and subglacial drainage systems, affecting the broader glacial ecosystem (Anesio et al., 2009; Onuma et al., 2023).

Variability of Cryoconite Material

Cryoconite material exhibits significant variability both between and within glaciers, influenced by local environmental conditions and processes (Kalińska et al., 2017). For example, in Longyearbyen, Svalbard, heavy rainfall and foehn winds contribute to the breakdown of granular cryoconite into loose, organic-rich sediment (Table 1). Conversely, observations from Austerdalsbreen in

Norway reveal a distinct absence of granules in the upper ablation zone, with the accumulation of large, black granules at the glacier's terminus. Observations have hinted at the prevalence of a particular morphological type, granules or loose sediment, across various samples and seasons within specific glaciers (Cook et al., 2015). The morphological classification and the understanding of cryoconite distribution provide critical insights into the dynamics and environmental interactions of these glacial micro-ecosystems.

Table 1: Categories of cryoconite material based on their morphology

Type	Description	Size Range	References
Granular Cryoconite	Large granules with smooth edges, oval or round	0.3 mm to 4.5 mm; can reach up to 1 cm	Rozwalak et al., 2022
Irregular Cryoconite	Resemble granular cryoconite, with irregular edges	Variable	Buda et al., 2020
Overgrown Granules	Overgrown due to the presence of filamentous *Cyanobacteria* on their surface	Variable	Wejnerowski et al., 2023
Variable (Chaotic) Granules	Granules of different shapes and sizes, exhibiting a mixture of morphological characteristics	Variable	Zawierucha et al., 2019
Loose Cryoconite Material	Mineral matter lacking typical bio-aggregations	Highly variable in size, colour and particle types	Langford et al., 2014
Fine-Grained Loose Cryoconite	Primarily contains sand particles	Smaller than 0.5 mm	Rozwalak et al., 2022
Coarse-Grained Loose Material	Includes coarse sand or gravel grains	0.5 to 64 mm	Samui et al., 2018

Geographic Distribution of Cryoconite Holes in Antarctica

Cryoconite holes are widespread across Antarctic glaciers. They are distributed across various regions, with significant concentrations in the McMurdo Dry Valleys, the Antarctic Peninsula, and coastal regions of East Antarctica.

1. **McMurdo Dry Valleys:** Among the various glaciers in the McMurdo Dry Valleys, the Taylor, Canada, and Commonwealth glaciers are among the most extensively studied sites for cryoconite holes. These valleys, characterized by

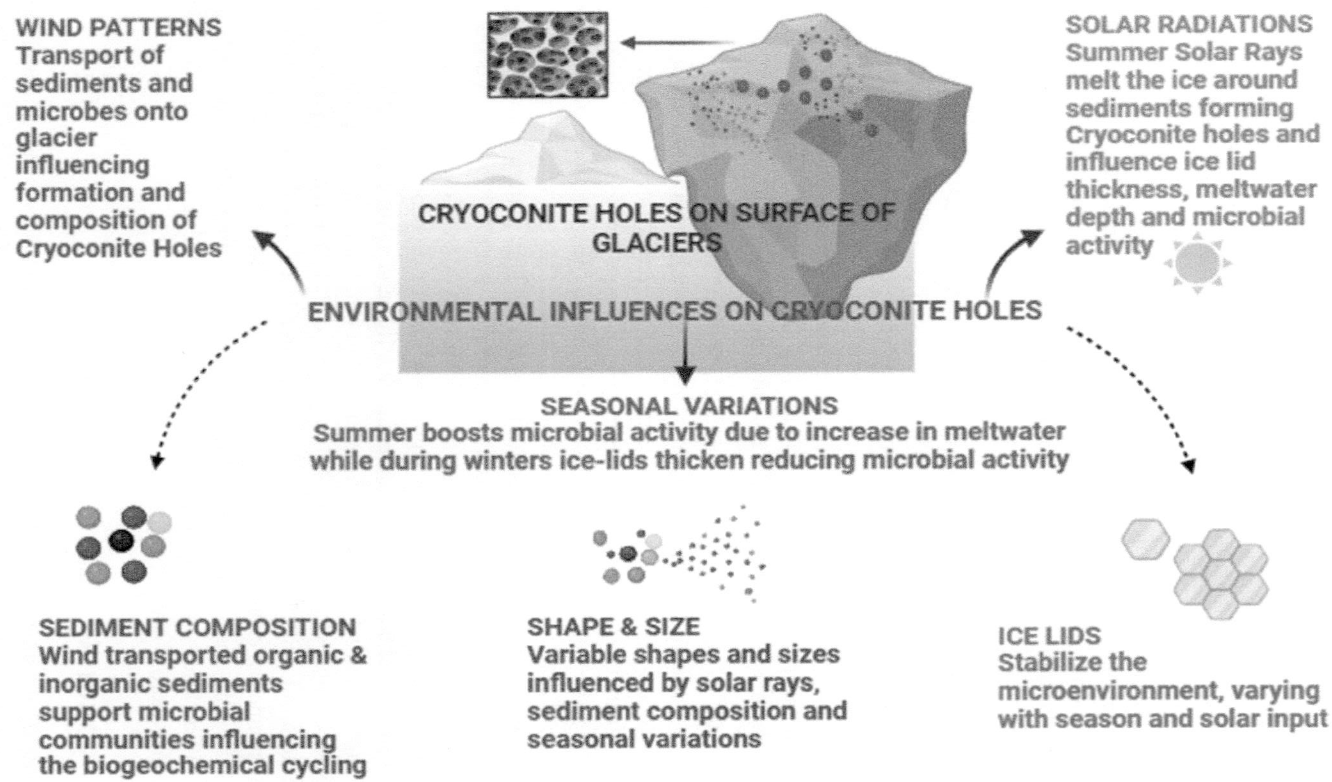

Figure 2: Environmental dynamics that influence the physical characteristics of cryoconite holes.

their cold desert climate and minimal precipitation, facilitate the formation of perennial ice-lidded cryoconite holes. The unique climatic conditions of this hyper-arid polar desert contribute to the stability and persistence of these microhabitats, making them ideal for long-term ecological and biogeochemical studies (Cameron et al., 2012; Sommers et al., 2019).

2. **Antarctic Peninsula:** The glaciers of the Antarctic Peninsula, such as those on James Ross Island and the Larsen Ice Shelf, also harbour cryoconite holes. The Peninsula experiences comparatively milder temperatures and higher precipitation than the interior of Antarctica, which influences the formation and evolution of cryoconite holes. Seasonal melt cycles play a significant role in the dynamics of these features, affecting their physical characteristics and the microbial communities they support (Sommers et al., 2020; Weisleitner et al., 2020).
3. **Coastal Regions of East Antarctica:** Glaciers in the areas around the Davis and Casey research stations host numerous cryoconite holes. These regions benefit from coastal climatic conditions, resulting in both seasonal and perennial cryoconite holes. Notable locations include the glaciers in Adelie Land and the Amery Ice Shelf, where unique climatic and glaciological conditions prevail, further contributing to the diversity of cryoconite hole ecosystems (Darcy et al., 2018; Hodson et al., 2008).

Specific Locations of Interest include Taylor Valley where three main glaciers – Canada, Taylor, and Commonwealth, exhibit distinct cryoconite hole communities. The microbial diversity and nutrient content of these holes vary significantly based on their proximity to the coast (Schmidt et al., 2022). Specifically, the Commonwealth Glacier is closest to the coast and demonstrates the highest bacterial diversity, whereas the inland Taylor Glacier has fewer nutrients and lower diversity (Darcy et al., 2018). This pattern highlights the influence of geographic location on the microbial composition of cryoconite holes within the valley. The Untersee Oasis in East Antarctica, particularly the Anuchin Glacier, presents another significant site for cryoconite hole research. These holes are characterized by perennial ice lids and sediment-filled interiors. There are substantial differences in microbial communities and biogeochemical parameters across various zones of the glacier, such as the medial moraine and glacier terminus (Stanish et al., 2013; Weisleitner et al., 2020). Figure 2 depicts the environmental changes that influence the physical characteristics of cryoconite holes. These regions provide valuable data for understanding microbial life in extreme conditions, contributing to the broader knowledge of cryoconite hole ecosystems across Antarctica.

3. Cryoconite Hole Ecosystems

Cryoconite holes are nutrient-rich microhabitats within glacier systems, functioning as ‘greenhouse’ environments that support diverse life forms. These holes host microorganisms such as bacteria, viruses, fungi, cyanobacteria, green algae, diatoms, rotifers, tardigrades, and nematodes, that are well-adapted to extreme conditions (Poniecka et al., 2020). Dominant bacterial groups within Antarctic cryoconite holes include *Cyanobacteria*, *Actinobacteria*, *Proteobacteria*, and *Bacteroidetes*, which play key roles in nutrient cycling and primary production by utilizing sunlight and organic matter to sustain the ecosystem (Pittino et al., 2023). Green algae and diatoms contribute to photosynthesis and can form visible biofilms on the sediment surface, particularly during the summer months (Ambrosini et al., 2017). Fungi contribute to the decomposition of organic matter, while viruses influence microbial population dynamics through infection and lysis of bacterial cells (Kaczmarek et al., 2015). Wind-transported organic and inorganic sediments support microbial communities, further influencing biogeochemical processes. The interaction between meltwater and sediment provides essential nutrients, with microorganisms photosynthesizing in summer to sustain the ecosystem (Hassan et al., 2024). This primary production makes cryoconite holes significant nutrient reservoirs. Microbial activity produces humic substances, forming cryoconite granule complexes of sediment and microbes that reduce sediment albedo and contribute to melting through biothermal energy. Together, this accounts for up to 10% of the energy within cryoconite holes, making them one of the largest nutrient reservoirs in glacier systems (Xu et al., 2010). The extent of material decomposition influences the optical properties of the cryoconite hole, with more decomposed material resulting in darker granules and a lower albedo due to the higher carbon content from microbial activity (Buda et al., 2021).

The study of cryoconite holes has extended to astrobiology. Their ability to harbour life in extreme conditions makes them models for potential habitats on other planets, such as Mars. Recent work has discussed how the conditions within cryoconite holes on Earth could provide insights into where and how life might exist on icy extraterrestrial bodies. Cryoconite holes have also been used to hypothesize about life on other planets and moons, given their ability to support diverse microorganisms in extreme conditions (Cook et al., 2015). Cryoconite holes are hotspots for carbon and nutrient cycling. Specifically, the microbial activity within these holes contributes to the carbon cycle by fixing carbon dioxide through photosynthesis (Stibal et al., 2020). Additionally, new methods in molecular biology are providing deeper insights into the microbial diversity and metabolic functions within cryoconite ecosystems (Telling et al., 2012). Climate change can impact cryoconite hole ecosystems, with rising temperatures potentially enhancing microbial activity and nutrient cycling due to more frequent hole formation and longer meltwater retention (Novotna et al., 2022). Figure 3 depicts the microbial diversity of Antarctic cryoconite holes. Cryoconite holes also influence meltwater drainage patterns on glaciers, acting as temporary storage and affecting overall hydrology (Samui et al., 2018).

Technological advances like remote sensing, drones, and high-resolution imaging have improved the study of cryoconite holes, allowing precise mapping and monitoring over large areas. Advances in molecular biology methodologies, including metagenomics and proteomics, further our understanding of microbial communities' metabolic pathways and functions. Cryoconite holes remain a key focus for understanding microbial life and glacial environments. Continued research will expand knowledge of these unique microhabitats, with implications for Earth and potentially other planetary systems.

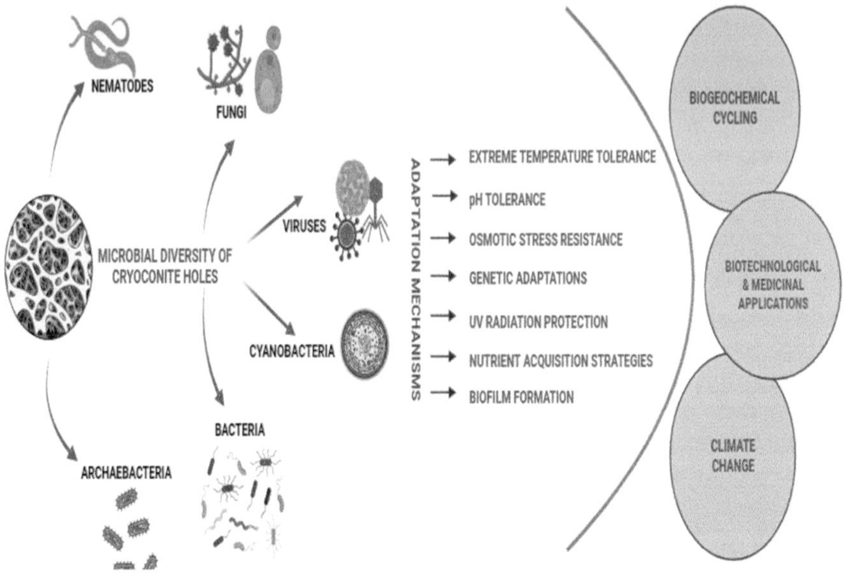

Figure 3: Microbial diversity of Antarctic cryoconite holes.

4. Microbial Diversity of Antarctic Cryoconite Holes

Eukarya

The eukaryotic assemblages in cryoconite holes are diverse and include various groups of algae, microfauna, and ciliates (Doytchinov & Dimov, 2022). Notably, the algae *Pleurastrum* sp. is dominant, especially in the Commonwealth Glacier (Sommers et al., 2019). Dominant microfauna include *Bdelloid rotifers*, *Rotaria* sp. and *Adineta* sp. and *tardigrades*, *Acutuncus* sp. and *Diphascon* sp. Sequences from a ciliate, *Stokesia* sp., and a flatworm related to *Phaenocora* sp. are also commonly detected (Sommers et al., 2020).

A recent study on eukaryotic diversity within Antarctic cryoconite holes used clone libraries of 18S rRNA genes. *Cercozoa* were found to constitute approximately one-third of the communities at each location, with Alveolata-related sequences consistently present. *Metazoans*, specifically *Tardigrada*, were also identified (Sohm et al., 2020). Across the five cryoconite communities, a total of 96 eukaryotic operational taxonomic units (OTUs) were identified, with around 15% common to multiple locations. These OTUs represented several taxa, including *Cercomonadida*, *Silicofilosea*, *Phaeodarea*, and the *Vampyrellidae* (Obbels et al., 2016). The small subunit rDNA molecules amplified and cloned from cryoconite DNA using Eukarya-specific primers revealed the presence of several eukaryotic taxa, nematodes (*Plectus aquatilis*), tardigrades (*Macrobiotus hufelandi*), rotifers (*Philodina acuticornis*), truffles (*Choiromyces meandriformis*), green algae (*Pleurastrium insigne*), and ciliates (*Spathidium* sp.) (Christner et al., 2003).

Cyanobacteria

The predominant photoautotrophic organisms in the investigated glaciers are algae and cyanobacteria, with prominent families like *Pseudanabaenaceae* and *Phormidiaceae*. Genera such as *Leptolyngbya*, *Pseudanabaena*, *Phormidium*, *Phormidesmis*, *Wilmottia*, and *Microcoleus*, are frequently observed (Kaczmarek et al., 2015; Wejnerowski et al., 2023). Morphological identification has revealed that *Leptolyngbya* and *Pseudanabaena* are common cyanobacterial genera in cryoconite samples. Complementing this, DNA sequencing data indicates that *Phormidesmis* is the most common cyanobacterial genus on glaciers (Segawa et al., 2017; Uetake et al., 2016). The sediments of different glaciers are predominantly colonized by cyanobacteria, with *Nostoc* sp. dominant in the Commonwealth Glacier, and *Chamaesiphon* sp. and *Phormidium* sp. prevalent in the Canada and Taylor Glaciers, respectively. These cyanobacterial OTUs, along with *Polaromonas* sp. are consistently among the top 10 most abundant OTUs. Phylogenetic analyses show close relationships between identified cyanobacteria and species from microbial aggregates in Lake Bonney and microbial mats in the Dry Valley region (Sohm et al., 2020; Webster et al., 2015).

Bacteria

In a recent study, the predominant bacterial phyla in cryoconite holes were *Proteobacteria* (26.54%), *Cyanobacteria* (25.45%) *Bacteroidetes* (23.49%), *Actinobacteria* (13.38%), and *Acidobacteria* (3.20%), cumulatively accounting for 92% of all bacterial counts. Notably, 12.05% could not be assigned to a known genus, and 1.98% remained unclassified at the phylum level (Pittino et al., 2023). This diversity was further highlighted by sequencing clone libraries of 16S rRNA genes, revealing autotrophic and heterotrophic bacteria from phyla like *Alphaproteobacteria*, *Betaproteobacteria*, *Deltaproteobacteria*, *Bacteroidetes*, *Cyanobacteria*, and *Actinobacteria* (Cameron et al., 2012; Millar et al., 2021). Another study utilized DNA preparations directly from cryoconite sediment for PCR assays, amplifying bacterial rDNA molecules. This revealed sequences from diverse bacterial lineages including *Acidobacteria*, *Actinobacteria*, *Cytophagales*, *Gemmimonas*, *Planctomycetes*, α- and γ-*Proteobacteria*, and *Verrucomicrobia* (Edwards et al., 2011). These findings were consistent with the dominance of *Proteobacteria*, *Cyanobacteria*, *Bacteroidetes*, *Actinobacteria*, and *Acidobacteria* in cryoconite holes, reflecting results from other Antarctic environments. The most abundant Proteobacteria sequences were identified as *Silanimonas* sp. *Sphingomonas* sp. and *Polaromonas* sp., each known from various Antarctic habitats (Obbels et al., 2016; Weisleitner et al., 2020).

The microbial community structure has been shown to vary with depth, where deeper cryoconite holes exhibit more stable conditions and reduced biotic inoculation, leading to more stable microbial communities. The bacterial communities in Antarctic cryoconite holes display significant diversity and adaptability, with the presence of both autotrophic and heterotrophic species. The site-specific diversity and unique community compositions in different cryoconite holes highlight the influence of local environmental conditions on bacterial assemblages.

Archaea

Archaea, particularly *Halobacteriales* and the *Halorubrum* genus, dominate cryoconite holes, constituting 65.48% of all archaeal reads. *Methanomicrobiales* represent 10.01% of the reads, while *Methanocellales*, *Methanosarcinales*, and *Thermoplasmatales* collectively account for less than 1%. These archaeal sequences primarily belong to *Halobacteria*, *Methanomicrobia*, and *Thermoplasmata* classes within *Euryarchaeota* (Christner et al., 2003). Recent PCR amplification studies targeting 16S rRNA genes in Antarctic cryoconite holes have revealed that 77% of the detected sequences are related to *Thaumarchaeota*, exhibiting limited diversity with clones sharing over 97% similarity (Cameron et al., 2012; Kaczmarek et al., 2015).

Fungi

Fungi isolated from Antarctic cryoconite holes primarily consist of psychrophilic and psychrotolerant species, with dominant genera including *Rhodotorula* (species such as *Rhodotorula psychrophenolica*, *R. svalbardensis*, and *R. glacialis*), *Cryptococcus*, *Debaryomyces*, *Torulopsis,* and *Candida* (Borzecka et al., 2022). The *Pezizales* order is represented by species like *Choiromyces meandriformis*, *Hydnotrya tulasnei*, and *Verpa bohemic*a, along with *Dothideomycetes* like *Alternaria* species and *Thelebolus* (Edwards et al., 2013). Further *micromycetes* identified include *Penicillium* spp. *Phialophora* sp. *Cladosporium* spp. and *Circinella* species, alongside members of *Chytridiomycota* and *Ascomycota* groups like *Acremonium, Articulospora*, *Ascochyta*, *Preussia*, *Pseudeurotium*, and *Varicosporium* genera (Uetake et al., 2016). These fungi are often related to plant endophytes and are presumed to be transported by wind from plant litter or soil into cryoconite holes. Basidiomycetous yeast such as *Mrakia* species (e.g., *Mrakia robertii*), *Varicosporium elodeae*, and *Glaciozyma watsonii* have been documented, primarily from glacial meltwater (Singh et al., 2016). *Chytridiomycota* are present in all samples, while *Rozellomycota* are found in sea water, lake water, and tap water, highlighting the diversity and adaptability of fungal communities (Kaczmarek et al., 2015). Notably, the fungal communities within and among glaciers are diverse and spatially heterogeneous. Fungal species *Choiromyces meandriformis* has been reported from Antarctica (Singh et al., 2016).

Viruses

Antarctic cryoconite holes host diverse viral communities, yet these remain largely uncharacterized (Bellas et al., 2013). A metagenomic analysis study aimed at identifying single-stranded DNA (ssDNA) viruses in Taylor Valley's cryoconite holes revealed that *Microviridae* are predominant, and also detected the presence of unclassified circular replication-associated protein (Rep)-encoding single-stranded (CRESS) DNA viruses. Notably, all detected circular molecules originated from cryoconite sediments, rather than bare ice (Sommers et al., 2019). As reported previously, there is a correlation between viral genome diversity and microbial diversity and abundance. *Microviruses* found in the samples were clustered within the subfamily *Gokushovirinae*, known for infective bacteria such as *Bdellovibrio*, *Chlamydia*, and *Spiroplasma*. This presence could potentially be linked to the microbiomes of gut micro-invertebrates or other bacteria within these ecosystems (Smeele et al., 2017).

Antarctic cryoconite holes host unique viral communities, and it is crucial to understand their role in extreme environments. While ssDNA viruses are widespread, polar viral genomic data remain limited, hindering comprehension of their biogeography, evolution, and ecological roles (Roux et al., 2019). Viruses have an important role in shaping microbial communities in these extreme environments, highlighting the need for further research to elucidate their ecological roles and evolutionary dynamics (Bellas et al., 2015). The discovery of unique viral genomes can contribute to the growing body of knowledge on Antarctic virology and pave the way for future studies on virus-host interactions in polar ecosystems.

5. Adaptation Mechanisms of Microorganisms

The microorganisms in cryoconite holes exhibit remarkable resilience to shifting environmental conditions (Singh et al., 2014). For example, cryoconite holes may fluctuate between oxygen saturation and anoxia; they can range from excessively dark to overly bright conditions; transition from hypersaline to low ionic strength states; shift between acidic and alkaline conditions; undergo repeated freeze-thaw cycles; and experience nutrient spikes or severe nutrient limitation (Poniecka et al., 2020). As such, it is anticipated that Antarctic microorganisms possess the resilience to endure harsher conditions than their Arctic counterparts (Núñez & Barrientos, 2018).

Metagenomic and molecular analyses of cryoconite hole communities have elucidated the extensive stress response genes and genetic potential for efficient nutrient scavenging and recycling (Edwards et al., 2013). These communities harbour genes encoding metabolic pathways characteristic of heterotrophic anoxygenic phototrophs and anaerobes, as well as enzymes capable of processing various organic carbon sources such as cellulose, chitin, and other polysaccharides, including Extracellular Polymeric Substances (EPS) (Margesin et al., 2002). The metagenomes of microbial assemblages on the Greenland Ice Sheet indicate a potential for resistance to, and degradation of, anthropogenic pollutants (Hauptmann et al., 2017). However, the genetic capabilities of Antarctic communities remain unexplored. The phenotypic diversity within these communities is crucial for ecosystem resilience and responsiveness to environmental changes.

pH Tolerance

Antarctic microbial isolates have demonstrated growth in media with a pH range from 4 to 10.5 (Margesin et al., 2002). No significant difference in pH tolerance has been observed between oxic and anoxic conditions or between yeasts and bacteria. Most yeast isolates were able to grow at pH ≥ 3. Svalbard yeasts demonstrated a pH growth range of 3–10. One bacterial isolate from Svalbard exhibited growth at pH 2.5, though other bacterial isolates did not survive in media with pH ≤ 4 (Poniecka et al., 2020). Further research corroborates these findings, emphasizing the adaptability of cryoconite hole microbes to extreme pH conditions. This adaptability is crucial for their survival and functionality in highly variable and often extreme pH conditions, characteristic of cryoconite holes, which undergo fluctuations due to factors such as meltwater input, organic matter decomposition, and mineral weathering (García et al., 2016).

The mechanisms underlying this tolerance have been extensively studied, revealing several strategies that these microorganisms employ to thrive across a wide pH spectrum. One primary mechanism is the ability of microbes to regulate their intracellular pH. Microorganisms maintain a stable internal pH through active transport systems like the proton pump ATPase, expelling excess H^+ ions to maintain homeostasis. This is crucial for optimal enzyme function and cellular integrity (Guan & Liu, 2020). Another key strategy is Extracellular Polymeric Substances (EPS) production, which is a mixture of biopolymers secreted by microbes, forming a protective matrix around the cells. EPSs buffer external pH changes and neutralize harmful ions, thus providing a stable microenvironment (Yin et al., 2019). Microbes produce acidophilic and alkaliphilic enzymes with unique amino acid compositions, enhancing their stability and functionality under extreme pH conditions. Importantly, these enzymes enable essential metabolic processes despite external pH stress (Rapuano & Graziano, 2022). As such, the regulation of gene expression in response to pH changes is another critical mechanism. Microbial cells upregulate or downregulate specific genes related to pH homeostasis, stress response, and metabolic adaptation. This allows rapid adaptation to changing pH conditions, involving genes encoding proton pumps, stress response proteins, and metabolic enzymes (Clarke et al., 2007). Adjustments in membrane lipid composition are also a significant adaptation mechanism. For example, microbial membranes can alter their lipid composition to maintain fluidity and functionality under varying pH conditions. Specific fatty acids can enhance membrane stability and permeability, which is crucial for maintaining cellular integrity (Singh et al., 2014).

Horizontal gene transfer (HGT) also plays a role in ensuring pH tolerance traits among microbial communities. HGT allows microbes to acquire genes conferring pH tolerance traits from other organisms, spreading stress response genes, proton pumps, and other adaptive mechanisms within the community, enhancing overall resilience (Hall et al., 2020).

Tolerance to Salinity

Microbial isolates from cryoconite holes exhibit remarkable tolerance to a wide range of salinity conditions. In a recent study, nearly all (34 out of 39) isolates demonstrated the ability to grow in environments with salinity levels up to 5%, 42000 $\mu S\ cm^{-1}$ (Poniecka et al., 2020). This salinity tolerance is crucial for survival in cryoconite holes, where salinity can fluctuate significantly due to melting and refreezing cycles, evaporation, and the concentration of solutes. Fungi isolated from Svalbard exhibited improved growth on halotolerant media, thriving at salinities up to 24% (Wang et al., 2017). This salinity resistance is likely an adaptation to other harsh conditions such as high UV exposure, dehydration, or freezing. In hypersaline ponds on the McMurdo Ice Shelf, cyanobacteria utilize organic osmolytes to mitigate osmotic stress, and substantial amounts of EPS have been found in sea ice brine channels (Pittino et al., 2023).

Microbes synthesize or accumulate compatible solutes, also known as osmoprotectants, such as trehalose, glycine betaine, and ectoine. These compounds help to balance the osmotic pressure between the inside and outside of the cell, preventing dehydration and maintaining cellular turgor (Roberts, 2005). Active ion transport mechanisms, including Na^+/H^+ antiporters and ATP-driven ion pumps, are employed to expel excess salt ions from the cell, thereby regulating internal ion concentration and maintaining osmotic balance. Adjustments in membrane lipid composition are also critical for salinity tolerance. The incorporation of specific lipids can enhance membrane fluidity and stability under high salinity conditions, ensuring proper membrane function and integrity (Hassan et al., 2020). Some microorganisms produce proteins and chaperones that protect cellular proteins from denaturation caused by high salt concentrations. These protective proteins ensure that essential enzymatic functions continue, even under osmotic stress (Godin et al., 2015).

Temperature Range

The temperature tolerance of cryoconite hole microbes is equally impressive. In a study examining the growth temperature ranges of various microbial isolates, all tested isolates were capable of growth at the lowest temperature of 1°C. Bacterial isolates demonstrated the ability to grow at temperatures above 22°C, showing a maximum growth temperature of 36°C. In contrast, yeast isolates were limited to growth at or below 22°C (Poniecka et al., 2020). Psychrophilic microbes produce cold-adapted enzymes that remain functional at low temperatures. These enzymes have higher catalytic efficiencies at low temperatures due to increased flexibility in their active sites, which compensates for the reduced kinetic energy in cold environments. The production of cryoprotectants, such as antifreeze proteins and exopolysaccharides, prevents ice crystal formation within cells (Hamid et al., 2022). At higher temperatures, microbes induce the expression of heat shock proteins (HSPs) that function as molecular chaperones. HSPs assist in the refolding of denatured proteins and prevent the aggregation of misfolded proteins, ensuring cellular functionality under thermal stress (Park & Seo, 2015). Microbes adjust their membrane lipid composition in response to temperature changes by incorporating unsaturated fatty acids at low temperatures to maintain fluidity and saturated fatty acids at higher temperatures to ensure membrane integrity. Temperature fluctuations also increase reactive oxygen species (ROS) production, which microbes counteract by upregulating antioxidant systems such as superoxide dismutase and catalase to neutralize ROS and protect cellular components (Zandi & Schnug, 2022).

Freeze-Thaw Cycles

Microbial isolates from cryoconite holes show varied resilience to freeze-thaw cycles. Some Antarctic isolates withstand multiple cycles without losing viability, while others do not fare as well. Yeast strains in particular demonstrate notable resilience, often surviving multiple freeze-thaw cycles without significant viability loss. Some even show increased cell counts after several cycles, indicating an adaptation to freeze-thaw stress and possibly transitioning from a non-culturable to a culturable state (Poniecka et al., 2020).

Bacterial isolates exhibit more variability in their response to freeze-thaw stress. Gram-positive bacteria, particularly *Actinobacteria*, exhibit higher resilience, often remaining viable through multiple freeze-thaw cycles. In contrast, Gram-negative bacteria typically lose culturability more rapidly under these conditions. The structural integrity of the cell envelope and the presence of carotenoids play crucial roles in protecting these microbes against reactive oxygen species generated during thawing. There are varying levels of adaptation to freeze-thaw stress among different microbial taxa (Akulava et al., 2022).

Microbes often produce cryoprotectants such as antifreeze proteins, trehalose, and exopolysaccharides. These substances help to prevent ice crystal formation inside cells and protect cellular structures from damage during freeze-thaw cycles. Specifically, antifreeze proteins bind to ice crystals and inhibit their growth, thereby reducing cellular injury. In addition, adjustments in cell membrane composition are crucial for freeze-thaw tolerance (Deming & Young, 2017). Heat shock proteins (HSPs) and cold shock proteins (CSPs) also play significant roles in freeze-thaw tolerance. These proteins act as molecular chaperones, assisting in the refolding of denatured proteins and preventing the aggregation of misfolded proteins during thermal stress (Margesin & Collins, 2019; Sajjad et al., 2020). Since freeze-thaw cycles can cause DNA damage, microorganisms with efficient DNA repair mechanisms are better equipped to survive such stress. These repair systems can correct DNA lesions caused by the formation and dissolution of ice crystals (De et al., 2014). Some microbes form biofilms, which provide a protective environment that enhances survival during freeze-thaw cycles. Biofilms can trap cryoprotectants and create a microenvironment that buffers against extreme temperature fluctuations (Wu et al., 2012).

Substrate Utilization

Microbial cultures from cryoconite holes exhibit remarkable versatility in utilising a wide range of carbon sources, including carboxylic acids, simple and complex carbohydrates, amino acids, alcohols, polyalcohols, and other complex substances (Siddiqui et al., 2012). Studies have shown that genera like *Flavobacterium* and *Cryobacterium* can metabolise nearly all tested substrates (Poniecka et al., 2020).

Antarctic isolates are unique in their ability to metabolize non-competitive substrates such as choline, betaine, and methylamine. Yeasts demonstrate a greater capability to utilize alcohols compared to bacteria, and are less likely to metabolisee amino acids (Wang et al., 2017). Cryoconite holes are rich in organic matter, providing a nutrient source for microorganisms. Studies have identified genes associated with biodegradation, enabling microbes to effectively decompose organic matter (Rozwalak et al., 2022). Microorganisms in cryoconite holes produce a variety of enzymes that allow them to metabolize different substrates, indicating their effectiveness in scavenging available sources (Buda et al., 2020). Isolates obtained under anoxic conditions tend to favour carbohydrates more than those obtained aerobically. Facultatively anaerobic microorganisms adjust their metabolism based on oxygen availability, often utilizing monomeric sugars as electron acceptors (Buda et al., 2021). There is considerable variability in the metabolic capabilities of closely related isolates, which can be attributed to differences in initial incubation conditions, culture fitness, and genome rearrangements in stressed and starving microorganisms.

Biochemical Cycling

Cryoconite holes harbour a diverse array of microorganisms that interact within intricate food webs, influencing vital biogeochemical processes. Bacterial communities, constituting the majority of the biomass, play a central role in decomposing organic matter and cycling nutrients (Porazinska et al., 2004). Studies have shown that bacterial taxa in cryoconite holes are highly diverse, and include representatives from phyla such as *Proteobacteria*, *Cyanobacteria*, *Bacteroidetes*, and *Actinobacteria* (Sanyal et al., 2018). One of the primary roles of microorganisms in cryoconite holes is the cycling of carbon. Autotrophic microbes, such as cyanobacteria and algae, perform photosynthesis, converting carbon dioxide into organic matter. This primary production forms the base of the food web, supporting heterotrophic bacteria that decompose organic matter and release CO_2 back into the environment through respiration. The balance between these processes regulates the carbon dynamics within cryoconite holes (Anesio et al., 2009).

Nitrogen cycling in cryoconite holes is primarily mediated by bacterial communities. Key processes include nitrogen fixation, nitrification, and denitrification. Nitrogen-fixing bacteria, such as certain *Cyanobacteria*, convert atmospheric nitrogen into ammonia, which can be used by other organisms. Nitrifying bacteria further convert ammonia to nitrite and then to nitrate, while denitrifying bacteria reduce nitrate back to nitrogen gas, completing the nitrogen cycle. These processes are crucial for maintaining nitrogen availability in nutrient-poor environments (Banerjee et al., 2023; Murakami et al., 2022). Phosphorus is another essential nutrient cycled by microbial communities in cryoconite holes. Phosphorus availability often limits primary production in these ecosystems. Microorganisms contribute to phosphorus cycling through the mineralization of organic phosphorus compounds, making inorganic phosphate available for uptake by autotrophs. This cycling is crucial for sustaining the productivity of cryoconite holes, especially given the limited external phosphorus inputs in Antarctic environments (Samui et al., 2018).

Sulphur cycling in cryoconite holes involves microbial-mediated transformations of sulphur compounds. Sulphate-reducing bacteria play a significant role in this process by converting sulphate to hydrogen sulphide under anaerobic conditions. Hydrogen sulphide can then be oxidized back to sulphate by sulphur-oxidizing bacteria, completing the sulphur cycle (Sommers et al., 2020). These transformations influence the redox state and the overall biogeochemistry of cryoconite holes. Viruses, particularly bacteriophages, are abundant in cryoconite holes and play a pivotal role in microbial dynamics and biogeochemical cycles. They regulate bacterial population sizes through lytic cycles, releasing cellular contents into the environment upon bacterial cell lysis. This viral lysis contributes to the recycling of nutrients and organic matter, influencing microbial community structure and function (Smeele et al., 2017).

Microorganisms in Antarctic cryoconite holes are integral to biogeochemical cycling, driving key processes that sustain these unique ecosystems. Through their roles in carbon, nitrogen, phosphorus, and sulphur cycles, these microbes maintain nutrient availability, influence microbial community dynamics, and impact broader environmental and climatic systems.

Role in Climate Change

Antarctic cryoconite holes are hubs of microbial diversity, wielding considerable influence on climate dynamics. Microbial activities within these holes impact biogeochemical cycles and alter glacier albedo, thereby shaping broader climate processes (Hassan et al., 2024). Microbial communities in cryoconite holes are pivotal in carbon cycling, a critical process which influences atmospheric CO_2 levels and, consequently, climate dynamics. Autotrophic microorganisms like *Cyanobacteria* and algae fix atmospheric CO_2 through photosynthesis, generating organic carbon. This organic carbon then supports heterotrophic bacteria, leading to the decomposition of organic matter and the release of CO_2 and methane (CH_4) through respiration and anaerobic digestion, respectively (Antony et al., 2024). Recent studies have highlighted the significant role of cryoconite hole microorganisms in greenhouse gas emissions, particularly methane production under anaerobic conditions. While the overall contribution of Antarctic cryoconite holes to global methane emissions is relatively small, their localized impact on the carbon budget of glacier surfaces can be substantial, contributing to the greenhouse effect and potentially accelerating regional warming and ice melt (Zwerschke et al., 2022).

Microorganisms in cryoconite holes influence glacier albedo by darkening the ice surface with their pigments, such as chlorophyll from *Cyanobacteria* and algae, and organic matter produced by microbial activity. This reduction in albedo leads to increased absorption of solar radiation, enhancing the melting of glacier ice. The resultant ice melt not only contributes to rising sea levels, but also exposes more ice to microbial colonization, creating a feedback loop that further accelerates melting (Pittino et al., 2023; Zwerschke et al., 2022). Microorganisms in cryoconite holes play a crucial role in nutrient cycling, which can have indirect effects on climate change. Nitrogen-fixing bacteria convert atmospheric nitrogen into bioavailable forms, supporting primary production in these nutrient-limited environments. Enhanced microbial activity and nutrient cycling can increase the biomass and metabolic activity of cryoconite communities, further influencing carbon dynamics and albedo (Porazinska et al., 2004).

Understanding the role of microorganisms in cryoconite holes is essential for improving polar climate models. Their contribution to carbon cycling, greenhouse gas emissions, glacier albedo reduction, and nutrient cycling must be incorporated into climate models to accurately predict the impacts of climate change on polar regions and global sea levels. Future research should focus on quantifying the contributions of cryoconite hole microorganisms to greenhouse gas emissions and understanding their interactions with environmental factors such as temperature and nutrient availability. Integrating microbial processes into climate models will be crucial for accurately predicting the impacts of climate change on polar regions and global sea levels (Stokes et al., 2022).

Biotechnological Applications

Extremophiles found within cryoconite holes possess novel biochemical properties and metabolic pathways that have garnered significant scientific interest due to their possible biotechnological and medical applications (Esclapez et al., 2018). Microorganisms in cryoconite holes are adapted to extreme conditions, and these adaptations have led to the evolution of unique enzymes that function optimally under such harsh conditions. Cold-adapted enzymes, or psychrophilic enzymes, from these microorganisms have sparked interest for their potential industrial applications, including biocatalysis in low-temperature processes. These enzymes are often characterized by high catalytic efficiency at low temperatures, reduced thermal stability, and unique structural properties that differentiate them from their mesophilic counterparts (Sahay et al., 2013; Hamid et al., 2022). For example, cold-active proteases, lipases, and amylases isolated from Antarctic microorganisms have potential applications in the food industry for processing at low temperatures, with broader impacts related to preserving nutritional quality and reducing storage-energy consumption. Additionally, these enzymes can be used in bioremediation processes to degrade pollutants in cold environments, offering sustainable solutions for environmental management (Sanyal et al., 2018; Tarfeen et al., 2022).

Microbial communities in Antarctic cryoconite holes produce a variety of secondary metabolites with potential pharmaceutical applications. These metabolites, often produced as survival strategies under extreme conditions, include antibiotics, antifungals, and antiviral compounds. The unique biochemical pathways utilized by these extremophiles provide a rich source of novel compounds that could be developed into new drugs (Parvizpour et al., 2021). Recent studies have identified antimicrobial peptides and other bioactive compounds from Antarctic bacteria and fungi that show activity against a range of pathogens, including antibiotic-resistant bacteria. These findings accentuate the potential of cryoconite hole microorganisms as sources of new antimicrobial agents to combat the growing threat of antibiotic resistance (Wang et al., 2017).

Microorganisms in cryoconite holes produce cryoprotectants, substances that protect biological tissues from freezing damage. These include antifreeze proteins, exopolysaccharides, and compatible solutes. Cryoprotectants from Antarctic microorganisms have potential applications in medicine and biotechnology, particularly in the preservation of cells, tissues, and organs for transplantation, and in the development of freeze-tolerant crops (Singh et al., 2014; Kuddus et al., 2024). Antifreeze proteins can be utilized in cryopreservation techniques, improving the viability of preserved biological materials. Additionally, exopolysaccharides with cryoprotective properties have potential applications in the food industry as thickeners and stabilizers, particularly in frozen products (Hamid & Mohiddin, 2018; Ali et al., 2020). Microorganisms in cryoconite holes exhibit metabolic diversity, including pathways for pollutant degradation and metal transformation. These capabilities hold promise for bioremediation in cold environments, where traditional methods are less effective. They have demonstrated the ability to degrade hydrocarbons like those found in oil spills, and convert heavy metals into less toxic forms, offering sustainable solutions for environmental cleanup in polar regions and beyond (Ferrario et al., 2017).

Advances in high-throughput sequencing and metagenomics have provided detailed insights into the genetic potential of cryoconite hole microorganisms. Genomic studies have revealed the presence of genes involved in stress response, secondary metabolite production, and novel metabolic pathways. These genetic resources can be explored for biotechnological applications, including the synthesis of novel biomolecules and the engineering of microorganisms for specific industrial processes (Kumari et al., 2021; Hamid, 2021). The study of viral communities in cryoconite holes has also highlighted the potential for phage therapy, an emerging approach to combat bacterial infections using bacteriophages. Phages isolated from these environments may possess unique properties due to their co-evolution with extremophilic bacteria, providing new tools for the development of alternative antimicrobial therapies (Bhatia et al., 2020; Parvizpour et al., 2021). Continued research and exploration of these unique microbial communities will likely yield further discoveries and applications, contributing to advancements in biotechnology and medicine.

6. Conclusion

The enigmatic microbial ecosystems within Antarctic cryoconite holes present an unparalleled frontier for scientific exploration, with profound implications for our knowledge of life in extreme environments and the biotechnological potential. Advancing our understanding of cryoconite microbiomes necessitates the integration of high-resolution metagenomics with functional genomics. By employing next-generation sequencing technologies and sophisticated bioinformatics tools, we can decode the complex genetic blueprints of these communities. Functional genomics in particular can illuminate the specific roles of genes and metabolic pathways, enabling the identification of novel enzymes and bioactive compounds with potential industrial and medical applications. Future research should focus on isolating and characterizing psychrophilic enzymes, antimicrobial agents, and cryoprotectants. These unique biomolecules could revolutionize various industries, from bioremediation and food processing to medical therapeutics and organ preservation. Investigating the interactions between cryoconite hole microorganisms and climate change remains a critical research avenue. Elucidating the ecological interactions and evolutionary adaptations of microorganisms in cryoconite holes can provide insights into their survival strategies and ecosystem dynamics. Addressing the complexity and interdisciplinary nature of cryoconite hole microbiology requires collaborative efforts across various scientific disciplines. For example, integrating microbiology, climatology, biogeochemistry, and biotechnology will foster a holistic understanding of these systems. Moreover, international collaborations and data sharing can be pivotal in advancing the field, enabling researchers to tackle the multifaceted challenges and leverage diverse expertise. Innovations in single-cell genomics, proteomics, metabolomics, advanced imaging, and *in situ* monitoring tools can further enhance our ability to study these ecosystems. Through these interdisciplinary efforts, we can unlock the mysteries of cryoconite microbiology and gain insights into the limits of life on Earth and beyond.

References

Akulava, V., Miamin, U., Akhremchuk, K., Valentovich, L., Dolgikh, A., & Shapaval, V. 2022. Isolation, physiological characterization, and antibiotic susceptibility testing of fast-growing bacteria from the sea-affected temporary meltwater ponds in the Thala Hills Oasis (Enderby Land, East Antarctica). *Biology*, 11(8), 1143.

Ali, P., Shah, A.A., Hasan, F., Hertkorn, N., Gonsior, M., Sajjad, W., & Chen, F. 2020. A glacier bacterium produces high yield of cryoprotective exopolysaccharide. *Frontiers in Microbiology*, 10, 3096.

Ambrosini, R., Musitelli, F., Navarra, F., Tagliaferri, I., Gandolfi, I., Bestetti, G. et al. 2017. Diversity and assembling processes of bacterial communities in cryoconite holes of a Karakoram Glacier. *Microbial Ecology*, 73(4), 827–837.

Anesio, A., Hodson, A., Friitz, A., Psenner, R., & Sattler, B. 2009. High microbial activity on glaciers: Importance to the global carbon cycle. *Global Change Biology*, 15, 955–960.

Antony, R., Mongad, D., Sanyal, A., Dhotre, D., & Thamban, M. 2024. Holed up, but thriving: Impact of multitrophic cryoconite communities on glacier elemental cycles. *The Science of the Total Environment*, 933, 173187.

Banerjee, A., Sarangi, C., Rashid, I., Vijay, S., Najar, N., & Chandel, A. 2023. A scaling relation for cryoconite holes. *Geophysical Research Letters*, 50.

Bellas, C., Anesio, A., Telling, J., Stibal, M., Tranter, M., & Davis, S. 2013. Viral impacts on bacterial communities in Arctic cryoconite. *Environmental Research Letters*, 8, 45021–45029.

Bellas, C.M., Anesio, A.M., & Barker, G. 2015. Analysis of virus genomes from glacial environments reveals novel virus groups with unusual host interactions. *Frontiers in Microbiology*, 6, 656.

Bendia, A.G., Moreira, J.C.F., Ferreira, J.C.N. Romano, R.G., Ferreira, I.G.C., Franco, D.C. et al. 2023. Insights into Antarctic microbiomes: Diversity patterns for terrestrial and marine habitats. *Anais Da Academia Brasileira De Ciencias*, 95(suppl 3), e20211442.

Bhatia, R., Ullah, S., Hoque, M., Ahmad, I., Bhatt, A., Yang, Y.-H., & Bhatia, S. 2020. Psychrophiles: A source of cold-adapted enzymes for energy efficient biotechnological industrial processes. *Journal of Environmental Chemical Engineering*, 9.

Borzęcka, J., Suchodolski, J., Dudek, B., Matyaszczyk, L., Spychała, K., & Ogórek, R. 2022. The first comprehensive biodiversity study of culturable fungal communities inhabiting cryoconite holes in the Werenskiold Glacier on Spitsbergen (Svalbard Archipelago, Arctic). *Biology*, 11(8), 1224.

Buda, J., Łokas, E., Pietryka, M., Richter, D., Magowski, W., Iakovenko, N.S. et al. 2020. Biotope and biocenosis of cryoconite hole ecosystems on Ecology Glacier in the maritime Antarctic. *The Science of the Total Environment*, 724, 138112.

Buda, J., Poniecka, E., Rozwalak, P., Ambrosini, R., Bagshaw, E., Franzetti, A. et al. 2021. Is oxygenation related to the decomposition of organic matter in cryoconite holes? *Ecosystems*, 25.

Cameron, K.A., Hodson, A.J., & Osborn, A.M. 2012. Structure and diversity of bacterial, eukaryotic and archaeal communities in glacial cryoconite holes from the Arctic and the Antarctic. *FEMS Microbiology Ecology*, 82(2), 254–267.

Cameron, K., Hodson, A., & Osborn, A. 2012. Carbon and nitrogen biogeochemical cycling potentials of supraglacial cryoconite communities. *Polar Biology*, 35, 1375–1393.

Christner, B., Kvitko, B., & Reeve, J. 2003. Molecular identification of bacteria and Eukarya inhabiting an Antarctic cryoconite hole. *Extremophiles: Life under Extreme Conditions*, 7, 177–183.

Clarke, A., Johnston, N.M., Murphy, E.J., & Rogers, A.D. 2007. Introduction. Antarctic ecology from genes to ecosystems: The impact of climate change and the importance of scale. *Philosophical Transactions of the Royal Society B: Biological Sciences*, 362(1477), 5–9.

Convey, P., & Biersma, E. 2022. *Antarctic Ecosystems*. Elsevier.

Cook, J., Hodson, A., Gardner, A., Flanner, M., Tedstone, A., Williamson, C. et al. (2017). Quantifying bioalbedo: A new physically-based model and critique of empirical methods for characterizing biological influence on ice and snow albedo. In: *The Cryosphere Discussions*. Copernicus Publications, pp. 1–29.

Cook, J.M., Edwards, A., Takeuchi, N., & Irvine-Fynn, T. 2015. Cryoconite: The dark biological secret of the cryosphere. *Progress in Physical Geography*, 40.

Darcy, J., Gendron, E., Sommers, P., Porazinska, D., & Schmidt, S. 2018. Island biogeography of cryoconite hole bacteria in Antarctica's Taylor Valley and around the world. *Frontiers in Ecology and Evolution*, 6.

De Maayer, P., Anderson, D., Cary, C., & Cowan, D.A. 2014. Some like it cold: Understanding the survival strategies of psychrophiles. *EMBO Reports*, 15(5), 508–517.

Deming, J.W., & Young, J. 2017. The role of exopolysaccharides in microbial adaptation to cold habitats. In: *Psychrophiles: From Biodiversity to Biotechnology*. Second Edition. Springer Nature, pp. 259–284.

Di Mauro, B., Baccolo, G., Garzonio, R., Giardino, C., Massabò, D., Piazzalunga, A. et al. 2017. Impact of impurities and cryoconite on the optical properties of the Morteratsch Glacier (Swiss Alps). *The Cryosphere*, 11, 2393–2409.

Doytchinov, V.V., & Dimov, S.G. 2022. Microbial community composition of the antarctic ecosystems: Review of the bacteria, fungi, and archaea identified through an NGS-based metagenomics approach. *Life* (Basel, Switzerland), 12(6), 916.

Edwards, A., Anesio, A.M., Rassner, S.M., Sattler, B., Hubbard, B., Perkins, W.T. et al. 2011. Possible interactions between bacterial diversity, microbial activity, and supraglacial hydrology of cryoconite holes in Svalbard. *The ISME Journal*, 5(1), 150–160.

Edwards, A., Douglas, B., Anesio, A., Rassner, S., Irvine-Fynn, T., Sattler, B., & Griffith, G.W. 2013. A distinctive fungal community inhabiting cryoconite holes on glaciers in Svalbard. *Fungal Ecology*, 6, 168–176.

Edwards, A., Rassner, S., Anesio, A., Worgan, H., Irvine-Fynn, T., Williams, H. et al. 2013. Contrasts between the cryoconite and ice-marginal bacterial communities of Svalbard glaciers. *Polar Research*, 32, 19468.

Esclapez, J., Bautista, V., Torregrosa-Crespo, J., Vegara, A., Camacho, M., Pire, C. et al. 2018. Extremophile Enzymes and Biotechnology. In: *Extremophiles*. doi: 10.1201/9781315154695-11.

Ferrario, C., Pittino, F., Tagliaferri, I., Gandolfi, I., Bestetti, G., Azzoni, R.S. et al. 2017. Bacteria contribute to pesticide degradation in cryoconite holes in an Alpine glacier. *Environmental Pollution* (*Barking, Essex: 1987*), 230, 919–926.

Foreman, C., Sattler, B., Mikucki, J., Porazinska, D, & Priscu, J. 2007. Metabolic activity and diversity of cryoconites in the Taylor Valley, Antarctica. *J. Geophys. Res.*, 112, 4–32.

Fountain, A., Tranter, M., Nylen, T., Lewis, K., & Mueller, D. 2004. Evolution of cryoconite holes and their contribution to meltwater runoff from glaciers in the McMurdo Dry Valleys, Antarctica. *Journal of Glaciology*, 50.

García-López, E., Alcázar, P., Postigo, M., & Cid, C. 2016. The effect of climate change on microbial communities from glaciers. In: *Glaciers: Formation, Climate Change, and Their Effects*. Nova Science Publishers.

Godin-Roulling, A., Schmidpeter, P.A.M., Schmid, F.X., & Feller, G. 2015. Functional adaptations of the bacterial chaperone trigger factor to extreme environmental temperatures. *Environmental Microbiology*, 17(7), 2407–2420.

González-Herrero, S., & Vasallo, F. 2019. Antarctic climates. In: *Reference Modules in Earth Systems and Environmental Sciences*. doi: 10.1016/B978-0-12-409548-9.11876-7.

Guan, N., & Liu, L. 2020. Microbial response to acid stress: Mechanisms and applications. *Applied Microbiology and Biotechnology*, 104(1), 51–65.

Hall, R.J., Whelan, F.J., McInerney, J.O., Ou, Y., & Domingo-Sananes, M.R. 2020. Horizontal gene transfer as a source of conflict and cooperation in prokaryotes. *Frontiers in Microbiology*, 11, 1569.

Hamid, B., & Mohiddin, F.A. 2018. Cold-active enzymes in food processing. In: *Enzymes in Food Technology*. Springer, Singapore, pp. 383–400.

Hamid, B. 2021. Potent biotechnological applications of psychrozymes. In: *Microbiomes and the Global Climate Change*. Springer, Singapore, pp. 349–364.

Hamid, B., Bashir, Z., Yatoo, A.M., Mohiddin, F., Majeed, N., Bansal, M. et al. 2022. Cold-active enzymes and their potential industrial applications: A review. *Molecules*, 27(18), 5885.

Hassan, N., Anesio, A.M., Rafiq, M., Holtvoeth, J., Bull, I., Haleem, A. et al. 2020. Temperature driven membrane lipid adaptation in glacial psychrophilic bacteria. *Frontiers in Microbiology*, 11, 824.

Hassan, S., Mushtaq, M., Ganiee, S.A., Zaman, M., Yaseen, A., Shah, A.J., & Ganai, B.A. 2024. Microbial oases in the ice: A state-of-the-art review on cryoconite holes as diversity hotspots and their scientific connotations. *Environmental Research*, 252(Pt 2), 118963.

Hauptmann, A., Sicheritz-Ponten, T., Cameron, K., Baelum, J., Plichta, D., Dalgaard, M., & Stibal, M. 2017. Contamination of the Arctic reflected in microbial metagenomes from the Greenland ice sheet. *Environmental Research Letters*, 12, 074019.

Hodson, A., Anesio, A., Tranter, M., Fountain, A., Osborn, A., Priscu, J. et al. 2008. Glacial ecosystems. *Ecological Monographs*, 78, 41–67.

Jiang, X., Van Horn, D.J., Okie, J.G., Buelow, H.N., Schwartz, E., Colman, D.R. et al. 2022. Limits to the three domains of life: Lessons from community assembly along an Antarctic salinity gradient. *Extremophiles: Life Under Extreme Conditions*, 26(1), 15.

Kaczmarek, Ł., Jakubowska, N., Celewicz, S., & Zawierucha, K. 2015. The microorganisms of cryoconite holes (algae, Archaea, bacteria, cyanobacteria, fungi, and Protista): A review. *Polar Record*, 52, 1–28.

Kalińska, E., Lamsters, K., Karušs, J., Krievāns, M., Rečs, A., & Meija, R. 2017. Fine-grained quartz from cryoconite holes of the Russell Glacier, southwest Greenland: A scanning electron microscopy study. *Baltica*, 30, 63–73.

Kang, S., Zhang, Y., Qian, Y., & Wang, H. 2020. A review of black carbon in snow and ice and its impact on the cryosphere. *Earth-Science Reviews*, 210, 103346.

Kuddus, M., Roohi, Bano, N., Sheik, G.B., Joseph, B., Hamid, B., Raveendran Sindhu, & Aravind Madhavan 2024. Cold-active microbial enzymes and their biotechnological applications. *Microbial Biotechnology*, 17, e14467.

Kumari, M., Padhi, S., Sharma, S., Phukon, L.C., Singh, S.P., & Rai, A.K. 2021. Biotechnological potential of psychrophilic microorganisms as the source of cold-active enzymes in food processing applications. *3 Biotech*, 11(11), 479.

Langford, H., Hodson, A., Banwart, S., & Bøggild, C. 2010. The microstructure and biogeochemistry of Arctic cryoconite granules. *Annals of Glaciology*, 51(56), 87–94.

Langford, H., Irvine-Fynn, T., Edwards, A., Banwart, S., & Hodson, A. 2014. A spatial investigation of the environmental controls over cryoconite aggregation on Longyearbreen glacier, Svalbard. *Biogeosciences*, 11, 5365–5380.

Lewandowski, M., Kusiak, M., Werner, T., Barzycka, B., Laska, M., Nawrot, A. et al. 2020. Seeking the sources of dust: Geochemical and magnetic studies on "Cryodust" in glacial cores from Southern Spitsbergen (Svalbard, Norway). *Atmosphere*, 11, 1325.

Łokas, E., Zawierucha, K., Cwanek, A., Szufa, K., Gaca, P., Mietelski, J.W., & Tomankiewicz, E. 2018. The sources of high airborne radioactivity in cryoconite holes from the Caucasus (Georgia). *Scientific Reports*, 8(1), 10802.

Lyons, W., Welch, K., Levy, J., Fountain, A., & Mcknight, D. 2016. Solute and sediment fluxes from rivers and streams in the McMurdo Dry Valleys, Antarctica. In: *Source-to-Sink Fluxes in Undisturbed Cold Environments*. Cambridge University Press, pp. 260–272.

MacDonell, S., & Fitzsimons, S. 2008. The formation and hydrological significance of cryoconite holes. *Progress in Physical Geography*, 32, 595–610.

MacDonell, S., Sharp, M., & Fitzsimons, S. 2016. Cryoconite hole connectivity on the Wright Lower Glacier, McMurdo Dry Valleys, Antarctica. *Journal of Glaciology*, 1, 1–11.

Mallet, M., Humphries, R., Fiddes, S., Alexander, S., Altieri, K., Angot, H. et al. 2023. Untangling the influence of Antarctic and Southern Ocean life on clouds. *Elementa: Science of the Anthropocene*, 11.

Margesin, R., & Collins, T. 2019. Microbial ecology of the cryosphere (glacial and permafrost habitats): Current knowledge. *Applied Microbiology and Biotechnology*, 103(6), 2537–2549.

Margesin, R., Zacke, G., & Schinner, F. 2002. Characterization of heterotrophic microorganisms in Alpine Glacier cryoconite. *Arctic, Antarctic, and Alpine Research*, 34, 88.

Millar, J., Bagshaw, E.A., Edwards, A. Poniecka, E., & Jungblut, A. 2021. Polar cryoconite associated microbiota is dominated by hemispheric specialist genera. *Frontiers in Microbiology*, 12.

Murakami, T., Takeuchi, N., Mori, H., Hirose, Y., Edwards, A., Irvine-Fynn, T. et al. 2022. Metagenomics reveals global-scale contrasts in nitrogen cycling and cyanobacterial light-harvesting mechanisms in glacier cryoconite. *Microbiome*, 10(1), 50.

Novotna Jaromerska, T., Ambrosini, R., Richter, D., Pietryka, M., Niedzielski, P., Souza-Kasprzyk, J. et al. 2022. *Spatio-temporal Changes in Cryoconite Community, Isotopic, and Elemental Composition over the Ablation Season of an Alpine Glacier*. doi:10.5194/bg-2022-198.

Núñez-Montero, K., & Barrientos, L. 2018. Advances in antarctic research for antimicrobial discovery: A comprehensive narrative review of bacteria from antarctic environments as potential sources of novel antibiotic compounds against human pathogens and microorganisms of industrial importance. *Antibiotics*, 7(4), 90.

Obbels, D., Verleyen, E., Mano, M.-J., Namsaraev, Z., Sweetlove, M., Tytgat, B. et al. 2016. Bacterial and eukaryotic biodiversity patterns in terrestrial and aquatic habitats in the

Sør Rondane Mountains, Dronning Maud Land, East Antarctica. *FEMS Microbiology Ecology*, 92(6), fiw041.

Onuma, Y., Fujita, K., Takeuchi, N., Niwano, M., & Aoki, T. 2023. *Modelling the Development and Decay of Cryoconite Holes in Northwest Greenland*. doi:10.5194/tc-17-3309-2023.

Park, C.-J., & Seo, Y.-S. 2015. Heat shock proteins: A review of the molecular chaperones for plant immunity. *The Plant Pathology Journal*, 31(4), 323–333.

Parvizpour, S., Hussin, N., Shamsir, M.S., & Razmara, J. 2021. Psychrophilic enzymes: Structural adaptation, pharmaceutical, and industrial applications. *Applied Microbiology and Biotechnology*, 105(3), 899–907.

Pearce, D. 2012. Extremophiles in Antarctica: Life at low temperatures In: *Adaptation of Microbial Life to Environmental Extremes*. Research Gate, pp. 87–118.

Pittino, F., Ambrosini, R., Seeger, M., Azzoni, R.S., Diolaiuti, G., Alviz Gazitua, P., & Franzetti, A. 2023. Geographical variability of bacterial communities of cryoconite holes of Andean glaciers. *Scientific Reports*, 13(1), 2633.

Poniecka, E.A., Bagshaw, E.A., Sass, H., Segar, A., Webster, G., Williamson, C. et al. 2020.. Physiological capabilities of cryoconite hole microorganisms. *Frontiers in Microbiology*, 11, 1783.

Porazinska, D., Fountain, A., Nylen, T., Tranter, M., Virginia, R., & Wall, D. 2004. The biodiversity and biogeochemistry of cryoconite holes from McMurdo Dry Valley Glaciers, Antarctica. *Arctic, Antarctic and Alpine Research*, 36, 84–91.

Ramírez, C.F., Cavieres, L.A., Sanhueza, C., Vallejos, V., Gómez-Espinoza, O., Bravo, L.A. et al. 2024. Ecophysiology of antarctic vascular plants: An update on the extreme environment resistance mechanisms and their importance in facing climate change. *Plants* (Basel, Switzerland), 13(3), 449.

Rapuano, R., & Graziano, G. 2022. Some clues about enzymes from psychrophilic microorganisms. *Microorganisms*, 10(6), 1161.

Roberts, M.F. 2005. Organic compatible solutes of halotolerant and halophilic microorganisms. *Saline Systems*, 1, 5.

Roux, S., Adriaenssens, E.M., Dutilh, B.E., Koonin, E.V., Kropinski, A.M., Krupovic, M. Kuhn et al. 2019. Minimum Information about an Uncultivated Virus Genome (MIUViG). *Nature Biotechnology*, 37(1), 29–37.

Rozwalak, P., Podkowa, P., Buda, J., Niedzielski, P., Kawecki, S., Ambrosini, R. et al. 2022. Cryoconite: From minerals and organic matter to bioengineered sediments on glaciers' surfaces. *The Science of the Total Environment*, 807(Pt 2), 150874.

Sahay, S., Hamid, B., Singh, P., Ranjan, K., Chauhan, D., Rana, R.S., & Chaurse, V.K. 2013. Evaluation of pectinolytic activities for oenological uses from psychrotrophic yeasts. *Letters in Applied Microbiology*, 57(2), 115–121.

Sajjad, W., Din, G., Rafiq, M., Iqbal, A., Khan, S., Zada, S. et al. 2020. Pigment production by cold-adapted bacteria and fungi: Colorful tale of cryosphere with wide range applications. *Extremophiles*, 24(4), 447–473.

Samui, G., Antony, R., & Thamban, M. 2018. Chemical characteristics of hydrologically distinct cryoconite holes in coastal Antarctica. *Annals of Glaciology*, 59, 1–8.

Sanyal, A., Antony, R., Samui, G., & Thamban, M. 2018. Microbial communities and their potential for degradation of dissolved organic carbon in cryoconite hole environments of Himalaya and Antarctica. *Microbiological Research*, 208, 32–42.

Schmidt, S., Johnson, B., Solon, A., Sommers, P., Darcy, J., Vincent, K. et al. 2022. Microbial biogeochemistry and phosphorus limitation in cryoconite holes on glaciers across the Taylor Valley, McMurdo Dry Valleys, Antarctica. *Biogeochemistry*, 158.

Segawa, T., Yonezawa, T., Edwards, A., Akiyoshi, A., Tanaka, S., Uetake, J. et al. 2017. Biogeography of cryoconite forming cyanobacteria on polar and Asian glaciers. *Journal of Biogeography*, 44.

Segawa, T., Takeuchi, N., Mori, H., Rathnayake, R.M.L.D., Li, Z., Akiyoshi, A. et al. 2020. Redox stratification within cryoconite granules influences the nitrogen cycle on glaciers. *FEMS Microbiology Ecology*, 96(11), fiaa199.

Siddiqui, K., Williams, T., Wilkins, D., Yau, S., Allen, M., Brown, M. et al. 2012. Psychrophiles. *Annual Review of Earth and Planetary Sciences*, 41.

Singh, P., Singh, S.M., & Dhakephalkar, P. 2014. Diversity, cold active enzymes, and adaptation strategies of bacteria inhabiting glacier cryoconite holes of High Arctic. *Extremophiles: Life Under Extreme Conditions*, 18(2), 229–242.

Singh, P., Roy, U., & Tsuji, M. 2016. Characterisation of yeast and filamentous fungi from Brøggerbreen glaciers, Svalbard. *Polar Record*, 1, 1–8.

Smeele, Z., Ainley, D., & Varsani, A. 2017. Viruses associated with Antarctic wildlife: From serology-based detection to identification of genomes using high throughput sequencing. *Virus Research*, 243.

Sohm, J.A., Niederberger, T.D., Parker, A.E., Tirindelli, J., Gunderson, T., Cary, S.C. et al. 2020. Microbial mats of the McMurdo Dry Valleys, Antarctica: Oases of biological activity in a very cold desert. *Frontiers in Microbiology*, 11, 537960.

Sommers, P., Darcy, J.L., Porazinska, D.L., Gendron, E.M.S., Fountain, A.G., Zamora, F. et al. 2019. Comparison of microbial communities in the sediments and water columns of frozen cryoconite holes in the McMurdo Dry Valleys, Antarctica. *Frontiers in Microbiology*, 10, 65.

Sommers, P., Fontenele, R.S., Kringen, T., Kraberger, S., Porazinska, D.L., Darcy, J.L. et al. 2019. Single-stranded DNA viruses in antarctic cryoconite holes. *Viruses*, 11(11), 1022.

Sommers, P., Porazinska, D.L., Darcy, J.L., Gendron, E.M.S., Vimercati, L., Solon, A.J. et al. 2020. Microbial species – Area relationships in antarctic cryoconite holes depend on productivity. *Microorganisms*, 8(11), 1747.

Stanish, L., Bagshaw, E., Mcknight, D., Fountain, A., & Tranter, M. 2013. Environmental factors influencing diatom communities in antarctic cryoconite holes. *Environmental Research Letters*, 8, 5006.

Stibal, M., Tranter, M., Benning, L.G., & Rehák, J. 2008. Microbial primary production on an Arctic glacier is insignificant in comparison with allochthonous organic carbon input. *Environmental Microbiology*, 10(8), 2172–2178.

Stibal, M., Schostag, M., Cameron, K.A., Hansen, L.H., Chandler, D.M., Wadham, J.L. et al. 2015. Different bulk and active bacterial communities in cryoconite from the margin and interior of the Greenland ice sheet. *Environmental Microbiology Reports*, 7(2), 293–300.

Stibal, M., Box, J., Cameron, K., Langen, P., Yallop, M., Mottram, R. et al. 2017. Algae drive enhanced darkening of bare ice on the Greenland ice sheet: Algae drive Greenland bare ice darkening. *Geophysical Research Letters*, 44.

Stibal, M., Bradley, J.A., Edwards, A., Hotaling, S., Zawierucha, K., Rosvold, J. et al. 2020. Glacial ecosystems are essential to understanding biodiversity responses to glacier retreat. *Nature Ecology and Evolution*, 4(5), 686–687.

Stokes, C.R., Abram, N.J., Bentley, M.J., Edwards, T.L., England, M.H., Foppert, A. et al. 2022. Response of the East Antarctic Ice Sheet to past and future climate change. *Nature*, 608(7922), 275–286.

Tarfeen, N., Nisa, K.U., Hamid, B., Bashir, Z., Yatoo, A.M., Zakir, A. et al. 2022. Microbial remediation: A promising tool for reclamation of contaminated sites with special emphasis on heavy metal and pesticide pollution: A review. *Processes*, 10(7), 1358.

Telling, J., Anesio, A., Tranter, M., Stibal, M., Hawkings, J., Irvine-Fynn, T. et al. 2012. Controls on the autochthonous production and respiration of organic matter in cryoconite holes on High Arctic glaciers. *Journal of Geophysical Research*, 117.

Uetake, J., Tanaka, S., Segawa, T., Takeuchi, N., Nagatsuka, N., Motoyama, H. et al. 2016. Microbial community variation in cryoconite granules on Qaanaaq Glacier, NW Greenland. *FEMS Microbiology Ecology*, 92, fiw127.

Varliero, G., Lebre, P.H., Adams, B., Chown, S.L. Convey, P., Dennis, P.G. et al. 2024. Biogeographic survey of soil bacterial communities across Antarctica. *Microbiome*, 12(1), 9.

Wang, M., Tian, J., Xiang, M., & Liu, X. 2017. Living strategy of cold-adapted fungi with the reference to several representative species. *Mycology*, 8(3), 178–188.

Webster-Brown, J.G., Hawes, I., Jungblut, A.D., Wood, S.A., & Christenson, H.K. 2015. The effects of entombment on water chemistry and bacterial assemblages in closed cryoconite holes on Antarctic glaciers. *FEMS Microbiology Ecology*, 91(12), fiv144.

Weisleitner, K., Perras, A.K., Unterberger, S.H., Moissl-Eichinger, C., Andersen, D.T., & Sattler, B. 2020. Cryoconite hole location in East-Antarctic Untersee Oasis shapes physical and biological diversity. *Frontiers in Microbiology*, 11, 1165.

Wejnerowski, Ł., Poniecka, E., Buda, J., Klimaszyk, P., Piasecka, A., Dziuba, M.K. et al. 2023. Empirical testing of cryoconite granulation: Role of cyanobacteria in the formation of key biogenic structure darkening glaciers in polar regions. *Journal of Phycology*, 59(5), 939–949.

Wu, Z., Kan, F., She, Y., & Walker, V. 2012. Biofilm, ice recrystallization inhibition, and freeze-thaw protection in an epiphyte community. *Prikladnaia Biokhimiia i Mikrobiologiia*, 48, 403–410.

Xu, Y., Simpson, A., Eyles, N., & Simpson, M. 2010. Sources and molecular composition of cryoconite organic matter from the Athabasca Glacier, Canadian Rocky Mountains. *Organic Geochemistry*, 41, 177–186.

Yin, W., Wang, Y., Liu, L., & He, J. (2019). Biofilms: The microbial "Protective Clothing" in extreme environments. *International Journal of Molecular Sciences*, 20(14), 3423.

Zandi, P., & Schnug, E. 2022. Reactive oxygen species, antioxidant responses, and implications from a microbial modulation perspective. *Biology*, 11(2), 155.

Zawierucha, K., Baccolo, G., Di Mauro, B., Nawrot, A., Szczuciński, W., & Kalińska, E. 2019. Micromorphological features of mineral matter from cryoconite holes on Arctic (Svalbard) and alpine (the Alps, the Caucasus) glaciers. *Polar Science*, 100482.

Zawierucha, K., Buda, J., Azzoni, R., Niśkiewicz, M., Franzetti, A., & Ambrosini, R. 2019. Water bears dominated cryoconite hole ecosystems: Densities, habitat preferences, and physiological adaptations of Tardigrada on an alpine glacier. *Aquatic Ecology*, 53.

Zwerschke, N., Sands, C.J., Roman-Gonzalez, A., Barnes, D.K.A., Guzzi, A., Jenkins, S. et al. 2022. Quantification of blue carbon pathways contributing to negative feedback on climate change following glacier retreat in West Antarctic fjords. *Global Change Biology*, 28(1), 8–20.

CHAPTER

8

Psychrophiles and Psychrozymes: Innovative and Potential Resources for Commercial Uses

Neesa Majeed[1*], Mawish Mahroof[2], Burhan Hamid[2], Ashiq Hussain Magrey[3] and Rihab Djebaili[4]

[1] Government Medical College, Baramulla – 193101, J&K, India
[2] Centre of Research for Development, University of Kashmir, Srinagar – 190006, India
[3] Centre for Scientific Research and Development, Peoples University, Bhopal, India
[4] Department of Life, Health and Environmental Sciences, University of L'Aquila, Via Vetoio, Coppito, 67100 L'Aquila, Italy

1. Introduction

Extreme environments can be found throughout the earth's biosphere, some of which are marked by extremely high or low temperatures (Shu et al., 2022). Microorganisms can be isolated from a variety of habitats, including those with extreme temperatures, high salinities, low water, and fluctuating pH levels. They are ubiquitous in Mother Nature (Waditee et al., 2022). Because they live in such hostile surroundings, microbes have evolved adaptive strategies to survive in harsh situations. Extremophiles are microorganisms that survive and multiply in these harsh circumstances; polyextremophiles are those that can endure and multiply in a variety of harsh environments (Alam & Tiwary, 2023). Cold-adapted microbes, including bacteria, fungus, archaea, protists, and unicellular algae, face a variety of challenges, and they have evolved a variety of physiological and molecular defense mechanisms (Chatragadda et al., 2021). Therefore, in terms of diversity, biomass, and dispersion, psychrophiles are the most commonly detected microorganisms. It has been documented that microorganisms can survive in harsh, frigid conditions; these organisms are known as psychrophiles, and they live in conditions like the Siberian permafrost,

*Corresponding author: loneneesa@gmail.com

DOI: 10.1201/9781003717003-8

which is thought to be unique due to its permanently icy ground, scarcity of organic matter, low water activity, and other characteristics (Bhatt, 2022). The capacity of psychrophiles to endure and multiply in frigid conditions confers upon these microorganisms a number of unique traits that make them a prime candidate for industrial applications aimed at producing various pharmaceutical, medicinal, agricultural, and biotechnological products (Liu et al., 2023). In this sense, strengthening the enzymatic reactions at low temperatures and the cooling or heating phase of the processes are crucial requirements for these applications. Previous research has demonstrated that psychrophiles may break down a variety of polymeric materials and can also produce a wide range of enzymes, including cellulases, lipase, dehalogenases, amylases, and oxidases (Hamid et al., 2022). The scientific community has been paying more interest in psychrozymes in order to better understand their molecular mechanisms for cold adaptation and investigate their potential for increased industrial use (Zhu et al., 2020). Since their reactions are slower at lower temperatures, psychrophiles create enzymes with a particular activity that is roughly ten times higher in this temperature range (Sharma et al., 2022). Because of the low stability of these proteins and their unfolding and inactivation at mild temperatures, the maximal activity of these enzymes is moved towards the low-temperature region. Because destabilization of the active site or even the entire protein permits the catalytic center to be more flexible at temperatures that tend to freeze molecular movements and biological activities at low temperatures, such high activity at low temperatures is possible (Bhatia et al., 2021). Psychrophilic enzymes increase the structural elements' flexibility in the catalytic cycle, which makes the activity substantially more heat-labile (Collins & Feller, 2023). For example, during the production of many foods, beverages, fine chemicals, and pharmaceuticals, low temperatures are necessary because they prevent contamination issues or involve volatile, heat-sensitive components that react unfavorably at high temperatures (Rabbani et al., 2023).

2. Habitat of Psychrophiles

The Greek words 'psychros' (cold) and 'philos' (loving, or cold-loving) are the source of the name 'psychrophile' (Kuddus et al., 2024). Other names for them include 'cryophiles' and 'rhigophiles'. Schmidt was the first to utilize both of these Greek-derived terms, which shared the same basic meaning of 'psychrophiles'. There are two categories into which psychophilic species have been divided.

- The ideal growth temperature for obligatory psychrophiles is around 20°C (Gracia et al., 2022).
- A temperature of more than 20°C is ideal for facultative psychrophile growth.

The two main groups of cold-adapted microorganisms are psychrophiles, which prefer the cold and grow best at temperatures below 15°C (Pathania et al., 2022), and psychrotrophs, which can withstand the cold and grow best at temperatures between 20 and 25°C (Furhan, 2020). A vast array of interconnected chemical reactions or metabolic pathways, such as photosynthesis, chemoautotrophy, and heterotrophy, provide the means of life for psychrophiles, or cold-loving microorganisms (Bhanu et al., 2024), which are classified into three realms of existence: bacteria, eukaryotes, and archaea. About 85% of our world is made up of various cold habitats, both natural and man-made (Bansode & Banpatte, 2021). A wide variety of psychrophilic microorganisms, mostly viruses, bacteria, fungi, yeasts, and archaea, grow in these habitats (Karadayi et al., 2021). Cold-active enzymes are thought to mostly originate from psychrophilic microbes, particularly bacteria. Approximately three-fourths of all organisms on the Earth are psychrophiles (Hamid et al., 2022), and they are commonly found in high-altitude soils, mountains, glacial provinces, fresh and marine deep-sea waters, and the Antarctic (Bhat et al., 2024). The majority of Earth's biosphere, which includes permafrost, non-polar zones, the Arctic and Antarctica, and deep oceans, is frigid. As low as 32° has been reported for life, and that too for bacteria with a high metabolic rate. There are several reports on psychrophile isolation from extremely low-temperature environments. *Planococcus halocryophilus* Orl (Rizzo & Lo, 2022), a bacterium isolated from Arctic permafrost, is now being cultured at the lowest temperature ever recorded –15°C which is also the lowest growth temperature that has been verified. These low-temperature environments have been found to be home to a wide variety of cold-loving microorganisms. These include gram-negative bacteria such as *Pseudoalteromonas* sp., *Moraxella* sp., *Psychrobacter* sp., *Polaromonas* sp., *Psychroflexus* sp., *Polaribacter* sp., *Moritella* sp., *Vibrio* sp., and *Pseudomonas* sp.; some are gram-positive bacteria such as *Arthrobacter* sp., *Bacillus* sp., and *Micrococcus* sp. (Kanekar & Kanekar, 2022); some archaea also belong to psychrophiles such as *Methanogenium* sp., *Methanococcoides* sp., and *Halorubrum* sp.; yeasts such as *Candida* sp. and *Cryptococcus* sp., and fungi such as *Penicillium* sp. and *Cladosporium* sp. have also been isolated from these low temperatures (Anwar et al., 2020). In light of their prospective uses in a variety of biotechnological contexts, the molecular mechanisms of psychrophilic enzymes have gotten adequate attention (Ramasamy et al., 2023). The basis for designing effective catalysts with desired properties for particular applications is provided by this understanding. Studying and examining the ways by which psychrophilic species adapt to the cold is therefore vitally essential. Numerous adaptation traits are needed for living in cold climates in almost every area of cell structure and function. The harsh physicochemical limitations that cold-adapted organisms must overcome make their environment unsuitable for survival. Enzyme activity, biochemical reaction rate, cellular membrane fluidity, molecular diffusion rate, water viscosity, and gas solubility all decrease and increase with decreasing temperature (Akindolire et al., 2022). Moreover, proteins are denatured and hydrogen bonds are reinforced. Furthermore, diverse ecological circumstances and restricting variables, such as excessive salinity, low nutrition levels, oxidative stress, freeze-thaw cycles, and low water activity, are the main characteristics of cold habitats (Bharali et al., 2024). As a result, the organisms have evolved their genotypic and phenotypic traits to let them grow in these incredibly hostile settings by overcoming the main obstacles posed by the frigid surroundings. Numerous investigations have been conducted to examine the tactics used by psychrophilic creatures to adapt to the cold. Some of their cellular constituents, such as the membrane, the machinery involved in protein synthesis, the systems that produce energy, and other physicochemical aspects of the cell structure, have undergone modifications (Sharma et al., 2022). To boost their activity at low temperatures, the psychrophilic enzymes have also undergone a variety of structural and molecular modifications. In fact, these enzymes' remarkable structural flexibility enhances their capacity to go through conformational changes during catalysis, which is essentially the source of their cold activity (Angelin & Kavitha, 2022).

3. Cryo-Defense Strategies of Psychrophiles

Psychrophiles possess a diverse array of distinct and versatile characteristics throughout nearly every step of cell construction and function, enabling them to endure harsh and frigid environments (Xu et al., 2024). They are believed to experience cold shock in three stages. Phase I is the "acclimation phase", and it begins as soon as you are exposed to the cold (Rathore et al., 2022). Different cold-shock proteins are created during this phase, and as a result, the membrane's fluidity decreases, slowing the rate of growth of these proteins (Li et al., 2020). The second stage, which is referred to as the "recovery phase," Cells are said to have "cold-adapted" at this phase (Gupta et al., 2023). At this stage, cells start producing proteins again in large amounts and keep growing. Known as the "stationary phase", phase III is characterized by a markedly slower rate of cell development due to the production of non-cold shock proteins. Psychrophiles usually attribute their ability to adjust to such conditions to crafty tactics. The largest availability of psychrophiles favourable to matter cycling is found in the vast biotope volume of the seas (Salwan & Sharma, 2020); nonetheless, psychrophiles and related biomolecules have already found numerous uses. Numerous aspects of their physiology, ecology, and molecular changes have been covered in earlier studies. Life in cold environments requires a broad variety of adaptive features at nearly every cell design and functional level (Suzuki & Ley, 2020). The physicochemical consequences of cold place significant constraints on living organisms. For example, low temperatures increase the viscosity of water, slow down biological processes, lower the rates of molecule diffusion, disrupt weak molecular recognition and interaction, and reinforce hydrogen bonds that preserve inhibitory nucleic acid structures (Grassmann et al., 2024). Additionally, it lessens the fluidity of cellular membrane systems, stabilizes hazardous metabolites, and enhances gas solubility.

4. Commercially Significant Cold-Adapted Enzymes

Psychrophilic microbes produce psychrozymes, which are enzymes resistant to low temperatures and other cold-related stresses. These enzymes are used in the pulp, paper, food, and pharmaceutical industries (Figure 1) (Gupta et al., 2023). Psychrozymes, or cold-adapted enzymes, can withstand temperatures ranging from 10 to 15°C. Psychrozymes are in great demand in businesses due to their capacity to withstand adverse conditions (Kuddus et al., 2014). Since they function best at room temperature, proteins extracted from bacteria that enjoy the cold are being employed more and more to increase the recovery of enzymatic reaction products. Psychrozymes are capable of degrading a variety of polymeric materials, including cellulose, amylose, pectin, β-galactoside, protein, and lipid (Debroy & George, 2023). Psychrozymes have drawn a lot of capital from all over the world due to their great promise. Psychrozymes that are required for industry application include those in the food technology sector (such as β-galactosidase and pectinase), textile product biopolishing, and detergent composition (Kumari et al., 2021). These psychrozymes are also used in several pollution control methods, such as bioremediation (oxidases), and biotransformations (aminotransferases and methylases) (Hamid, 2021).

Among the many benefits offered by psychrozymes is their high specific activity at low temperatures. Other advantages they can provide are energy savings, a decrease in volatile chemicals, contamination prevention, and the simple deactivation of enzymes. Psychrozymes are used by the majority of the food

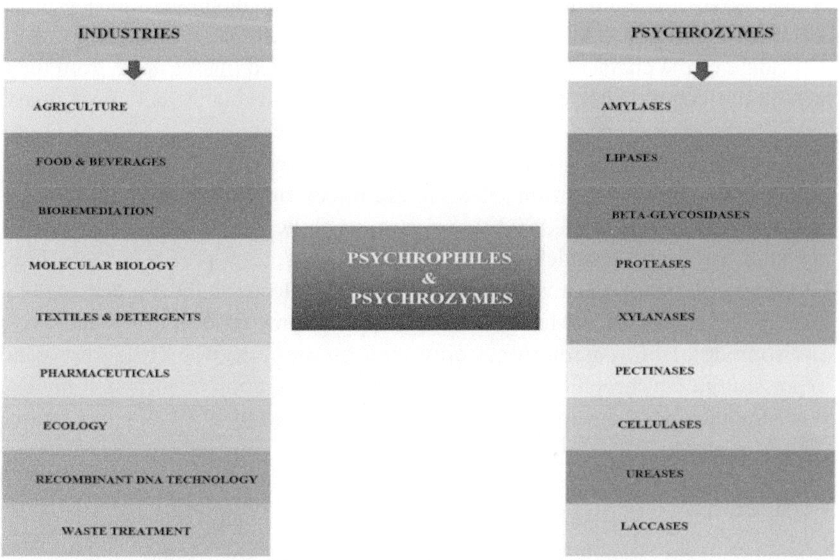

Figure 1: Psychrozymes and their commercial significance.

industry to preserve food quality throughout storage and transit. Psychrozymes are also frequently employed in the textile and detergent industries (Parvizpour et al., 2021). Fruit fluids are also clarified by pectinases and cellulases, while fish skin is removed by proteases. Psychrozymes are used in the food sector as well as low-temperature biodegradation. Bioremediation is the best substitute for physicochemical methods in the removal of solids and wastewater contaminated with hydrocarbons, lipids, and oils. Psychrozyme biodegradation is superior to other established conventional techniques in several ways. It has been found that treating polluted soil with psychrozymes is significantly more economical than storing, traditional incineration, and concentrating techniques (Miri et al., 2022). Most of the food sector favour using cold-adapted enzymes to treat items in order to maintain product quality during shipment and storage. The textile and detergent industries are two more that frequently use enzymes that are adapted to cold temperatures (Kumar et al., 2021).

Lipases

An enzyme called lipase aids in fat digestion by catalyzing the breakdown of triglycerides to release fatty acids and glycerol. Lipases, also known as triacylglycerol acyl hydrolases, are capable of hydrolyzing and producing fatty acid esters in both aqueous and non-aqueous environments (Sood et al., 2023). The canonical/hydralase fold-containing cold-adapted lipases from psychrophilic and psychrotrophic microbes isolated from deep-sea settings, polar regions, the Antarctic, and refrigerated food samples were compared to all other psychrotrophic lipases (Guo et al., 2021). One of the features that sets psychrophilic lipases apart is their structural alteration, which permits significant flexibility and substrate accommodation at low temperatures (Kochhar et al., 2022).

Some of the characteristics that set the cold-adapted lipases apart from their mesophilic counterparts, as demonstrated by the molecular structure of lipase from *Pseudomonas immobilis* and *P. fragis.* CLPs featureless proline residues, arginine residues relative to lysine residues, a weaker hydrophobic core, fewer salt bridges, and fewer aromatic-aromatic interactions than their mesophilic counterparts (Gupta et al., 2023). Since mesophilic and thermophilic lipases are more rigid and less flexible, psychophilic lipase is a preferable selection. Psychrophilic lipase has a wide range of uses in biotechnological processes like detergent formulation, molecular biology applications, food industry additives, environmental bioremediations, biotransformation, and heterologous gene expression in psychrophilic hosts to prevent the generation of inclusion bodies. It is a unique option due to its exceptional catalytic activity at lower temperatures (Vivek et al., 2022).

Proteases

Microorganisms found in the Arctic, frigid desert soil, polar regions, ice permafrost, deep sea, alpine regions, sub-Antarctic sediments, and other colder places of the earth are known to have psychophilic proteases (Hamid et al., 2024; Abirami et al., 2021). *Pedobacter cryoconitis* from glacier ice, *Penicillium chrysogenum* from the cold marine environment (Sarsan et al., 2024), *Pseudomonas* sp. from the deep sea, *Psychrobacter proteolyticus* from Antarctic krill *Euphausia superba,* and *Serratia* sp. from coastal waters are some of the sources of cold-active proteases that have been isolated from various microorganisms (Wang et al., 2021).

The mesophilic analogue alkaline protease from *P. aeruginosa* and the psychrophilic alkaline protease (PAP) from an *Antarctic Pseudomonas* species were compared with a marine protease from *Flavobacterium* YS-80 (Gupta et al., 2023). The psychrophilic alkaline protease was found to have a more flexible Zn^{2+}-Tyrosine-OH bond, which aids in substrate accessibility and preserves activity at low temperatures. In other words, the psychrophilic protease exhibits higher specific activity (4 times higher at 4 and 25°C), a multiple time shorter half-time of denaturation (10 times), and excellent catalytic efficiency at low temperatures (Arcus & Mulholland, 2020). Textiles can be laundered at room temperature, and the use of less toxic chemicals for dehairing hides and skins also saves energy. One disadvantage of cold proteases is their weakness at temperatures higher than 25°C (Yang et al., 2023). Recent studies have examined the molecular flexibility and heat stability of psychrophilic protease. Even now, our inadequate knowledge of psychrophilic protease is a setback (Rahban et al., 2022).

Amylases

Amylase is an enzyme that facilitates the cleavage of internal α-1,4-glycosidic bonds in starch, glycogen, and intermediate hydrolysis products, hence facilitating the creation of dextrin and glucose subunits. Based on their mode of action, amylases have been divided into three categories: α-, β-, and γ-amylases (Shad et al., 2023). 30% of the commercial enzymes produced worldwide are amylases. Bacteria, fungi, mammals, and plants all have amylases (Kumar & Rastogi, 2020). Feller et al. (1998) isolated and expressed the first psychrophilic amylase using the Antarctic bacteria *Alteromonas haloplanktis*. An amylase variant of *B. licheniformis* that shows enhanced enzyme activity at temperatures between 10 and 60°C was patented by Novozymes (Muras et al., 2021). *Pseudoalteromonas haloplanktis* amylase extract exhibits increased specificity, flexibility, and catalytic activity for its mesophilic counterpart (Parrilli et al., 2021). These industrial processes can be catalyzed by cold-active amylases, which reduces the danger of fouling and also helps to save energy. Because of the poor stability caused by oxidizing conditions at low temperatures, psychrophilic α-amylases are less useful for commercial applications (Paul et al., 2021).

Cellulases

Hydrolytic enzymes known as cellulases help break down the polysaccharide cellulose into monosaccharides and shorter polysaccharides, including glucose, and oligosaccharides (Ejaz et al., 2021). Due to their wide range of uses, these enzymes have a high economic value. Cellulases are widely used in the textile, paper, pulp, laundry, food and feed, pharmaceutical, and agricultural industries (Ranjan et al., 2023). They can be used, among other things, to process coffee, soften textiles, remove excess color from fabrics, pulp woody raw materials, bioremediate industrial wastes, and treat phytobezoars.

Protozoa, bacteria, and fungi are the main sources of cellulases. These microbial cellulases are made up of a catalytic module, a linker region, and a module for binding carbohydrates (Shi et al., 2023). Mesophilic cellulases, in contrast to psychrophilic cellulases, have a repeating, easily understood linker region that is often abundant in serine, glycine, proline, valine, and threonine residues (Annapure et al., 2022). In their linker region, they are rich in cysteines, asparagine, serine, glycine, aspartate, and threonine. They differ structurally from mesophilic cellulases, which makes them more versatile and active at low temperatures. These qualities enable them to work in the aforementioned fields and receive larger remuneration.

Pectinases

Plant cell walls include pectin, which is reduced to simpler molecules like galacturonic acids by the enzyme pectinase. Pectin lyases, polygalacturonases, and pectin esterases are the three categories into which they can be divided (Bonnin et al., 2020). The primary challenge confronting the commercial food and beverage industry is that mesophilic proteins typically require heating, which has an adverse effect on the organic qualities and nutritional functions of the finished product (Sun et al., 2022). Using cold-adapted pectinase reduces the thickness of the natural product juice at low temperatures and improves the mash's ability to squeeze, which breaks down the structure of the jam and makes it easier to extract more raw product juice (Gupta et al., 2023). The psychrophilic acidic pectinases enzyme compositions used in contemporary technology are used in retting in the textile industry, degumming fiber crops, and the fermentation, oil extraction, and wastewater treatment processes used to produce high-quality paper (Rehman et al., 2023). At 20°C, the pectinase XPect®, developed by Novozymes, breaks down 1,4-glycosidic bonds in polygalacturonic acid into water-soluble "simpler sugars" (Staack, 2023).

β-Galactosidase (Lactases)

When a sugar moiety in an oligosaccharide is switched from a glycosyl donor to an acceptor, the β-galactosidase (EC 3.2.1.23) enzyme catalyzes the hydrolysis of the β-galactosidic bonds in the oligosaccharides. Microorganisms originating from areas with constant low temperatures possess β-galactosidases that are suited to the cold (Mangiagalli et al., 2021). Maximum catalytic potential of psychrozymes at low temperatures indicates a significant feature of energy saving and desired characteristic of cold-active enzymes (Hamid et al., 2022).

An optimum shared by some cold-adapted enzymes and their thermophilic counterparts, however, is the same. Psychrophilic β-galactosidases have various advantages, including the capacity to catalyze lactose breakdowns at higher temperatures, accelerate the reaction, and reduce the overall cost of the process by utilizing less energy (Adapa et al., 2022). They not only reduce the risk of infection but also improve dairy products' flavor, solubility, sweetness, and digestibility (Amiri et al., 2021). Cold-adapted enzymes include pectinase, which lowers viscosity and turbidity in chilled juices, and β-galactosidase, which extracts lactose from chilled milk (Mutlu-Ingok et al., 2022). Hamid et al. (2013) reported two psychrophilic yeasts producing cold-active β-galactosidase which can be applied in lactose hydrolysis at low temperatures.

Laccases

Laccase (benzenediol: oxygen oxidoreductase, EC 1.10.3.2) is a multi-copper oxidase that is present in fungi, bacteria, and plants (Rajan, 2020). It is a member of the multi-copper oxidases superfamily. Among the fungi that break down wood, such as *Trametes ochracea*, *T. versicolor*, *T. hirsute*, *T. gallica*, and *T. villosa*, laccase is extensively dispersed (Gupta et al., 2023). While extracellular glycoproteins make up the majority of the laccases produced by fungi that degrade wood, intracellular laccases have also been found. Fungal laccases are involved in the morphogenesis, pathogenicity, and breakdown of lignin (Janusz et al., 2020). Bacterial laccase was initially discovered by the plant-root-associated bacterium *Azospirillum lipoferum* (Palaskar et al., 2022).

Most bacterial laccases that have been studied are intracellular, like those found in *A. lipoferum* and *B. subtilis*; extracellular laccases have also been found in *S. cyaneus* (Akram et al., 2022). The three sequentially arranged copper domains that make up the structure of all fungal laccases are identical. Apart from the typical three-domain laccases, little two-domain laccases have also been found in bacteria recently. The enzyme laccase exclusively targets substances that are phenolic. One of the three types of copper that laccases frequently contain gives them their characteristic blue colour (Khatami et al., 2022).

Alkaline Phosphate

The widely used enzyme alkaline phosphatase (AP; EC 3.1.3.1) catalyzes the hydrolytic cleavage of phosphate monoesters in alkaline conditions, releasing inorganic phosphate from a range of phosphate-containing materials (Le-Vinh et al., 2022). In molecular biology, alkaline phosphatase is used to remove pyrophosphate and dNTPs from polymerase chain reaction assays, dephosphorylate 5′-nucleic acid termini prior to polynucleotide kinase 5′-end labelling, and dephosphorylate plasmid DNA before cloning to prevent circularization (Ritte et al., 2008). In terms of catalytic efficiency (k_{cat}/K_m) and turnover number (kcat), it was found that metagenome-derived AP (mAP) performed better than Aps from the Antarctic, prawn, calf intestine, and *Sinorhizobium meliloti* as well as that from *S. meliloti* and *E. coli* (Gupta et al., 2023).

DNA Ligase

An enzyme known as DNA ligase requires either adenosine triphosphate (ATP) or oxidized nicotinamide adenine diphosphate (NAD^+) as a cofactor in order to form a phosphodiester link between the hydroxyl group at the 3′ end of one DNA strand and the 5′ phosphate of the opposing DNA strand (Williamson et al., 2019). This procedure does not change the structure of sugar-phosphate. Using the thermos-lability of a cold-adapted DNA ligase, a novel temperature-sensitive tularemia vaccine was recently produced. Since low temperature promotes base pairing between short sections of complementary nucleotides at the break site, it aids in bridging double-strand breaks with cohesive ends. Lower temperatures reduce ligase activity, but DNA overhangs can still base pair and stay annealed long enough for the ligase to connect them (Schlichting, 2021). Cold-adapted enzymes exhibit greater thermal denaturation susceptibility (k_{cat}/K_m) at low and moderate temperatures in comparison to their mesophilic counterparts (Collins et al., 2007).

Nucleases

There are many applications for nucleases in both industry and research. They are an important class of hydrolases that degrade nucleic acids. Nuclease from psychrophilic organisms will be synthesized and purified to provide an incredibly useful molecular biology tool (Maciejewska et al., 2019). The highly psychrophilic bacteria *Psychromonas ingrahamii,* whose nuclease was discovered and characterized, can grow exponentially at 5°C and thrive at temperatures as low as 12°C (Bruno et al., 2019).

5. Applications of Psychrophilic Enzymes

The enzymes exhibit high activity at low temperatures, therefore the predicted efficiency for the goal activity is provided by their reduced concentration. Therefore, it is possible to arrange cost-effective processes by reducing the quantity of costly enzymes. The capacity to heat the surroundings at the conclusion of a process to inactivate the enzymes' activity is made possible by their heat lability (Chen et al., 2024). Additionally, the utilization of enzymes derived from psychrotolerant organisms can serve as a beneficial means of obtaining catalysts for the effective processing of enzymatic reactions (Kumar et al., 2021). The cold-active enzymes are interesting candidates for a range of biotechnological applications because of these three crucial features. At high temperatures, undesired chemical processes can be inactivated due to the cold-adapted enzymes' heat lability. Because of this characteristic, the enzymes are especially appealing as candidates for biotechnological applications, since heat-sensitive substrate transformation necessitates cold-temperature reactions. Enzymes that like the cold actually reduce the possibility of unfavorable reactions at high temperatures (Socan et al., 2020).

Due to this characteristic, the enzymes are especially appealing as candidates for biotechnological applications, since reactions involving the transformation of heat-sensitive substrates must take place at low temperatures. Due to the necessity of preventing product spoiling and loss of flavour and nutritional value, this feature is especially important to the food and pharmaceutical industries (Firouz et al., 2021). In molecular biology applications, where it's required to deactivate each enzyme after its reaction in turn, cold-active enzymes can also be used. Enzymes that are heat-labile stop DNA from melting at high temperatures due to heat inactivation (Park et al., 2022). The ability of these enzymes to hydrolyze substrates has been shown to be beneficial for uses in industrial cleaning. Manufacturers of detergents have taken advantage of the relationship between low washing temperatures and increased energy use, wherein lowering the washing temperature results in lower electricity usage (Barasanti et al., 2024). These enzymes are widely used in many different industries, such as automated dishwashers, water treatment membrane cleaning, and cleaning brewing equipment, because of their efficient cleaning capabilities in detergents at low temperatures.

Owing to their exceptional qualities, psychrozymes have found extensive application in a variety of biotechnological domains. The sectors include baking, bioremediation, biotransformations, food industries, detergent formulation, biopolishing, and textile industries, among many more (Ghosh et al., 2021). Psychrophiles have surfactants in their cell membranes, which give them a special stability that allows them to live in frigid climates. This skill may be useful in a variety of fields, including medicine and pharmaceutical formulations (Dhole et al., 2024). As an illustration, surfactants have lately been utilized at low temperatures to dry, freeze, and preserve goods related to the food, pharmaceutical, and biological sectors.

Application of Psychrozymes in Food Industry

The use of mesophilic enzymes in the brewing and food sectors is not new, but replacing these enzymes with their psychrophilic counterparts is showing potentially positive outcomes (Qian et al., 2023). Because psychrozymes are very energy-efficient and contribute to energy conservation, their use in the food industry is proving to be a useful asset (Hamid et al., 2022). The dairy industry uses lactases; according to recent research, a comparative analysis of several commercially available β-galactosidases revealed that these enzymes are efficiently active in milk at lower temperatures that cause lactose hydrolysis. The production of commercially produced lactose-free products could be made more efficient and affordable by using psychrophilic β-galactosidases. *Pseudoalteromonas haloplanktis*, an Antarctic marine bacterium, has been shown to produce a cold-active β-galactosidase that is highly effective in hydrolyzing lactose when it is refrigerated; this characteristic will help with lactose intolerance (Reddy et al., 2023).

With a variety of biotechnological uses at low temperatures, the utilization of cold-active β-galactosidase (temperature optima 15–18°C) produced by psychrophilic microbes has opened up new research avenues in the dairy and food processing industries (Adapa et al., 2022). In the dairy industry, enzymes are used in the production of cheese and other dairy products. For example, lactase from *Kluyveromyces lactis* yeast is used to hydrolyze lactose, the sugar found in milk, in order to produce dairy products that are lactose-free (Li et al., 2023). Energy savings in industry could be greatly increased by using cold-active enzymes. The k_{cat} of cold-active enzymes at low temperatures is comparable to that of mesophilic enzymes, according to their kinetic characteristics (Starck et al., 2022). Without losing their activity, they work by reducing the reaction temperature. Additionally, at high temperatures, these enzymes stop unwanted chemical reactions from happening. With a range of substrates, the K_m value of cold-active hydrolases from diverse bacterial sources has been observed to be between 0.3–1.01 mg/mL k_{cat}. However, reports of its V_{max} range from 236 to 97,951 mmol/min/mg.

The food and beverage industries benefit greatly from low-temperature processing because it offers numerous advantageous outcomes, such as avoiding bacterial contamination and spoiling, averting unintended chemical reactions that may arise at higher temperatures, improving food quality, and preserving flavor because heat-labile flavoring compounds do not break down (Aaliya et al., 2021). Apart from these advantages, a key benefit of employing cold-adaptive enzymes in the food industry is their ease of separation from a reaction process through the simple use of increased temperatures. Therefore, proteases in meat tenderization, pectinases in fruit juice, amylases and xylanases in baking, and lactases in the dairy business are among the enzymes that find various applications in the food and beer industries (Table 1) (Sari et al., 2023).

Application of Psychrozymes in Detergent and Cleaning Industry

To replace dangerous chemicals with enzymes that can be utilized in the detergent industry, a sustainable strategy is required. Since they make up more than 60% of all enzyme sales in the global enzyme market, proteases extracted from microorganisms are employed in a variety of industries (Solanki et al., 2021). Because they are sustainable and beneficial to the environment, cold-active enzymes are widely employed in the commercial sector as detergent flavorings. Given their substantial market share in the enzyme industry, proteases are recognized to be important enzymes in detergents and washing powders. Numerous cold-active proteases have demonstrated unmatched detergent-compatible stability and activity throughout a broad range of alkaline

Table 1: Utilization of psychrophilic enzymes in the food industry

Enzymes	Application	Examples	References
Lipase Glucanases	Poorly degradable animal diets have improved and gained more nutrients and digestibility	*Candida Antarctica, Glaciozyma antarctica PI12*	Mohammadi et al., 2018
Protease	Utilized in the manufacturing of cheese, beer, and bakeries. Its capacity to hydrolyze soluble materials is advantageous for functional food components	*Bacillus cereus*	Kuddus, 2018
β-mannanase	Reduces food components' viscosity to relieve some of the difficulties associated with food processing	*Flavobacterium* sp.	Parvizpour et al., 2018
α-amylase	Used in the juice production process to lessen haze development and staling retardation in the baking sector	*Alteromonas haloplanktis*	Kuddus et al., 2011
Xylanase	Enhances the bread's volume and texture in addition to the dough's firmness and stability. It is useful for extracting juice, preventing the loss of nutrients, preventing dough from spoiling, and modifying the flavor of items.	*Pseudoalteromonas*	Qiu et al., 2017
Chitinase	Used to cause fungus food to deteriorate in freezing temperatures	*Glaciozyma antarctica* PI12	Jabeen and Qazi, 2014; Ramli et al., 2011
Laccase	It is imperative for the dairy industry to eliminate lactose from milk and convert it into glucose and galactose	*Cryptococcus neoformans*	Sharma et al., 2018

pH (Farooq et al., 2021). Additives to detergent for cold washing can be made from proteases that have been identified from *Acinetobacter* sp., *Bacillus* sp., *Planococcus* sp., *Pseudomonas aeruginosa*, and *Serratia marcescens* (Yang et al., 2023). Due to their greater cleaning effectiveness and environmental advantages over chemical agents, mesophilic enzymes have found application in a variety of industries, including the food, beverage, and detergent industries. Proteases have been widely used in the food processing, laundry, and cleaning agent industries for a very long time (Naveed et al., 2021). Enzymes that can adapt to cold temperatures represent a significant advancement in technology, as they offer numerous advantages while consuming less energy (Akindolire et al., 2022). Energy conservation is achieved to a considerable degree by the employment of cold-adaptive enzymes in cleaning procedures, which guarantee lower wash temperatures. According to a survey, reducing the wash temperature by just 10°C resulted in a 30% decrease in electricity consumption (Perfumo et al., 2020).

Various cold-active enzymes, including lipases, amylases, and proteases, have demonstrated exceptional efficacy in low-temperature cleaning procedures. Wipes or various formulations containing psychrozymes can be an easy way to clean solid things that cannot be heated for washing (Khan et al., 2023). The effectiveness of detergents or cleaning solutions in eliminating stains will be enhanced by the incorporation of lipases. Cold-active enzymes are perfect for cleaning and washing in the food industry since they significantly reduce energy costs and downtime (Yao et al., 2021). Thus, the psychrozymes also offer endless potential for use in cleaning and washing procedures; however, in order to extend their shelf life, genetic modifications must be undertaken (Table 2). In cold-active detergents, for example, subtilisins derived from Antarctic; *Bacillus* sp. were modified to lengthen their shelf life. The quality of products has increased with the use of cold-active enzymes in detergents (Al-Ghanayem et al., 2020).

Pharmaceutical Applications of Psychrophilic Enzymes

The pharmaceutical sector has drawn the attention of the scientific community to improve the quality of its products based on psychrozymes due to the aforementioned properties of these enzymes (Table 3). Thus, strains of bacteria, *Streptomyces*, and fungus have been used to develop a number of anti-fungal, anti-cancer, and anti-tumor medicines. Most cold-adapted enzymes can be rendered inactive by mild heat due to their thermolability (Tiwari et al., 2021). Consequently, the enzymes are not inactivated by substances derived from chemicals. Because of this characteristic, psychrophilic enzymes are a solid option for live vaccine development (Murthy et al., 2024). Additionally, mesophilic infections can be engineered to produce thermolabile homologs of vital enzymes, which make them sensitive to temperature. These strains are designed to lose their pathogenicity and become inactive at body temperature similar to mammals, all the while retaining their whole antigenic repertoire. In a study, *F. novicida* completely changed its lifestyle to avoid causing tularemia in mice by replacing the genes encoding its vital enzymes with those found in an Arctic bacterium (Mlynek et al., 2022).

The Duplantis team used a similar technique for both Gram-positive Mycobacterium and *Salmonella enterica* (*S. enterica*). They demonstrate the safety of temperature-sensitive *S. enterica* strains in analytical labs and their capacity to elicit an immune response that is protective (Shin et al., 2022). The utilization of *Candida antarctica's* cold-adapted lipases is still useful for a variety of tasks, such as improving the quality of cosmetics, synthesizing optically active drug

Table 2: Application of psychrophilic enzymes in cleaning and detergent industry

Enzymes	Application	Examples	References
Lipase	In order to minimize the energy usage during cold washing and minimize the deterioration of textile fibers	*Photobacterium lipolyticum, Psychrobacter* sp.	Andualema and Gessesse, 2012
Protease	Because of its strong activity towards a range of substrates in cold environments and stability over a broad pH range, this enzyme is valuable in the formulation of detergents	*Pseudoalteromonas arctica*	Park et al., 2018
Cellulase	Utilized in the process of removing cellulose microfibrils from textiles that are created during the production and cleaning of clothing made of cotton. The color of agents can also be softened and brightened with the help of cellulases. At the moment, the detergents are the focus of industry which is producing cellulases for use in cold weather washing and lowering water usage	*Pseudoalteromonas arctica*	Kasana and Gulati, 2011
Esterase	Helps in getting rid of fat stains by breaking down into more hydrophilic materials that are easily removed by washing. Energy consumption is decreased when it is used in formulations for cold-temperature stain removal of fat. Additionally, it is crucial for the decrease of unfavourable substances in detergents	*Streptomyces coelicolor*	Brault et al., 2012

intermediates, altering sugars and their derivatives, and resolving racemates, like amines and alcohols, during the production of fragrance esters, pharmaceuticals (like iso-propylmyristate), cosmetic products, and fragrance esters (Mhetras et al., 2021). Tagatose serves as an antihyperglycemic and prebiotic, while the psychrophilic enzyme β-galactosidase is employed in the manufacture of galacto-oligosaccharides (Morales-Contreras et al., 2021).

Table 3: A summary of psychrophilic enzymes utilized in pharmaceuticals

Enzymes	Application	Examples	References
Protease	In order to prevent periodontal disease, remove plaque, reduce the infectivity of viruses, and reduce glabellar lines in cosmeceutical applications, barriers to tissue regeneration and healing	*Euphausia superba*	Fornbacke and Clarsund, 2013
Lipase	Used in the production of pharmaceuticals, including arylaliphatic glycolipids, ethyl esterification of docosahexaenoic acid to ethyl docosahexaenoic (EtDHA), synthesis of single isomer chiral drugs, formation of optically active amines as intermediates, and synthesis of citronellol laurate from citronellol and lauric acid lovastatin as a medication with lower serum cholesterol levels	*Candida antarctica, Pseudomonas* sp.	Gotor-Fernández et al., 2006; Kavitha, 2016
Laccase	Carries potential use in the synthesis of anticancer medications. Non-steroidal anti-inflammatory medicines (NSAIDs) are also used to treat prostaglandins while preventing cyclooxygenases from being produced	*Pseudomonas putida*	Duarte et al., 2018
Esterase	Chiral resolution plays a crucial role in enhancing the potency and range of medicines	*Alcanivorax dieselolei B-5*	Kashif et al., 2017
Amylase	An important enzyme in the synthesis of cyclodextrins, which are utilized extensively in the pharmaceutical sector. It functions as a complexing agent to improve the stability, aqueous solubility, and bioavailability of poorly soluble medications	*Alteromonas haloplanctis*	Hamid, 2015
Chitinase	Significantly used as an ingredient in antifungal creams and lotions, where it hydrolyzes chitin to chitosan	*Glaciozyma antarctica PI12*	Ramli et al., 2011
Lysozyme	Used in pharmaceuticals as an antibacterial agent	*Debaryomyces hansenii*	Sotelo-Mundo et al., 2007

6. Conclusion

Microorganisms may survive in a variety of environments and are found everywhere. Due to the peculiarities of their metabolic processes, microbiologists and researchers from around the globe are becoming interested in microorganisms, such as bacteria, fungus, and yeast, that live in cold climates on Earth. Considering their potential and qualities that are much needed in industries, it is anticipated that more cold-active and cold-adapted microbial enzymes will be found soon. The food industry can benefit greatly from cold-active enzymes derived from psychrophilic microorganisms due to their numerous benefits over other sources of enzymes. Many industrial uses are served by the cold-active enzymes and their psychrophilic microbial sources. Different sources of cold-active enzymes are added to detergents for cold washing and to food products for usage in baking, cheese production, meat tenderization, and fermentation. The food industry can benefit greatly from cold-active enzymes derived from psychrophilic microorganisms due to their numerous benefits over other sources of enzymes. Many industrial uses are served by the cold-active enzymes and their psychrophilic microbial sources. Cold-active enzymes derived from microorganisms are finding intriguing uses in both biotechnology and industry. These enzymes have a few potential uses, some of which are included here (like β-galactosidase), which is used to hydrolyze lactose in milk, and cellulases for bio-polishing and stone washing of textile items. Microbial cold-active enzymes have enormous potential and are used in a variety of food Industries. Owing to their remarkable stability at low temperatures, they are garnering increased interest and require more studies.

References

Aaliya, B., Sunooj, K.V., Navaf, M., Akhila, P.P., Sudheesh, C., Mir, S.A., & George, J. 2021. Recent trends in bacterial decontamination of food products by hurdle technology: A synergistic approach using thermal and non-thermal processing techniques. *Food Res. Int.*, 147, 110514.

Abirami, B., Manigundan, K., Radhakrishnan, M., Gopikrishnan, V., Bhaskar, P.V., Shanmugasundaram, T., & Dastager, S.G. 2021. Diversity and bioprospecting potentials of antarctic (polar) microbes. In: *Microbiome-Host Interactions.* CRC Press, pp. 349–365.

Adapa, V., Ramya, L.N., & Pulicherla, K.K. 2022. Cold-active enzymes: Enabling nonthermal processing in food industry. In: *Microbial Extremozymes.* Academic Press, pp. 39–53.

Akindolire, M.A., Rama, H., & Roopnarain, A. 2022. Psychrophilic anaerobic digestion: A critical evaluation of microorganisms and enzymes to drive the process. *Renewable and Sustain. Energy Rev.*, 161, 112394.

Akram, F., Ashraf, S., Haq, I.U., Shah, F.I., & Aqeel, A. 2022. Eminent industrial and biotechnological applications of laccases from bacterial source: A current overview. *Appl. Biochem. Biotechnol.*, 1–21.

Alam, M., & Tiwary, B.K. (Eds.). 2023. *Extremophiles: Diversity, Adaptation, and Applications*. Bentham Science Publishers.

Al-Ghanayem, A.A., & Joseph, B. 2020. Current prospective in using cold-active enzymes as eco-friendly detergent additive. *Appl. Microbial. Biotechnol.*, 104(7), 2871–2882.

Amiri, S., Aghamirzaei, M., Mostashari, P., Sarbazi, M., Tizchang, S., & Madahi, H. (2021). The impact of biotechnology on dairy industry. In: *Microbial Biotechnology in Food and Health.* Academic Press, pp. 53–79.

Andualema, B., & Gessesse, A. 2012. Microbial lipases and their industrial applications. *Biotechnol.* (Faisalabad), 11(3), 100–118.

Angelin, J., & Kavitha, M. 2022. Molecular mechanisms behind the cold and hot adaptation in extremozymes. In: *Extremozymes and Their Industrial Applications*. Academic Press, pp. 141–176.

Annapure, U.S., & Pratisha, N. 2022. Psychrozymes: A novel and promising resource for industrial applications. In: *Microbial Extremozymes.* Academic Press, pp. 185–195.

Anwar, U.B., Zwar, I.P., & de Souza, A.O. 2020. Biomolecules produced by extremophiles microorganisms and recent discoveries. In: *New and Future Developments in Microbial Biotechnology and Bioengineering*. Elsevier, pp. 247–270.

Arcus, V.L., & Mulholland, A.J. 2020. Temperature, dynamics, and enzyme-catalyzed reaction rates. *Ann. Rev. of Biophys.*, 49, 163–180.

Bansode, S., & Banpatte, A. (2021). Case study of Global Warming. *Adn. Envi. Was. Mana. Rec.*, 4(3), 197–206.

Barsanti, M., Yilmaz, S., & Binder, C.R. 2024. Beyond the average consumer: Mapping the potential of demand-side management among patterns of appliance usage. *Energy Res. Soc. Sci.*, 111, 103463.

Bhanu, L.M., & Chatterjee, S. 2024. Cellulases from psychrophilic and psychrotrophic microorganisms and their potential applications. In: *Harnessing Microbial Potential for Multifarious Applications.* Springer Nature Singapore, pp. 549–573.

Bharali, P., Gogoi, B., Sorhie, V., Acharjee, S.A., Walling, B., Alemtoshi, ... & Shah, M.P. 2024. Autochthonous psychrophilic hydrocarbonoclastic bacteria and its ecological function in contaminated cold environments. *Biodegradation*, 35(1), 1–46.

Bhat, A.M., Hussain, A., Hassan, Q.P., & Bhat, A. 2024. Culturable Streptomyces spp. from high-altitude, oligotrophic North Western Himalaya: A comprehensive study on the diversity, bioactivity and insights into the proteome of potential species. *FEMS Microbiology Ecology*, 100(4), fiae026.

Bhatia, R.K., Ullah, S., Hoque, M.Z., Ahmad, I., Yang, Y.H., Bhatt, A.K., & Bhatia, S.K. 2021. Psychrophiles: A source of cold-adapted enzymes for energy efficient biotechnological industrial processes. *J. Environ.Chem. Eng.*, 9(1), 104607.

Bhatt, S.A. 2022. Psychrophiles: Distribution, ecology, physiology, metabolism, cold adapted enzymes, and proteins. In: *Physiology, Genomics, and Biotechnological Applications of Extremophiles.* IGI Global, pp. 94–122.

Bonnin, E., & Pelloux, J. 2020. Pectin degrading enzymes. *Pectin: Technological and Physiological Properties*, pp. 37–60.

Brault, G., Shareck, F., Hurtubise, Y., Lépine, F., & Doucet, N. 2012. Isolation and characterization of EstC, a new cold-active esterase from Streptomyces coelicolor A3 (2). *PLOS One*, 7(3), e32041.

Bruno, S., Coppola, D., di Prisco, G., Giordano, D., & Verde, C. 2019. Enzymes from marine polar regions and their biotechnological applications. *Marine Drugs*, 17(10), 544.

Chatragadda, R., & Dufossé, L. 2021. Ecological and biotechnological aspects of pigmented microbes: A way forward in development of food and pharmaceutical grade pigments. *Microorganisms*, 9(3), 637.

Chen, Y., Tian, Q., Wang, H., Ma, R., Han, R., Wang, Y., ... & Qin, Y. 2024. A manganese-based metal–organic framework as a cold-adapted nanozyme. *Advanced Materials*, 36(10), 2206421.

Collins, T., D'Amico, S., Marx, J.C., Feller, G., & Gerday, C. 2007. Cold-adapted enzymes. In: *Physiology and Biochemistry of Extremophiles*. ASM Press, pp. 165–179.

Collins, T., & Feller, G. 2023. Psychrophilic enzymes: Strategies for cold-adaptation. *Essays Biochem.*, 67(4), 701–713.

Debroy, A., & George, N. 2023. Psychrophiles and psychrozymes: Structural adaptations and applications. In: *AIP Conference Proceedings*. AIP Publishing, 2558(1).

Dhole, N.P., Dar, M.A., Ullah, K., Xie, R., Sun, J., Shahnawaz, M., & Pandit, R.S. 2024. A review on the bioprospection of psychrophilic bacteria and their potential applications for sustainability. In: *Soil Microbiome of the Cold Habitats*. CRC Press, pp. 127–146.

Duarte, A.W.F., Dos Santos, J.A., Vianna, M.V., Vieira, J.M.F., Mallagutti, V.H., Inforsato, F.J., ... & Durães Sette, L. 2018. Cold-adapted enzymes produced by fungi from terrestrial and marine Antarctic environments. *Crit. Rev. Biotechnol.*, 38(4), 600–619.

Ejaz, U., Sohail, M., & Ghanemi, A. 2021. Cellulases: From bioactivity to a variety of industrial applications. *Biomimetics*, 6(3), 44.

Farooq, S., Nazir, R., Ganai, S.A., & Ganai, B.A. 2021. Isolation and characterization of a new cold-active protease from psychrotrophic bacteria of Western Himalayan glacial soil. *Scientific Reports*, 11(1), 12768.

Feller, M.B. 1999. Spontaneous correlated activity in developing neural circuits. *Neuron.*, 22(4), 653–656.

Firouz, M.S., Mohi-Alden, K., & Omid, M. 2021. A critical review on intelligent and active packaging in the food industry: Research and development. *Food Res. Int.*, 141, 110113.

Fornbacke, M., & Clarsund, M. 2013. Cold-adapted proteases as an emerging class of therapeutics. *Infectious Diseases and Therapy*, 2(1), 15–26.

Furhan, J. 2020. Adaptation, production, and biotechnological potential of cold-adapted proteases from psychrophiles and psychrotrophs: Recent overview. *J. Genet. Eng. Biotechnol.*, 18(1), 36.

García-Descalzo, L., García-López, E., & Cid, C. 2022. Comparative proteomic analysis of psychrophilic vs. mesophilic bacterial species reveals different strategies to achieve temperature adaptation. *Front. Microbiol.*, 13, 841359.

Ghosh, M., & Pulicherla, K.K. 2021. Psychrophiles as the source for potential industrial psychrozymes. In: *Recent Developments in Microbial Technologies*. Springer, pp. 355–366.

Gotor-Fernández, V., Brieva, R., & Gotor, V. 2006. Lipases: Useful biocatalysts for the preparation of pharmaceuticals. *J. Mol. Catal. B. Enzym.*, 40(3–4), 111–120.

Grassmann, G., Miotto, M., Desantis, F., Di Rienzo, L., Tartaglia, G.G., Pastore, A., ... & Milanetti, E. 2024. Computational approaches to predict protein–protein interactions in crowded cellular environments. *Chemical Reviews*, 124, 3645-4540.

Guo, C., Zheng, R., Cai, R., Sun, C., & Wu, S. 2021. Characterization of two unique cold-active lipases derived from a novel deep-sea cold seep bacterium. *Microorganisms*, 9(4), 802.

Gupta, V., Bhaskar, P., Thoudam, J., Bisht, S., Sharma, A., & Tripathi, R. 2023. Psychrophiles as a novel and promising source of cold-adapted industrial enzymes. *Appl. Biol. Chem. J.*, 4(2), 54–68.

Hamid, B., Singh, P., Mohiddin, F.A., & Sahay, S. 2013. Partial characterization of cold-active β-galactosidase activity produced by *Cystofilobasidium capitatum* SPY11 and *Rhodotorula mucilaginosa* PT1. *J. Endocytobiosis Cell Res.*, 24, 23–26.

Hamid, B. 2015. Cold-active α-amylase from psychrophilic and psychrotolerant yeast. *J. Glob. Biosci.*, 4, 2670–2677.

Hamid, B. 2021. Potent biotechnological applications of psychrozymes. In: *Microbiomes and the Global Climate Change*. Springer, Singapore, pp. 349–364.

Hamid, B., Bashir, Z., Yatoo, A.M., Mohiddin, F., Majeed, N., Bansal, M., ... & Alfaifi, M.Y. 2022. Cold-active enzymes and their potential industrial applications: A review. *Molecules*, 27(18), 5885.

Hamid, B., Baba, Z.A., Sheikh, T.A., Perveen, K., & Sayyed, R.Z. 2024. Investigating potential protease activity of psychrotrophic bacteria from a municipal landfill for solid waste management. *Biomass Conversion and Biorefinery*. https://doi.org/10.1007/s13399-024-05621-2.

Jabeen, F., & Qazi, J.I. 2014. Potential of bacterial chitinases and exopolysaccharides for enhancing shelf life of food commodities at varying conditions. *Intl. Res. J. Environ. Sci.*, 3, 87–93.

Janusz, G., Pawlik, A., Świderska-Burek, U., Polak, J., Sulej, J., Jarosz-Wilkołazka, A., & Paszczyński, A. 2020. Laccase properties, physiological functions, and evolution. *Int. J. of Mol. Sci.*, 21(3), 966.

Kanekar, P.P., & Kanekar, S.P. 2022. Psychrophilic, psychrotrophic, and psychrotolerant microorganisms. In: *Diversity and Biotechnology of Extremophilic Microorganisms from India.* Springer Nature Singapore, pp. 215–249.

Karadayi, G., Alaylar, B., Karadayi, M., Gürkök, S., Aksu, Ş., Egamberdieva, D., & Güllüce, M. 2021. Biodiversity, ecological, and commercial importance of psychrophilic microorganisms. *Microbial Communities and their Interactions in the Extreme Environment*. Springer, pp. 225–247.

Kasana, R.C., & Gulati, A. 2011. Cellulases from psychrophilic microorganisms: A review. *J. Basic Microbiol.*, 51(6), 572–579.

Kashif, A., Tran, L.H., Jang, S.H., & Lee, C. (2017). Roles of active-site aromatic residues in cold adaptation of Sphingomonas glacialis Esterase EstSP1. *ACS Omega*, 2(12), 8760–8769.

Kavitha, M. 2016. Cold active lipases: An update. *Front. Life Sci.*, 9(3), 226–238.

Khan, A., Punetha, A., Kukreti, B., Sharma, R.S., Panthari, D., Naudiyal, N. & Singh, A.V. 2023. Microbial community dynamics of Antarctica: Their ecological potential and industrial importance. In: *Microbial Bioactive Compounds: Industrial and Agricultural Applications.* Springer Nature Switzerland, pp. 261–289.

Khatami, S.H., Vakili, O., Movahedpour, A., Ghesmati, Z., Ghasemi, H., & Taheri-Anganeh, M. 2022. Laccase: Various types and applications. *Biotechnol. Appl. Biochem.*, 69(6), 2658–2672.

Kochhar, N., Shrivastava, S., Ghosh, A., Rawat, V.S., Sodhi, K.K., & Kumar, M. 2022. Perspectives on the microorganism of extreme environments and their applications. *Current Research in Microbial Sciences*, 3, 100134.

Kuddus, M., Arif, J.M., & Ramteke, P.W. 2011. An overview of cold-active microbial α-amylase: Adaptation strategies and biotechnological potentials. *Biotechnology*, 10(3), 246–258.

Kuddus, M. 2018. Cold-active enzymes in food biotechnology: An updated mini review. *J. Appl. Bio. Biotechnol.*, 6(3), 58–63.

Kuddus, M., Roohi, Bano, N., Sheik, G.B., Joseph, B., Hamid, B., ... & Madhavan, A. 2024. Cold-active microbial enzymes and their biotechnological applications. *Microbial Biotechnology*, 17(4), e14467.

Kumar, A., Mukhia, S., & Kumar, R. 2021. Industrial applications of cold-adapted enzymes: Challenges, innovations, and future perspective. *3 Biotech.*, 11(10), 426.

Kumar, D., & Rastogi, S. 2020. Industrial application of amylase: A review. *Journal of Natural Remedies*, 20, 23–27.

Kumari, M., Padhi, S., Sharma, S., Phukon, L.C., Singh, S.P., & Rai, A.K. 2021. Biotechnological potential of psychrophilic microorganisms as the source of cold-active enzymes in food processing applications. *3 Biotech*, 11(11), 479.

Le-Vinh, B., Akkuş-Dağdeviren, Z.B., Le, N.M.N., Nazir, I., & Bernkop-Schnürch, A. 2022. Alkaline phosphatase: A reliable endogenous partner for drug delivery and diagnostics. *Advanced Therapeutics*, 5(2), 2100219.

Li, A., Zheng, J., Han, X., Yang, S., Cheng, S., Zhao, J., ... & Lu, Y. 2023. Advances in low-lactose/lactose-free dairy products and their production. *Foods*, 12(13), 2553.

Li, C., Murugaiyan, J., Thomas, C., Alter, T., & Riedel, C. 2020. Isolate specific cold response of Yersinia enterocolitica in transcriptional, proteomic, and membrane physiological changes. *Front. Microbiol.*, 10, 3037.

Liu, Y., Jia, K., Chen, H., Wang, Z., Zhao, W., & Zhu, L. 2023. Cold-adapted enzymes: Mechanisms, engineering, and biotechnological application. *Bioprocess Biosyst. Eng.*, 46(10), 1399–1410.

Maciejewska, N., Walkusz, R., Olszewski, M., & Szymańska, A. 2019. New nuclease from extremely psychrophilic microorganism Psychromonas ingrahamii 37: Identification and characterization. *Mol. Biotechnol.*, 61, 122–133.

Mangiagalli, M., & Lotti, M. 2021. Cold-active β-galactosidases: Insight into cold adaptation mechanisms and biotechnological exploitation. *Marine Drugs*, 19(1), 43.

Mhetras, N., Mapare, V., & Gokhale, D. 2021. Cold active lipases: Biocatalytic tools for greener technology. *Appl. Biochem. Biotechnol.*, 193(7), 2245–2266.

Miri, S., Davoodi, S.M., Robert, T., Brar, S.K., Martel, R., & Rouissi, T. 2022. Enzymatic biodegradation of highly p-xylene contaminated soil using cold-active enzymes: A soil column study. *J. Hazard. Mater.*, 423, 127099.

Mlynek, K.D., & Bozue, J.A. 2022. Why vary what's working? Phase variation and biofilm formation in Francisella tularensis. *Front. Microbiol.*, 13, 1076694.

Mohammadi, S., Parvizpour, S., Razmara, J., Abu Bakar, F.D., Illias, R.M., Mahadi, N.M., & Murad, A.M. 2018. Structure prediction of a novel exo-β-1, 3-glucanase: Insights into the cold adaptation of psychrophilic yeast Glaciozyma antarctica PI12. *Interdisciplinary Sciences: Computational Life Sciences*, 10, 157–168.

Morales-Contreras, J.A., Rodríguez-Pérez, J.E., Álvarez-González, C.A., Martínez-López, M.C., Juárez-Rojop, I.E., & Ávila-Fernández, Á. 2021. Potential applications of recombinant bifidobacterial proteins in the food industry, biomedicine, process innovation, and glycobiology. *Food Sci. Biotechnol.*, 30, 1277–1291.

Muras, A., Romero, M., Mayer, C., & Otero, A. 2021. Biotechnological applications of Bacillus licheniformis. *Crit. Rev. Biotechnol.*, 41(4), 609–627.

Murthy, P.S., Vedashree, M., Sneha, H.P., & Prakash, I. 2024. Extremophiles as a source of biotechnological products. In: *Research Anthology on Bioinformatics, Genomics, and Computational Biology*. IGI Global, pp. 606–632.

Mutlu-Ingok, A., Kahveci, D., Karbancioglu-Guler, F., & Ozcelik, B. 2022. Applications of extremozymes in the food industry. In: *Microbial Extremozymes*. Elsevier, pp. 197–206.

Naveed, M., Nadeem, F., Mehmood, T., Bilal, M., Anwar, Z., & Amjad, F. 2021. Protease—A versatile and ecofriendly biocatalyst with multi-industrial applications: An updated review. *Catalysis Letters*, 151, 307–323.

Palaskar, R.S., Kate, S.A., Khandagale, M.S., Namekar, S.B., Aher, S.B., Nale, A.R., & Tandale, A.J. 2022. Screening of laccase producer from soil and its applications. *J. Adv. Sci. Res.*, 13(01), 311–318.

Park, H.J., Lee, C.W., Kim, D., Do, H., Han, S.J., Kim, J.E., ... & Yim, J.H. 2018. Crystal structure of a cold-active protease (Pro21717) from the psychrophilic bacterium, Pseudoalteromonas arctica PAMC 21717, at 1.4 Å resolution: Structural adaptations to cold and functional analysis of a laundry detergent enzyme. *PLOS One*, 13(2), e0191740.

Park, S., Shin, Y.K., Yoon, J.Y., Nam, K.H., Munashingha, P.R., Park, S., ... & Chang, I. 2022. Computational design of a thermolabile uracil-DNA glycosylase of Escherichia coli. *Biophys. J.*, 121(7), 1276–1288.

Parrilli, E., Tedesco, P., Fondi, M., Tutino, M.L., Giudice, A.L., de Pascale, D., & Fani, R. 2021. The art of adapting to extreme environments: The model system Pseudoalteromonas. *Phys. Life Rev.*, 36, 137–161.

Parvizpour, S., Razmara, J., Ramli, A.N.M., Md Illias, R., & Shamsir, M.S. 2014. Structural and functional analysis of a novel psychrophilic β-mannanase from Glaciozyma antarctica PI12. *J. Comput..Aided Mol. Des.*, 28, 685–698.

Parvizpour, S., Hussin, N., Shamsir, M.S., & Razmara, J. 2021. Psychrophilic enzymes: Structural adaptation, pharmaceutical and industrial applications. *Appl. Microbial. Biotechnol.*, 105, 899–907.

Pathania, S., Solanki, P., Putatunda, C., Bhatia, R.K., & Walia, A. 2022. Adaptation to cold environment: The survival strategy of psychrophiles. In: *Survival Strategies in Cold-adapted Microorganisms*. Springer, Singapore, pp. 87–111.

Paul, J.S., Gupta, N., Beliya, E., Tiwari, S., & Jadhav, S.K. 2021. Aspects and recent trends in microbial α-amylase: A review. *Appl. Biochem. Biotechnol.*, 193, 2649–2698.

Perfumo, A., Freiherr von Sass, G.J., Nordmann, E.L., Budisa, N., & Wagner, D. 2020. Discovery and characterization of a new cold-active protease from an extremophilic bacterium via comparative genome analysis and *in vitro* expression. *Front. Microbial.*, 11, 518462.

Qian, Y.F., Yu, J.Y., Xie, J., & Yang, S.P. 2023. A mini-review on cold-adapted enzymes from psychrotrophic microorganisms in foods: Benefits and challenges. *Current Research in Biotechnology*, 100162.

Qiu HaiYan, Q.H., Li ZhongYuan, L.Z., Wang Hui, W.H., Zhang HaiYing, Z.H., Li Shuang, L.S., Luo XueGang, L.X., ... & Zhang TongCun, Z.T. 2017. Molecular and biochemical characterization of a novel cold-active and metal ion-tolerant GH10 xylanase from frozen soil. *Biotechnology and Biotechnological Equipment*, 31(5).

Rabbani, G., Ahmad, E., Ahmad, A., & Khan, R.H. 2023. Structural features, temperature adaptation, and industrial applications of microbial lipases from psychrophilic, mesophilic, and thermophilic origins. *Int. J. Bio. Macromol.*, 225, 822–839.

Rahban, M., Zolghadri, S., Salehi, N., Ahmad, F., Haertlé, T., Rezaei-Ghaleh, N.,... & Saboury, A.A. 2022. Thermal stability enhancement: Fundamental concepts of protein engineering strategies to manipulate the flexible structure. *Int. J. Bio. Macromol.*, 214, 642–654.

Rajan, K. 2020. *Discovery of Industrially Relevant Oxidoreductases*. Doctoral dissertation, Dublin City University.

Ramasamy, K.P., Mahawar, L., Rajasababapathy, R., Rajeshwari, K., Miceli, C., & Pucciarelli, S. 2023. Comprehensive insights on environmental adaptation strategies in Antarctic bacteria and biotechnological applications of cold adapted molecules. *Front. Microbiol.*, 14, 1197797.

Ramli, A.N., Mahadi, N.M., Rabu, A., Murad, A.M., Bakar, F.D., & Illias, R.M. 2011. Molecular cloning, expression and biochemical characterisation of a cold-adapted novel recombinant chitinase from Glaciozyma antarctica PI12. *Microbial Cell Factories*, 10, 1–13.

Ranjan, R., Rai, R., Bhatt, S.B., & Dhar, P. 2023. Technological road map of Cellulase: A comprehensive outlook to structural, computational, and industrial applications. *Biochem. Eng. J.*, 109020.

Rathore, N., Kumar, P., Mehta, N., Swarnkar, M.K., Shankar, R., & Chawla, A. 2022. Time-series RNA-Seq transcriptome profiling reveals novel insights about cold acclimation and de-acclimation processes in an evergreen shrub of high altitude. *Scientific Reports*, 12(1), 15553.

Reddy, L.J., Kumar, P.S., Pandrangi, S.L., Chikati, R., Srinivasulu, C., John, A., ... & Pamuru, R.R. 2023. A review on psychrophilic β-D-Galactosidases and their potential applications. *Appl. Biochem. Biotechnol.*, 195(4), 2743–2766.

Rehman, H.U. 2023. Applications of pectinolytic enzymes in process industries. In: *Utilization of Pectin in the Food and Drug Industries*. IntechOpen.

Rittié, L., & Perbal, B. 2008. Enzymes used in molecular biology: A useful guide. *J. Cell Communication and Signaling*, 2, 25–45.

Rizzo, C., & Lo Giudice, A. 2022. Life from a snowflake: Diversity and adaptation of cold-loving bacteria among ice crystals. *Crystals*, 12(3), 312.

Salwan, R., & Sharma, V. (Eds.). 2020. *Physiological and Biotechnological Aspects of Extremophiles*. Academic Press.

Sari, T.P., Gaur, V.K., Tarafdar, A., Sirohi, R., Sindhu, R., & Rai, A.K. 2023. Microbial lipases: Production and application. In: *Microbial Enzymes and Metabolites for Health and Well-Being*. CRC Press, pp. 75–90.

Sarsan, S., Rodhe, A.V., Roy, K.V.V., & Jagavati, S. 2024. Biotechnological potential of cold-adaptive extremozymes. *Microbial Essentialism: An Industrial Prospective*. Academic Press, pp. 265–299.

Schlichting, N.C. 2021. Optimization and automation of the ligase cycling reaction.

Shad, M., Hussain, N., Usman, M., Akhtar, M.W., & Sajjad, M. 2023. Exploration of computational approaches to predict the structural features and recent trends in α-amylase production for industrial applications. *Biotechnol. Bioeng.*, 120(8), 2092–2116.

Sharma, K.K., Singh, D., & Rawat, S. 2018. Molecular dynamics simulation studies suggests unconventional roles of non-secretary laccases from enteropathogenic gut bacteria and Cryptococcus neoformans serotype D. *Comput. Bio. Chem.*, 73, 41–48.

Sharma, S., Chaturvedi, U., Sharma, K., Vaishnav, A., & Singh, H.B. 2022. An overview of survival strategies of psychrophiles and their applications. In: *Survival Strategies in Cold-adapted Microorganisms*. Springer, Singapore, pp. 133–151.

Sharma, S., Sharma, V., Chatterjee, S., & Kumar, S. 2022. Psychrophilic enzymes. In: *Extremophiles: General and Plant Biomass Based Biorefinery*. CRC Press, 166 pp.

Shi, Q., Abdel-Hamid, A.M., Sun, Z., Cheng, Y., Tu, T., Cann, I., ... & Zhu, W. 2023. Carbohydrate-binding modules facilitate the enzymatic hydrolysis of lignocellulosic biomass: Releasing reducing sugars and dissociative lignin available for producing biofuels and chemicals. *Biotechnol. Adv.*, 108126.

Shin, H., La, T.M., Lee, H.J., Kim, T., Song, S.U., Park, E., ... & Lee, S.W. 2022. Evaluation of immune responses and protective efficacy of a novel live attenuated Salmonella enteritidis vaccine candidate in chickens. *Vaccines*, 10(9), 1405.

Shu, W.S., & Huang, L.N. 2022. Microbial diversity in extreme environments. *Nat. Rev. Microbiol.*, 20(4), 219–235.

Sočan, J., Purg, M., & Åqvist, J. 2020. Computer simulations explain the anomalous temperature optimum in a cold-adapted enzyme. *Nature Communications*, 11(1), 2644.

Solanki, P., Putatunda, C., Kumar, A., Bhatia, R., & Walia, A. 2021. Microbial proteases: Ubiquitous enzymes with innumerable uses. *3 Biotech.*, 11(10), 428.

Sood, A., Kaur, M., & Gupta, R. 2023. Lipases and their applications in biomedical field. *Curr. Biotechnol.*, 12(1), 25–36.

Sotelo-Mundo, R.R., Lopez-Zavala, A.A., Garcia-Orozco, K.D., Arvizu-Flores, A.A., Velazquez-Contreras, E.F., Valenzuela-Soto, E.M., ... & Kanost, M.R. 2007. The lysozyme from insect (Manduca sexta) is a cold-adapted enzyme. *Protein and Peptide Letters*, 14(8), 774–778.

Staack, L. 2023. *Modulating Poultry Microbiota to Promote Gut Function by Enzymatic Tailoring of Feed.* DTU Research Database.

Stark, C., Bautista-Leung, T., Siegfried, J., & Herschlag, D. 2022. Systematic investigation of the link between enzyme catalysis and cold adaptation. *Elife*, 11, e72884.

Sun, W., Shahrajabian, M.H., & Lin, M. 2022. Research progress of fermented functional foods and protein factory-microbial fermentation technology. *Fermentation*, 8(12), 688.

Suzuki, T.A., & Ley, R.E. 2020. The role of the microbiota in human genetic adaptation. *Science*, 370(6521), eaaz6827.

Tiwari, B.R., Rouissi, T., Brar, S.K., & Surampalli, R.Y. 2021. Critical insights into psychrophilic anaerobic digestion: Novel strategies for improving biogas production. *Waste Management*, 131, 513–526.

Vivek, K., Sandhia, G.S., & Subramaniyan, S.J.B.A. 2022. Extremophilic lipases for industrial applications: A general review. *Biotechnol. Adv.*, 60, 108002.

Waditee-Sirisattha, R., & Kageyama, H. 2022. Extremophilic cyanobacteria. In: *Cyanobacterial Physiology*. Academic Press, pp. 85–99.

Wang, F., Sheng, J., Chen, Y., & Xu, J. 2021. Microbial diversity and dominant bacteria causing spoilage during storage and processing of the Antarctic krill, Euphausia superba. *Pol. Bio.*, 44, 163–171.

Williamson, A., & Leiros, H.K.S. 2019. Structural intermediates of a DNA–ligase complex illuminate the role of the catalytic metal ion and mechanism of phosphodiester bond formation. *Nucleic Acids Research*, 47(14), 7147–7162.

Xu, H., Xu, D., & Liu, Y. 2024. Molecular biology applications of psychrophilic enzymes: Adaptations, advantages, expression, and prospective. *Appl. Biochem. Biotechnol.*, 1–25.

Yang, Z., Huang, Z., Wu, Q., Tang, X., & Huang, Z. 2023. Cold-adapted proteases: An efficient and energy-saving biocatalyst. *Int. J. Mol. Sci.*, 24(10), 8532.

Yao, W., Liu, K., Liu, H., Jiang, Y., Wang, R., Wang, W., & Wang, T. 2021. A valuable product of microbial cell factories: Microbial lipase. *Front. Microbiol.*, 12, 743377.

Zhu, D., Adebisi, W.A., Ahmad, F., Sethupathy, S., Danso, B., & Sun, J. 2020. Recent development of extremophilic bacteria and their application in biorefinery. *Front. Bioengin. and Biotechnol.*, 8, 483.

CHAPTER

9

Role of Psychrophiles and Psychrozymes in Microbial Remediation of Cold Environments and Their Applications

Monika Chaudhary[1], Dilfuza Jabborova[2*], Amar P. Garg[3] and Giuseppe Graziano[4]

[1] School of Biotechnology & Life Sciences, Shobhit Institute of Engineering & Technology, Meerut – 250110, India

[2] Institute of Genetics and Plant Experimental Biology, Kibray 111208, Uzbekistan

[3] Swami Vivekanand Subharti University, Subhartipuram, NH-58, Delhi – 250005, India

[4] Department of Science and Technology, University of Sannio-82100, Benevento, Italy

1. Introduction

The unique characteristics of cold and high-altitude environments include their substantial contributions to the biological succession of various habitats and their conflicting impacts on soil physicochemical properties and biodiversity. High-altitude habitats significantly affect biodiversity due to a variety of climatic and nutritional variables, including exposure to intense UV radiation, low partial pressure, low temperature, high wind velocities, soft water, and nutrient availability (Gupta et al., 2023). Because microbes can produce specialized enzymes that carry out specific biological activities, they are able to colonise and maintain metabolic activity at temperatures below freezing in spite of unfavourable environmental circumstances. They found that unforgiving environments are ideal for isolating effective new enzymes or genes that might be used for crucial industrial uses (Kumar et al., 2018; Thakur et al., 2018). These psychrophilic bacteria constitute a possible reservoir of cold-adapted enzymes, known as psychrozymes, because of their ability to preserve enzyme performance and stability even under severe settings (D'Amico et al., 2006).

*Corresponding author: dilfuzajabborova@yahoo.com

DOI: 10.1201/9781003717003-9

The influence of temperature on the structure and function of cells makes it a crucial factor in the survival of an organism. It is now discovered that the extremely cold, previously thought to be dead habitats are home to numerous, varied microbial populations (Anesio et al., 2017).

The three types of bacteria that are separated and recognized from cold settings are psychrophilic, psychrotolerant, and psychrotrophic microbes. Renowned microbiologists have defined a psychrophilic organism as one that has three cardinal temperatures: 0°C at lowest growth, 15°C at ideal growth, and 20°C at maximum growth. This description is widely recognized. From an ecological point of view, frigid settings are home to both psychrophilic and psychrotrophic (sometimes called psychrotolerant) species. However, psychrotrophs may survive in colder climates; in fact, a temperature ranging from 15 to 20°C is ideal for optimum development. It's important to remember that laboratory temperatures in the United States normally fall between 21 and 22°C (Richard Morita, 1975). Two names that are often used in the field of animal physiology are eurypsychrophiles and stenopsychrophils. This usage of the term 'psychrophilic' refers to any bacterium that can survive in conditions below 5°C. Researchers have come to understand the importance of microorganisms in cold environments during the past several decades. *Ascomycota*, *Actinobacteria*, *Basidiomycota*, *Chloroflexi*, *Chlamydiae*, *Cyanobacteria*, *Euryarchaeota*, *Firmicutes*, *Gemmatimonadetes*, *Mucoromycota*, *Nitrospirae*, *Planctomycetes*, *Proteobacteria*, *Thaumarchaeota* and *Verrucomicrobia* are just a few of the phyla that have been found to love the cold (Yadav et al., 2018; Kumar et al., 2022).

The production of anti-freezing proteins, antibiotics, and a variety of other cold-adapted enzymes that promote plant development by microbes that thrive in low-temperature conditions has potential applications for these organisms in the agricultural, medical, and pharmaceutical sectors. Psychrophilic microbe-derived enzymes are very useful for industrial uses. Numerous enzymes, such as amylase, cellulase, chitinase, protease, laccase, lipase, pectinase, β-galactosidase, β-glucosidase, and xylanase, can be produced by these cold-adapted bacteria. Species from genera including *Arthrobacter*, *Clostridium*, *Bacillus*, *Flavobacterium*, *Acinetobacter*, *Marinomonas*, *Moraxella*, *Paenibacillus*, *Pseudomonas*, *Shewanella*, *Vibrio*, and *Xanthomonas* are examples of psychrophilic microorganisms that are known to manufacture these enzymes (Singh et al., 2016). Moreover, cold-adapted bacteria can create anti-freezing proteins, which facilitate low-temperature adaptations. Potential uses for these anti-freezing substances include cryopreserving commercially significant microorganisms and improving the quality of frozen food (Yadav et al., 2017).

The fields of agriculture, pharmaceuticals, medicine, and allied fields may now more widely use advantageous cold-adapted bacteria that are obtained from a variety of cold environments thanks to developments in biotechnology (Baba et al., 2021; Hamid et al., 2024). The richness of psychrophilic microorganisms and the range of ways in which they are used in agriculture to support farmers financially are well understood (Afrouz et al., 2023; Sheikh et al., 2022). These microorganisms do a number of tasks for plants, including nitrogen fixation, vitamin solubilization, and the induction of physiologically active substances that improve plant metabolic pathways. Farmers can use psychrophilic bacteria as biocontrol agents to combat plant pathogenic germs in low-temperature conditions. There is a lot of promise for improving agricultural sustainability under various eco-climatic circumstances through the employment of cold-adapted bacteria in agriculture as biofertilizers, bioremediation agents, bio-inoculants, and biocontrol agents. Psychrophilic microorganisms have garnered increasing attention from both scientific communities and private-sector organizations due to their notable contributions to revenue-generating industries and their diverse capacities in stimulating plant development and creating cold-active enzymes (Kumar et al., 2022).

2. Biodiversity of Psychrophiles

The prevalence of psychrophilic microorganisms that thrive in a wide range of extremely cold conditions has been thoroughly studied by researchers in recent decades. Scientists have efficiently identified these cold-loving microbes, which belong to three primary domains: fungus, eubacteria, and archaea, using culture-dependent approaches. Represented by 17 distinct phyla—*Actinobacteria*, *Ascomycota*, *Bacteroidetes*, *Basidiomycota*, *Chlamydiae*, *Chloroflexi*, *Cyanobacteria*, *Euryarchaeota*, *Firmicutes*, *Gemmatimonadetes*, *Mucoromycota*, *Nitrospirae*, *Planctomycetes*, *Proteobacteria*, *Spirochaetes*, *Thaumarchaeota*, and *Verrucomicrobia*—are among the varied phyla that these microbes originate from. In the beginning, different genera, including *Arthrobacter*, *Alcaligenes*, *Achromobacter*, *Brevibacterium*, *Bacillus*, *Cryobacterium*, *Clostridium*, *Flavobacterium*, *Microbacterium*, *Micrococcus*, *Pseudomonas*, *Shewanella,* and *Vibrio*, were used to classify distinct species of psychrophilic microorganisms. Numerous frigid habitats, such as Antarctica, glaciers, subglacial lakes, the Himalayan cold deserts, and subalpine areas, have been reported to support the growth of these creatures (Saxena et al., 2016; De Pascale et al., 2012).

3. Adaptive Features of Cold Adapted Enzymes

The temperature of the environment has a direct impact on the activity of enzymes. Researchers have long known that at lower temperatures, cold-adapted enzymes (CAEs) have better specific activity due to their increased flexibility. The structural flexibility of CAEs has been investigated using a range of methods, including time-resolved fluorescence, electron paramagnetic resonance (EPR) spectroscopy, neutron scattering, X-ray crystallography, and fluorescence quenching. We now have better knowledge of how CAEs function at low temperatures thanks to recent investigations. For instance, it appears that the flexibility of psychrophilic enzymes is mostly shown in the enzyme's periphery rather than in its active area based on both experimental and computational data, including MDS. The disintegration of intramolecular bonds that stabilize the growing polypeptide chain and facilitate protein folding is the main source of this flexibility. Moreover, it has been noted that a reduction in the amount of disulphide bridges, hydrogen bonds, and hydrophobic and electrostatic contacts increases the flexibility of enzymes that have adapted to cold conditions (Arcus et al., 2020; Kumar et al., 2021).

Conformational flexibility can be improved by adjustments to the amino acid composition, such as a decrease in the ratio of arginine to lysine, an increase in glycine, a decrease in proline in loops, the promotion of more alpha-helices, and an increase in non-polar residues on the surface of the protein. Furthermore, a key factor in cold adaptation is the creation of salt bridges between amino acids, which reduce electrostatic interactions. In addition, the stability of proteins is also maintained by the number of hydrogen bonds (Al-Ghanayem & Joseph, 2020). Psychrophiles have developed unique genetic and protein adaptation mechanisms to maintain their metabolic activity and ensure their survival in extremely cold conditions. The aforementioned adaptations encompass alterations in the transcription and translation processes, modifications to the fluidity of the membrane, increased expression of cold-shock proteins (CSPs), synthesis of proteins that nucleate ice or antifreeze, synthesis of compatible solutes, exopolysaccharides, and the creation of biocatalysts that can function at low temperatures. Many biomolecules, such as different proteins and enzymes, are produced as a result of the essential adaptation to cold environments. These biomolecules may find use in biotechnological applications in the future and present. Investigating the unique adaptations, structural traits, metabolic pathways, enzymatic capabilities, and genetic composition of psychrophiles, as well as their biodiversity in cold climates, is essential to making the most use of cold-adapted enzymes (Koh et al., 2017; Turchetti et al., 2020; Bhatia et al., 2021; Rapuano & Graziano, 2022).

Membrane Fluidity

Microbes that survive in freezing settings use a variety of adaptation techniques, most of which are centred around controlling and modifying membrane fluidity in order to mitigate the deleterious effects of extreme temperatures. These tactics include increasing membrane fluidity, triggering membrane-associated sensors, and stiffening the membrane. Low temperatures cause gel-phase transitions, reduce fluidity, and disrupt cellular processes, all of which have a negative impact on the physical characteristics and functionality of membranes. In response, microbes alter the lipid fatty acyl chains in their membranes to preserve the ideal level of fluidity. Desaturases, an enzyme that is triggered by cold shock, are essential to this process because they change saturated acyl fatty acids into unsaturated acyl chains. Moreover, methyl-branched fatty acid concentration increases help microbiological cells endure hostile environments. Enhancing fluidity requires changing the composition of the membrane, which is accomplished by adding steric restrictions to decrease interactions and change the packing order in the membrane structure (Shivaji & Prakash, 2010; Pathania et al., 2022).

Anti-Freezing Proteins

Psychrophiles produce special proteins called antifreeze proteins (AFPs) in reaction to low temperatures. These proteins attach to ice and stop it from growing and recrystallizing (Munoz et al., 2017). All cell components must have these proteins in order for them to operate properly at low temperatures. Psychrophiles with genetic alterations are able to produce enzymes that are active in cold environments. Antifreeze proteins found in a wide range of animals, such as fungus, bacteria, plants, insects, and crustaceans, have been shown to possess varying structures and characteristics. To improve their quick reaction time and stability, these proteins are frequently lipidated and/or glycosylated (Casillo et al., 2017). Antifreeze activity has been shown by a number of bacteria, including *Bacillus aquamarinus*, *Psychrobacter* sp., *Enterobacter agglomerans*, *Pseudomonas fluorescens*, *Micrococcus cryophilus*, *Sphingomonas* sp., *Halomonas* sp., *Pseudoalteromonas* sp., *Stenotrophomonas maltophilia, and Marinomonas protea* (Santagio et al., 2016).

Cold Shock Proteins

When an organism is exposed to low temperatures, several genes are regulated, which causes the organism to produce cold-shock proteins (Csps). These proteins are essential for the processes of translation, transcription, folding of proteins, and control of membrane permeability and fluidity (Phadtare, 2004). Translocation and translation are impacted by cold temperatures because they alter protein folding efficiency, membrane fluidity, and the synthesis of secondary structures in DNA and RNA. Protein stress is increased when bacteria produce cold-induced proteins (Cips) in response to a fast drop in temperature (Jung et al., 2010). Csps are tiny, mostly 65–75 amino acid long proteins that bind nucleic acids. They are highly conserved, however their thermostability varies. As chaperones, they inhibit mutations that alter RNA structure, prevent RNA misfolding to preserve functional conformation, and are linked to the production of persister cells and biofilms, which help bacteria adapt to environmental stress. For example, *Psychromonas arctica*, a psychrophilic organism, produces a CspA that aids in cell survival in cold environments. Important in this regard are also the cold-adaptive chaperones Cpn60 and Cpn10, which Agilent Technologies markets as Arctic Express (Vancuren & Hill, 2019).

Synthesis of Crypprotectants

Osmotic shrinkage and dehydration of the cell interior brought on by cryoinjuries cause enzyme deactivation and adverse effects on the survival and function of the injured cell. Psychrophiles, on the other hand, offset these effects by generating a variety of unique cryoprotectants, such as biosurfactants, EPS, ice-nucleating proteins, and suitable solutes (Bar-Dolev et al., 2016). Low molecular weight and non-toxic cryoprotectants such as betaine, glycine, glycerol, mannitol, sarcosine, sorbitol, sucrose, and trehalose are examples of organic osmolytes synthesized by psychrophiles. These organic osmolytes build up inside the cell to stop shrinkage and stop the solution from freezing inside the membrane, protecting the cell from cryoinjuries when it's freezing (Sun et al., 2016).

Protein Folding

The bulk of enzymatic processes in microorganisms are catalyzed by enzymes, which function differently depending on temperature. Metabolic flow in mesophiles and psychrophiles is similar, suggesting that the latter have evolved to counterbalance the chilling influence on response rates. Psychrophiles are organisms that prefer cold temperatures. Studies on the structure have demonstrated a connection between particular alterations in the content of amino acids and adjustments in the thermostability of psychrophilic enzymes. Even at temperatures below the freezing point of water, substrates can be accommodated because of the heightened flexibility in the molecular structure (Pathania et al., 2022).

4. Psychrozymes and Their Importance

Compared to mesophilic enzymes, cold-adapted enzymes exhibit higher catalytic activity at low temperatures because of their active site changes and structural flexibility, which make substrate binding simpler. Studies conducted on cold-adapted enzymes have shown a number of unique characteristics that set them apart from their mesophilic and thermophilic cousins (Al-Maqtari et al., 2019). These features include decreased hydrophobicity of the core, increased hydrophobicity of the surface, and decreased electrostatic interactions, fewer disulfide bridges, more glycine residues, increased development of loop structures, and less proline residues (Siddiqui et al., 2013).

Usually thermolabile, cold-active enzymes can be rendered inactive by a little amount of heat after a reaction. This property is especially important in sequential reactions where it's required to deactivate an enzyme once it's done its job while maintaining the right circumstances for other enzymes that are engaged in the process. Within the α/β hydrolase fold family, esterase is among the most extensively researched classes of enzymes. Unlike lipases, which prefer long-chain triglycerides, esterases typically hydrolyze soluble esters with short acyl chains (less than 10 carbons). A catalytic trio consisting of Cys, His, and Asp has been found within the active site of esterases by structural research. Their structure is characterized by amphiphilic α-helices at the periphery and centre hydrophobic β-sheets, and they belong to the α/β hydrolase fold family (Al-Maqtari et al.,

2019). Cold-active enzymes offer an additional benefit in the food industry by preventing bacterial contamination and undesirable chemical side reactions that may occur at high temperatures. Because of their advantageous characteristics, cold-active enzymes hold promise for various industrial applications and have garnered significant interest in recent times (Hamid et al., 2022; Kuddus et al., 2024).

Lipase

The enzyme lipase, which is essential for the digestion of fats, breaks down triglycerides into fatty acids and glycerol. Triacylglycerol acyl hydrolases, which are also referred to as lipases, are capable of carrying out this hydrolysis in conditions that include or do not contain water. Cold-adapted lipases from microbes from Antarctica, the Polar Regions, deep sea settings, and chilled food samples have been studied by researchers. These lipases were contrasted with other psychrotrophic lipases and lipases that were cold-adapted and had the canonical hydrolase fold. Psychrophilic lipases are unique in that they can tolerate substrates at low temperatures due to structural alterations that provide them great flexibility (Gerday et al., 1997).

Cold-adapted lipases (CLPs) typically have fewer proline residues, lesser arginine residues relative to lysine residues, a weaker hydrophobic core, a smaller number of salt bridges, and fewer aromatic-aromatic interactions than their counterparts, mesophilic lipases, which are adapted to moderate temperatures. By comparison, psychrophilic lipases show reduced ion pair counts, restricted charge-dipole interactions inside helices, and increased concentration and buildup of glycine residues (which improve local mobility). Because of these structural features, psychrophilic lipases can catalyze hydrolysis even in low-water conditions and efficiently accept substrates at low temperatures. Because of their greater flexibility and adaptability, psychrophilic lipases are preferred over their mesophilic and thermophilic counterparts, which typically have lower flexibility and more stiffness (Joseph et al., 2008).

Buchon and colleagues (2000) highlighted the significant potential of psychrophilic lipases for various applications. These enzymes show promise in addressing pollution in cold environments contaminated with fats, as well as in the synthesis of active compounds under cold conditions. They are valuable in diverse fields including detergent formulation, molecular biology research, food industry additives, environmental cleanup, biotransformation processes, and utilizing psychrophilic hosts for gene expression to prevent the formation of inclusion bodies. Waste disposal by lipase biotechnology-based bioremediation is a promising strategy. The potential of cold-adapted organophosphorus acid anhydrolases to effectively detoxify pesticides and nerve agents has been investigated. Furthermore, the production of active chemicals in cold settings, bioremediation of fat-contaminated cold environments, and wastewater treatment are all areas where cold-adapted lipases exhibit great potential (Joseph et al., 2008).

Cold-adapted microorganisms found naturally in cold locations, such as Alpine soils, aid in the biodegradation of petroleum hydrocarbons. These microbes are able to break down pollutants by generating enzymes that are active in the cold. CLPs have great promise for a number of uses, such as the bioremediation of oil-contaminated cold settings, the synthesis of active chemicals in cold environments, and the treatment of wastewater. Research has revealed that psychrotrophic bacteria, such *Acinetobacter* sp., produce cold-active lipases that effectively hydrolyze soybean oil's triglycerides at low temperatures. Furthermore, a number of strains that can degrade oil at low temperatures have been identified by researchers, suggesting the possibility of in situ bioremediation or bio-augmentation in cold habitats polluted by oil (Kavitha, 2016).

Protease

Cold-active proteases are essential to many technical and industrial operations. They are especially useful in the food sector, where they act as a microbial substitute for rennet in the ripening of cheeses that need low temperature and low moisture. These enzymes are also used in the processing of seafood to take the skin and scales off of fish and to extract the carotenes from shellfish. They also aid in the preparation of meat by giving chilled goods a soft texture without sacrificing flavour (Huston, 2008). Numerous cold-active proteases that come from both psychrophilic and psychrotolerant yeasts have been found and examined. Maximum activity of these enzymes is usually observed in situations with somewhat low pH values, usually in the range of 3.5–4.5 (Ray et al., 1992). An Antarctic strain of *Cryptococcus humicola* (now known as *Vanrija humicola*) produced a cold-active aspartyl proteinase with a molecular mass of 36 kDa, which is a noteworthy finding from the early 1990s. This enzyme exhibits activity from 0 to 45°C, a broad temperature range, with its highest activity being measured at 37°C. It is astonishing because even at 0°C, it maintains around 15% of its maximal activity.

Three different proteases were compared: a mesophilic alkaline protease from *Pseudomonas aeruginosa*, a psychrophilic alkaline protease (PAP) from an Antarctic *Pseudomonas* strain, and a marine protease from *Flavobacterium* YS-80. The investigation showed that in the psychrophilic alkaline protease, the Zn2+-Tyrosine-OH bond displays increased flexibility. This flexibility keeps the enzyme active at low temperatures and improves substrate accessibility. As a result, at 4°C and 25°C, the specific activity of the psychrophilic protease is four times higher than that of its mesophilic equivalents. Additionally, the psychrophilic protease shows exceptional catalytic effectiveness at low temperatures and a much shorter half-time of denaturation—roughly 10 times quicker. Waste management in cold locations is best served by enzymes that are tolerant of low temperatures or that are acclimated to cold temperatures. This is especially helpful because the freezing temperatures in these settings reduce the ability of local microbes to degrade materials (Joshi & Satyanarayana, 2013).

Amylase

Because of their extensive biotechnological uses in a variety of industrial processes, including starch liquefaction, textiles, food preparation, and baking, amylases are extremely important enzymes. Exo-amylases and endo-amylases are the two main categories into which they may be divided based on the differences in their catalytic processes. Of all the endo-amylases, α-amylase (EC 3.2.1.1) is of special significance as it is an industrial enzyme that can hydrolyze substrates' internal α-1,4 glycosidic bonds to generate glucose, maltose, and dextrin while maintaining the α-anomeric conformation in the end products (Gupta et al., 2003).

When it comes to water treatment and bioremediation in cold regions, an amylase enzyme that is suited to cold conditions is developed from Glaciozyma antarctica PI12. Similar enzymes have been found to be synthesized by a number of psychrophilic bacteria, including *Schwanniomyces alluvius*, *Lipomyces kononenkoae*, *Candida antarctica*, *Trichosporon pullulans,* and *C. flavus*. Cold-active amylases are essential for catalyzing industrial processes in cold environments, which lowers the danger of fouling and saves energy. However, oxidative conditions at low temperatures restrict the efficacy of psychrophilic α-amylases, leading to poor stability and limiting their economic value (Ramli et al., 2013; Hamid et al., 2014).

Cellulase

Hydrolytic enzymes called cellulases are essential for converting the polysaccharide cellulose into shorter polysaccharides, oligosaccharides, and monosaccharides like glucose. Their wide range of uses in industries like textiles, agriculture, medicine, food and feed production, pulp and paper, detergents, and laundry contribute to their economic worth. These enzymes are used in a variety of applications, including as the softening of textiles, the biopolishing of denim, the processing of coffee, the creation of wood pulp, the bioremediation of industrial waste, and the treatment of phytobezoars. The main producers of cellulases are bacteria, protozoa, and fungus. A linker region, a carbohydrate-binding module, and a catalytic module are the usual components of microbial cellulases. Psychrophilic cellulases differ from mesophilic cellulases in that they have a unique linker region with a repeating sequence that is rich in serine, glycine, proline, valine, and threonine residues. In their linker region, they also show increased amounts of cysteines, asparagine, serine, glycine, aspartate, and threonine. In contrast to their mesophilic counterparts, these structural variations increase the psychrophilic cellulases' flexibility and enzymatic activity at low temperatures (Pulicherla et al., 2011).

β-Galactosidase

The majority of cold-active β-D-galactosidases are derived from psychrophilic and psychrotolerant bacteria, with the majority of these enzymes being recovered from Arthrobacter sp. This preference for Arthrobacter sp. is explained by the fact that many of these strains are derived from soil samples, and soil environments are often the source of Arthrobacter sp. Thus far, only two eukaryotic microorganisms' cold-active β-D-galactosidases—the yeast-like fungus *Guehomyces pullulans*, in particular—have been discovered, isolated, and biochemically characterized. One of the bacterial strains known to produce cold-active β-D-galactosidases that has been isolated from cold environments is Arthrobacter sp. C2-2, which was found in soil in a fell field (Song et al., 2010; Nakagawa et al., 2006). The capacity of psychrophilic β-galactosidases to effectively catalyze lactose breakdown, even at lower temperatures, speeds up processes and lowers total process costs by conserving energy, is only one of their many benefits. They also help to improve dairy products' flavour, solubility, sweetness, and digestibility while reducing the chance of contamination. The enzymes in question are demonstrated by their ability to eliminate lactose from chilled milk and, by means of pectinase, to decrease viscosity and turbidity in chilled juices.

Nuclease

Enzymes that degrade nucleic acids, known as nucleases, are a significant family of hydrolases with a wide range of uses in both research and industry. A nuclease from psychrophilic organisms may be synthesized and purified, which has great potential as a molecular biology tool. It has been possible to find and characterize a nuclease from the psychrophilic bacteria Psychromonas ingrahamii, which grows exponentially at 5°C and can survive at temperatures as low as 12°C. After cloning and expressing the putative endonuclease I gene in Pichia pastoris, the nucleolytic characteristics of the purified recombinant protein were studied.

There are now three nucleases that are sold commercially: Benzonase® (Merck), HL-SAN (ArcticZymes), and CryonaseTM (Takara, Japan). Whereas Benzonase® comes from the mesophilic bacteria *Serratia marcescens*, CryonaseTM is derived from the psychrotrophic *Shewanella* sp. strain AC10 (Yang et al., 2011; Pan et al., 2013; Mangiagalli et al., 2020).

Xylanses

Enzymes that break down the polysaccharide beta 1,4-xylan and hemicellulose, which is the main building block of plant cell walls, are known as xylanase enzymes and glycoside hydrolases. Xylanases are frequently found in industrial dough conditioners, which are used to improve the quality of bread. A new family of xylanases has been identified from a bacteria found in the Antarctic. These xylanases help create big, soft, elastic loaves of bread with a fluffy, yet robust, dough by breaking down the insoluble hemicellulose in the dough into soluble sugars prior to baking. This increased efficiency is thought to be related to xylanase's unique method of xylan hydrolysis as well as its psychrophilic activity during the dough's obligatory cold temperature resting phase (Feller, 2013). Three psychrophilic xylanases from *Flavobacterium* sp. MSY-2 were shown to be more effective than mesophilic xylanases from *Aspergillus aculeatus* and *B. subtilis* in a recent study. *P. haloplanktis* TAH3A and a bacterium of unknown origin markedly enhanced the final bread volume and dough quality (up to 28%) (Maciejewska et al., 2019).

Pectinase

Pectin is a component of plant cell walls that is reduced to simpler molecules like galacturonic acids by the enzyme pectinase. Pectin lyases, polygalacturonases, and pectin esterases are the three categories of pectinases. One of the biggest problems facing the commercial food and beverage business is that mesophilic proteins are typically heated, which can negatively impact the finished product's organic qualities and nutritional value (Garg et al., 2016). Using cold-adapted pectinase increases fruit mash extraction capacity, aids in the thinning of natural juice products at low temperatures, and breaks down jam structure to produce higher yields of raw juice products. With today's technology, psychrophilic acidic pectinases are used in many different sectors. These include retting in textiles, degumming fibre crops, enhancing paper quality through fermentation, supporting oil extraction procedures, and cleaning wastewater (Margesin, 2005).

Uracil-DNA Glycosylase

Psychozymes, such as uracil-DNA N-glycosylase (UNG) and double-strand-specific DNase, are crucial components of molecular biology. Uracil is extracted from DNA that contains this nucleotide by UNG. This enzyme is very helpful in avoiding cross-contamination when using methods like SNP genotyping, RT-PCR, PCR amplification, and site-directed mutagenesis. However, DNase, which was created specially to break down double-stranded DNA, does so without damaging single-stranded DNA, including primers and probes. DNase eliminates genomic DNA from RNA preparations, aiding in the detoxification of PCR mixes. In order to ensure that these enzymes' effects are eradicated in later phases of investigation, it is possible to enhance their selective and irreversible heat inactivation by using psychrophilic versions (Sarmiento et al., 2015).

PinNuc Enzyme

PinNuc, an enzyme that was recently found, demonstrates characteristics of nonselective endonucleases. After PinNuc was isolated from Pichia pastoris and homogeneously purified, its activity was evaluated with a range of substrates that represented different nucleic acid kinds and shapes. Remarkably, even in solutions with low ion strengths and little magnesium ion present, the enzyme exhibits substantial activity at room temperature. Furthermore, the enzyme, which has six cysteine residues that are probably involved in disulfide bridges, may be efficiently deactivated with a modest amount of a reducing agent. Only when PinNuc is oxidized does it become active. As an alternative to currently available commercially accessible counterparts, this disclosed nuclease is predicted to expand the variety of bioactive compounds usable in cold conditions and has exciting prospective applications in molecular biology and the biotechnology industry (Awazu et al., 2011).

5. Industrial and Biotechnological Application of Psychrophiles

Waste Management

Yeasts with high performance in biological waste treatment technologies, such as *Saccharomyces cerevisiae*, *Candida boidinii*, and *Yarrowia lipolytica*, have been evaluated for their potential in treating in-house waste fluids and manufacturing or industrial wastewater. In particular, water samples taken from Lake Vanda in Antarctica have been found to include Candida species that are acclimated to freezing temperatures. This discovery presents a potential answer for handling dissolved organic compounds at low temperatures (Goncalves et al., 2009). For the purpose of bioremediating solids and wastewater polluted with hydrocarbons, oils, and lipids, cold-adapted microorganisms, such as Candida species, and their cold-active enzymes, such as oxidases, peroxidases, and catalases, have been offered as alternatives to physico-chemical processes. In addition to being essential in several molecular biology applications, these psychrozymes and the microbes that produce them are also important in environmental bioremediation processes including oil breakdown, composting, and digesting. These microorganisms aid in the conversion of garbage into compost by secreting enzymes that break down particular substances or change them into non-toxic forms. This provides a long-term solution for waste management (Margesin et al., 2002).

Yeasts are important sources of this enzyme because they have been subjected to screening and selection across a broad variety of yeast species based on their phytase activity. Psychrophilic yeasts have proven useful in the biofuel and environmental biotechnology sectors. A class of putative biocatalysts known as phytoses helps reduce ecological phosphorus pollution by facilitating the release of minerals, calcium, and phosphorus from phytic acid. It has been noted in earlier studies that yeast phytases, especially those from *Candida krusei,* are consistent and that cold-adapted yeasts that produce cold-active phytases have potential for reducing phosphorus contamination in cold conditions. Thus, phytotases can be very important in controlling phosphate contamination. Furthermore, lipases from psychrophiles are extremely desirable for the synthesis of biofuels, especially since they may be used for low-temperature trans-esterification catalysis. This emphasizes how crucial cold-active lipases are to the synthesis of biofuels since they support effective conversion activities even in cold environments (Hamid, 2021).

Biomedicines, Pharmaceuticals, and Cosmetics

A large number of drugs, fine chemicals, flavourings, fragrances, and active pharmaceutical ingredients (APIs) must be synthesized at low temperatures, where cold-adapted enzymes function best, due to their volatility and heat sensitivity. Furthermore, it has been suggested that the high degree of structural flexibility exhibited by cold-adapted enzymes allows them to function effectively in low-water environments, such as aqueous/organic and non-aqueous solvent systems, which are widely employed in the chemical synthesis of complex molecules (Karan et al., 2012).

Many different types of enzymes, such as hydrolases, oxidoreductases, lyases, transferases, reductases, and carboxylases, are being employed more and more in this market sector, even though only a limited number of cold-adapted enzymes have been thoroughly explored as mentioned in Figure 1. Among them, lipases and esterases are the most often used cold-adapted enzymes, particularly for the synthesis of optically pure intermediate compounds that are important to industry. It's important to note that lipases—primarily those derived from *Candida antarctica* (CAL B)—are used in a broad range of chemical compound production processes. A few of the remarkably wide range of applications they find themselves in include modification of sugars and sugar-related compounds, desymmetrization of complex prochiral drug intermediates, and resolution of racemic alcohols and amines during the synthesis of various pharmaceuticals (e.g., calcium antagonists for antihypertensive drugs, NK1/NK2 antagonists for the treatment of asthma), cosmetics (e.g., iso-propylmyristate, a skin emollient), flavours, and fragrance esters. Moreover, cold-adapted β-galactosidases are ideally suited for the synthesis of galacto-oligosaccharides (prebiotics) and tagatose (an

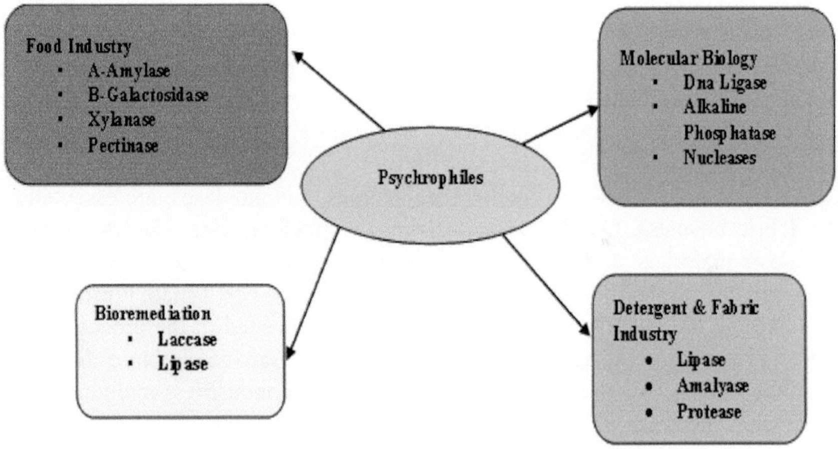

Figure 1: Application of cold adapted enzymes.

anti-hyperglycemic medication). Furthermore, these enzymes have demonstrated their ability to catalyze the synthesis of hetero-oligosaccharides from lactose, such as lactulose (which is used to treat constipation and hepatic encephalopathy), galactosyl-xylose (which is used in diagnostics), glycosylated salicin (an anti-inflammatory agent), and alkyl glycosides (foaming agents) (Kirk & Christensen, 2002; Suen et al., 2004).

Based on research on psychrophilic bacteria from Antarctic species including Colwellia and Shewanella, a considerable number of bacteria suited to life in sea ice have been identified to generate polyunsaturated fatty acids (PUFA), such as eicosapentaenoic acid (EPA) and docosahexaenoic acid (DHA). These long-chain polyunsaturated fatty acids (LPUFA) are beneficial to health in many ways, including the treatment and prevention of diseases including diabetes, atherosclerosis, and high blood pressure. Nevertheless, the disadvantages of getting LPUFA from fish oil exceed its benefits, especially when taking into account the difficulties involved in bulk manufacturing and purification. Furthermore, it's unclear how long commercial fish populations will be available. Although algal-derived oils provide a different way to obtain LPUFA, they are more expensive and technologically demanding than bacterial fermentation, even though bacteria normally contain less lipids than algae. In bacteria, PUFAs are normally present in the cell membrane as parts of certain phospholipids. Generally speaking, LPUFAs) are made up of a carbon chain (usually C20–C22) with four to six methylene-interrupted double bonds. The synthesis of a single PUFA by bacteria is a noteworthy benefit over the complicated combination derived from fish or algae oils (Pulicherla et al., 2011).

Leather and Textile Sectors

Cold-active enzymes have attracted attention in a number of areas, such as low-water biocatalysis, the textile and food industries, bioremediation, and cleaning procedures. Their use is multifaceted; for example, proteases are used in businesses to cause leather or skin items to flake. Psychrozymes may effectively complete this activity under tap water, in contrast to the traditional procedure that calls for heating to 37°C in order for mesophilic biocatalysts to grow. Cold-active enzymes, including as cellulases, laccases, and amylases, are essential to the fabric business. Quick textile desizing or starch removal, bio-finishing of cellulosic materials, and less aggressive enzymatic stonewashing and bleaching of denim are among the applications for which they are best suited. Furthermore, by lowering the quantity of harsh chemicals often needed in the process, cold-active proteases or keratinases are essential to the low-temperature dehairing of animal skins. All things considered, a variety of industries, including food processing, detergents, leather and textiles, beverages, cosmetics, and baking, heavily rely on cold-active enzymes. Among the many benefits they offer are more production and less environmental impact (Sarmiento et al., 2015).

Biotechnology

It has been demonstrated that biotechnological activities that take place at low temperatures are supported by psychozymes. Yeasts are frequently used as model organisms in gene cloning, protein purification, and the study of several biological processes. By changing the characteristics of enzymes to boost their activities, genetic modification can provide a number of advantages. For example, a mutant of *Candida antarctica's* cold-tolerant Lipase B showed higher stability and activity. Furthermore, an Antarctic bacterium's heat-labile alkaline phosphatase has been identified and isolated. These enzymes are used in molecular biology and recombinant technology for a variety of tasks, such as eliminating phosphate groups from DNA vectors before cloning to stop re-circularization, eliminating phosphate groups from 5′-nucleic acid termini before polynucleotide kinase labels the 5′ end, and preventing dNTPs and pyrophosphate from PCR reactions. Additionally, a heat-sensitive uracil-DNA N-glycosylase from the cold-adapted Atlantic cod (*Gadus morhua)* has been used to eliminate DNA impurities in successive polymerase chain reaction (PCR) procedures (Leiros et al., 2003; Goodchild et al., 2004; Jill, 2007).

6. Application of Psychrophiles in Remediation of Cold Environment

Bioremediation

Environmental contamination is always increasing as a result of human activity. The removal of pollutants from the environment by means of organisms or their products, such as enzymes, is known as bioremediation, and it is imperative to address this contamination. Microbial enzymes are thought to be a cost-effective bioremediation tool because of their great efficacy in eliminating contaminants. However, because these enzymes have poor activity in cold settings, bioremediation utilizing mesophilic or thermophilic enzymes is unsuccessful under such conditions. Because of their innate ability to work well in cold temperatures, cold-active enzymes, or CAEs, are useful agents for bioremediation in cold settings as mentioned in Figure 2 (Miri et al., 2019; Okino-Delgado et al., 2019; Bhandari et al., 2021).

There are difficulties with bioremediation in cold climates that are not present in tropical or temperate areas. Remedial techniques are hampered in these areas by elements including low oxygen levels, low temperatures, poor hydrocarbon bioavailability, nutritional scarcity, and the existence of native microbial communities. The particular circumstances that exist in cold settings determine how successful the remediation procedures that are now accessible. Therefore, careful site analysis and taking into account influencing factors are essential for creating effective bioremediation strategies. Before introducing non-indigenous microorganisms into Antarctic regions, regulatory limits should

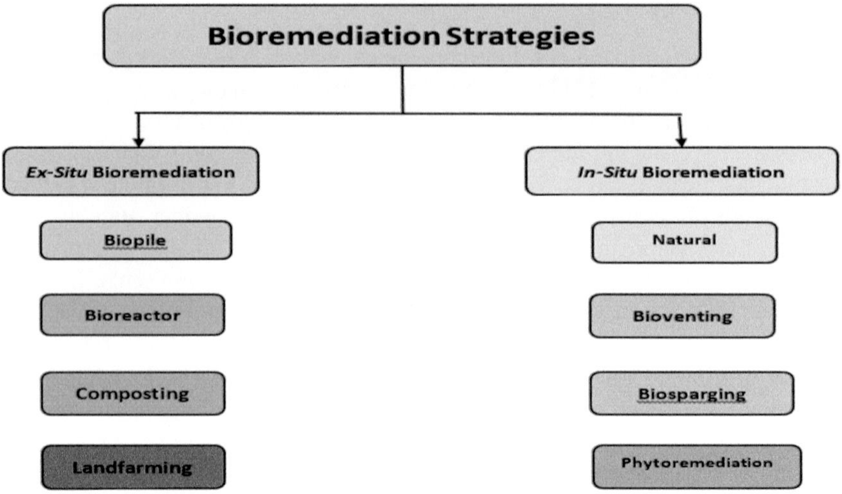

Figure 2: Strategies of bioremediation in a cold environment.

be considered, even if biostimulation and various bioaugmentation methods may be useful instruments.

Re-inoculation with native psychrophilic bacteria that can break down oil is necessary for effective cleanup of contaminated areas. Therefore, one of the most important steps in bioremediation operations is to evaluate the population of psychrophiles that degrade oil. Comparatively speaking to mesophilic bacteria, real psychrophilic bacteria are more difficult to use. Bioaugmentation by itself is still debatable and could not be a very effective tactic. Therefore, the best remediation option for cold climates is to combine bioaugmentation and biostimulation. Psychrophilic oil-degrading bacteria are used in conjunction with bioengineering approaches to overcome obstacles and optimize bioremediation efforts in cold conditions (Chaudhary & Kim, 2019).

Degradation of Petroleum Hydrocarbons in Cold Environment

The capacity of microorganisms to break down petroleum hydrocarbons and alkanes at low temperatures as well as their adaptation to cold conditions. Microorganisms that have evolved to the cold include cyanobacteria, fungus, bacteria, and psychoactive archaea. Even at 4°C, these organisms have been shown to effectively break down polycyclic aromatic hydrocarbons (PAHs) and petroleum hydrocarbons. These bacteria' capacity to break down hydrocarbons is aided by an increase in cell membrane fluidity at low temperatures. *Colwellia psychrerythraea* 34H, for instance, modifies the ratio of trans- to cis-esterified fatty acids in phospholipids, improving the fluidity of the membrane at low temperatures. It has been shown that anteiso-C15 fatty acid content is greater in psychrophilic Bacillus species, which also aids in membrane fluidity.

However, because of lower enzymatic activity rates, the biodegradation rate of petroleum hydrocarbons tends to decrease with lowering temperatures. Furthermore, oil's increased viscosity at low temperatures restricts its ability to spread in soil and aquatic ecosystems. The volatilization of short-chain alkanes is further slowed down by oil's decreased solubility in water at low temperatures. Psychrophilic microbes generate hydrocarbon-degrading enzymes, which are expressed and degrade hydrocarbons at low temperatures thanks to genes encoding these enzymes. These microbes also generate cryoproteins, which have an impact on the low-temperature folding and synthesis of proteins. Certain cold-adapted microbes have hydrocarbon-degrading enzymes outside of their cytoplasm, and they use sigma-70 transcription factors to control the production of extracellular polysaccharides under cold conditions. Overall, cold-adapted microorganisms have evolved various mechanisms to efficiently degrade hydrocarbons in cold environments, which is significant for bioremediation efforts in cold regions (Rajkumari et al., 2019).

Biocontrol of *Penicillium expansum* in Apples

Effective biocontrol capabilities are exhibited by the Antarctic yeast strain Candida sake 41E, especially in lowering the incidence of blue mould on apples during cold storage. Its ability to regulate biological processes continues to work at room temperature, allowing fruits to be stored for up to five days. Furthermore, this strain has the capacity to reduce the content of patulin in apple juice by almost 80% at room temperature—all without just binding the mycotoxin to its cell walls. There is a theory that the mycotoxin is biologically degraded within cells. According to these results, strain 41E may be useful as a biocontrol agent against blue mould and may also help to lessen patulin contamination, particularly when total inhibition of *P. expansum* rot is not attained. Furthermore, its possible application as a detoxifying agent for apple trash polluted with patulin may help turn such waste into environmentally beneficial feed sources (Alvarez et al., 2019).

Biocontrol and Growth Promotion in Wheat at Low Temperatures Using *Eurypsychrophilic pseudomonas* Spp.

Isolates of *Erypsychrophilic pseudomonas* spp. that were kept for long periods of time in tropical glacier ice have shown amazing capacities to support wheat plantlet germination and development, even at low temperatures. They also successfully protect young plantlets from *P. ultimum*, a phytopathogenic oomycete, and its assaults. Their generation of phytohormones like IAA and the release of volatile and soluble hazardous compounds are responsible for these advantageous effects. Making use of this unanticipated microbiological resource might result in the creation of bio-inoculants that are incredibly effective. These bio-inoculants not only fight phytopathogens and promote plant development, but they also continue to function in cold climates. Their possible use has great potential to advance ecologically friendly mountain agriculture methods (Rondón et al., 2019).

Controlling Ice-Water Phase Changes in Frost-Susceptible Soils Using Psychrphiles

A study done by Rahman et al. (2024) using extracellular proteins from psychrophilic bacteria to prevent ice from growing in soil that is sensitive to frost. Samples of soil combined with water or porous media showed a thermal hysteresis of 4.62°C, freezing values of –4.59°C, and thawing points of 0.03°C. The outcomes showed that freezing point depression was significantly increased by protein solutions made from psychrophilic bacteria. With the lowest freezing point of –8.85°C on day 21 at a protein content of 1.55 mg/ml, *Polaromonas hydrogenivorans* demonstrated the highest control over freezing point depression among the studied strains. *Polaromonas hydrogenivorans* had the lowest temperature range of –1.14 to –1.32°C among all bio-treated soils, indicating that thawing had completed below 0°C. When compared to water-soil mixes, the extracellular solutions of psychrophilic bacteria significantly increased thermal hysteresis activity in frost-susceptible soils, with values ranging from 7.04 to 7.71°C. These results imply that ice-binding proteins (IBPs), such as CSPs or AFPs, or both, may be present in the protein solutions. These proteins alter the characteristics of water and ice, which affects how the soil freezes and thaws. Characterizing these psychrophilic IBPs may provide information on their role in the dynamics of soil freezing and thawing in the future.

Psychrophilic Bacterial Phosphate Biofertilizers for Sustainable Crop Production in Cold Environments

Significant crop losses occur worldwide as a result of abiotic stressors, especially those that are associated with low temperatures and their adverse impact on the physiological activities, composition, and structure of soil micro biomes. Psychrophilic and psychrotolerant phosphate-solubilizing bacteria (PSB), on the other hand, are interesting reservoirs of low-temperature-tolerant bacteria that can survive and continue to function in severe cold environments. To efficiently resist cold stress, these bacteria generate antifreeze proteins and turn on stress-responsive genes. Their capacity to improve plant development by providing phosphorus, a crucial nutrient, especially in soil lacking in phosphorus, is one of their noteworthy functional characteristics. Furthermore, the overall performance of stressed plants is enhanced by cold-tolerant rhizosphere microorganisms. *Pseudomonas*, *Acinetobacter*, *Bacillus*, and *Stenotrophomonas* are just a few of the genera of cold-active phosphate-solubilizing bacteria that have great promise for converting nutrient-deficient soils into nutrient-rich soils under low temperatures. By using these cold-adapted PSB strains, it may be possible to reduce the need for chemical fertilizers in intensive agricultural systems by increasing the production of vital food crops including cereals, vegetables, and legumes in colder climates. To understand the molecular processes behind these cold-active PSB strains' capacity to promote development and to clarify how they preserve their functional characteristics in the face of extreme cold, more investigation is required. More crop yield in colder climates might result from the development of ecologically acceptable cold-active phosphate biofertilizers, creating new opportunities for sustainable agriculture (Rizvi et al., 2021).

7. Conclusion

Our knowledge of the physiology and molecular adaptations of psychrophilic species, as well as their mode of existence, has been substantially expanded by studying the microbial communities present in extremely cold conditions. The use of psychrophilic enzymes through methods like protein engineering and evolution will benefit from this understanding. There's growing interest in mining difficult environments for effective biocatalysts as the need for affordable, sustainable, and eco-friendly substitutes for harsh chemicals in various industrial processes develops. Even with the many benefits, there are obstacles in the way of cold-adapted enzyme technology's transfer from lab to commercial use. It is anticipated that more cold-active and cold-adapted microbial enzymes will be found in the near future due to their potential and highly desirable characteristics for industrial applications. Cold-active enzymes generated from psychrophilic bacteria have several advantages over other types of enzymes, especially in the food industry. They are employed in many different industries for a broad variety of purposes. In the food sector, cold-active enzymes from different microbial sources are used for baking, cheese-making, fermenting, and softening meat, among other operations. These enzymes are also helpful constituents in cold-washing detergents. These enzymes originate from psychrophilic bacteria and have intriguing biotechnological and commercial possibilities.

References

Afrouz, M., Sayyed, R.Z., Fazeli-Nasab, B., Piri, R., Almalki, W.H., & Fitriatin, B.N. 2023. Seed bio-priming with beneficial *Trichoderma harzianum* alleviates cold stress in maize. *PeerJ.*, 11, e15644. http://doi.org/10.7717/peerj.15644.

Al-Ghanayem, A.A., & Joseph, B. 2020. Current prospective in using cold active enzymes as eco-friendly detergent additive. *Appl. Microbiol. Biotechnol.*, 104, 2871–2882.

Al-Maqtari, Q.A., Waleed, A.A., & Mahdi A.A. 2019. Cold-active enzymes and their applications in industrial fields: A review. *Int. J. Res. Agric. Sci.*, 6(4), 2348–2399.

Alvarez, A., Gelezoglo, R., & Garmendia, G. 2019. Role of Antarctic yeast in biocontrol of *Penicillium expansum* and patulin reduction of apples. *Environmental Sustainability*, 2, 277–283.

Anesio, A.M., Lutz, S., Chrismas, N.A.M., & Benning, L.G. 2017. The microbiome of glaciers and ice sheets. *NPJ Biofilms Microbiomes*, 3, 1–11.
Awazu, N., Shodai, T., Takakura, H., Kitagawa, M., Mukai, H., & Kato, I. 2011. *Microorganism-derived Psychrophilic Endonuclease*. United States Patent US8034597B2.
Baba, Z.A., Hamid, B., Sheikh, T.A., Alotaibi, S., Enshasy, H.E., Ansari, M.J. et al. 2021. Psychrotolerant *Mesorhizobium* sp. isolated from temperate and cold desert regions solubilize potassium and produces multiple plant growth promoting metabolites, *Molecules*, 26, 5758. https://www.mdpi.com/1420-3049/26/19/5758/htm.
Bar Dolev, M., Braslavsky, I., & Davies, P.L. 2016. Ice-binding proteins and their function. *Annual Review of Biochemistry*, 85, 515–542.
Bhandari, S., Poudel, D.K., & Marahatha, R. 2021. Microbial enzymes used in bioremediation. *J. Chem.*, 2021, 1–17.
Bhatia, R.K., Ullah, S., Hoque, M.Z., Ahmad, I., Yang, Y.H., Bhatt, A.K., & Bhatia, S.K. 2021. Psychrophiles: A source of cold-adapted enzymes for energy efficient biotechnological industrial processes. *Journal of Environmental Chemical Engineering*, 9(1), 104607.
Buchon, L., Laurent, P., Gounot, A.M., & Guespin-Michel, J.F. 2000. Temperature dependence of extracellular enzymes production by psychrotrophic and psychrophilic bacteria. *Biotechnol. Lett.*, 22, 1577–1581.
Casillo, M., Ziaco, B., Lindner Parrilli, D., & Schwudke. 2017. Unusual lipid A from a cold-adapted bacterium: Detailed structural characterization. *Chem. Bio. Chem.*, 18, 1845–1854.
Chaudhary, D.K., & Kim, J. 2019. New insights into bioremediation strategies for oil-contaminated soil in cold environments. *International Biodeterioration and Biodegradation*, 142, 58–72.
D'Amico, S., Collins, T., Marx, J.C., Feller, G., Gerday, C., & Gerday, C. 2006. Psychrophilic microorganisms: Challenges for life. *EMBO Rep.*, 7(4), 385–389.
De Pascale, D., De Santi, C., Fu, J., & Landfald, B. 2012. The microbial diversity of Polar environments is a fertile ground for bioprospecting. *Marine Genomics*, 8, 15–22.
Feller, G. 2013. Psychrophilic enzymes: From folding to function and biotechnology. *Scientifica* (*Cairo*), 2013, 512840.
Garg, G., Singh, A., Kaur, A., Singh, R., Kaur, J., & Mahajan, R. 2016. Microbial pectinases: An ecofriendly tool of nature for industries. *3 Biotech.*, 6, 1–3.
Gerday, C., Aittaleb, M., Arpigny, J.L., Baise, E., & Chessa, J.P. 1997. Psychrophilic enzymes: A thermodynamic challenge. *Biochimica Biophysica Acta*, 1342(2), 119–131.
Goodchild, A., Saunders, N.F.W., Ertan, H., Raftery, M., Guilhaus, M., Curmi, P.M.G., & Cavicchioli, R. 2004. A proteomic determination of cold adaptation in the Antarctic archaeon, Methanococcoides burtonii. *Mol. Microbiol.*, 53, 309–321.
Gupta, R., Gigras, P., Mohapatra, H., Goswami, V.K., & Chauhan, B. 2003. Microbial α-amylases: Biotechnological perspective. *Process Biochem.*, 38, 1599–1616.
Gupta, V., Bhaskar, P., Thoudam, J., Bisht, S., Sharma, A., & Tripathi, R. 2023. Psychrophiles as a novel and promising source of cold-adapted industrial enzymes. *T. Appl. Biol. Chem. J.*, 4(2), 54–68.
Hamid, B., Rana, R.S., Chauhan, D., Singh, P., Mohiddin, F.A., Sahay, S., & Abidi, I. 2014. Psychrophilic yeasts and their biotechnological applications: A review. *African J. Biotechnol.*, 13(22), 2188–2197.

Hamid, B. 2021. Potent biotechnological applications of psychrozymes. In: Lone, S.A., & Malik A. (Eds.), *Microbiomes and the Global Climate Change*. Springer, Singapore, pp. 349–364.

Hamid, B., Bashir, Z., Yatoo, A.M., Mohiddin, F., Majeed, N., Bansal, M. et al. 2022. Cold-active enzymes and their potential industrial applications: A review. *Molecules*, 27(18), 5885.

Hamid, B., Baba, Z.A., Sheikh, T.A., Perveen, K., & Sayyed, R.Z. 2024. Investigating potential protease activity of psychrotrophic bacteria from a municipal landfill for solid waste management. *Biomass Conversion and Biorefinery*. https://doi.org/10.1007/s13399-024-05621-2.

Jill, B.K. 2007. Microorganisms: applications in molecular biology. In: *Encyclopedia of Life Sciences*. Wiley, Hoboken, pp. 1–8.

Joseph, B., Ramteke, P.W., & Thoma, G. 2008. Cold active microbial lipases: Some hot issues and recent developments. *Biotechnology Advances*, 26(5), 457–470.

Joshi, S., & Satyanarayana, T. 2013. Biotechnology of cold-active proteases. *Biology* (Basel), 2(2), 755–783.

Jung, Y.H., Yi, J.Y., Jung, H.J., Lee, Y.K., & Lee, H.K. 2010. Overexpression of cold shock protein of a *Psychromonas arctica* KOPRI 22215 confers cold resistance. *Protein J.*, 29, 136–142.

Karan, R., Capes, M.D., & Dassarma, S. 2012. Function and biotechnology of extremophilic enzymes in low water activity. *Aquat. Biol.*, 8, 1–5.

Kavitha, M. 2016. Cold active lipases: An update. Frontiers in *Life Science*, 9(3), 226–238.

Kirk, O., & Christensen, M.W. 2002. Lipases from *Candida antarctica*: Unique biocatalysts from a unique origin. *Org. Process Res. Dev.*, 6(4), 446–451.

Koh, H.Y., Park, H., Lee, J.H., Han, S.J., Sohn, Y.C., & Lee, S.G. 2017. Proteomic and transcriptomic investigations on cold-responsive properties of the psychrophilic Antarctic bacterium Psychrobacter sp. PAMC 21119 at subzero temperatures. *Environmental Microbiology*, 19(2), 628–644.

Kuddus, M., Roohi, Bano, N., Sheik, G.B., Joseph, B., Hamid, B., Raveendran, Sindhu, & Aravind, Madhavan. 2024. Cold-active microbial enzymes and their biotechnological applications. *Microbial Biotechnology*, 17, e14467.

Kumar, A., Mukhia, S., & Kumar, R. 2021. Industrial applications of cold-adapted enzymes: Challenges, innovations, and future perspective. *3 Biotech.*, 11, 1–18.

Kumar, S., Sravani, B., Korra, T., Behera, L., Datta, D., Dhakad, P.K. & Yadav, M.K. 2022. Psychrophilic microbes: Biodiversity, beneficial role, and improvement of cold stress in crop plants. In: *New and Future Developments in Microbial Biotechnology and Bioengineering*. Elsevier, pp. 177–198.

Kumar, V., Thakur, V., Kumar, S., & Singh, D. 2018. Bioplastic reservoir of diverse bacterial communities revealed along altitude gradient of Pangi-Chamba trans-Himalayan region. *FEMS Microbiol. Lett.*, 365(14), fny144.

Leiros, I., Moe, E., Lanes, O., Smalas, A.O., & Willassen, N.P. 2003. The structure of uracil-DNA glycosylase from Atlantic cod (*Gadus morhua*) reveals cold-adaptation features. *Acta Crystallogr. D. Biol. Crystallogr.*, 59, 1357–1365.

Maciejewska, N., Walkusz, R., Olszewski, M., & Szymańska, A. 2019. New nuclease from extremely psychrophilic microorganism *Psychromonas ingrahamii* 37 identification and characterization. *Mol. Biotechnol.*, 61, 122–133.

Mangiagalli, M., Brocca, S., Orlando, M., & Lotti, M. 2020. The "cold revolution": Present and future applications of cold-active enzymes and ice-binding proteins. *New Biotechnol.*, 55, 5–11.

Margesin, R., Fauster, V., & Fonteyne, P.A. 2005. Characterization of cold-active pectate lyases from psychrophilic Mrakia frigida. *Lett. Appl. Microbiol.*, 40(6), 453–459.

Margesin, R., Zacke, G., & Schinner, F. 2002. Characterization of heterotrophic microorganisms in alpine glacier cryoconite. *Arctic, Antarctic, and Alpine Research*, 34(1), 88–93.

Miri, S., Naghdi, M., Rouissi, T., Kaur Brar, S., & Martel, R. 2019. Recent biotechnological advances in petroleum hydrocarbons degradation under cold climate conditions: A review. *Critical Reviews in Environmental Science and Technology*, 49(7), 553–586.

Muñoz, P.A., Márquez, S.L., González-Nilo, F.D., Márquez-Miranda, V., & Blamey, J.M. 2017. Structure and application of antifreeze proteins from Antarctic bacteria. *Microbial Cell Factories*, 16, 1–3.

Nakagawa, T., Ikehata, R., Uchino, M., Miyaji, T., Takano, K., & Tomizuka, N. 2006. Cold-active acid β-galactosidase activity of isolated psychrophilic-basidiomycetous yeast *Guehomyces pullulans*. *Microbiol Res.*, 161, 75–79.

Okino-Delgado, C.H., Zanutto-Elgui, M.R., do Prado, D.Z., Pereira, M.S., & Fleuri, L.F. 2019. Enzymatic bioremediation: Current status, challenges of obtaining process, and applications. In: *Microbial Metabolism of Renobiotic Compounds*. pp. 79–101.

Pan, Y., Xiao, L., Li, A.S., Zhang, X., Sirois, P., Zhang, J., & Li, K. 2013. Biological and biomedical applications of engineered nucleases. *Mol. Biotechnol.*, 55, 54–62.

Pathania, S., Solanki, P., Putatunda, C., Bhatia, R.K. & Walia, A. 2022. Adaptation to cold environment: The survival strategy of psychrophiles. In: *Survival Strategies in Cold-adapted Microorganisms*. pp. 87–111.

Phadtare, S. 2004. Recent developments in bacterial cold-shock response. *Curr. Issues Mol. Biol.*, 6(2), 125–136.

Pulicherla, K.K., Ghosh, M., Kumar, P.S., & Rao, K.R. 2011. Psychrozymes – The next generation industrial enzymes. *J. Mar. Sci. Res. Dev.*, 1(2), 102.

Rahman, R., Bheemasetti, T.V., & Govil, T. 2024. Psychrophiles to control ice-water phase changes in frost-susceptible soils. *Sci. Rep.*, 14, 47.

Rajkumari, J., Bhuyan, B., & Das, N. 2019. Environmental applications of microbial extremophiles in the degradation of petroleum hydrocarbons in extreme environments. *Environmental Sustainability*, 2, 311–328.

Ramli, A.N., Azhar, M.A., Shamsir, M.S., Rabu, A., Murad, A.M., Mahadi, N.M., & Illias, R.M. 2013. Sequence and structural investigation of a novel psychrophilic α-amylase from *Glaciozyma antarctica* PI12 for cold-adaptation analysis. *J. Mol. Model.*, 19, 3369–3383.

Rapuano, R., & Graziano, G. 2022. Some clues about enzymes from psychrophilic microorganisms. *Microorganisms*, 10(6), 1161.

Ray, M.K., Devi, K.U., Kumar, G.S., & Shivaji, S. 1992. Extracellular protease from the Antarctic yeast *Candida humicola*. *Applied and Environmental Microbiology*, 58(6), 1918–1923.

Richard Morita, Y. 1975. Psychrophilic Bacteria. *Bacteriologic. Rev.*, 39(2), 144–167.

Rizvi, A., Ahmed, B., Khan, M.S., Umar, S., & Lee, J. 2021. Psychrophilic bacterial phosphate-biofertilizers: A novel extremophile for sustainable crop production under cold environment. *Microorganism*, 9(12), 2451.

Rondón, J.J., Ball, M.M., & Castro, L.T. 2019. Eurypsychrophilic *Pseudomonas* spp. isolated from Venezuelan tropical glaciers as promoters of wheat growth and biocontrol agents of plant pathogens at low temperatures. *Environmental Sustainability*, 2, 265–275.

Santiago, M., Ramírez-Sarmiento, C.A., Zamora, R.A., & Parra, L.P. 2016. Discovery, molecular mechanisms, and industrial applications of cold-active enzymes. *Frontiers in Microbiology*, 7, 222937.

Sarmiento, F., Peralta, R., & Blamey, J.M. 2015. Cold and hot extremozymes: Industrial relevance and current trends. *Front. Bioeng. Biotechnol.*, 3, 148.

Saxena, A.K., Yadav, A.N., Rajawat, M.V.S., Kaushik, R., Kumar, R., Kumar, M. et al. 2016. Microbial diversity of extreme regions: An unseen heritage and wealth. *Indian J. Plant Genet. Resour.*, 29(3), 246–248.

Sheikh, T., Baba, Z., Hamid, B., Iqbal, S., Yatoo, A., Fatima, S. et al. 2022. Extracellular polymeric substances in psychrophilic cyanobacteria: A potential bioflocculant and carbon sink to mitigate cold stress. *Biocatalysis and Agricultural Biotechnology*, 42, 102375. https://doi.org/10.1016/j.bcab.2022.102375.

Shivaji, S., & Prakash, J. 2010. How do bacteria sense and respond to low temperature? *Arch. Microbiol.*, 192, 85–95.

Siddiqui, K.S., Williams, T.J., & Wilkins, D. 2013. Psychrophiles. *Annu. Rev. Earth Planet Sci.*, 41, 87–115.

Singh, R.N., Gaba, S., Yadav, A.N., Gaur, P., Gulati, S., Kaushik, R., & Saxena, A.K. 2016. First high quality draft genome sequence of a plant growth promoting and cold active enzyme producing psychrotrophic *Arthrobacter agilis* strain L77. *Standards in Genomic Sciences*, 11, 1–9.

Song, C., Liu, G.L., Xu, J.L., & Chi, Z.M. 2010. Purification and characterization of extracellular β-galactosidase from the psychrotolerant yeast *Guehomyces pullulans* 17–1 isolated from sea sediment in Antarctica. *Process. Biochem.*, 45, 954–960.

Suen, W.C., Zhang, N., Xiao, L., Madison, V., & Zaks, A. 2004. Improved activity and thermostability of *Candida antarctica* lipase B by DNA family shuffling. *Protein Eng. Des. Sel.*, 17(2), 133–140.

Sun, M.L, Zhao, F., Shi, M., Zhang, X.Y., Zhou, B.C., Zhang, Y.Z., & Chen, X.L. 2015. Characterization and biotechnological potential analysis of a new exopolysaccharide from the Arctic marine bacterium *Polaribacter* sp. SM1127. *Scientific Reports*, 5(1), 18435.

Thakur, V., Kumar, V., Kumar, S., & Singh, D. 2018. Diverse culturable bacterial communities with cellulolytic potential revealed from pristine habitat in Indian trans-Himalaya. *Can. J. Microbiol.*, 64(11), 798–808.

Turchetti, G., Marconi, C., Sannino, P., Buzzini, E., & Albertini. 2020. DNA methylation changes induced by cold in psychrophilic and psychrotolerant Naganishia yeast species. *Microorganisms*, 8, 296.

Vancuren, S.J., & Hill, J.E. 2019. *Update on CPN DB: A Reference Database of Chaperonin Sequences*. Database 033.

Yadav, A.N., Kumar, V., Prasad, R., Saxena, A.K., & Dhaliwal, H.S. 2018. Microbiome in crops: Diversity, distribution and potential role in crops improvements. In: Prasad, R., Gill, S.S., & Tuteja, N. (Eds.), *Crop Improvement through Microbial Biotechnology*. Elsevier Enfield, pp. 305–332.

Yadav, A.N., Verma, P., Kumar, V., Sachan, S.G., & Saxena, A.K. 2017. Extreme cold environments: A suitable niche for selection of novel psychrotrophic microbes for biotechnological applications. *Adv. Biotechnol. Microbiol.*, 2(2), 1–4.

Yang, W. 2011. Nucleases: Diversity of structure, function, and mechanism. *Q. Rev. Biophys.*, 44(1), 1–93.

CHAPTER

10

Biodegradation of Environmental Contaminants by Psychrophiles and Psychrozymes

Manuraj P, Neha Bharti, Priti Raj and Garima Kaushik*

Department of Environmental Sciences, Central University of Rajasthan, NH-8, Bandarsindri, Kishangarh, Ajmer, 305817, India

1. Introduction

Pollutants in the environment can exist in various media, including air, soil, dust, food, water, and consumer products (Foster et al., 2017). Wastewater treatment involves hydrolysis, photolysis, and microbial breakdown of pollutants. Recent research has identified byproducts formed when artificial sweeteners like sucralose and acesulfame are broken down by light-photolysis (Lewis & Tzilivakis, 2021). Additionally, scientists were able to pinpoint oxypurinol, a compound resulting from the degradation of the medication allopurinol, in wastewater (Funke et al., 2015). These byproducts can occur at higher concentrations than the original molecules and are stable in the environment. In streams, rivers, and groundwater sources, a maximum of 22.6 μg/L of the microbe-derived product oxypurinol was found following its production in wastewater treatment (Funke et al., 2015). Per- and poly-fluoroalkyl substances (PFAS) are frequently detected in water for consumption, environmental water bodies, biota, and human blood (Kurwadkar et al., 2022). PCBs (polychlorinated biphenyls) are environmentally stable and can accumulate in food chains, possibly damaging both people and ecosystems (Hu et al., 2019).

Pharmaceuticals, illicit drugs, hormonal substances, and antibacterial agents are among the significant pollutants in environmental waters, contaminating waters and potentially affecting human health and ecosystems (Naseem et al., 2025; Chaturvedi et al., 2021). A new generation of flame retardants has emerged to

*Corresponding author: garimakaushik@curaj.ac.in

DOI: 10.1201/9781003717003-10

replace PBDEs (polybrominated diphenyl ethers). These alternatives encompass a variety of chemical classes, including organophosphate esters (both chlorinated and non-chlorinated), brominated aromatic compounds (such as brominated phenyl esters, bromophenols, and bromophthalates), and halogenated cyclic aliphatic substances (containing bromine and chlorine). Global production of brominated flame retardants ranges from 100,000 to 180,000 tons annually. Research (Janna et al., 2011) has also identified environmental contaminants like Toly triazoleenzotriazole in water used for drinking, streams, and sewage, with dishwashing tablets and powders being a significant major source of pollution for both the environment and waterways. Dioxane is challenging to eliminate from water and is not biodegradable, with concentrations exceeding 2,900 μg/L detected in polluted groundwater (Isaacson et al., 2006).

Bacteria can live in various environments, including frigid locales like snow-covered lakes, icy seas, permafrost soils, cloud droplets, rocky environments, and glaciers (Margesin & Collins, 2019). *Halocryophilus* strain Or1 has been isolated from high-polar frozen ground and can develop and replicate at –15°C, with active metabolism observed in a microcosm at –25°C (Mykytczuk et al., 2013). The proliferation of *E. coli* in lakes, water, and sand is correlated with algae growth, with an ideal temperature range for *E. coli* proliferation between 25–35°C, like summer temperatures near a lake's shore (Meyers et al., 2022). Bacteria exhibit various behaviors in response to temperature changes, including heat- or cold-shock reactions (Zhang & Gross, 2021). Streambed silt can serve as a bacterial reservoir, with bacterial quantities often significantly higher than in the water column. Bed sediment can contain several times more *E. coli* and *Cryptosporidium oocysts* than the water column (Pandey et al., 2012).

2. Psychrophiles

The term 'extremophiles' refers to organisms that can survive in harsh environments with unfavorable conditions related to pressure, pH, temperature, salinity, and water availability. Extremophiles are primarily found in bacteria and archaea, although some have been identified in eukaryotes (Zeldes et al., 2015). The microbes known as psychrophiles are fascinating organisms that can grow below 15°C. They are primarily found in cold environments such as glaciers, mountainous regions, polar zones, ocean depths, upper atmosphere, underground systems, and Arctic and Antarctic ice beds. Psychrophiles belong to the Archaea, Bacteria, and Eukarya kingdoms and have various metabolic routes, including photosynthesis, heterotrophic, and chemoautotrophic processes (Sarmiento et al., 2015). Psychrophilic organisms are classified into two groups based on their growing temperature: psychrophiles, which have an ideal growth temperature below 15°C, and psychrotrophs or psychrotolerants, which grow best at temperatures between 20–25°C (Kanekar & Kanekar, 2022). Genuine psychrophiles are found in persistently frigid environments, while psychrotolerant species are more prevalent in habitats experiencing temperature changes.

However, it has been noted that over 50% of microorganisms in cold conditions are not psychrophilic, indicating that psychrophiles are relatively rare (Kanekar & Kanekar, 2022). Many psychrophilic organisms, like halo-psychrophiles, are well-adapted to survive multiple environmental stresses. These halo-psychrophiles thrive in the harsh conditions of sea ice, where they encounter both low temperatures and high salt concentrations, and piezo-psychrophiles, which survive in deep waters with low temperatures and high pressure (Dheeran & Kumar, 2022). These organisms can develop at freezing temperatures and withstand temperatures as low as –20°C to 10°C. For example, *Planococcus halocryophilus* Orl, isolated from extremely cold Arctic frozen ground, demonstrated optimal development and fertilization at a temperature of –15°C (Mykytczuk et al., 2013). Psychrophilic microbes face several obstacles in cold environments, including reduced enzyme activity, modified transport systems, membrane fluidity issues, and protein freezing-denaturation (Raval et al., 2018). To survive, they have evolved adaptations such as producing unsaturated fatty acids and shorter acyl-chain fatty acids to increase membrane fluidity (Sarmiento et al., 2015). The majority of psychrophile research focuses on areas such as anti-freeze proteins, bioremediation processes, medicinal and pharmaceutical uses, and enzymes in process industries, including food technology (Białkowska et al., 2020; Kuddus et al., 2024). Psychrophilic microbes are incredibly diverse and prevalent in cold environments, showcasing unique adaptations to thrive in extreme conditions (Raval et al., 2018).

3. Psychrozymes

Microbial biocatalysts are increasingly sought for their ability to withstand harsh processing conditions in industrial applications. Mesophiles are the primary source of enzymes in modern industry, but their activity is limited by factors such as temperature, pH, and ionic strength (Nadaroglu et al., 2022). This limitation becomes particularly challenging in processes involving heating or cooling, which are costly. Improving the yield of enzymatic reaction products is crucial for cost-cutting measures. Proteins from cold-loving bacteria are gaining attention for their potential to address these challenges (Garcia-Lopez et al., 2021). Additionally, enzymes such as family 8 xylanases from thermophiles (*Clostridium thermocellum*), mesophiles (*Streptomyces* sp.), psychrophiles (*P. holoplanktons*), and other cold-adapted species have shown remarkable adaptation to low temperatures. They also exhibit thermal and chemical stabilities, thermal inactivation, temperature dependence of activity, and conformational flexibility. These properties make them valuable candidates for industrial applications (Hamid et al., 2022, 2024).

4. Psychrozymes and Its Stability

Psychrozymes exhibit lower levels of weak contacts, including ion pairs, aromatic interactions, hydrogen bonds, and helix dipoles, which influence their stability. The hydrophobicity of non-polar core clusters in psychrozymes is weaker, leading to a less compact protein interior. Proline residues, limiting an enzyme's rotation ability, and arginine residues, which are less effective in forming hydrogen bonds and salt bridges, contribute to this trend. In contrast, glycine residue clusters, with their minimal side chains, enhance chain mobility. Also, weakly binding stabilizing cofactors are observed to help tighten loose or relaxed protein extremities. In multimeric enzymes, reducing the quantity and strength of contacts contributing to association leads to decreased cohesion between monomers (Parvizpou et al., 2021).

5. Types of Major Emerging Environmental Contaminants

In the present scenario, the production and unlimited use of synthetic materials in industry has gained importance, as well as many pharmaceuticals and cosmetic products. However, the poor biodegradability of these compounds is a significant obstacle to their widespread use. Among all the natural resources, water is the most crucial part of life. Currently, over 780 million individuals worldwide lack access to safe drinking water, leading to health issues. The existence of these pollutants in water sources poses a greater danger and can result in harmful effects. Organic contaminants, also referred to as emerging contaminants (ECs), have recently drawn attention from the public due to evidence that they not only significantly deteriorate water quality but also present significant challenges to the effectiveness of current water treatment systems (Rathi et al., 2021).

Emerging contaminants refer to both man-made and naturally occurring substances, as well as microorganisms, that aren't routinely tracked in the environment but can still find their way into it, posing potential risks to ecological balance and human health, whether confirmed or suspected. Due to their poor biodegradability, it becomes a downside associated with heavy reliance. Removing a growing class of environmental pollutants (emerging contaminants), including antibiotics (used in medicine and agriculture), hormones (synthesized for various applications), illegal drugs (derived from various plants or synthetically manufactured), cosmetics, personal care products (used for personal hygiene and appearances), pesticides (employed in pest control), surfactants present in detergents and cleaning products, products from various industries microplastics (both primary and secondary), particles engineered at atomic or molecular levels (nanoparticles), poses a significant challenge due to the inadequacy of conventional wastewater treatment systems designed to address these substances. Additionally, these contaminants are frequently present in trace amounts within complex organo-mineral mixtures, further complicating their removal process (Figure 1).

Pharmaceuticals and Personal Care Products

These are generally found in low concentrations in water bodies in trace

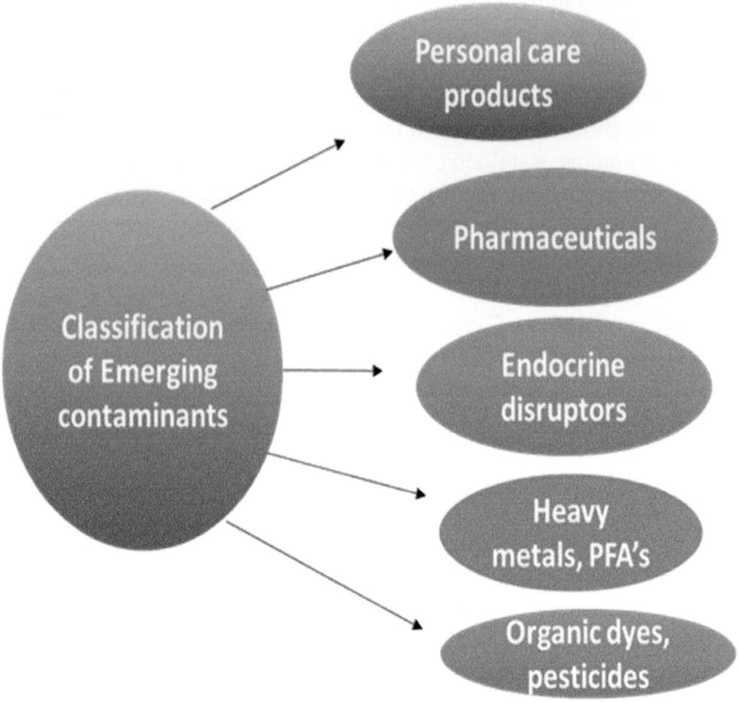

Figure 1: Classification of emerging contaminants.

quantities, due to which they are unable to be detected and have many harmful environmental and health effects from long-term exposure. The chemicals in the environment come from many sources, the most important of these are wastewater from sewage treatment plants, water bodies, and agriculture, and some personal care items, like synthetic fragrances and antimicrobial agents. For example, enzymes like oxide-reductase can facilitate the breakdown of potentially harmful substances, including phenol, dichloroethane, dyes (natural and synthetic), pesticides, including herbicides/fungicides, pharmaceuticals, and personal care products (Bilal et al., 2017).

Per- and Poly-Fluoroalkyl Substances (PEAS)

These are chemicals known for their water- and grease-resistant properties, as they can resist heat, oil, stains, grease, and water. Due to their persistence in nature, they have raised concerns about their removal. A study by Baker and Horden (2013) found the presence of illicit drugs like cocaine and 3,4-methylenedioxymethamphetamine (MDMA) in river water, with concentrations ranging from 18 to 25 ng/l. Similarly, a study by Kasprzyk-Hordern et al. (2009) reported the presence of 4-benzophenone, a sunscreen ingredient, in higher concentrations (3600–5790 ng/L) compared to pharmaceuticals and personal care products in the same water source. Over the last few years, the increasing demand for veterinary antibiotics has resulted in accelerated growth and emergence. Adding to the list of hazardous contaminants, Perfluorinated compounds (PFCs) like Perfluorooctanoic sulfonate (PFO) pose a significant health risk. These compounds have been linked to the development of various cancers, including hepatic adenoma, benign testicular tumor cells, and malignant pancreatic tumors (Chang et al., 2014).

Endocrine Disruptors

Chemicals that affect hormonal balance, when they interact with the endocrine system. Examples involve bisphenol A and phthalates, found in plastic food packaging and personal care products.

Cyanotoxins

Due to eutrophication, the growth of cyanobacterial blooms is increasing in surface water. The fertilizers obtained from agricultural lands increase the quantity of nutrients such as nitrogen and phosphorus, and products like detergents. These blooms can release toxins that pollute water and endanger the health of humans and wildlife.

Industrial Chemicals

Disposal of wastewater containing dyes used in paper and textile industries leads to many environmental impacts; for this we have many chemical and physical methods for treatment, but instead of using them, we need to switch towards enzyme related methods, emerging as the most effective the most efficient and ecofriendly solution bioremediation by the biodegradation of pollutants (Ahmed et al., 2017).

Microplastics

Particles size less than 5 mm, which are also an example of emerging contaminants due to their harmful effects, and sometimes they are manufactured as microbeads in personal care products. Eliminating these types of contaminants is a major focus for researchers and ecologists to guarantee the availability of safe drinking water to the public (Khan et al., 2020).

Artificial Sweeteners

Many sweetening agents are used in different kinds of food and beverage products, such as sucralose and saccharin, in water bodies.

Methyl Tert-Butyl Ether (MTBE)

They act as a significant additive used in gasoline and have the potential to become a new pollutant known as an emerging contaminant. MBTBE exposure causes damage to various organs and systems, such as the immune system and the central nervous system, as well as the liver and kidneys. It can also cause neurotoxicity (Ekpenyong et al., 2017).

Nanomaterials

They are considered a new class of emerging contaminants due to their tremendous influence on health and the environment. According to the researchers, some emerging contaminants are carcinogenic and exhibit developmental toxicity. Other health problems related to diseases caused by nanomaterials include cancer formation in the lungs and reproductive disorders (Khan et al., 2022).

6. Degradation Methods

Degradation processes enclose the modification or decomposition of substances into simpler forms, usually leading to decreased concentrations or conversion into less detrimental compounds. These changes can occur through diverse mechanisms, including physical, chemical, and biological processes. However, certain micropollutants present in wastewater, including emerging contaminants, exhibit properties like recalcitrance to some extent to conventional wastewater treatment processes, so we need some methods to degrade the contaminants to reduce their harmful impacts. The discharged effluents from different facilities, such as household and industrial, always contain a heterogeneous mixture of substances. Many efforts have been made for the elimination of substances like PPCPs in drinking and wastewater (Yang et al., 2017). These studies have extensively analyzed their effects on human health and have explored diverse methods for their removal.

Physical Degradation Process

In contrast to chemical degradation, which is the breaking of chemical bonds, physical degradation processes involve changes in the size, shape, or phase of the substance rather than a change in its molecular structure. Physical degradation involves the breakdown or modification of a substance's physical characteristics without altering its chemical composition. Common physical degradation techniques include mechanical processes such as grinding or crushing, which reduce the size of particles, and dissolution, where a solid substance disperses in a liquid solvent to form a solution.

Adsorption Oriented Process for Removal of Emerging Contaminants

It is one of the most important physical processes used for the removal of contaminants. One important way to remove pollutants from wastewater or drinking water sources is through liquid-solid adsorption using commercial activated carbons. This industrial pollutant removal method can produce high-quality water and is both economically and technologically feasible. Activated carbon offers a significant advantage in pharmaceutical removal as it doesn't produce toxic or pharmacologically active byproducts. Research literature indicates that activated carbons typically exhibit a high adsorption capacity for pharmaceuticals. Rivera-Utrilla et al. (2009) conducted a comprehensive investigation into antibiotics (nitro-imidazoles).

Utilizing commercial activated carbons for liquid-solid adsorption represents a pivotal approach in eliminating contaminants from both wastewater and drinking water sources. This method not only yields high-quality water but also stands out for its simplicity in technology and economic viability. Commercially available activated carbon, whether granular or powdered, demonstrates efficacy across various contaminants, such as pesticides, pharmaceuticals, and organic matter for reducing chemical and biochemical oxygen demand, decolorizing water, and addressing problems with odor or taste (Khan et al., 2020). Although active carbons are widely recognized to have a high adsorption capacity, there are significant drawbacks due to rapid saturation and regeneration. The regeneration process of saturated carbon is expensive, complex, and results in the decline of the absorbent. Thus, despite the widespread recognition of active carbon technology's efficacy, it is important to acknowledge its associated drawbacks. Active carbons can be highly costly; the higher the quality, the more expensive they are.

Removal of Pharmaceuticals Using Cyclodextrin Bead Polymers

In recent decades, there has been significant research devoted to utilizing cyclodextrin polymers for producing complex compounds suited for environmental purposes (Cova et al., 2021). These commercial polymers have gained popularity for their effectiveness in both detecting and eliminating trace contaminants. Cyclodextrins are carbohydrates produced from starch and can encapsulate compounds in their cavity (Wüpper et al., 2021).

The procedure for eliminating emerging pollutants using cyclodextrin beads operates through a mechanism known as adsorption (Figure 2). Cyclodextrins, cyclic oligosaccharides consisting of glucose units arranged in a ring structure, have a unique architecture facilitating the creation of inclusion complexes with various organic compounds, including emerging contaminants like pharmaceuticals, personal care products, and pesticides.

Cyclodextrin beads typically involve immobilizing cyclodextrin molecules onto a solid support material such as silica or polymer beads. This immobilization enhances the stability and ability to reuse cyclodextrin beads. Upon introduction into polluted water, cyclodextrin beads selectively trap the target contaminants via host-guest interactions. The pollutants are trapped within the cyclodextrin molecule's cavities, effectively extracting them from the water. A key advantage

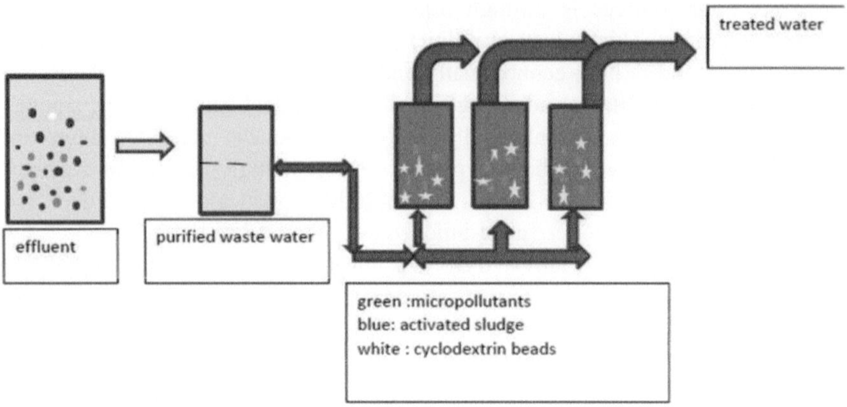

Figure 2: Purification is done by using cyclodextrin beads polymer.

of utilizing cyclodextrin beads for removing pollutants is their high selectivity and affinity for specific substances, enabling efficient removal even at low concentrations. Additionally, cyclodextrin beads can be regenerated and reused multiple times, providing a cost-effective water treatment solution. In conclusion, employing cyclodextrin beads to eliminate emerging contaminants offers a promising approach that integrates effectiveness, selectivity, and sustainability into water treatment practices.

Metal Organic Framework for Removal of Emerging Contaminants

Over the last few years, metal-organic frameworks (MOFs) have emerged as promising materials for effectively removing emerging contaminants from effluents. This can be accomplished through a range of techniques including adsorption, catalytic degradation, and membrane processes, as evidenced by several studies (Dhaka et al. (2018). Metal-organic frameworks are crystalline materials characterized by permanent porosity, where more than 45% of their crystal volume consists of interconnected metal ions or clusters bonded by multifunctional organic ligands (Dhaka et al. (2018). Like cyclodextrin molecules, these frameworks possess cavities that enable specific host-guest interactions, facilitating the adsorption and/or degradation of pollutants. MOFs exhibit exceptionally high pore volumes and surface areas ranging from 1,000 to 10,000 m^2/g, exceeding conventional materials like carbons. Their physicochemical properties are easily adjustable, and various synthesis methodologies, purification, and activation steps are outlined in reviews by (Bedia et al., 2019). A notable example of an MOF is chromium terephthalate MIL-101, comprising trimeric chromium (III) triangular cluster complexes linked by linear terephthalate linkers. This substance possesses a highly porous three-dimensional structure with large pores (>32 Å), high surface areas (>2900 m^2/g), and significant cell volumes (e.g., 705,000 $Å^3$) (Rojas & Horcajada, 2020). In water and wastewater treatment, MOFs serve as effective adsorbents for removing harmful substances, which include pharmaceuticals, personal care products, artificial sweeteners, agricultural products, organic dyes, xylene compounds, pesticides, alkylphenols, products from the photographic industry, and plasticizers (Mon et al., 2018).

Degradation of Emerging Contaminants by Dielectric Barrier Discharge (DBD) Method

DBD mainly relies on physical processes, where high-voltage electrical discharges occur between electrodes separated by a dielectric barrier. This generates plasma containing reactive species such as electrons, ions, free radicals, and UV radiation. These reactive components physically interact with contaminants, leading to their molecular breakdown and altering their structure. While there might be some chemical reactions occurring within the plasma, the primary impact of DBD is perceived as predominantly physical, rather than chemical or biological. The intense plasma produced in DBD can trigger chemical reactions, breaking down organic contaminants into less complex and less harmful forms. Reactive species like ozone, hydroxyl radicals, and other oxygen-based compounds generated within plasma can interact with the contaminants, resulting in their decomposition. Another process, like the UV radiation emitted during the discharge, can induce photolysis, breaking chemical bonds in the contaminants directly. Also, Reactive oxygen species (ROS) generated in the plasma, such as ozone and hydroxyl radicals, can oxidize the contaminants, leading to their degradation into non-toxic products (Morin-Crini et al., 2021)

Adsorption and Photodegradation of Organic Contaminants by Silver Nanoparticles

Silver nanoparticles (Ag-NPs) are a type of advanced material that has been recognized for their promising capabilities as effective adsorbents and photocatalysts (Da Silva et al., 2021). It has a high tendency for adsorption with a large surface area. It has a high affinity towards positively and negatively charged contaminants. Because of their high level of thermal stability, they can be used for high-temperature adsorption. Ag-NPs have been found to have several unique advantages, such as a high sensitivity to even visible light (Jaast & Grewal 2021), high antimicrobial properties (Rodrigues et al., 2024), and ease of reusability. Ag-NPs can produce ROS like superoxide radicals and hydrogen peroxide in specific situations. These ROS can engage with contaminants, resulting in their degradation.

Membrane Filtration Method

Membrane technology is a type of physical treatment method that uses a membrane to filter out contaminants based on their size and characteristics. The primary factor for this filtration is the need for hydrostatic force (Ricceri et al., 2019). Based on the size of pores in the membrane, this filtration process is of different types. For instance, microfiltration has pores that range from 0.001 to 0.1 micrometers; ultrafiltration has pores that are smaller than 0.001 micrometers; nano filtration has pores that range from 1 to 10 nanometers, and so on. Moreover, during reverse osmosis (RO), the use of a semipermeable membrane helps to eliminate particles smaller than 1 nm (Anis et al., 2019). Nanofiltration is effective for eliminating pharmaceuticals and naturally occurring hormones like anti-inflammatory drugs (Dhangar et al., 2020).

Chemical Process for Degradation

Chemical degradation of emerging contaminants involves altering their chemical makeup through various reactions, leading to reduced concentrations or conversion into less harmful substances. This is crucial for cleaning up polluted environments and protecting human health and ecosystems. The most used processes are hydrolysis, reduction, and oxidation, and one of them is catalytic degradation. Catalytic degradation is the chemical process that involves treating contaminants by using a catalyst for breakdown (Lama et al., 2022). Catalytic degradation is a process that breaks down emerging contaminants by reacting with the catalyst and breaks down into simpler, less harmful substances. The catalysts can be either homogeneous (liquid-phase catalysts) or heterogeneous (solid-phase catalysts) (Argyle & Bartholomew, 2015).

Biological Degradation Process

Biological degradation utilizes living organisms like bacteria, fungi, or enzymes to decompose organic or inorganic substances. The aim of this process is to create a mechanism for the effective management of decomposition byproducts. Biological treatment is commonly employed due to its superior cost efficiency compared to chemical or physical methods (Samer, 2015). This approach relies on natural cellular processes and involves the utilization of nematodes, bacteria, or other organisms in the degradation of organic waste.

Use of *Rhodococcus actinobacteria* for Degradation of Pharmaceuticals

Rhodococcus is seen as a promising choice for the degradation of pharmaceutical pollutants because of its notable adaptability, wide-ranging abilities, robust catalytic performance even under adverse conditions, and capability to break down organic compounds from diverse recognized groups. Research has demonstrated that Rhodococcus can break down a wide range of chemically complex medications (Gauthier et al., 2010).

Enzyme Assisted Biodegradation

Enzymes categorized as oxidoreductases, such as LiPs, MnPs, HRPs, laccases, and tyrosinases, are extensively utilized as environmentally friendly catalysts due to their wide application in environmental contexts. Oxidoreductases show significant promise in the bioremediation and breakdown of various harmful substances like phenols and their derivatives, bisphenols, natural or synthetic dyes, and pharmaceuticals. However, it's important to note that each type of biocatalyst varies in its three-dimensional structures, types of cofactors, and modes of action (Zdarta et al., 2018).

Bioremediation

Bioremediation, an eco-friendly method, has been employed in remediating sites contaminated with organic compounds. This approach utilizes microorganisms to degrade contaminants, potentially leading to the full conversion of organic pollutants into harmless substances (Safdari et al., 2018). The effectiveness of in situ bioremediation can be enhanced through bio-stimulation, which involves stimulating native bacteria by providing additional nutrients, oxygen, and other supplements, as well as bioaugmentation, which entails introducing specific bacteria to boost the degradation of contaminants. Bio stimulation or bioaugmentation can be employed to expedite site remediation when environmental conditions are unfavorable for natural bioremediation processes, such as the absence of electron acceptors, nutrients, and bacteria capable of degrading contaminants (Shahsavari et al., 2019).

Anaerobic Membrane Bioreactors

Anaerobic membrane bioreactors (MBRs) are advanced wastewater treatment systems that blend anaerobic biological processes with membrane filtration technology. In these setups, microorganisms decompose organic pollutants without oxygen, generating biogas, primarily methane, as a byproduct. The membrane filtration unit segregates biomass and suspended solids, permitting treated water to pass through, resulting in high-quality effluent suitable for reuse or discharge. MBRs offer multiple advantages, including effective removal of organic substances, reduced sludge production, and the potential for energy recovery from biogas production. They find widespread use in treating diverse wastewater types, including industrial and municipal sewage, thereby advancing sustainable water management practices.

7. Role of Psychrophiles and Psychrozymes in Biodegradation of Environmental Contaminants

The polar and alpine environments spanning Polar Regions, mountain peaks, and ocean abysses harbor unique life forms-psychrophiles adapted to survive as well as flourish in these extreme environments where that would cripple most organisms (Sharma et al., 2022). Unfortunately, these environments face a grave threat, such as pollution. Unsolicited guests like Plastic, Persistent Organic Pollutants (POPs), petroleum hydrocarbons, and crude oil disrupt the delicate equilibrium of these frigid havens (Saeed et al., 2022, 2023; Depledge et al., 2013).

Remediation in cold environments presents a unique set of challenges where conventional physical and chemical methods often prove ineffective or environmentally deleterious. The low temperature retards the degradation rates and alters the physicochemical properties of pollutants, impairing their bioavailability. In these conditions, nature unveils its remarkable solution, 'psychrophiles'. These microorganisms have mastered the art of life in the cold, their very structure and function adapted to thrive in temperatures deemed inhospitable by most. Psychrophiles and psychrotolerant bacteria exhibit a significant contrast in their optimal growth temperature ranges. Psychrophiles typically prefer temperatures around 15°C for optimal growth, with a maximum growth temperature of approximately 20°C and a minimum temperature of 0°C or even below. In contrast, psychrotolerant bacteria are adapted to withstand fluctuations in temperature (Kanekar & Kanekar, 2022). They are usually found in glaciers, deep ocean layers, and alpine environments. Some psychrophiles can even withstand subzero conditions. It is the most diverse form of extremophiles (Sharma et al., 2022). These psychrophiles wield a potent weapon, 'psychrozymes'. Cold-adapted enzymes serve as powerful agents of biodegradation due to their distinct thermodynamic and kinetic characteristics, enabling them to operate effectively in low-temperature environments where mesophilic enzymes are typically ineffective (Pathania et al., 2022). Their adaptation is particularly crucial in regions contaminated with emerging pollutants, as it supports both their survival and enzymatic activity. As psychrophiles producing enzymes have the ability to function optimally at low temperatures, they contribute to the breakdown of various pollutants, including emerging contaminants. Psychrozymes exhibit reduced activation energy and enhanced catalytic efficiency under low-temperature conditions, resulting in lower energy requirements and reduced environmental impact (Furhan et al., 2020). This section delves into the world of these cryophilic warriors, exploring their crucial role in biodegradation within cold systems. Each subheading will shed light on the specific ways these microorganisms and their enzymes tackle distinct contaminants.

Psychrophiles and Psychrozymes: Powerhouses for POP Biodegradation in Cold Environments

POPs refer to organic substances that endure in the environment over a prolonged period, which are highly toxic and defy degradation to varying extents (Bajaj et al., 2015). They exhibit lipophilic and hydrophobic characteristics, biomagnifying in adipose tissues across the food web and posing a global risk due to their volatility (FAO & UNEP, 2021). POPs have been discovered in every region in the world, including remote cold habitats. The occurrence of Persistent Organic Pollutants in frigid environments raises specific concerns, given that the low temperature impacts the rate of biodegradation by modifying the physical nature of pollutants, including decreased bioavailability increased viscosity and reduced volatilization (Cabrerizo et al., 2019). Moreover, the harsh frigid environmental circumstances impact functioning of microbes by diminishing the rate of reactions, and the remote locations of cold environments pose challenges for bioremediation. The restricted evaporation in cold environments results in the buildup of pollutants and makes these habitats ultimate sinks (Gai et al., 2014). Contamination in cold environments can result from diverse sources such as research activities, tourism, migrating fauna such as birds and mammals. However atmospheric transport is identified as a prominent contributor (Xie et al., 2022), volatilization of Persistent Organic Pollutants at ambient temperature allows them to spread globally for instance the organochlorine insecticides like DDT, hexachlorobenzene (HCB), heptachlor, endrin, aldrin, chlordane have been recorded on the Antarctic continent despite not being used or produced there (Bhardwaj et al., 2020; Kim et al., 2021).

Global distillation theory, illustrated in Figure 3, asserts that the organochlorine compounds with reduced volatility undergo condensation in polar regions, thereby intensifying pollutant accumulation in such remote areas (Wania & Mackay, 1993). Several investigations have identified that the widespread dissemination of these contaminants has greatly impacted both the polar regions and the high-latitude alpine regions. Due to the slow mineralization in cold habitats and hydrophobic

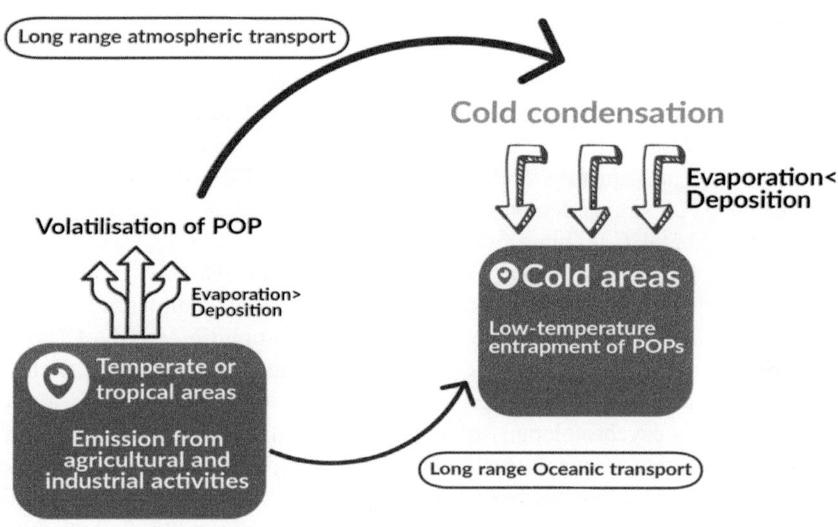

Figure 3: Global distillation of persistent organic pollutants.

nature and bio-magnify causing adverse health effects such as hormonal imbalance, compromised immune function, reduced fertility, and death in cold-adapted species (Ali et al., 2021; Morris et al., 2018).

Biodegradation of persistent organic pollutants is particularly important because of slow degradation and an increase in the amount in recent times. The traditional methods, like soil washing (utilizing water to purify contaminants through separation techniques), incineration (thermal waste treatment), are usually burdensome on wallets and the environment, with limited effectiveness, and the cleanup of cold regions poses further challenges, as most of them are remote locations. In these cases, microbial degradation is a promising alternative. The effectiveness of this process is dependent upon the biology of microbes, since the cold temperature severely inhibits the activity of microbes, the use of psychrophiles is of prime importance. The POPs can be grouped into three main categories: lindane, which covers the pesticides, PCBs, Dioxins, and furans, which are industrial products as well as byproducts.

Pesticides

Lindane or γ-hexachlorocyclohexane is an organochlorine insecticide, even though its use and production have been regulated due to its longevity and toxicity; however, it is currently manufactured for domestic use and overseas trade. γ-HCH undergoes degradation under both aerobic and anaerobic conditions, typically undergoing mineralization in aerobic settings (Zhang et al., 2020). *Sphingomonas* has demonstrated remarkable efficiency in dehalogenation and converting HCH. Even though most of the studies on bacterial degradation of lindane emphasized mesophilic temperatures (Zheng et al., 2011). Several Sphingomonas strains, including *Sphingomonas indicum* B90A, *Sphingomonas japonicum* UT26, and *Sphingomonas francense* sp., are capable of degrading γ-hexachlorocyclohexane (γ-HCH) under cold conditions, even though the efficiency of degradation was lower at 4°C compared to 30°C. However, a larger initial dose of the bacteria (inoculum size) reversed this trend (Zheng et al., 2011). Consequently, *Sphingomonas indicum* B90A and *Sphingomonas japonicum* UT26 strains demonstrated a significant increase in the degradation rate of γ-HCH at 4°C within a five-day period.

A study by Verma et al. (2021) identified *Ochrobactrum* sp. M1D, a soil isolates from Himachal Pradesh, India, capable of complete malathion degradation at a low temperature, indicating psychrophilic activity. GC-MS analysis revealed diverse metabolites (methyl phosphate, diethyl maleate, diethyl 2-mercaptosuccinate), suggesting multiple degradation pathways. This is the first report of a psychrotolerant strain exhibiting malathion degradation across cold and moderate temperatures, with potential applications in cold-climate bioremediation and enzymatic biotechnology. Another study by Verma et al. (2021) isolated a cold-tolerant (psychrophilic) bacterium, *Rahnella* sp. PFF2, from apple orchard soil in Himachal Pradesh, India. This bacterium effectively degraded profenofos, a harmful insecticide, completely (100%) at both cold (15°C) and moderate temperatures. Analysis using GC-MS and HPLC techniques identified breakdown products (metabolites) of profenofos, suggesting a potential pathway for its degradation. This study is the first to report a psychrotolerant *Rahnella* sp. from the Western Himalayas that can degrade profenofos under both cold and moderate temperature conditions.

Polychlorinated Biphenyls (PCBS)

PCBs are a cluster of Organic Pollutants that are persistent in nature and are employed in various technical applications such as coolants, thermal management systems, paints, lubricants, and inks. It consists of 209 congeners, and the degree of chlorination is such that its persistence increases in the environment. Several studies have been done by employing bacterial strains isolated from polluted cold sites.

An investigation into Antarctic freshwater ecosystems revealed the presence of psychrophilic bacteria capable of PCB oxidation (Papale et al., 2022). Enrichment cultures derived from four coastal lakes yielded 192 isolates. Of these, 57 demonstrated the ability to utilize Aroclor 1242 (a PCB mixture) as their sole carbon source. Analysis via PCR for the bphA gene, a marker for potential PCB degradation, identified 37 positive isolates. These isolates were predominantly affiliated with the genera *Salinibacterium*, *Arthrobacter* (*Actinobacteria*), and *Pusillimonas* (*Betaproteobacteria*). The cold-adapted strains from Arctic soil displayed an increased capacity for degrading both Mono and dichlorobenzenes at 7°C (Mohn et al., 1997). In this study, it has been observed that the *Pseudomonas Cam-1* adapted to cold conditions demonstrated proficiency in eliminating several tetras, tri, di, and mono-chlorobiphenyls. Certain isolates from Antarctica classified under the *Arthrobacter*, *Pseudoalteromonas genus, and Psychrobacter* displayed the ability to metabolize organic toxins. The activity remained unaffected at low temperature, and congeners were eliminated at almost the same rates at 4°C for the Arthrobacter as well as the *Pseudoalteromonas genera. Psychrobacter* active only at 15°C and *Arthrobacter* sp. exhibited higher removal efficiency for tri, tetra, penta-chlorobiphenyls at 4°C compared to 15°C.

Chlorobenzoates (CBAS) and Chlorobenzenes (CBS)

The degradation process results in the release of metabolites, including chlorobenzoates (CBAs), which are organic compounds derived from chlorinated benzoic acid, and chlorobenzenes, volatile aromatic compounds containing chlorine atoms. The breakdown of PCBs often encounters a bottleneck at the stage involving chlorobenzoates (CBAs), recognized as a step that significantly influences the overall rate of the process. *Rhodococcus erythropolis* S-7, adapted to cold conditions, demonstrated complete degradation of 3-chlorobenzoate at 10°C (Qi et al., 2007). Another strain, *Rerythropolis MS11,* thriving between 5°C and 30°C, was identified to degrade di, tri, and tetrachlorobenzenes.

Furthermore, the organism generated biosurfactants, enhancing the availability of water-insoluble contaminants for microbial breakdown (Rapp et al., 2003).

Furans and Dioxins

Industrial byproducts often include furans and dioxins; they lack any known purpose and are commercially produced. The major sources of emissions are vehicular exhaust, burning of municipal wastes, and coal. Currently, very little knowledge is available about the degradation of these in cold environments (Bajaj & Singh, 2015). A soil bacterium thriving in moderate climates, capable of withstanding cold temperatures, identified as *Pseudomonas* sp. C3211, employs dibenzofuran (DBF) as its sole source of carbon for growth, with acetone acting as a secondary substrate. Notably, DBF is a known environmental pollutant, an undesired byproduct of coal gasification, which was degraded by the strain across various temperatures (4°C, 10°C, 22°C, and 25°C). Although degradation occurred at all temperatures, the optimal temperature was found to be 20°C–21°C, with the lowest rates observed at 4°C (Jensen et al., 2003).

8. Psychrophiles and Psychrozymes: Powerhouses for Petroleum Hydrocarbon Biodegradation in Cold Environment

Hydrocarbon contamination presents a serious environmental challenge, especially in cold environments like alpine and polar regions. While natural systems do contribute to contamination, a significant portion originates from human activities such as oil extraction, transport, and storage, which frequently result in spills and pose serious environmental risks (Chaudhary et al., 2019). In Alpine areas, leaks from tanks and accidents during transportation on roads and ski resorts can result in contamination of soil and groundwater. The high-water solubility of BTEX (benzene, toluene, ethylbenzene, and xylene) allows for their downward movement, consequently contaminating groundwater. Polar Regions face a double threat from petroleum hydrocarbons. Distant sources, carried by ocean currents and melting ice, contaminate these sites. Areas with a confluence of scientific research platforms (e.g., research vessels and support facilities) and nearshore commercial fishing operations and tourist sites pose further risk (Camenzuli et al., 2015). Melting polar icecaps have paved the way for new trade routes through the Arctic, and the heightened shipping of crude oil from Russia along Norway's northern coast has increased the potential threat of accidental spills, posing a grave threat to the region's pristine environment (Carayannis et al., 2020). Existing and abandoned human settlements, including military bases and oil rigs, add a layer of concern due to their potential for spills and leaks. The devastating Exxon Valdez disaster in 1989 serves as a sober reminder of the disastrous outcomes such spills can bring (Carls et al., 2016). Petroleum hydrocarbons, a broad category of pollutants, encompass various types: linear or branched chain aliphatic hydrocarbons, including n-alkanes (straight-chain alkanes) with carbon chain lengths typically between 6 and 22 carbons. Additionally, petroleum fuels like diesel, aromatic hydrocarbons, known for their BTEX compounds (benzene, toluene, ethylbenzene, and xylenes), and polycyclic aromatic hydrocarbons (PAHs) such as naphthalene, anthracene, and pyrene, all with multiple aromatic rings, are included (Yap et al., 2021).

Upon entering the environment, petroleum hydrocarbons are subjected to a natural breakdown process driven by various physical and chemical factors known as weathering. These include spreading of oil across the surface, the volatilization of lighter components, microdispersion of water within the oil phase (water in oil emulsification), dissolution of water into the oil, and photo oxidation (significantly aiding the natural reduction (or degradation) of the oil contaminants in the environment) (Guidice et al., 2010). Decreased (low) temperatures can significantly alter the physical characteristics and chemical makeup of petroleum (Yap et al., 2021). Lower temperatures can change the physical properties and composition of petroleum. This includes an increase in the oil's viscosity, which slows down the evaporation (volatilization) of short-chain alkanes (hydrocarbons with carbon chains less than 10 atoms). This has two consequences: it increases the toxicity of these alkanes to microorganisms and affects their solubility in water, subsequently increasing their microbial toxicity (Aislabie et al., 2006). Despite adverse conditions, a diverse range of autochthonous microbes demonstrates the ability to utilize a broad spectrum of n-alkanes and aromatic hydrocarbons as sole carbon and energy sources under psychrophilic conditions, therefore emphasizing their importance in *in situ* hydrocarbon biodegradation. Polar and alpine soils, along with the Arctic and Antarctic marine environments, have attracted most of the research on cold-adapted microorganisms, and the most studied degraders of petroleum hydrocarbons are aerobic bacteria, utilizing oxygen in their metabolism to break down hydrocarbons (Giudice et al., 2010). However, there are also anaerobic bacteria that can degrade these compounds through alternative metabolic pathways, while limited information exists about fungi and yeasts capable of degrading petroleum hydrocarbons. Research, for instance the work of (Eriksson et al., 2003), have shown the potential of enriched Arctic soil cultures to break down PAHs. These experiments were conducted under anoxic conditions (without oxygen) and with nitrate-reducing conditions at 7°C. Interestingly, Psychrophilic (or psychrotrophic) bacteria with hydrocarbon degradation capabilities, isolated from frigid (or cold) environments often consist of alkane degraders, specifically those capable of degrading alkanes with chain lengths ranging from 22 to 49 carbon atoms. Following an initial contamination event, these bacteria exhibit remarkable persistence, remaining at high concentrations for extended periods, even up to decades (Aislable et al., 2006). Unlike strictly cold-loving (psychrophilic) bacteria, these strains demonstrate a broader tolerance (e.g., psychrotolerant) for environmental conditions. They are primarily affiliated with the genera *Pseudomonas*, *Sphingomonas*, and *Rhodococcus*, which are adapted to cold conditions as well as synthesize bio-surfactants associated with their cell structures, contributing to the enhancement of hydrocarbon biodegradation. Deppe et al. (2005) reported the cultivation of a microbial community (consortium) from Svalbard, Norway, capable of effectively degrading crude oil at a cold temperature of 4°C. The consortium was derived from the drift ice and Marine water. After a period of 28 days (about 4 weeks) of incubation, n-alkanes (C_{1-30}), isoprenoids, phytane, and pristane were completely degraded. During the first 2 weeks, the consortium displayed a preference to metabolize short-chain n-alkanes (C_{1-6}). Subsequently, n-alkanes ranging from (C_{7-30}) were simultaneously attacked along with two isoprenoids. Majorly nine bacterial strains were identified within the mixed culture through DNA sequencing. The isolated strains belonged to the following bacterial genera: *Pseudomonas*, *Pseudoalteromonas*, *Shewanella*, *Marinobacter*, *Psychrobacter*, and *Agreia*. In the study, Murphy et al. (2021) metagenomics sequencing uncovered previously unrecognized alkane biodegradation capabilities in the genus *Paraperlucidibaca* and *Cycloclasticus*. Analyses using both whole-genomic and amplicon sequencing techniques indicated that the bacterial genera *Oleispira* and *Thalassolituss* possess the capability to degrade alkanes found in diesel fuel. Additionally, *Zhongshania* and a newly discovered lineage named PGZG01 contributed to the biodegradation of alkanes found in crude oil. The study also noted that diesel and crude oil spills, even at low concentrations (0.1% and 1%) quickly use up the essential nutrients in the surrounding environment. This depletion limits the effectiveness of biodegradation.

A recent study by Gregson et al. (2020) identified the psychrophilic bacterium, *Oleispira antartcica* RB-8, a dominant member in the cold environments following oil spills flourishes on n-alkanes and produces proteins that exhibited a significant increase in production during n-alkane breakdown at low temperatures (4°C–16°C). The study also suggests that this bacterium might strengthen its colder environments by enhancing its motility and tolerance to stress. In another study, Fergusen et al. (2020) investigated natural attenuation in Arctic soils by studying the Romulus C-42 well site (Ellesmere Island, Canada) was decommissioned in 1972 following an oil spill. Bacterial assemblages in contaminated soil were quite different from nearby uncontaminated areas after four decades. Notably, *Rhodanobacter* sp. (*Alphaprotobacteria*) present in contaminated soil and gas chromatography analysis revealed that moderate biodegradation of spilled hydrocarbons occurred at low temperature.

The biodegradation efficiency in cold climates can be increased by using cold-active enzymes. The enzymatic treatment of pollutants offers multiple benefits, such as shorter treatment periods and greater specificity in comparison to microbial processes. The initial stage involved the addition of one oxygen atom to the alkyl side chains, resulting in the formation of key intermediate compounds at the center of the molecule. Depending on their chemical structure, the intermediates can be categorized as either containing a catechol group (possessing a specific dihydroxybenzene structure) or non-catecholic. Subsequently, these central intermediates undergo de-aromatization through ortho, meta, or para cleavage, a crucial step in the detoxification process. Notably, this cleavage is mediated by catechol dioxygenases for catecholic intermediates, highlighting the specific enzymatic pathway involved in their biodegradation. This two-stage process forms the foundation for the biodegradation of most individual BTEX compounds under aerobic conditions (Tsai et al., 2007). Monooxygenases and dioxygenases play a crucial role as key enzymes in the majority of individual BTEX biodegradation pathways (Yavas et al., 2018). The instability of oxygenases presents a significant hurdle in their widespread practical and commercial application. To overcome this challenge, biotechnological techniques like immobilization are employed. Miri et al. (2021) found that attaching enzymes to microscopic and nanoscopic biochar-chitosan mixtures significantly improved their stability during storage, maintaining over 50% retained activity after 30 days (about 4 and a half weeks) at 4±1°C, and they degraded over 80% of BTEX in groundwater and soil.

9. Psychrophiles and Psychrozymes: Powerhouses for Phenol Biodegradation in Cold Environments

Phenol is a byproduct frequently associated with many industrial processes such as oil refineries, pesticides, insecticides, and pharmaceutical manufacturing, which has been listed as a priority hazardous pollutant by the US Environmental Protection Agency (Fatimah et al., 2024). Even though small concentrations of phenol possess the ability to inhibit the rate of microbial growth, they can also inhibit the rate of degradation of contaminant compounds, resulting in substantial harm to the ecosystem and living organisms. Instances of phenol pollution in cold environments, resulting from oil spills such MS Explorer incident, a 2009 oil spill near Carline station, King George Island, Antarctica involving 1,200 liters (about 317.01 gal) of petrol, 24,000 liters (about 6340.13 gal) of lubricant oil, and 190,000 liters (about 50,192.68 gal) of marine gas. These accidents create the dispersion of pollutants like phenol to other areas through various mechanisms such as dispersant application, coastal erosion, and vaporization. Despite this process, these toxic substances retain their potential to cause harm (Subramaniam et al., 2020). Besides this, the terrestrial spills from an increasing number of research stations and daily water discharges from coastal stations result in an increased level of phenol pollution.

Phenol, being a persistent pollutant, became a topic of concern due to the significant effect of cold temperature on the biodegradation rate by modifying its physical properties, such as a rise in viscosity, decline in volatilization, and diminished bioavailability (Kebede et al., 2021). There are several methods employed for the clean-up of phenol, such as dig and haul, low temperature thermal incineration, solvent extraction, and adsorption. But most of the processes are impractical on a large scale in cold environments due to higher cost, unavailability of equipment, and hindrances in the logistics due to the remote location of cold environments. Considering these disadvantages, research widely uses biological methods such as bioremediation to degrade environmental contaminants. While the feasibility of bioremediation for phenol removal from the environment has gained widespread acknowledgment, the limiting factor lies in the occurrence of low temperatures. In these conditions, the use of psychrophiles is essential for effective degradation. Numerous studies have investigated sediments, soil, and marine environments to identify such strains. After years of research, various microorganisms capable of degrading phenols and their derivatives have been identified, as outlined in the Table. 1. In the case of Antarctica, the Antarctic treaty prohibits the importation of extraneous organisms into the region, thus permitting the use of only native microbes for the biodegradation of phenols (Subramaniam et al., 2020).

Bacteria possess the capability to transform phenols into toxic intermediates, a process that can take place under both aerobic and anaerobic conditions. The

Table 1: Notable microorganisms involved in the biodegradation of phenol

Microorganisms	Origin	References
Pseudomonas Putida LY 1 [Bacteria]	Iran Binaloud northeastern mountains at an altitude of 1000–2000 m	Sepehr et al., 2019
Rhodotorula creatinivora [Yeast]	Shale mine and railway area in Austria	Krallish et al., 2006
Arthrobacter psychrophenolicus [Bacteria]	Ice caves in Alpine	Margesin et al., 2004
Arthrobacter Bambuse (AQ5-003) [Bacteria]	Contaminated soil of Antarctica	Ahmad et al., 2018
Basidiomycetous, Ascomycetous (Yeast isolates –41)	Contaminated Antarctic soil	Lee et al., 2018
Penicillium sp. *Aspergillus* sp. [Fungi]	Livingston, Antarctic soil sample	Alexieva et al., 2012
Rhodococcus sp. (AQ5-14) [Bacteria]	King George Island, Antarctica	Tengku et al., 2020
Pseudomonas sp. ATR2008 and Stenotrophomonas spp. [Bacteria]	Binaloud mountains, Northeastern Iran	Sepehr et al., 2019
Arthrobacter sulfureus BZ73 [Bacteria]	Contaminated soil from Bozen, South Tyrol, Italy	Margesin et al., 2013
Basidomycetes, *Urediniomycetes* [Yeast]	Alpine habitats	Bergauer et al., 2005
Trichosporon dulcitum (Basidiomycetous) [Yeast]	Alpine soil in Austria	Margesin et al., 2005
Candida subhashii (strain $A01_1$) *Candida oregonenis* (strain $B02_1$) *Schizoblastosporion starkeyi-henricii* (strain $L01_2$) [Yeast]	Rucianka peatland, Northeastern Poland	Filipowicz et al., 2020

beginning step of aerobic pathways involves the enzyme phenol hydroxylase, which converts phenol into catechol. Following this, catechol is completely degraded by two pathways, ortho or meta cleavage by the enzyme catechol dioxygenase. In the ortho-cleavage pathway, catechol is metabolized into cis, cis-meconate, which is then further degraded into 3-oxoadipyl CoA. This compound can function as a substrate for the tricarboxylic acid cycle. On the other hand, catechol is metabolized into 2-hydroxymuconic semialdehyde and pyruvate. Both compounds can enter the tricarboxylic acid cycle (Subramaniam et al., 2020). Most microorganisms in Antarctica have been documented to employ aerobic pathways for metabolism. The bacterial degradation of phenol is a complex process that involves multiple enzymes such as oxygenases, hydroxylases, peroxidases, tyrosinases, and oxidases. Phenol hydroxylases have been purified from numerous bacterial species. In summary, the degradation of phenol is a complex process involving multiple enzymes, ultimately leading to the production of substrates for the tricarboxylic acid cycle.

A recent study by Lee et al. (2018) identified three isolates (Aq5-05, AQ5-06, AQ5-07) capable of achieving complete degradation of 0.5 g/l phenol within 120 hours at a temperature of 10°C. The isolates identified as *Arthrobacter* sp. and *Rhodococcus* spp. demonstrated psychrotolerant traits with optimum temperature for phenol degradation between 10°C–15°C. In the study by Bergauer et al. (2005), yeast strains also demonstrated phenol-degrading potential. Specifically, 12 strains belonging to the *Microbotryomycetidae* group, identified as true psychrophiles and isolated from Alpine environments, showed the ability to degrade phenols. There has been growing interest in using mixed microbial cultures due to their advantages over single-species systems. These benefits arise from various biological interactions, including harmless co-existence (commensalism), cooperation (synergism), elimination of inhibitory compounds (relief of substrate inhibition), and co-metabolism, where one organism utilizes a primary substrate to enable the breakdown of a secondary one that otherwise remains undegraded. Collectively, these interactions lead to significantly enhanced degradation rates compared to monocultures. For instance, the study by Zeng et al. (2014) revealed that bacterial strains of *Flavobacterium* sp. and *Acetobacter* sp. exhibited accelerated and more efficient phenol degradation than each strain working independently.

10. Psychrophiles and Psychrozymes: Powerhouses for Synthetic Plastic Biodegradation in Cold Environments

Plastics, ubiquitous in our daily lives, are not without their risks. They can contain harmful substances like bisphenol A, phthalates, and polychlorinated biphenyls (Bryant et al., 2016). There are several types of plastics, including thermoplastics, thermosetting plastics, artificial plastics, naturalistic plastics, and partially synthetic plastics (Kumar et al., 2013). Many plastics undergo biodegradation when microorganisms break them down, such as bacteria like *Pseudomonas alcaligenes* and *D. nigrificans* (Begum et al., 2015). These bacteria produce various enzymes, both extracellular and intracellular, that break down plastic polymers, helping to prevent plastics from contaminating the environment (Okamoto et al., 2003). Some bacteria can even break down the C-C bonds in plastics directly, which is considered more efficient (Wei & Zimmermann, 2017). Biodegradation involves specific enzymes breaking down specific substrates (Adamcová & Vaverková, 2014). Biodeterioration is a process where abiotic variables and various bacteria work together to break down polymers into smaller pieces, leading to depolymerization (Helbling et al., 2006). This process involves chemical, physical, and mechanical alterations caused by certain microorganisms on the surface of plastic (Bostan et al., 2025).

Certain enzymes play crucial roles in biodegradation. The three principal lignolytic enzymes are laccase, manganese peroxidase, and lignin peroxidase (Thomas et al., 2023; Hofrichter et al., 2001). Another important enzyme is phenol oxidase or laccase (Thomas et al., 2020; Deepa et al., 2020; Dashtban et al., 2010). Polyethylene, the most common plastic, is a linear hydrocarbon polymer made of long chains of ethylene monomers (C2H4). It is difficult for microbes to break down, but *Actinobacter* sp. have been shown to partially decompose smaller molecular weight PE oligomers (MW = 600–800) (Ghosh et al., 2013). However, the process of PE biological degradation is slow (Yamada-Onodera et al., 2001). Polypropylene (PP) can be degraded by microbial communities, including fungi like *Aspergillus niger* and bacteria like *Pseudomonas* and *Vibrio*. During degradation, there is a reduction in viscosity and the production of new groups, specifically carboxyl and carbonyl (Arutchelvi et al., 2008). Polystyrene (PS) is a hydrophobic artificial polymer with a high molecular weight that cannot be biodegraded. Polyethylene terephthalate (PET) is a semi-crystalline polymer that is stable both chemically and thermally, with a molecular weight ranging from 30,000 to 80,000 g mol^{-1} (Webb et al., 2012). Microbial degradation can impact the crystalline structure of PET, as observed using scanning electron microscopy (SEM) micrographs (Sharon & Sharon, 2012).

11. Psychrophiles and Psychrozymes: Powerhouses for P-Xylene Biodegradation in Cold Environments

Petroleum product use has led to widespread contamination in cold climates, affecting both land and water. P-xylene, a highly volatile aromatic hydrocarbon, is a common contaminant due to its use in solvents for industries like leather, printing, and transportation fuels. While surface-level p-xylene may evaporate quickly, any remaining residue can contaminate deeper soil layers and pose a persistent environmental hazard (Fay et al., 2007).

Over the decades, several approaches have been implemented to address the issue, including physical, chemical, and thermal treatments. Physical methods like contaminated soil excavation and chemical methods, especially in remote and northern sites, prove expensive due to high mobilization and onsite costs. Operational limitations encompass the accessibility of tools, the seasonal aspects of the tasks, and the occurrence of intermediate pollutants. Bioremediation emerges as the preferred method for cleaning up pollution in cold environments because of its superior efficacy, reduced cost, and diminished environmental footprint in comparison to alternative choices. The biodegradation of p-xylene is quite difficult, even though it exhibits a similar aromatic structure, as this compound exhibits both persistence in the environment and inability to serve as a growth substrate for microorganisms, although some bacterial strains have shown the ability to break it down. In the investigation conducted by Miri et al. (2021), three psychrophilic Pseudomonas strains (*Pseudomonas putida* SZTR-01, *Pseudomonas syntanta* S2TR-20, and *Pseudomonas azotoformans* S2TR-09) were compared. Only *P. azotoformans* S2TR-09 displayed the capacity to break down 200 mg/L p-xylene within a 60-hour period at 15°C. Gene expression analysis highlighted the xylE gene as a bottleneck in p-xylene biodegradation., *P. azotoformans* S2TR-09 exhibited elevated performance of oxygenase enzymes such as xylene monooxygenase and catechol breakdown Catechol 2,3-dioxygenase upon exposure to the contaminant, with the presence and activity of the xylE gene and its associated regulatory protein can serve as a reliable indicator (biomarker) for effective degradation of p-xylene (Alcalde et al., 2006).

The use of degradative enzymes instead of whole-cell microorganisms has gathered attention in recent times due to its shorter treatment periods, safety for the environment and humans, absence of toxic by-products, high specificity, and minimal environmental risk (Alcalde et al., 2006). Although these enzymes offer numerous benefits for environmental applications, their high production cost and susceptibility to instability limit their practical implementation. The study by Miri et al. (2022) introduces a fresh approach by proposing the utilization of investigating the feasibility of utilizing economic and readily available byproducts, such as waste generated during crustacean processing, as growth media for enzyme production.

Researchers are currently exploring enzymatic biodegradation specifically designed using psychrozymes, p-xylene sequential oxygenation reactions for the biodegradation of soil with a high pollutant burden (Initial amount: 13,000 mg/kg). The enzymes for p-xylene biodegradation were produced through the mixed culture of 2 psychrophilic strains, *Pseudomonas synxantha* S2TR-26 and *Pseudomonas mandelii* S2TR-08, and the subsequent characterization of enzymes was done through tandem liquid chromatography-mass spectrometry (LC-MS/MS) to identify and characterize the enzymes.

To facilitate *in situ* application, the effectiveness of p-xylene bioremediation was assessed through soil column experiments by injecting different concentrations of enzyme cocktails. Different enzyme concentrations were tested: the original concentration (x), a five-fold dilution (x/5), and a ten-fold dilution (x/10). The original concentration (x) contained approximately 11 U/mL of xylene monooxygenase (XMO) and 21 U/mL of catechol 2,3-dioxygenase (2,3-DO). The results demonstrated a removal rate of 92–94% for p-xylene. enzymes concentration x and x/5 treated soil column. However, the rate of removal of p-xylene dipped below 30% in x/10 concentration, and in untreated soil, it is recorded as 22.2%.

Another recent study by Miri et al. (2021) instigated the use of psychrozymes for biodegrading p-xylene in contaminated groundwater. They isolated two cold-adapted *Pseudomonas* strains containing complementary enzymes: xylene monooxygenase and catechol 2,3-dioxygenase. Merging these enzymes in a particular ratio significantly accelerated p-xylene degradation, attaining a complete breakdown within 48 h (about 2 days) at 15°C. This breakdown demonstrates the potential of psychrozymes to overcome the temperature limitation of conventional bioremediation.

12. Biotechnological Advancements and Prospects

Biotechnological development, as noted by Kastenhofer (2007), represents a transformative shift in scientific enthusiasm and its societal role. This evolution has led to the exploration of innovative techniques like plant-based remediation and bioremediation, as highlighted by Conesa et al. (2012). These methods utilize living organisms, such as plants or microorganisms, to mitigate the effects of contaminants, offering a promising solution for pollution control (Cardenas et al., 2008). The process of bioremediation involves the introduction of specific microbes to contaminated sites, enhancing favorable physicochemical conditions, and accelerating degradation (Kumar et al., 2011). Selecting and designing bacterial strains with enhanced bioremediation capacities, as discussed by Singh et al. (2011), can lead to a secure and environmentally friendly remediation process.

Ex-situ biological remediation methods, including hydroponic farming, composting, and soil bio piles (Rayu et al., 2012), are effective in treating contaminants. Composting increases the region of interaction between compost and soil, promoting microbial growth (Kour et al., 2021; Abdusamatov et al., 2025; Duong et al., 2013). Air infusion rate is critical for biodegradation effectiveness in heavily polluted regions, as demonstrated by Machackova et al. (2012). Increased air sparging amplification leads to higher biodegradation rates, emphasizing the importance of air infusion.

Bioaugmentation, the process of adding microbial cultures to contaminated sites, accelerates the breakdown of pollutants (Tyagi et al., 2011). Fungal species like *Aspergillus niger*, *Aureobasidium pullulans, and Penicillium* sp. (Say et al., 2003) have been effective in bioremediation, along with *Rhizopus arrhizus* and *Trametes versicolor*, which can extract heavy metals (Norstrom et al., 2004). Microorganisms such as cyanobacteria and algae have also been evaluated for treating aquatic ecosystem contamination (Tripathi et al., 2008). For example,

Pb (II) can be effectively removed using *Cladophora fascicularis* as a biosorbent (Deng et al., 2007).

Biological stimulation activity can be promoted through the provision of nutrients, acceptors of electrons, and adding microbes with catalytic capabilities (Baldwin et al., 2008). Additionally, nematodes like *Caenorhabditis elegans* have shown promise in reducing the toxicity of acid mine drainage (Pagnanelli et al., 2012). Despite advancements in bioremediation techniques, landfill disposal and sea disposal remain widely used due to limitations in spatial capacity, costs, and environmental sustainability (Adriaens et al., 2006; Bortone et al., 2004). *In situ* capping and controlled aquatic disposal face challenges regarding long-term stability under different environmental conditions (Juwarkar et al., 2010).

13. Conclusion

The comprehensive review dives into the intriguing domain of cold bioremediation, examining the degrading potential of psychrophiles and psychrozymes of environmental contaminants in cold environments. POPs, petroleum hydrocarbons, phenol, synthetic plastics, and recently, the various emerging contaminants such as microplastics, pharmaceuticals, and personal products (Ilyas et al., 2023). These various pollutants originating from various sources exert detrimental effects on fragile ecological balances and pose significant risks to indigenous flora and fauna in cold regions like the Alps, Antarctica, and the Arctic. Their recalcitrance further amplifies the problem, demanding innovative solutions specifically designed to address the unique complexities of cold environments. Promisingly, psychrophiles and psychrozymes provide a glimpse of optimism. These interesting microorganisms and their enzymes thrive in frigid conditions, and they exhibit unique thermodynamic and kinetic properties that enable them to degrade contaminants at low temperatures, which sets them apart from their mesophilic counterparts, making them essential resources for bioremediation in cold climates. The chapter further delves into promising developments in harnessing the biodegradation potential of these cold-adapted microorganisms. Immobilization techniques, production of biosurfactants, mixed culture methods, and comobilization techniques exemplify the ongoing endeavors to enhance biodegradation efficiency, increase substrate specificity, and ensure long-term sustainability in cold environments. These approaches underscore the ongoing progress towards optimizing biodegradation efficiency in real-world scenarios. The significance of psychrophiles and psychrozymes in preserving and restoring the ecological integrity of cold environments cannot be overstated. As climate change intensifies, the threats posed by environmental contaminants increase, and the urgency of the challenges grows exponentially. In these challenging times, these biocatalysts offer a compelling path forward for bioremediation efforts. Integrating them with advanced technologies like engineered bioremediation systems could unlock synergic effects for even greater performance.

References

Abdusamatov, S., Jabborova, D., Azimova. N., Alimov, J.E., Djalolova, B., Alhewairini, S.S. et al. 2025. Eco-Friendly biodegradation of plant biomass using multienzyme-producing *Trichoderma harzianum* for sustainable feed improvement. *Waste and Biomass Valorization*. https://doi.org/10.1007/s12649-025-03132-3

Adamcová, D., & Vaverková, M. 2014. Degradation of biodegradable/degradable plastics in municipal solid-waste landfill. *Pol. J. Environ. Stud.*, 23(4).

Adriaens. P., Li, M.Y.,& Michalak, A.M. 2006. Scaling methods of sediment bioremediation processes and applications. *Eng. Life Sci.*, 6(3), 217–227. https://doi.org/10.1002/elsc.200520127.

Ahmad, Siti Aqlima, Asokan, G., Yasid, Nur, Mohammad Nawawi, Norazah, Subramaniam, Kavilasni et al. 2018. Effect of heavy metals on biodegradation of phenol by antarctic bacterium: Arthrobacter bambusae strain AQ5-003. *Malaysian Journal of Biochemistry and Molecular Biology*, 21, 47–51.

Ahmed, I., Iqbal, H.M., & Dhama, K. 2017. Enzyme-based biodegradation of hazardous pollutants:An overview. *J. Exp. Biol. Agric. Sci.*, 5(4), 402–411.

Aislabie, J., Foght, J., & Saul, D. 2000. Aromatic hydrocarbon-degrading bacteria from soil near Scott Base, Antarctica. *Polar Biol.*, 23, 183–188.

Aislabie, J., Saul, D.J., & Foght, J.M. 2006. Bioremediation of hydrocarbon-contaminated polar soils. *Extremophiles*, 10, 171–179.

Alcalde, M., Ferrer, M., Plou, F.J., & Ballesteros, A. 2006. Environmental biocatalysis: From remediation with enzymes to novel green processes. *Trends Biotechnol.*, 24(6), 281–287.

Alcazar, P., & Cid, C. 2021. Identification of biomolecules involved in the adaptation to the environment of cold-loving microorganisms and metabolic pathways for their production. *Biomol.*, 11(8), 1155.

Alexieva, Z., Yemendzhiev, H., Tossi, S., Krumova, E., Angelova, M., Terziyska, A. et al. 2012. Growth of fungal strains isolated from Livingston Island on phenolic compounds-biodegradation potential. In: *Microbes in Applied Research: Current Advances and Challenges*. World Scientific Publishing, pp. 131–134.

Ali, A.M., Langberg, H.A., Hale, S.E., Kallenborn, R., Hartz, W.F., Mortensen, Å.K. et al. 2021. The fate of poly-and perfluoroalkyl substances in a marine food web influenced by land-based sources in the Norwegian Arctic. *Environ. Sci.: Process. Impacts*, 23(4), 588–604.

Anis, S.F., Hashaikeh, R., & Hilal, N. 2019. Microfiltration membrane processes: A review of research trends over the past decade. *Journal of Water Process Engineering*, 32, 100941. https://doi.org/10.1016/j.jwpe.2019.100941

Argyle, M.D., & Bartholomew, C.H. 2015. Heterogeneous catalyst deactivation and regeneration: A review. *Catalysts*, 5(1), 145–269. https://doi.org/10.3390/catal5010145

Arutchelvi, J., Sudhakar, M., Arkatkar, A., Doble, M., Bhaduri, S., & Uppara, P.V. 2008. Biodegradation of polyethylene and polypropylene. *Indian J. Biotechnol.*, 7, 9–22.

Bajaj, S., & Singh, D.K. 2015. Biodegradation of persistent organic pollutants in soil, water, and pristine sites by cold-adapted microorganisms: Mini review. *Int. Biodeterior.*, 100, 98–105.

Baker, D.R., & Kasprzyk-Hordern, B. 2013. Spatial and temporal occurrence of pharmaceuticals and illicit drugs in the aqueous environment and during wastewater treatment: New developments. *Sci. Total Environ.*, 454, 442–456.

Baldwin, B.R., Peacock, A.D., Park, M., Ogles, D.M., Istok, J.D., McKinley, J.P. et al. 2008. Multilevel samplers as microcosms to assess microbial response to biostimulation. *Groundwater*, 46(2), 295–304. https://doi.org/10.1111/j.1745-6584.2007.00411.x.

Begum, M.A., Varalakshmi, B., & Umamagheswari, K. 2015. Biodegradation of polythene bag using bacteria isolated from soil. *Int. J. Curr. Microbio. and App. Sci.*, 4(11), 674–680.

Bedia, J., Muelas-Ramos, V., Peñas-Garzón, M., Gómez-Avilés, A., Rodríguez, J.J., & Belver, C. 2019. A review on the synthesis and characterization of metal organic frameworks for photocatalytic water treatment. *Catalysts*, 9(1), 52. https://doi.org/10.3390/catal9010052

Bergauer, P., Fonteyne, P.A., Nolard, N., Schinner, F., & Margesin, R. 2005. Biodegradation of phenol and phenol-related compounds by psychrophilic and cold-tolerant alpine yeasts. *Chemosphere*, 59(7), 909–918.

Bhardwaj, L., & Jindal, T., 2020. Persistent organic pollutants in lakes of Grovnes Peninsula at Larsemann Hill Area, East Antarctica. *Earth Syst. Environ.*, 4, 349–358.

Białkowska, A., Majewska, E., Olczak, A., & Twarda-Clapa, A. 2020. Ice binding proteins: Diverse biological roles and applications in different types of industry. *Biomolecules*, 10(2), 274. https://doi.org/10.3390/biom10020274.

Bilal, M., Adeel, M., Rasheed, T., Zhao, Y., & Iqbal, H.M. 2019. Emerging contaminants of high concern and their enzyme-assisted biodegradation: A review. *Environ. Int.*, 124, 336–353.

Bostan, N., Ilyas, N., Saeed, M., Debnath, A., Akhtar, N., Tanveer, S. et al. 2024. An *in vitro* phytotoxicity assessment of UV-enhanced biodegradation of plastics for spinach cultivation. *Frontiers of Env. Sci. and Engg.*, 19(2), 17. https://doi.org/10.1007/s11783-025-1937-3.

Bryant, J.A., Clemente, T.M., Viviani, D.A., Fong, A.A., Thomas, K.A., Kemp, P. et al. 2016. Diversity and activity of communities inhabiting plastic debris in the North Pacific Gyre. *MSystems*, 1(3), e00024-16. https://doi.org/10.1128/mSystems.00024-16.

Cabrerizo, A., Muir, D.C., Teixeira, C., Lamoureux, S.F., & Lafreniere, M.J. 2019. Snow deposition and melting as drivers of polychlorinated biphenyls and organochlorine pesticides in Arctic rivers, lakes, and ocean. *Environ. Sci. Technol.*, 53(24), 14377–14386.

Camenzuli, D., & Freidman, B.L. 2015. On-site and *in situ* remediation technologies applicable to petroleum hydrocarbon contaminated sites in the Antarctic and Arctic. *Polar Res.*, 34(1), 24492.

Carayannis, E., Ilinova, A., & Chanysheva, A. 2020. Russian Arctic offshore oil and gas projects: Methodological framework for evaluating their prospects. *JKEC*, 11, 1403–1429.

Carls, M.G., Holland, L., Irvine, G.V., Mann, D.H., & Lindeberg, M. 2016. Petroleum biomarkers as tracers of Exxon Valdez oil. *Environmental Toxicology and Chemistry*, 35(11), 2683–2690. https://doi.org/10.1002/etc.3454

Chaturvedi, P., Shukla, P., Giri, B.S., Chowdhary, P., Chandra, R., Gupta, P., & Pandey, A. 2021. Prevalence and hazardous impact of pharmaceutical and personal care products, and antibiotics in environment: A review on emerging contaminants. *Environ. Res. J.*, 194, 110664.

Chang, E.T., Adami, H.O., Boffetta, P., Cole, P., Starr, T.B., & Mandel, J.S. 2014. A critical review of perfluorooctanoate and perfluorooctanesulfonate exposure and cancer risk in humans. *Critical Reviews in Toxicology*, 44(sup1), 1–81. https://doi.org/10.3109/10408444.2014.905767

Chaudhary, D., & Kim, J. 2019. New insights into bioremediation strategies for oil-contaminated soil in cold environments. *Int. Biodeterior. Biodegradation*, 142, 58–72. 10.1016/j.ibiod.2019.05.001.

Cova, T.F., Murtinho, D., Aguado, R., Pais, A.A., & Valente, A.J. 2021. Cyclodextrin polymers and cyclodextrin-containing polysaccharides for water remediation. *Polysaccharides*, 2(1), 16–38. https://doi.org/10.3390/polysaccharides2010002

Conesa, H.M., Evangelou, M.W.H., Robinson, B.H., & Schulin, R. 2012. A critical view of current state of phytotechnologies to remediate soils: Still a promising tool? *Sci. World J.*, 1–10. https://doi.org/10.1100/2012/173829.

Dhaka, S., Kumar, R., Deep, A., Kurade, M.B., Ji, S., & Jeon, B. 2018. Metal–organic frameworks (MOFs) for the removal of emerging contaminants from aquatic environments. *Coordination Chemistry Reviews*, 380, 330–352. https://doi.org/10.1016/j.ccr.2018.10.003

Dashtban, M., Schraft, H., Syed, T.A., & Qin, W. (n.d.). Fungal biodegradation and enzymatic modification of lignin. *International Journal of Biochemistry and Molecular Biology*, 1(1), 36.

Deepa, T., Gangwane, A., Sayyed, R.Z., Jadhav, H.P., & Mehjabeen. 2020. Optimization and scale-up of laccase production by Bacillus sp. isolated from the soap industry waste site. *Environmental Sustainability*, 3(4), 471–479. https://doi.org/10.1007/s42398-020-00135-9.

Deng, L., Su, Y., Su, H., Wang, X., & Zhu, X. 2007. Sorption and desorption of lead (II) from wastewater by green algae Cladophora fascicularis. *J. Hazard. Mater.*, 143(1–2), 220–225. https://doi.org/10.1016/j.jhazmat.2006.09.009.

Depledge, M.H., Galgani, F., Panti, C., Caliani, I., Casini, S., & Fossi, M.C. 2013. Plastic litter in the sea. *Marine Environmental Research*, 92, 279–281.

Deppe, U., Richnow, H.H., Michaelis, W., & Antranikian, G. 2005. Degradation of crude oil by an arctic microbial consortium. *Extremophiles*, 9, 461–470.

Dhangar, K., & Kumar, M. 2020. Tricks, and tracks in removal of emerging contaminants from the wastewater through hybrid treatment systems: A review. *Sci. Total Environ.*, 738, 140320.doi: 10.1016/j.scitotenv.2020.140320.

Dheeran, P., & Kumar, S. 2022. *Extremophiles: General and Plant Biomass Based Biorefinery* (1st Edn.). CRC Press. https://doi.org/10.1201/9781003014188.

Dubey, S.K., Dubey, J., Mehra, S., Tiwari, P., & Bishwas, A.J. (2011). Potential use of cyanobacterial species in bioremediation of industrial effluents. *African Journal of Biotechnology*, 10(7), 1125–1132.

Duong, T.T.T., Verma, S.L., Penfold, C., & Marschner, P. 2013. Nutrients are released from compost into the surrounding soil. *Geoderma*, 195–196, 42–47. https://doi.org/10.1016/j.geoderma.2012.11.010.

Ekpenyong, C.E., & Asuquo, A.E. 2017. Recent advances in occupational and environmental health hazards of workers exposed to gasoline compounds. *Int. J. Occup. Environ. Health*, 30(1), 1–26.

Eriksson, M., Sodersten, E., Yu, Z., Dalhammar, G., & Mohn, W.W. 2003. Degradation of polycyclic aromatic hydrocarbons at low temperature under aerobic and nitrate-reducing conditions in enrichment cultures from northern soils. *Appl. Environ. Microbiol.*, 69(1), 275–284.

FAO and UNEP. 2021. *Global Assessment of Soil Pollution: Report*. Rome. https://doi.org/10.4060/cb4894en.

Farrell, R.L., Rhodes, P.L., & Aislabie, J. 2003. Toluene-degrading Antarctic Pseudomonas strains from fuel-contaminated soil. *Biochem. Biophys. Res. Commun.*, 312(1), 235–240.

Fatimah, N., Ashraf, S., Nayana, R.U., Anju, P.B., Showkat, M., Perveen, K. et al. 2024. Evaluation of suitability and biodegradability of the organophosphate insecticides to mitigate insecticide pollution in onion farming. *Heliyon*, 10, e32580. https://doi.org/10.1016/j.heliyon.2024.e32580.

Ferguson, D.K., Li, C., Jiang, C., Chakraborty, A., Grasby, S.E., & Hubert, C.R. 2020. Natural attenuation of spilled crude oil by cold-adapted soil bacterial communities at a decommissioned high Arctic oil well site. *Science of the Total Environment*, 722, 137258.

Foster, W.G., Evans, J.A., Little, J., Arbour, L., Moore, A., Sauve, R. et al. 2017. Human exposure to environmental contaminants and congenital anomalies: A critical review. *Crit. Rev. Toxicol.*, 47(1), 59–84. https://doi.org/10.1080/10408444.2016.1211090.

Funke, J., Prasse, C., Lütke Eversloh, C., & Ternes, T.A. 2015. Oxypurinol: A novel marker for wastewater contamination of the aquatic environment. *Water Res.*, 74, 257–265. https://doi.org/10.1016/j.watres.2015.02.007.

Furhan, J. 2020. Adaptation, production, and biotechnological potential of cold-adapted proteases from psychrophiles and psychrotrophs: Recent overview. *J. Genet. Eng. Biotechnol.*, 18, 1–13.

Gai, N., Pan, J., Tang, H., Chen, S., Chen, D., Zhu, X. et al. 2014. Organochlorine pesticides and polychlorinated biphenyls in surface soils from Ruoergai high altitude prairie, east edge of Qinghai-Tibet Plateau. *Sci. Total Environ.*, 478, 90–97.

Garcia-Lopez, E., Alcazar, P., & Cid, C. 2021. Identification of biomolecules involved in the adaptation to the environment of cold-loving microorganisms and metabolic pathways for their production. *Biomol.*, 11(8), 1155.

Gauthier, H., Yargeau, V., & Cooper, D.G. 2010. Biodegradation of pharmaceuticals by Rhodococcus rhodochrous and Aspergillus niger by co-metabolism. *The Science of the Total Environment*, 408(7), 1701–1706. https://doi.org/10.1016/j.scitotenv.2009.12.012

Ghosh, S.K., Pal, S., & Ray, S. 2013. Study of microbes having potential for biodegradation of plastics. *Environ. Sci. Pollut. Res.*, 20(7), 4339–4355. https://doi.org/10.1007/s11356-013-1706-x.

Giudice, A.L., Bruni, V., De Domenico, M., & Michaud, L. 2010. Psychrophiles-cold-adapted hydrocarbon-degrading microorganisms. In: *Handbook of Hydrocarbon and Lipid Microbiology* (pp. 1897–1921). Springer, Berlin, Heidelberg.

Gregson, B.H., Metodieva, G., Metodiev, M.V., Golyshin, P.N., & McKew, B.A. 2020. Protein expression in the obligate hydrocarbon-degrading psychrophile Oleispira antarctica RB-8 during alkane degradation and cold tolerance. *Environ. Microbiol.*, 22(5), 1870–1883.

Hamid, B., Bashir, Z., Yatoo, A.M., Mohiddin, F., Majeed, N., Bansal, M. et al. 2022. Cold-active enzymes and their potential industrial applications: A review. *Mol.*, 27(18), 5885.

Hamid, B., Baba, Z.A., Sheikh, T.A., Perveen, K., & Sayyed, R.Z. 2024. Investigating potential protease activity of psychrotrophic bacteria from a municipal landfill for solid waste management. *Biomass Conversion and Biorefinery*. https://doi.org/10.1007/s13399-024-05621-2.

Helbling, C., Abanilla, M., Lee, L., & Karbhari, V.M. 2006. Issues of variability and durability under synergistic exposure conditions related to advanced polymer composites in the civil infrastructure. *Composites Part A: Appl. Sci. Manuf.*, 37(8), 1102–1110. https://doi.org/10.1016/j.compositesa.2005.05.039.

Hofrichter, M., Lundell, T., & Hatakka, A. 2001. Conversion of milled pine wood by manganese peroxidase from Phlebia radiata. *Appl. Environ. Microbiol.*, 67(10), 4588–4593. https://doi.org/10.1128/AEM.67.10.4588-4593.2001.

Hu, J., Wu, J., Xu, C., Zha, X., Hua, Y., Yang, L. et al. 2019. Preliminary investigation of polychlorinated dibenzo-p-dioxin and dibenzofuran, polychlorinated naphthalene, and dioxin-like polychlorinated biphenyl concentrations in ambient air in an industrial park at the northeastern edge of the Tibet–Qinghai Plateau, China. *Sci. Total Environ.*, 648, 935–942. https://doi.org/10.1016/j.scitotenv.2018.08.241.

Ilyas, N., Bostan, N., Akhtar, N., Mehmood, S., Saman, R.U., Sayyed, R.Z. et al. 2023. Toxicity assessment of microplastic (MPs) on major components of food chain (soil, plant, human). *Environmental Research*, 234, 116523. https://doi.org/10.1016/j.envres.2023.116523.

Isaacson, C., Mohr, T.K.G., & Field, J.A. 2006. Quantitative determination of 1,4-dioxane and tetrahydrofuran in groundwater by solid phase extraction GC/MS/MS. *Environ. Sci. Technol.*, 40(23), 7305–7311. https://doi.org/10.1021/es0615270.

Janna, H., Scrimshaw, M.D., Williams, R.J., Churchley, J., & Sumpter, J.P. 2011. From dishwasher to tap? Xenobiotic substances benzotriazole and tolyltriazole in the environment. *Environ. Sci. Technol.*, 45(9), 3858–3864. https://doi.org/10.1021/es103267g.

Jensen, A.M., Finster, K.W., & Karlson, U. 2003. Degradation of carbazole, dibenzothiophene, and dibenzofuran at low temperature by Pseudomonas sp. strain C3211. *Environ. Toxicol. Chem.*, 22(4), 730–735.

Juwarkar, A.A., Singh, S.K., & Mudhoo, A. 2010. A comprehensive overview of elements in bioremediation. *Rev. Environ. Sci. Biotechnol.*, 9(3), 215–288. https://doi.org/10.1007/s11157-010-9215-6.

Kanekar, P.P.,& Kanekar, S.P.2022. Psychrophilic, psychrotrophic, and psychrotolerant microorganisms. In: Kanekar, P.P., &Kanekar, S.P. (Eds), *Diversity and Biotechnology of Extremophilic Microorganisms from India. Microorganisms for Sustainability*. Springer, Singapore.

Kasprzyk-Hordern, B., Dinsdale, R.M., & Guwy, A.J. 2009. The removal of pharmaceuticals, personal care products, endocrine disruptors, and illicit drugs during wastewater treatment and its impact on the quality of receiving waters. *Water Res.*, 43(2), 363–380.

Kastenhofer, K. 2007. Converging epistemic cultures: A discussion drawing on empirical finding. *Innovation: Eur. J. Soc. Sci.*, 20(4), 359–373. https://doi.org/10.1080/13511610701767908.

Kebede, G., Tafese, T., Abda, E.M., Kamaraj, M., & Assefa, F. 2021. Factors influencing the bacterial bioremediation of hydrocarbon contaminants in the soil: Mechanisms and impacts. *J. Chem.* 2021, 1–17.

Khan, S., Naushad, M., Govarthanan, M., Iqbal, J., & Alfadul, S.M. 2022. Emerging contaminants of high concern for the environment: Current trends and future research. *Environ. Res. J.*, 207, 112609.

Kim, J.T., Choi, Y.J., Barghi, M., Kim, J.H., Jung, J.W., Kim, K. et al. 2021. Occurrence, distribution, and bioaccumulation of new and legacy persistent organic pollutants in an ecosystem on King George Island, maritime Antarctica. *J. Hazard. Mater.*, 405.

Kour, D., Kaur, T., Devi, R., Yadav, A., Singh, M., Joshi, D. et al. 2021. Beneficial microbiomes for bioremediation of diverse contaminated environments for environmental sustainability: Present status and future challenges. *Environmental Science and Pollution Research*, 28, 24917–24939. https://doi.org/10.1007/s11356-021-13252-7.

Krallish, I., Gonta, S., Savenkova, L., Bergauer, P., & Margesin, R. 2006. Phenol degradation by immobilized cold-adapted yeast strains of Cryptococcus terreus and Rhodotorula creatinivora. *Extremophiles*, 10, 441–449.

Kuddus, M., Roohi, Bano, N., Sheik, G.B., Joseph, B., Hamid, B., Raveendran Sindhu, & Aravind Madhavan. 2024. Cold-active microbial enzymes and their biotechnological applications. *Microbial Biotechnology*, 17, e14467.

Kumar, R., Acharya, C., & Joshi, S.R. 2011. Isolation and analyses of uranium tolerant Serratia marcescens strains and their utilization for aerobic uranium U(VI) bioadsorption. *Res. J. Microbiol.*, 49(4), 568–574. https://doi.org/10.1007/s12275-011-0366-0.

Kumar, S., Das, M.P., Rebecca, L.J., & Sharmila, S. 2013. Isolation and identification of LDPE degrading fungi from municipal solid waste. *J. Chem. Pharm. Res.*, 5(3), 78–81.

Kurwadkar, S., Dane, J., Kanel, S.R., Nadagouda, M.N., Cawdrey, R.W., Ambade, B. et al. 2022. Per- and polyfluoroalkyl substances in water and wastewater: A critical review of their global occurrence and distribution. *Sci. Total Environ.*, 809, 151003.

Lama, G., Meijide, J., Sanromán, A., & Pazos, M. 2022. Heterogeneous advanced oxidation processes: Current approaches for wastewater treatment. *Catalysts*, 12(3), 344. https://doi.org/10.3390/catal12030344

Lee, G.L.Y.,Ahmad, S.A.,Yasid, N.A.,Zulkharnain, A., Convey, P., Wan Johari, W.L. et al. 2018. Biodegradation of phenol by cold-adapted bacteria from Antarctic soils. *Polar Biology*, 41, 553–562.

Leonardo da Silva, W., Moro Druzian, D., Rodrigues Oviedo, L., Cristine Ladwig Muraro, P., & Rodrigues Oviedo, V. 2021. Silver nanoparticles for photocatalysis and biomedical applications. IntechOpen. doi: 10.5772/intechopen.95922

Lewis, K., & Tzilivakis, J. 2021. Review and synthesis of data on the potential environmental impact of artificial sweeteners. *EFSA Supporting Publications*, 18(10), 6918E.

Ma, Y., Wang, L., & Shao, Z. 2006. Pseudomonas, the dominant polycyclic aromatic hydrocarbon-degrading bacteria isolated from Antarctic soils and the role of large plasmids in horizontal gene transfer. *Environ. Microbiol.*, 8(3), 455–465.

Mac Cormack, W.P.,& Fraile, E.R. 1997. Characterization of a hydrocarbon degrading psychrotrophic Antarctic bacterium. *Antarctic Science*, 9(2), 150–155.

Machackova, J., Wittlingerova, Z., Vlk, K., & Zima, J. 2012. Major factors affecting *in situ* biodegradation rates of jet-fuel during large-scale biosparging project in sedimentary bedrock. *J. Env. Sc. Health A Tox. Hazard. Subst. Env. Eng.*, 47(8), 1152–1165.

Margesin, R., Schumann, P., Spröer, C., & Gounot, A.M. 2004. Arthrobacter psychrophenolicus sp. nov., isolated from an alpine ice cave. *Int. J. Syst. Evol. Microbiol.*, 54(6), 2067–2072.

Margesin, R., Fonteyne, P.A., & Redl, B. 2005. Low-temperature biodegradation of high amounts of phenol by Rhodococcus spp. and basidiomycetous yeasts. *Res. J. Microbiol.*, 156(1), 68–75.

Margesin, R., Moertelmaier, C., & Mair, J. 2013. Low-temperature biodegradation of petroleum hydrocarbons (n-alkanes, phenol, anthracene, pyrene) by four actinobacterial strains. *Int. Biodeterior. Biodegradation*, 84, 185–191.

Margesin, R., & Collins, T. 2019. Microbial ecology of the cryosphere (glacial and permafrost habitats): Current knowledge. *Appl. Microbiol.Biotechnol.*, 103, 2 537–2549.

Meyers, B.C., & McLellan, S.L. 2022. Influence of nutrients and the native community on E. coli survival in the beach environment. *Appl. Environ. Microbiol.*, 88(21), e01043-22.

Miri, S., Perez, J.A.E., Brar, S.K., Rouissi, T., & Martel, R. 2021. Sustainable production and co-immobilization of cold-active enzymes from Pseudomonas sp. for BTEX biodegradation. *Environ. Pollute.*, 285, 117678.

Miri, S., Rasooli, A., Brar, S.K., Rouissi, T., & Martel, R. 2022. Biodegradation of p-xylene: A comparison of three psychrophilic Pseudomonas strains through the lens of gene expression. *Environmental Science and Pollution Research*, 29, 21465–21479, https://doi.org/10.1007/s11356-021-17387-5

Mohn W., Westerberg, K., Cullen, W., & Reimer, K. 1997. Aerobic biodegradation of biphenyl and polychlorinated biphenyls by arctic soil microorganisms. *Appl. Environ. Microbiol.*, 63, e3378-e3384.

Mohr, T.K., & DiGuiseppi, W.H. 2020. Case studies of 1, 4-dioxane releases, treatment, and drinking water contamination. In: *Environmental Investigation and Remediation*. CRC Press, pp. 405–456.

Mon, M., Bruno, R., Tiburcio, E., Leguvano, C., & Armentano, D. 2018. Metal–organic framework technologies for water remediation: Towards a sustainable future. *Journal of Materials Chemistry A*, 6(12), 4912–4947. https://doi.org/10.1039/C8TA00264A

Morin-Crini, N., Lichtfouse, E., Fourmentin, M., Ribeiro, A.R.L., Noutsopoulos, C., Mapelli, F. et al. 2021. Remediation of emerging contaminants. In: *Environmental Chemistry for a Sustainable World*, pp. 1–106. https://doi.org/10.1007/978-3-030-69090-8_1

Morris, A.D., Muir, D.C., Solomon, K.R., Teixeira, C.F., Duric, M.D., & Wang, X. 2018. Bioaccumulation of polybrominated diphenyl ethers and alternative halogenated flame retardants in a vegetation–caribou–wolf food chain of the Canadian Arctic. *Environ. Sci. Technol.*, 52(5), 3136–3145.

Murphy, S.M., Bautista, M.A., Cramm, M.A., & Hubert, C.R. 2021. Diesel and crude oil biodegradation by cold-adapted microbial communities in the Labrador Sea. *Appl. Environ. Microbiol.*, 87(20), e00800-21.

Mykytczuk, N.C.S., Foote, S.J., Omelon, C.R., Southam, G., Greer, C.W., & Whyte, L.G. 2013. Bacterial growth at −15°C; molecular insights from the permafrost bacterium Planococcus halocryophilus Or1. *The ISME J.*, 7(6), 1211–1226. https://doi.org/10.1038/ismej.2013.8.

Nadaroglu, H., & Polat, M.S. 2022. Microbial extremozymes: Novel sources and industrial applications. In: *Microbial Extremozymes*. Academic Press, pp. 67–88.

Naseem, A., Alneghery, L.M., Al-Zaharani, M., Nasr, F.A., Jawad, S.S., Sumer, M. et al. 2025. An insight into the impacts of pharmaceutical pollutants on the ecosystem, and the potential role of bioremediation in mitigating pharmaceutical pollutants. *International Journal of Pharmaceutics*, 680, 125791. https://doi.org/10.1016/j.ijpharm.2025.125791.

Norström, A., Larsdotter, K., Gumaelius, L., La Cour Jansen, J., & Dalhammar, G. 2004. A small-scale hydroponics wastewater treatment system under Swedish conditions. *Water Sci. Technol.*, 48(11–12), 161–167. https://doi.org/10.2166/wst.2004.0830.

Okamoto, K., Izawa, M., & Yanase, H. 2003. Isolation and application of a styrene-degrading strain of Pseudomonas putida to biofiltration. *J. Biosci. Bioeng.*, 95(6), 633–636.

Pagnanelli, F., Cruz Viggi, C., Cibati, A., Uccelletti, D., Toro, L., & Palleschi, C. 2012. Biotreatment of Cr (VI) contaminated waters by sulphate reducing bacteria fed with ethanol. *J. Hazardous Materials*, 199–200, 186–192.

Pandey, P.K., Soupir, M.L., & Rehmann, C.R. 2012. A model for predicting resuspension of Escherichia coli from streambed sediments. *Water Res.*, 46(1), 115–126. https://doi.org/10.1016/j.watres.2011.10.019.

Papale, M., Lo Giudice, A., Rappazzo, A.C., Azzaro, M., & Rizzo, C. 2022. A first glimpse on cold-adapted PCB-oxidizing bacteria in Edmonson Point Lakes (Northern Victoria Land, Antarctica). *Water J.*, 14(1), 109.

Parvizpour, S., Hussin, N., Shamsir, M.S., & Razmara, J. 2021. Psychrophilic enzymes: Structural adaptation, pharmaceutical and industrial applications. *Appl. Microbiol. Biotechnol.*, 105, 899–907.

Pathania, S., Solanki, P., Putatunda, C., Bhatia, R.K., & Walia, A. 2022. Adaptation to cold environment: The survival strategy of psychrophiles. In: *Survival Strategies in Cold-adapted Microorganisms*, Springer Singapore, pp. 87–111.

Qi, Y., Zhao, L., Ojekunle, Z.O., & Tan, X. 2007. Isolation and preliminary characterization of a 3-chlorobenzoate degrading bacteria. *Journal of Environmental Sciences*, 19(3), 332–337. https://doi.org/10.1016/s1001-0742(07)60054-0

Rapp, P., & Gabriel-Jurgens, L. 2003. Degradation of alkanes and highly chlorinated benzenes, and production of biosurfactants, by a psychrophilic Rhodococcus sp. and genetic characterization of its chlorobenzene dioxygenase. *J. Microbiol.*, 149, e2879-e2890.

Raval, V.H., Bhatt, H.B., & Singh, S.P. 2018. Adaptation strategies in halophilic bacteria. In: *Extremophiles*. CRC Press, pp. 137–164.

Rayu, S., Karpouzas, D.G., & Singh, B.K. 2012. Emerging technologies in bioremediation: Constraints and opportunities. *Biodegradation*, 23(6), 917–926. https://doi.org/10.1007/s10532-012-9576-3.

Ricceri, F., Giagnorio, M., Farinelli, G., Blandini, G., Minella, M., Vione, D., & Tiraferri, A. 2019. Desalination of produced water by membrane distillation: Effect of the feed components and of a pre-treatment by fenton oxidation. *Scientific Reports*, 9(1), 14964. https://doi.org/10.1038/s41598-019-51167-z

Rivera-Utrilla, J., Prados-Joya, G., Sánchez-Polo, M., Ferro-García, M.A., & Bautista-Toledo, I. 2009. Removal of nitroimidazole antibiotics from aqueous solution by adsorption/bioadsorption on activated carbon. *Journal of Hazardous Materials*, 170(1), 298–305. https://doi.org/10.1016/j.jhazmat.2009.04.096

Rodrigues, A.S., Batista, J.G.S., Rodrigues, M.Á.V., Thipe, V.C., Minarini, L.A.R., Lopes, P.S., & Lugão, A.B. 2024. Advances in silver nanoparticles: A comprehensive review on their potential as antimicrobial agents and their mechanisms of action elucidated by proteomics. *Frontiers in Microbiology*, 15, 1440065. https://doi.org/10.3389/fmicb.2024.1440065

Rojas, S., & Horcajada, P. 2020. Metal-organic frameworks for the removal of emerging organic contaminants in water. *Chemical Reviews*, 120(16), 8378–8415. https://doi.org/10.1021/acs.chemrev.9b00797

Saeed, M., Ilyas, N., Jayachandran, K., Shabir, S., Akhtar, N., Shahzad, A. et al. 2022. Advances in biochar and PGPR engineering system for hydrocarbon degradation:

A promising strategy for environmental remediation. *Environmental Pollution*, 35, 199282. https://doi.org/10.1016/j.envpol.2022.119282.

Saeed, M., Ilyas, N., Bibi, F., Shabir, S., Jayachandran, K., Sayyedd, R.Z. et al. 2023. Adsorption kinetics of a bioengineered microbiome based biochar system for *in situ* remediation of total petroleum hydrocarbons (TPHs) contaminated soil. *Chemosphere*, 324, 138311. https://doi.org/10.1016/j.chemosphere.2023.138311.

Safdari, M.S., Kariminia, H.R., Rahmati, M., Fazlollahi, F., Polasko, A., Mahendra, S. et al. 2018. Development of bioreactors for comparative study of natural attenuation, biostimulation, and bioaugmentation of petroleum-hydrocarbon contaminated soil. *Journal of Hazardous Materials*, 342, 270–278.

Samer, M. 2015. *Biological and Chemical Wastewater Treatment Processes*. InTech. doi: 10.5772/61250

Sarmiento, F., Peralta, R., & Blamey, J.M. 2015. Cold and hot extremozymes: Industrial relevance and current trends. *Front. Bioeng. Biotechnol.*, 3, 148.

Saul, D.J., Aislabie, J.M., Brown, C.E., Harris, L., & Foght, J.M. 2005. Hydrocarbon contamination changes the bacterial diversity of soil from around Scott Base, Antarctica. *FEMS Microbiol. Ecol.*, 53(1), 141–155.

Say, R., Yilmaz, N., & Denizli, A. 2003. Removal of heavy metal ions using the fungus Penicillium canescens. *Adsorpt. Sci. Technol.*, 21(7), 643–650. https://doi.org/10.1260/026361703772776420.

Sepehr, S., Shahnavaz, B., Asoodeh, A., & Karrabi, M. 2019. Biodegradation of phenol by cold-tolerant bacteria isolated from alpine soils of Binaloud Mountains in Iran. *J. Environ. Sci. Health Part A*, 54(4), 367–379.

Shahsavari, E., Mansur, A.A., Aburto-Medina, A., Haleyur, N., Jones, N., & Ball, A.S. 2019. Microbial remediation of contaminated soils. In: *CABI eBooks*, pp. 517–530. https://doi.org/10.1079/9781786394675.0517

Sharma, S., Chaturvedi, U., Sharma, K., Vaishnav, A., & Singh, H.B. 2022. An overview of survival strategies of psychrophiles and their applications. In: *Survival Strategies in Cold-adapted Microorganisms*. Springer, Singapore, pp. 133–151.

Sharon, C., & Sharon, M. 2012. Studies on biodegradation of polyethylene terephthalate: A synthetic polymer. *J Microbiol Biotechnol Res.*, 2(2), 248–257.

Singh, A., Nigam, P.S., & Murphy, J.D. 2011. Mechanism and challenges in commercialization of algal biofuels. *Bioresource Technology*, 102(1), 26–34. https://doi.org/10.1016/j.biortech.2010.06.057.

Subramaniam, K., Ahmad, S.A., & Shaharuddin, N.A. 2020. Mini review on phenol biodegradation in Antarctica using native microorganisms. *Asia Pac. J. Mol. Biol. Biotechnol.*, 28, 77–89.

Tengku-Mazuki, T.A., Subramaniam, K., Zakaria, N.N., Convey, P., Abdul Khalil, K., Lee, G.L.Y. et al. 2020. Optimization of phenol degradation by Antarctic bacterium Rhodococcus sp. *Antarct. Sci.*, 32(6), 486–495. doi:10.1017/S0954102020000358.

Thomas, D., Gangwane, A., Sayyed, R.Z., Sapna, Enshasy H.E., & Zaidel, D.N.A. 2020. UV induced mutagenesis elevates the production of laccase in *Enterobacter cloacae*. *Journal of Scientific and Industrial Research*, 79, 442–448. http://nopr.niscpr.res.in/handle/123456789/54695.

Tripathi, R.D., Dwivedi, S., Shukla, M.K., Mishra, S., Srivastava, S., Singh, R. et al. 2008. Role of blue green algae biofertilizer in ameliorating the nitrogen demand and fly-ash stress to the growth and yield of rice (Oryza sativa L.) plants. *Chemosphere*, 70(10), 1919–1929. https://doi.org/10.1016/j.chemosphere.2007.07.038.

Tsai, S.-C., & Li, Y.-K. 2007. Purification and characterization of a catechol 1,2-dioxygenase from a phenol degrading Candida albicans TL3. *Arch. Microbiol.*, 187(3), 199–206.

Tyagi, M., Da Fonseca, M.M.R., & De Carvalho, C.C.C.R. 2011. Bioaugmentation and biostimulation strategies to improve the effectiveness of bioremediation processes. *Biodegradation*, 22(2), 231–241.

Verma, S., Singh, D., & Chatterjee, S. 2021. Malathion biodegradation by a psychrotolerant bacteria Ochrobactrum sp. M1D and metabolic pathway analysis. *Letters in J. Appl. Microbiol.*, 73(3), 326–335.

Wania, F., & Mackay, D. 1993. Global fractionation and cold condensation of low volatility organochlorine compounds in polar regions. *Ambio*, 22, 10–18.

Webb, H., Arnott, J., Crawford, R., & Ivanova, E. 2012. Plastic degradation and its environmental implications with special reference to poly (ethylene terephthalate). *Polym. J.*, 5(1), 1–18. https://doi.org/10.3390/polym5010001.

Wei, R., & Zimmermann, W. 2017. Microbial enzymes for the recycling of recalcitrant petroleum-based plastics: How far are we? *Microb. Biotechnol.*, 10(6), 1308–1322. https://doi.org/10.1111/1751-7915.12710.

White, C., Sayer, J.A., & Gadd, G.M. 1997. Microbial solubilization and immobilization of toxic metals: Key biogeochemical processes for treatment of contamination. *FEMS Microbiol. Rev.*, 20(3–4), 503–516. https://doi.org/10.1111/j.1574-6976.1997.tb00333.x.

Whyte, L.G., Bourbonniere, L., & Greer, C.W. 1997. Biodegradation of petroleum hydrocarbons by psychrotrophic Pseudomonas strains possessing both alkane (alk) and naphthalene (nah) catabolic pathways. *J. Appl. Microbiol.*, 63(9), 3719–3723.

Wilkin, R. 2022. Per- and polyfluoroalkyl substances in water and wastewater; A critical review of their global occurrence and distribution. *Sci. Total Environ.*, 809, 151003.

Wüpper, S., Lüersen, K., & Rimbach, G. 2021. Cyclodextrins, natural compounds, and plant bioactives: A nutritional perspective. *Biomol.*, 11(3), 401.

Xie, Z., Zhang, P., Wu, Z., Zhang, S., Wei, L., Mi, L. et al. 2022. Legacy and emerging organic contaminants in the polar regions. *Sci. Total Environ.*, 835, 155376.

Yakimov, M.M., Gentile, G., Bruni, V., Cappello, S., D'Auria, G., Golyshin, P.N. et al. 2004. Crude oil-induced structural shift of coastal bacterial communities of rod bay (Terra Nova Bay, Ross Sea, Antarctica), and characterization of cultured cold-adapted hydrocarbonoclastic bacteria. *FEMS Microbiol. Ecol.*, 49(3), 419–432.

Yamada-Onodera, K., Mukumoto, H., Katsuyaya, Y., Saiganji, A., & Tani, Y. 2001. Degradation of polyethylene by a fungus, Penicillium simplicissimum YK. *Polym. Degrad. Stab.*, 72(2), 323–327. https://doi.org/10.1016/S0141-3910(01)00027-1.

Yap, H.S., Zakaria, N.N., Zulkharnain, A., Sabri, S., Gomez-Fuentes, C., & Ahmad, S.A. 2021. Bibliometric analysis of hydrocarbon bioremediation in cold regions and a review on enhanced soil bioremediation. *J. Biol.*, 10(5), 354.

Yavas, A., & Icgen, B. 2018. Aerobic bacterial degraders with their relative pathways for efficient removal of individual BTEX compounds. *Clean J.*, 46(11), 1800068.

Zakaria, N.N., Ahmad, S.A., Yin, G.L.L., Yasid, N.A., Manogaran, M., Nawawi, N.M. et al. 2018. Biodegradation of phenol by Antarctic bacterium Rhodococcus baikonurensis strain AQ5-001 in the presence of heavy metals. *Malays. J. Biochem. Mol. Biol.*, 21, 29–36.

Zdarta, J., Meyer, A.S., Jesionowski, T., & Pinelo, M. 2018. A general overview of support materials for enzyme immobilization: Characteristics, properties, practical utility. *Catalysts*, 8(2), 92. https://doi.org/10.3390/catal8020092

Zeldes, B.M., Keller, M.W., Loder, A.J., Straub, C.T., Adams, M.W.W., & Kelly, R.M. 2015. Extremely thermophilic microorganisms as metabolic engineering platforms for production of fuels and industrial chemicals. *Front. Microbiol.*, https://doi.org/10.3389/fmicb.2015.01209.

Zeng, H.Y., Cao, X.L., Xiong, L.B., Cai, X.L., Huang, L.Q., Zhang, C.Y. et al. 2014. Microbiological degradation of phenol using two co-aggregating bacterial strains. *Environ. Earth Sci.*, 71, 1339–1348.

Zhang, W., Lin, Z., Pang, S., Bhatt, P., & Chen, S. 2020. Insights into the biodegradation of lindane (γ-hexachlorocyclohexane) using a microbial system. *Front. Microbiol.*, 11, 523963.

Zhang, Y., & Gross, C.A. 2021. Cold shock response in bacteria. *Annu. Rev. Genet.*, 55, 377–400.

Zhang, Y., Lu, W., Wang, J., Chen, M., Zhang, W., Lin, M. et al. 2021. Characterization of EstDR4, a novel cold-adapted insecticides-metabolizing esterase from deinococcus radiodurans. *Appl. Sci.*, 11(4), 1864.

Zheng, G., Selvam, A., & Wong, J. 2011. Rapid degradation of lindane (g-hexachlorocyclohexane) at low temperature by Sphingobium strains. *Int. Biodeterior. Biodegrad.*, 65, 612e618.

CHAPTER
11

Role of Cryophiles Towards Sustainability of Agro-Ecosystem

Zakir Amin[1], F.A. Mohiddin[2*], Burhan Hamid[3], Liliana Aguilar Marcelino[4] and Mohd Sajjad Ahmad Khan[5]

[1] ICAR-Central Institute of Temperate Horticulture, Srinagar – 191132, India
[2] Section of Plant Pathology, MRCFC, Khudwani, SKUAST-K – 192101, India
[3] Centre of Research for Development, University of Kashmir, Srinagar – 190006, India
[4] Cenid-Salud Animal E Inocuidad, Inifap., Carretera Federal Cuernavaca-Cuautla No. 8534, Col. Progreso, Jiutepec, Morelos, México
[5] Department of Basic Sciences, Deanship of Preparatory Year and Supporting Studies, Imam Abdulrahman Bin Faisal University, P.O. Box 1982, Dammam 34212, Saudi Arabia

1. Introduction

Microscopic life forms are present across a wide range of natural settings, including those with extreme environmental conditions like extremely high or low temperatures, high salt concentrations, limited water availability, and varying acidity levels. Through the process of adaptation over time, these microorganisms have developed specialized mechanisms that let them endure and even flourish in such harsh and inhospitable situations (Tarfeen et al., 2022). Those microorganisms that live and flourish in these harsh environments are termed extremophiles, and if they can endure and thrive in numerous challenging environments, they are referred to as polyextremophiles. In cold environments, various microorganisms, including archaea, protists, bacteria, fungi, and unicellular algae face multiple stresses and, in response, have established abundant molecular and physiological tactics to cope with such challenging conditions (Vallesi et al., 2020; De Maayer et al., 2014; Collins

*Corresponding author: famohiddin@rediffmail.com

DOI: 10.1201/9781003717003-11

& Margesin, 2019). Cryo-environments are characterized by either continuous or predominant sub-zero temperatures, commonly found in polar and alpine regions. The extreme cold environments mentioned are characterized by visually striking large-scale geological and atmospheric features. These include glaciers, sea ice, permanently frozen lakes, and the distinct polygonal patterns on the ground associated with permafrost. Additionally, the cold conditions extend into the atmosphere, dominating in clouds and airborne particulate matter. Research conducted both in these freezing environments on Earth as well as in controlled laboratory conditions has provided enough evidence of the widespread existence of microbial life thriving under such frigid conditions. Microorganisms thriving in challenging cold environments, known as psychrophiles, have been periodically reported. Psychrophiles are inhabitants of unique environments like the Siberian permafrost, characterized by various factors including low water activity, limited organic matter, enduringly icy lands, etc. (Margesin & Collins, 2019; Rivkina et al., 2018; Bakermans, 2017). Psychrophilic microbes particularly those included in fungal and bacterial genera have received comprehensive isolation and investigation, whereas yeasts, although constituting a smaller portion, also possess the potential to produce essential psychrozymes (Margesin & Miteva, 2011; Mangiagalli & Lotti, 2021). Cold-active enzymes, including psychrozymes, have significant potential and versatile applications in food biotechnology (Kuddus, 2018).

Cold-adapted microorganisms, initially regarded as biological curiosities, have evolved into valuable sources of diverse tools applicable across a wide range of innovative technologies. Thriving in permanently low-temperature environments, these microbes have undergone a sophisticated range of adaptations in its development at various cellular levels in order to tolerate the harshness intrinsic to cooler environments. These adaptations include modifications in the cell envelope, enzyme activities, cryoprotectant and chaperone production, as well as novel metabolic capabilities (Collins & Margesin, 2019).

Cold-tolerant bacteria possess a unique capacity to promote plant growth by making the essential nutrients available in nutrient-deficient soils (Baba et al., 2021). Conversely, plants subjected to stressful conditions benefit from the presence of cold-hardy rhizosphere bacteria, which can bolster their effectiveness. In this regard, the application of psychrophilic (cold-loving) formulations has proven advantageous for maximizing crop yields under conditions of temperature stress (Rizvi et al., 2021). The majority of research on microbial phosphorus biofertilizers has focused on their impact on plant growth under standard cultivation conditions (Afrouz et al., 2023; Sheikh et al., 2022). While psychrophilic bacteria and fungi have demonstrated impressive adaptations for survival in cold environments, there has been relatively little research investigating the plant growth-promoting capabilities of cold-tolerant phosphate-solubilizing bacteria (PSB) under low-temperature conditions (Rizvi et al., 2021). This lack of knowledge serves as the foundation for the current work, which aims to explore the potential of introducing these cold-hardy microbes into competitive agricultural settings. The manuscript explores the mechanisms of mineral solubilization/mineralization by these bacteria, their release of bioactive molecules, molecular engineering approaches to enhance their mineralizing or solubilizing potential and host range. There is a need to expand the projections of the biofertilizers based on these psychro-tolerant/psychrophilic microbes, leveraging their valuable and economical potential to promote plant growth, mitigate stress, and boost the crop production for rapidly increasing populations. Cold-adapted phosphorus biofertilizer production and their utilization can help to ensure long-term agricultural sustainability in cold-climate agroecosystems.

Extensive basic research has provided valuable understanding into the possible modes of survival of these microorganisms to flourish in challenging conditions, forming the base for development of several knowledge-driven advanced biotechnological tools (Rafeeq et al., 2023). The mechanisms that enable these microorganisms to adapt to extreme cold environments involve various cellular components and processes. These include modifications to the fatty acid configuration of cell membranes, the presence of pigments in membranes, alterations to the peptidoglycan layer of the cell wall, changes to the lipopolysaccharide constituent of the outer membrane, the accumulation of well-matched solutes, the production of specialized proteins that prevent freezing or induce ice nucleation, the secretion of extracellular polymeric substances and biosurfactants, the action of chaperone proteins, the storage of reserve materials like polyhydroxy-alkanoates and cyanophycins, and adjustments to their overall metabolism. Recent assessments predict substantial growth in the enzyme marketplace, with the global enzyme market projected to reach USD 13–14 billion by 2025. The growth can be ascribed to the unique functioning associated with these enzymes as well as the rising demand from different industries. Enzymes, these biological catalysts, possess a wide range of applications across various biotechnological and industrial domains. They play crucial roles in agricultural sustainability, molecular biology techniques, the production of detergents, food and beverage processing operations, textile manufacturing processes, as well as in the medical field and bioremediation endeavours aimed at environmental clean-up and restoration (Wackett, 2019; Santiago et al., 2016). The widespread use of enzymes across these industries underscores their importance and the immense potential of cold-active enzymes from psychrophiles in contributing to various technological advancements. This chapter explores the existing understanding of the adaptation approaches employed by cold-active microorganisms, and also explores their potential applications in agriculture and across various industries.

2. Cryophiles: Habitat and Applications

The exploration of microbial life thriving in extreme environmental conditions has been a focal point of research in recent times. Cryophiles, encompassing cold-loving microorganisms, span the three domains of eukaryotes, archaea, and bacteria. They exploit the complex randomness inherent in interconnected chemical reactions and metabolic pathways, engaging in processes such as chemoautotrophy, photosynthesis, and heterotrophy. The spectrum of cold habitats, constituting approximately 85% of Earth's surface, includes both natural and artificial settings, harbouring a diverse array of psychrophilic microorganisms (Figure 1). The microbial life found in these extreme cold environments consists primarily of various groups of microorganisms, including bacteria, archaea, yeasts, fungi, and viruses. These microbes can be further categorized on the basis of their capability to thrive in low-temperature conditions. Psychrophiles are those that not only tolerate but actively thrive in cold temperatures, while psychrotrophs are those that can endure and function under cold conditions, although they may prefer more moderate temperatures (Margesin & Miteva, 2011). Among these, bacteria, particularly psychrophilic ones, are recognized as significant producers of cold-active enzymes. Despite the myriad challenges posed by cold environments, life not only survives but thrives, displaying a remarkable microbial biodiversity primarily consisting of bacteria, fungi (especially yeasts), and microalgae. Among the bacteria identified in these cold environments, Gram-negative species having their place in the alpha, beta, and gamma groups of the phylum proteobacteria are frequently reported. These include various species from genera such as *Pseudomonas* and *Vibrio*. Additionally, bacteria from the phylum Cytophaga–Flavobacterium–Bacteroides are also commonly found in these extreme cold habitats (D'Amico et al., 2006). In addition to these, Gram-positive bacterial species are also prevalent in these extreme cold environments. Frequently encountered Gram-positive genera include *Coryneforms*, *Arthrobacter*, and *Micrococcus*. While bacteria generally outnumber and exhibit more diversity than Archaea, certain environments like deep-sea waters show equivalent numbers, with *Methanogenium* and *M. ethanococcus* being frequently cited genera. Antarctic habitats often feature dominant cyanobacteria like *Oscillatoria*, *Phormidium*, and *Nostoc* (Pandey et al., 2004). Cold-loving yeast species, especially those belonging to the genus *Cryptococcus*, are commonly isolated from soil samples collected from frigid regions, with some researchers emphasizing their crucial role these psychrophilic yeasts play as a dominant form of life in the arid soils found in Antarctica (Touchette et al., 2022).

Microorganisms exhibit differentiation based on the cold habitats they inhabit, categorizing them into two main groups. One category of microorganisms found in these extreme cold environments is known as psychrophiles. These organisms exhibit optimal growth and metabolic activity at remarkably low temperatures, typically around 15°C or lower. While their maximum growth temperature may reach around 20°C, they are capable of surviving and proliferating even at temperatures of 0°C or below, which is their minimum growth threshold (Troncoso et al., 2017; Robinson, 2001). Psychrophiles, constituting approximately three-fourths of the biosphere, are commonly found in diverse environments such as mountains, glacial regions, deep-sea waters (both freshwater and marine), and high-altitude soils. In contrast, the second group encompasses psychrotolerant

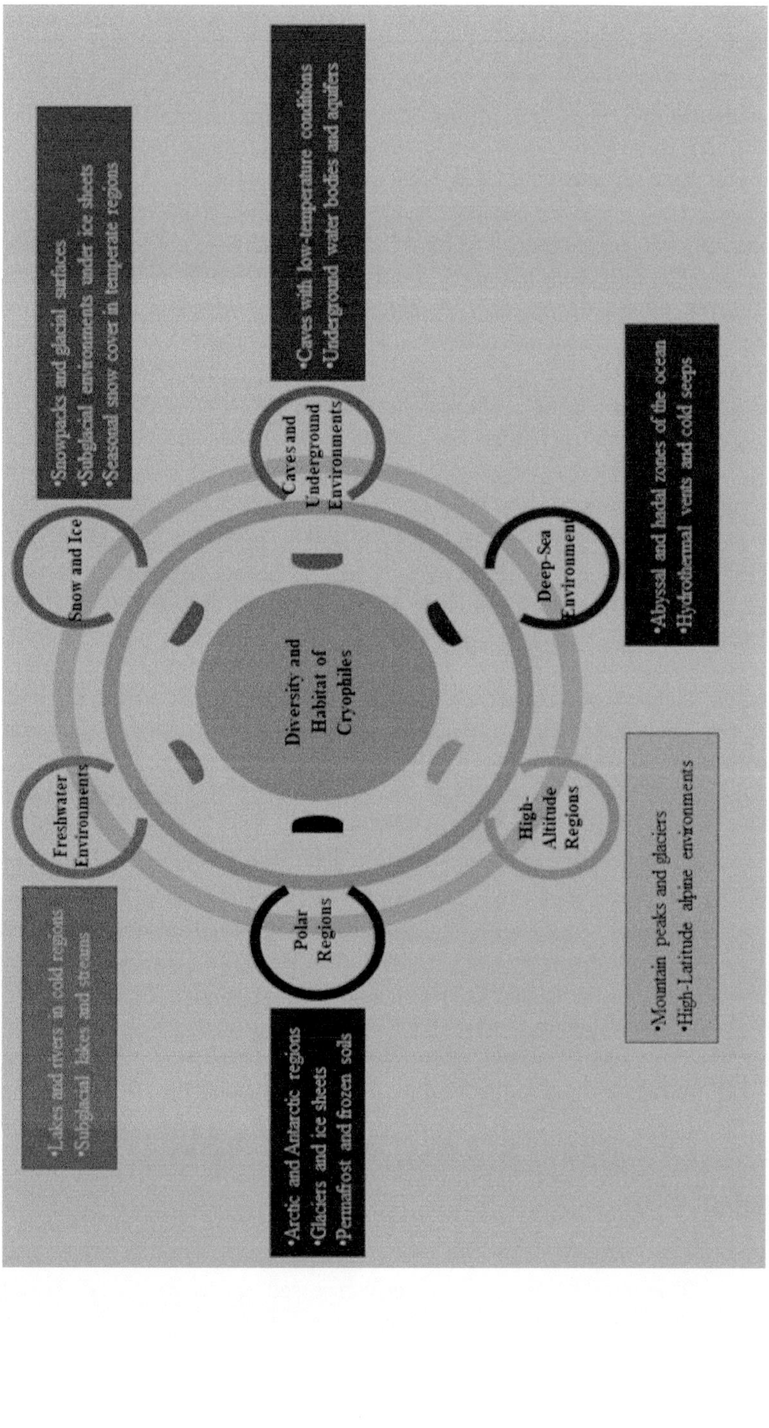

Figure 1: Diversity and habitat of cryophiles.

microorganisms. With a maximum growth temperature of over 20°C, these microbes require a temperature of roughly 15°C or higher. The differentiation between psychrophiles and psychrotolerant microorganisms is crucial in understanding their ecological niches and adaptations to varying temperature ranges (Collins & Margesin, 2019). The prevalence of psychrophiles in numerous environments underscores their significance in shaping microbial ecosystems across different ecosystems, including extreme and challenging habitats such as the Antarctic, high-altitude soils, and deep-sea environments.

The Earth's biosphere encompasses a substantial portion characterized by cold conditions, including regions such as the Arctic and Antarctica, permafrost areas, non-polar regions, and deep oceans (Hamid et al., 2022; Buzzini & Margesin, 2014). Temperatures as low as –32°C have been shown to support life, and metabolically active bacteria are able to survive in these conditions (Bakermans & Skidmore, 2011). Psychrophiles, isolated from extreme low-temperature ecosystems, exhibit remarkable adaptability to cold environments. For instance, *Planococcus halocryophilus* strain Orl, derived from freezing permafrost, has demonstrated growth at an exceptionally low temperature of –15°C, representing the lowermost temperatures to support growth to date (Mykytczuk et al., 2013). Another example is the cold-adapted yeast *Cystofilobasidium capitatum* SPY11, isolated from the soil of the Northern Province of India, Kashmir valley, showing growth up to 20°C (Hamid et al., 2012). Due to their habitat in cold environments, psychrophiles hold immense potential for producing the enzymes of agricultural and industrial importance. These enzymes exhibit resistance to the typical slowdown in chemical reaction rates that occurs at lower temperatures. Despite the challenges posed by frigid conditions, these biological catalysts maintain a moderate level of efficiency and catalytic activity, enabling the organisms to sustain vital metabolic processes (D'Amico et al., 2002). Recent reviews have highlighted the sources of important cold-adapted protease producers in addition to their molecular-level adaptations that enable these proteases to function effectively in low-temperature environments. Additionally, they have explored the promising agricultural and industrial applications and potential of utilizing these cold-active protease enzymes across various sectors (Hamid et al., 2024; Furhan, 2020). Extremophiles, particularly psychrophilic microorganisms, are increasingly recognized for their valuable contributions to bio-catalysis of proteins, organic acids, antimicrobial agents and bioactive compounds (Archna et al., 2015; Verma et al., 2015; Kumar et al., 2016; Kumar et al., 2017; Kaur et al., 2017).

3. Factors Affecting the Biological Activity of Cryophiles

The key factors that influence and shape the diverse array of microbial life present in extreme cold environments are frequently assessed and characterized based on the temperature dependency exhibited by psychrophilic organisms. However, soil nutrition is equally influential in growth determination, impacting substrate availability, metabolism, and enzymatic rates (Chauhan et al., 2023). Certain strains of *Pseudomonas*, through effective mineral absorption tactics, exhibit root invasion, rapid colonization, and establish host-microorganism interactions, enabling them to thrive in nutrient-deficient locations. This strategic defensive pathway facilitates the degradation of various compatible solutes via metabolic pathways such as the Entner-Doudoroff and β-ketoadipate pathways (Craig et al., 2021; Cabrera et al., 2022). Psychrophiles, exemplified by *P. psychrophila* MTCC12324 and *P. frederiksbergensis*, demonstrate the ability to tolerate broader temperature ranges than their ambient isolation conditions, growing between 4°C–37°C. Similarly, pH emerges as a crucial factor influencing microbial survival, with microorganisms from extreme temperature environments exhibiting growth across a broad pH spectrum, from extreme acidity to extreme alkalinity (Dhakar & Pandey, 2016; Abraham et al., 2020; Cui et al., 2023; Panicker et al., 2002; Ravi et al., 2022). Cold temperatures set a finite boundary for the viability of life in the majority of Earth's ecosystems and are a crucial factor in determining habitability, not only on Earth but also in the exploration for life on Mars and other celestial bodies in our solar system (Fairén et al., 2010; McKay et al., 2011; Doyle et al., 2012). Presently, –12°C and –32°C are among the lowest observed temperatures allowing microbe development and metabolic processes (Anitori, 2012). Biological processes face multiple constraints in sub-freezing temperatures. One major limiting factor is the scarcity of liquid water, which is essential for life. Additionally, the molecular motion and energy levels required for chemical reactions are significantly reduced at such low temperatures. Furthermore, any pockets of liquid water that may be present in these otherwise frozen habitats often have high concentrations of dissolved solutes, imposing osmotic stress on living organisms. The interplay of these factors severely restricts biological activity in sub-zero environments (Mykytczuk et al., 2013). Despite the extreme conditions, microorganisms inhabiting these freezing environments possess the remarkable ability to sustain metabolic activities and undergo cell division at sub-zero temperatures. This is made possible by the presence of minute quantities of liquid water, even if it exists only as thin films. As long as this essential requirement of liquid water availability is met, these resilient microbes can remain biologically active and proliferate within these harsh, sub-freezing habitats (Goordial, 2021). Organisms capable of thriving at refrigerator temperatures play a crucial role in maintaining the quality of dairy products. It is acknowledged that various organisms, optimized for growth at room temperatures or higher, may still proliferate in the range of 5°C–10°C. However, the most significant impact comes from true cryophiles, which exhibit their optimal growth within this specific temperature range. Despite the widespread practice of storing dairy products at low temperatures, bacteriology laboratory analyses and quality tests often utilize incubation temperatures ranging from 20°C–37°C (Dortey et al., 2020).

The primary challenge faced by cold-adapted microbes is the adverse impact of low temperatures on reaction rates. According to the Arrhenius Law, all reactions, including those catalyzed by enzymes, experience an exponential decrease in rate with decreasing temperature (Arrhenius, 1889). While most enzymes not adapted to cold conditions exhibit a significant reduction in activity when transitioning from 37°C to 0°C, enzymes from psychrophilic organisms are specially tailored to sustain high specific activities, ensuring appropriate rates of metabolism in cold environments (Collins et al., 2012). The majority of cold-adapted enzymes that have been studied exhibit distinct properties that enable them to function optimally in extreme cold conditions. These enzymes typically display heightened catalytic efficiency at relatively low to modest temperatures in comparison to their mesophilic (moderate temperature-adapted) and thermophilic (heat-adapted) equals. Additionally, the temperature at which these cold-adapted enzymes exhibit peak activity is shifted towards lower values. However, this enhanced low-temperature activity often comes at the cost of reduced structural stability when compared to enzymes adapted to warmer environments. These "cold enzymes" achieve this by reducing activation energy (ΔG^*), temperature dependence, and activation enthalpy (ΔH^*), implying fewer or weaker enthalpic interactions during the catalytic cycle (Lonhienne et al., 2000). This results in a more disordered enzyme-substrate (ES) ground state, leading to increased catalytic rates. Various studies have explored the structural flexibility or disorder in psychrophilic enzymes. Although protein flexibility is a complex parameter, it is generally accepted that psychrophilic enzymes exhibit increased structural flexibility, overcoming the low-temperature challenge by enhancing plasticity in specific regions or the entire protein structure. This flexibility facilitates the conformational changes necessary for low-temperature activity but often comes at the expense of decreased stability. The modifications contributing to increased flexibility include a reduction in stabilizing interactions like H bonds, salt bridges, aromatic interactions, and disulphide bonds, as well as a weakening of the hydrophobic core. Cold-adapted enzymes are also observed to exhibit alterations in their amino acid composition, specifically with regards to the prevalence of proline and glycine residues. These changes tend to be concentrated in the loop regions of the protein structure. Additionally, research has indicated that these cold-active enzymes often display a greater degree of exposure of hydrophobic amino acid residues on their surface (Collins & Gerday, 2017). The intrinsic properties exhibited by enzymes from psychrophilic (cold-loving) organisms, such as their ability to maintain high catalytic activity at low to moderate temperatures, coupled with their relatively lower structural stability, render these enzymes advantageous for a diverse range of applications. The unique combination of enhanced low-temperature performance and decreased stability offers practical benefits that can be leveraged across various agricultural practices and industrial processes. They are utilized to enhance the efficiency and economics of processes requiring lower temperatures, thereby improving environmental impact. Cold enzymes find applications in industrial sectors such as food, feed, and technical processes, with notable successes in organic synthesis, bread quality improvement, low-temperature cleaning detergents, and molecular biology techniques (Sarmiento et al., 2015).

Organisms adapted to thrive in extremely cold environments possess the remarkable capability of producing enzymes that exhibit exceptionally high levels of catalytic activity within their optimal temperature range. Remarkably, the specific activity, which quantifies the enzymatic efficiency per unit amount of enzyme, can be up to ten times greater for these cold-adapted enzymes when compared to their counterparts from organisms adapted to more moderate temperature conditions. This heightened activity compensates for the typically sluggish response rates at low temperatures (Santiago et al., 2016; Mangiagalli & Lotti, 2021). Because these proteins are not very stable and have a propensity to unfold and become inactive at lower temperatures, the maximum activity of these enzymes is moved toward the lower temperature range. Destabilization of the active site, or potentially the entire protein structure, is responsible for the very high activity at low temperatures. In low temperatures where molecular movements and biological processes would normally freeze, this instability makes the catalytic centre more flexible (Santiago et al., 2016). It is notable that recognized cryophilic bacteria are also halophilic, signifying a close connection between sub-zero growth and osmo-tolerance (Anitori, 2012). Osmo-tolerance expands the limits for microbial survival at sub-zero temperatures (Chin et al., 2010). Thus, the ideal creatures to explore the limitations of life at low temperatures and the processes necessary to sustain metabolic activity in sub-zero habitats are those found in Earth's cryo-environments. Permafrost cryo-environments, prevalent in Arctic regions, consistently experience sub-zero temperatures ranging from 10 to –20°C. These environments host diverse microbial communities (Steven et al., 2006; Margesin & Miteva, 2011). These communities have been found in permafrost, but their metabolic activity below 0°C has only lately been shown, both in pure culture (Bakermans et al., 2003; Panikov & Sizova, 2007) and in soil microcosms (Rivkina et al., 2000; Panikov & Sizova, 2007; Steven et al., 2006, 2007). As demonstrated by ice brine veins, the sub-zero brine veins enclosing soil particles are probably home to actively metabolizing microorganisms within permafrost (Junge et al., 2004; Bakermans & Skidmore, 2011). Determining the function of cold-active permafrost microbes in biogeochemical cycles throughout the melting of permafrost requires an understanding of their sub-zero metabolic activity (Singh et al., 2010; Vishnivetskaya et al., 2011). Nevertheless, the literature on permafrost bacteria is still scarce, and there are few draft genomes and studies of their cold-adaptive processes available. These include creatures such as *Exiguobacterium sibiricum* (Rodrigues et al., 2008), *Psychrobacter arcticus* 273-4 (Bergholz et al., 2009; Ayala-del-Río et al., 2010), and *Psychrobacter cryohalolentis* K5 (Bakermans et al., 2006). Elevation is a significant determinant of microbial growth in specific climates. Research in the Colorado Rocky Mountains reveals that soil microbiome diversity declines linearly with increasing elevation (Bryant et al. 2008). Edaphic factors, such as soil organic carbon, total nitrogen content, and moisture, significantly impact microbial diversity, with fluctuations observed at mid-elevational gradients (Bhattacharya, Tiwari, Rai et al., 2022). Research conducted in the Indian Himalayan area and on the post-deglaciation patterns of the Gangotri glacier highlights the significance of edaphic conditions on the composition and metabolism of microbial groups (Bhattacharya, Tiwari, Talukdar et al., 2022; Selvakumar et al., 2009). In mountain ecosystems, the direction of the slope affects the bioactivities shown by microorganisms that promote plant development, implying a coordinated pattern of temperature, height, and species variety (Nottingham et al., 2018). Along elevation gradients, the enzymes that break down organic materials align up with microbial transition patterns. Therefore, the diversity of microbes in these cold conditions is shaped by a combination of temperature, soil nutrition, pH, and elevation. These factors interact to influence the colonization and supremacy of explicit microbial inhabitants in mountain ecosystems, impacting both plant and microbial diversity.

4. Exploration of Cryophilic Microorganisms: Multifaceted and Promising Contenders for Agriculture and Other Diverse Applications

Applications in Agriculture

Promotion of Plant Growth

The promotion of plant growth (PGP) is a multifaceted phenomenon that can be facilitated through diverse mechanisms, including both direct and indirect pathways. Indirect mechanisms are closely associated with the management of plant diseases through biological inputs, relying on the synthesis of various compounds such as biocidal metabolites, siderophores, lysozymes, and antibiotics (Pandey & Yarzábal, 2019). Additionally essential to the development of plant defensive responses are microbial secondary metabolites that produce phytohormones, release bioavailable nitrogen, and solubilize phosphate (Pandey & Yarzábal, 2019). The synthesis of siderophores, HCN, phytohormones, IAA, ACC deaminase activity, and conversion of vital nutrients constitute key components of plant growth promotion (Balcazar et al., 2015; Ortiz-Ojeda et al., 2017; Kaur et al., 2021). The exploration of psychrophiles in this context introduces a novel dimension to sustainable agriculture, particularly in hilly provinces (Pandey & Yarzábal, 2019; Yadav et al., 2019; Torracchi C. et al., 2020). Cold-active microorganisms, or psychrophiles, exhibit the capability to produce phytohormones, including gibberellins, cytokinins, and auxins, which are pivotal regulators of physiological and molecular plant responses. The attention towards development of microbe based bioformulations as biopesticides and biofertilizers underscores the potential of psychrophiles in promoting sustainable agricultural practices in cold environments. However, challenges arise in the application of exogenous microorganisms in field trials due to their low survival rates in soils from different locations, particularly in the face of temperature variations (Feller & Gerday, 2003). To address this issue, the idea is to obtain and choose microorganisms from their natural microbial communities for use as bioinoculants (Velázquez et al., 2016). Field research in the Mamlay watershed of the Sikkim Himalaya in mountain habitats, emphasized the importance of native microorganisms and the requirement for PGP bacteria that are acclimated to the cold (Pandey & Palni, 1998). This study also demonstrated the potential of cold-adapted *Pseudomonas* (*P. corrugata*) as bioinoculants for maize cultivation at higher altitudes (Kumar, Trivedi, & Pandey, 2007). The genetic diversity and PGP activities of Pseudomonas species isolated from various mountain locations have been extensively investigated (Negi et al., 2011; Kour et al., 2020). Psychrotolerant PGP bacteria have demonstrated the dominance of *Pseudomonas* and *Bacillus* strains with PGP activities. These bacteria were isolated from the rhizosphere of the Andean tuber crop, *Lepidium meyenii* Walp. (maca) (Ortiz-Ojeda et al., 2017; Chica et al., 2019). The identification of specific strains, such as ILQ215 (*P. silesiensis*) and JUQ307 (*P. plecoglossicida*), from the Peruvian Andean Plateau region has further highlighted their potential for enhancing plant growth and serving as bioinoculants (Chumpitaz-Segovia et al., 2020). Continued research and development of carrier-based formulations tailored for cold environments will be crucial for the widespread application and commercialization of this technology (Trivedi, Pandey, & Palni, 2005; Du Jardin, 2015). Agriculture must simultaneously minimize its negative effects on the environment while providing food for a growing world population; the exploration of psychrophiles in plant growth promotion emerges as a valuable avenue for future agricultural innovation (Figure 2). Table 1 represents the role of cryophiles towards sustainable agriculture.

Plant Disease Management

The cold-active microbial control agents employ various mechanisms to combat plant pathogens and herbivores. These mechanisms, often redundant, involve direct and indirect biotic interactions. Negative direct mechanisms include the production of toxic compounds, synthesis of hydrolytic enzymes, and hyper parasitism (Köhl et al., 2019; Yarzábal et al., 2018). Indirect mechanisms encompass niche overlap, quorum quenching, and selection of biological control agents (BCA) by plant hosts. In terms of direct biological control, BCAs utilize metabolites or enzymes that impede pathogen growth or exhibit hyper-parasitic behaviour. For instance, microbial secondary metabolites, such as antibiotics, play a crucial role in inhibiting plant pathogenic microorganisms (Karlovsky, 2008). Cold-active Pseudomonads, especially CA-Pseudomonads, are well-studied PGPM and BCA bacteria from cold environments. They produce antimicrobial compounds like biosurfactants, iron-chelating chemicals, hydrogen cyanide (HCN), and antimicrobial compounds that cause plant systemic resistance (Ogata-Gutiérrez et al., 2017; Ortiz-Ojeda et al., 2017). By producing soluble and volatile compounds, Antarctic *Pseudomonas* strains and psychrotolerant

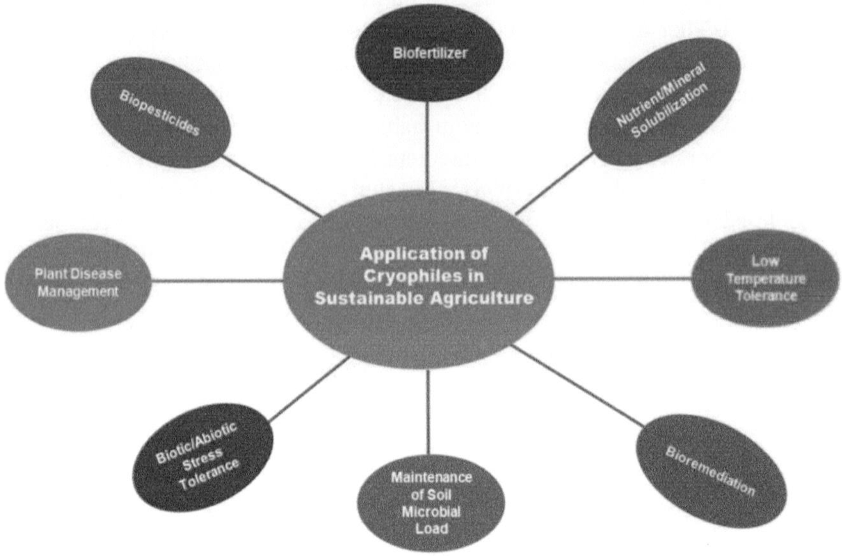

Figure 2: Application of cryophiles in sustainable agriculture.

yeasts are known to suppress the phytopathogenic fungal populations (Yarzábal et al., 2018; Vero et al., 2013). Bacteria producing gluconic acid, such as gluconic-acid-producing Pseudomonas, demonstrate biocontrol effectiveness against various plant pathogens (Kaur et al., 2006; Nieto-Penalver et al., 2014). Lactobacilli, specifically *Lactobacillus plantarum*, exhibit broad-spectrum ability to protect fruit crops from pathogens through acidification caused by lactic acid accumulation (Daranas et al., 2019). Cold-active microorganisms, including those producing pigments, play a role in antagonistic ecological interactions. Pigments like violacein and carotenoid-like compounds produced by bacterial and fungal species inhibit the growth of bacteria, fungi, and protozoa (Marizcurrena et al., 2019; Pandey et al., 2018).

Soil Health

Low temperature and drought are prominent abiotic stresses impacting agricultural productivity, with around 20% of the Earth's surface affected by frozen soil and permafrost. Microbes play a vital role in managing low temperature stress, offering genetic diversity and resilience (Yadav, 2021). In the case of drought, isolated microbes, once thoroughly studied, can be harnessed to control this stress. Inoculating plants with diverse PGP microbiomes is a promising approach to alleviate drought in arid regions. Water flooding stress, prevalent in rice-producing areas, affects a significant portion of global rice land annually. Effective strategies are needed to mitigate the impact of volatile flash floods on rice crops in these regions. Overall, understanding microbial roles

Table 1: Role of various cryophile microbes in sustainable agriculture

Microbe Group	Cold Tolerant Microbe	Utility	References
Cold Active Bacteria	*Pseudomonas* sp.	Reduce the impact that cold weather and other abiotic stressors place on plants.	Ogata-Gutiérrez et al., 2017; Mishra et al., 2008; Ortiz-Ojeda et al., 2017
		Encourage systemic resistance in plants.	Ogata-Gutiérrez et al., 2017
		Encourage development and germination in cold temperatures.	Yarzábal et al., 2018
		Effective against *F. oxysporum* f. sp. *cucumerinum* and *Verticillium dahlia.*	Liu et al., 2019
	Pseudomonas Fredericksbergensis	Uses a secreted CA chitinase to prevent Botrytis cinerea from growing its mycelial growth.	Melo et al., 2016
	Bacillus subtilis and *Bacillus amyloliquefaciens*	Used to suppress infections from certain fruit trees in addition to other crops like potatoes, quinoa, and maize.	Torracchi C. et al., 2020; Ortuño et al., 2013; Ortuño, Miranda, and Claros, 2013
	Planococcus versutus	Interference with the Quorum Sensing signalling system.	See-Too et al., 2017; See-Too et al., 2018
Cold Active Fungi	*Trichoderma* spp.	Used to manage some of the fungal diseases that affect maize, vegetables, and ornamental plants.	Ortuño, Miranda, and Claros 2013; Ortuño et al., 2013
	Trichoderma gamsii	Antagonistic toward *F. oxysporum* f. sp. *lycopersici* and protective against Fusarium wilt in tomato plants.	Shanmugam, Chugh, and Sharma 2015
	Mortierella sp.	This fungus is often found in Antarctic soils and works by baiting, injecting, or inoculating the soil to kill wax moth and housefly larvae.	Edgington et al., 2014

(Contd.)

(*Contd.*)

Microbe Group	Cold Tolerant Microbe	Utility	References
	Monacrosporium ellipsosporum	Pulls in nematodes by releasing sticky conidia that cling to their food and germinate.	Gray 1985; Gray, Wyborn, and Smith, 1982
	Lecanicillium muscarium	Fruit deterioration-causing mould was actively inhibited by the strong chitinase that this fungus generated.	Fenice, 2016
Yeasts	*Leucosporidium scottii*	Inhibits the growth of rot causing fungi in apples by generating certain kinds of metabolites.	Arrarte et al., 2017
	Candida sake	Repressed the progress of five fungal spoilers of apples (*Alternaria alternata*, *Penicillium expansum*, *Botrytis cinerea*, *Alternaria tenuissima*, and *Alternaria arborescens*).	
	Cystofilobasidium infirmominiatum PL1	Protect harvested fruit from *Botrytis cinerea* and *Penicillium expansum* rot.	Vero et al., 2013
	Candida albidus		
	Pichia membranifaciens		
	Vishniacozyma victoriae		
	Metschnikowia fructicola	Shields tuber crops including sweet potatoes, carrots, oranges, etc. against *Sclerotinia sclerotium*, *Penicillium digitatum*, *B. cinerea*, *Fusarium* sp., and *Rhizopus stolonifera*.	Blachinsky et al., 2007
	Rhodotorula mucilaginosa	Suppressed *Penicillium expansum* development, responsible for pear fruit blue mould at 4°C in a density-dependent way.	Hu et al., 2015

	Aureobasidium pullulans	It is applied to pome fruit to prevent storage rots. It shows efficient results against *B. cinerea*, *P. expansum*, *Neofabraea* sp., and *Monilinia fructigena*.	Zajc et al., 2020; Spadaro and Droby, 2016
Nematodes	*Heterorhabditis bacteriophora*	Effectiveness against vine weevil (*Otiorhynchus sulcatus)* at low temperatures.	Ehlers et al., 2005
	Heterorhabditis downesi	Suitable for controlling black vine weevil (*Otiorhynchus sulcatus*), citrus longhorn beetle (*Anoplophora chinensis*), giant pine weevil (*Hylobiusabietis* sp.), and spruce bark beetle (*Ips typographus*) in ornamentals and soft fruit crops.	Torracchi C. et al., 2020
	Steinernema kraussei	Capable of penetrating vine weevil larvae (*Otiorhynchus sulcatus*) through natural openings.	

and implementing innovative strategies can enhance agricultural resilience against various abiotic stresses (Yadav, 2021). Beyond biotechnological benefits, cold-loving microorganisms contribute to various ecological progressions in Himalayan ecologies. Their roles in biodegradation and agriculture, including plant growth promotion and biocontrol under colder conditions, support environmental sustainability (Pandey & Yarzábal, 2019; Tribelli & López, 2018). Microorganisms in these regions contribute to the degradation of complex hydrocarbons, such as petroleum products, with *Acinetobacter*, *Bacillus*, *Pseudomonas*, *Rhodococcus* and *Sphingobium* playing significant roles (Bisht et al., 2014; Mahajan et al., 2019; Margesin, 2007; Margesin & Schinner, 2001). Microbial changes and the number of efficient degraders within endemic populations reflect the impact of specific hydrocarbon compounds on the environment, as exemplified by a petroleum-contaminated colder area in China–Russia (Yang et al., 2016).The acetylene reduction assay to elucidate biological N_2 fixation activity is typically conducted under specific conditions. Bacterial cells, pre-cultured, are introduced into a nitrogen-free or nitrogen-scarce liquid medium that includes a carbon source, often d-glucose or d-mannitol (Stowers, 1985) at a concentration of 0.5–2.0%. The exposure lasts for 24 hours or less and is carried out at a distinctive room temperature, commonly around 25°C (Barraquio et al., 1983). However, this method faces limitations when dealing with free-living, microaerobic nitrogen-fixing bacteria, which are notorious for being challenging to culture. To address this issue, Döbereiner and her team developed a soft gel technique (Day & Döbereiner, 1976) that uses 0.2% agar as the medium's gel matrix. With the help of a vertical gradient in the concentration of dissolved oxygen created by these micro-aerobes, an optimal level of dissolved oxygen is secured at a specific depth inside the medium (Vardharajula et al., 2017). Nitrogen-fixing micro-aerobes, especially those on the rhizoplane of monocotyledonous agricultural plants, have been shown to have noteworthy activity in acetylene reduction experiments (e.g., *Azospirillum* and *Herbaspirillum* sp.) (Van et al., 1997). As a basic technique for screening free-living or difficult-to-cultivate nitrogen-fixing bacteria, thus far these 0.2% agar-solidified soft gel media have found widespread usage (Jeffrey et al., 2000; Barraquio et al., 1983).

Food Additives

Enzymes from psychrophilic bacteria find uses as additions in cleaning goods, flavouring agents in dairy and bread products, molecular biology and biotransformation processes, bioremediation as well as in pharmaceutics and medical sciences (Collins & Margesin, 2019; Feller et al., 1996; Bruno et al., 2019; Gupta et al., 2020). Fortunately, the psychrophiles discovered on chilled and frozen foods do not seem to generate toxins or harmful substances. However, there is a noteworthy apprehension after the revelation that spores of *Clostridium botulinum* type E can undergo germination, leading to the vegetative growth

2013; Sharma et al., 2010). The region also boasts a prevalence of multicopper enzymes, particularly laccase, with reported potential from cold-adapted microorganisms in glacial sites and hot springs of the Himalayas (Kaira et al., 2015; Gupta et al., 2017). Psychrophilic enzymes play a crucial role in industrial processes that require low temperatures due to the sensitivity of heat-sensitive or volatile components (Hamid et al., 2022). Additionally, these enzymes are beneficial in processes where undesirable chemical side-reactions occur at high temperatures, and avoiding contamination problems is essential. This is particularly relevant in the production of various items such as foods, beverages, fine chemicals, and pharmaceuticals (Mangiagalli & Lotti, 2021). Beyond temperature considerations, many industrial processes involve extremes of pH, pressure, salinity, and the presence of detergents, non-aqueous solvents, among other challenging conditions (Wackett, 2019). The versatility of psychrophilic enzymes makes them valuable tools in addressing the specific requirements and challenges encountered in these diverse industrial settings.

Enzyme Sources

Cold-adapted microorganisms are acknowledged as a promising source of cold-active enzymes and bioactive compounds (Margesin & Feller, 2010). In the current industrial landscape, there is a shift towards green chemistry, utilizing enzyme-based reactions in lieu of chemicals, particularly in sectors like textiles (Razzaq et al., 2019). Working with enzymes offers numerous advantages, including minimal interference with byproducts, reduced energy consumption, negligible pollution, and uncompromised product quality. Cold-active enzymes work well in catalytic processes because of their low kinetic energy in cold conditions. They are defined by their low activation enthalpy and structural flexibility (Cavicchioli et al., 2011). Cold-adaptive microorganisms generate enzymes including lignin-degrading enzymes, amylase, cellulase, lipase, invertase and protease which have a wide range of uses in sectors like detergents, food, and pharmaceuticals (Gomes & Steiner, 2004). Antifreeze proteins from these microorganisms also play crucial roles in physiological and biotechnological realms, controlling ice crystal shape and size, enabling biochemical activities in freezing temperatures. The production of antifreeze proteins from various organisms reflects convergent evolution due to extreme cold pressures, with potential applications in other industrial sectors (Kim et al., 2017; Venketesh & Dayananda, 2008). Florczak et al. (2013) discovered a cold-adapted lipase, LipG7, derived from the Antarctic filamentous fungus *Geomyces* sp. P7. This lipase exhibited unique properties, retaining 15% of its maximum activity at 0°C while maintaining 100% catalytic activity after 1 hour of incubation at 100°C. The Antarctic psychrophilic ciliate *Euplotes focardii* expressed three superoxide dismutases (SOD) belonging to two families, demonstrating tolerance to high temperatures (Tm in the range of 50–70°C) in conjunction with cold adaptation (Pischedda et al., 2018). *Planococcus halocryophilus* strain Or1, an aerobic Gram-positive bacterium isolated from Arctic permafrost in Canada, exhibited survival capabilities at –15°C (Mykytczuk et al., 2016). A psychrophilic laccase associated with copper oxide nanoparticles, entangled in a single-walled nanotube, demonstrated activity at low temperature (4°C) and enhanced activity at higher temperature (80°C) (Mukhopadhyay et al., 2015). Another significant study emphasized the thermostability of a cold-active transglutaminase, linking it to a versatile active site, a relatively stiff framework, large stable valine and lysine residues, coupled with strong electrostatic contacts at both N- and C-terminals (Zhang et al., 2020).

5. Conclusion and Future Perspectives

The exploration of the role of cryophiles in the sustainability of agro-ecosystems has unveiled a fascinating interplay between extremophiles and agricultural environments. This chapter has delved into the complicated relationships between cold-adapted microorganisms and the various components of agro-ecosystems, shedding light on their potential contributions to sustainable agriculture. The significance of cryophiles lies not only in their ability to thrive in extreme cold conditions but also in their capacity to positively impact soil health, plant growth, and overall agro-ecosystem resilience. The intricate interactions between cryophiles and plants, particularly in the rhizosphere, emphasize the potential for harnessing these microorganisms to enhance nutrient uptake, improve stress tolerance, and protect against pathogens. Understanding and leveraging these interactions could pave the way for innovative and sustainable agricultural practices. Moreover, the role of cryophiles in nutrient cycling and soil organic matter decomposition cannot be overstated. As key players in the breakdown of organic materials in cold environments, these microorganisms contribute to the formation of nutrient-rich soils. Harnessing their enzymatic capabilities for organic matter degradation could offer sustainable solutions for soil fertility enhancement, reducing the reliance on chemical fertilizers and minimizing environmental impacts. The potential applications of cryophiles in bioremediation have also been explored, highlighting their ability to adapt to cold environments contaminated with various pollutants. By tapping into the metabolic versatility of cryophiles, there is a promising avenue for developing eco-friendly strategies to mitigate the adverse effects of pollutants on agro-ecosystems. As we strive for sustainable agricultural practices to meet the growing global demand for food, the insights gained from this exploration of cryophiles open new doors for innovation. The integration of cryophiles into agricultural management practices holds the promise of enhanced resilience in the face of climate change, reduced environmental impact, and more importantly increased crop productivity. However, it is essential to acknowledge the challenges and gaps in our current understanding of cryophiles and their interactions within agro-ecosystems. Further research is needed to unravel the complexities of these microbial communities and to develop practical applications that can be seamlessly integrated into mainstream agricultural practices.

References

Abraham, Wilson Peter, Sajith Raghunandanan, Vipin Gopinath, Karthika Suryaletha, & Sabu Thomas. 2020. Deciphering the cold adaptive mechanisms in Pseudomonas psychrophila MTCC12324 isolated from the Arctic at 79 N. *Current Microbiology*, 77, 2345–2355.

Afrouz, M., Sayyed, R.Z., Fazeli-Nasab, B., Piri, R., Almalki, W.H., & Fitriatin, B.N. 2023. Seed bio-priming with beneficial *Trichoderma harzianum* alleviates cold stress in maize. *PeerJ.*, 11, e15644. http://doi.org/10.7717/peerj.15644.

Aislabie, Jackie, David J. Saul, & Julia M. Foght. 2006. Bioremediation of hydrocarbon-contaminated polar soils. *Extremophiles*, 10, 171–179.

Anitori, Roberto Paul. 2012. *Extremophiles: Microbiology and Biotechnology*. Caister Academic Press.

Archna, Suman, Verma, Priyank, Yadav, A.N., & Saxena, A.K. 2015. Bioprospecting for extracellular hydrolytic enzymes from culturable thermotolerant bacteria isolated from Manikaran thermal springs. *Research Journal of Biotechnology*, 10(4).

Arrarte, E., Garmendia, G., Rossini, C., Wisniewski, M., & Vero, S. 2017. Volatile organic compounds produced by Antarctic strains of Candida sake play a role in the control of postharvest pathogens of apples. *Biological Control*, 109, 14–20.

Arrhenius, Svante. 1889. Über die Reaktionsgeschwindigkeit bei der Inversion von Rohrzucker durch Säuren. *Zeitschrift* für *Physikalische Chemie*, 4(1), 226–248.

Ayala-del-Río, Héctor L., Patrick S. Chain, Joseph J. Grzymski et al. 2010. The genome sequence of Psychrobacter arcticus 273-4, a psychroactive Siberian permafrost bacterium, reveals mechanisms for adaptation to low-temperature growth. *Applied and Environmental Microbiology*, 76(7), 2304–2312.

Baba, Z.A., Hamid, B., Sheikh, T.A., Alotaibi, S., Enshasy, H.E., Ansari, M.J. et al. 2021. Psychrotolerant *Mesorhizobium* sp. isolated from temperate and cold desert regions solubilize Potassium and produces multiple plant growth promoting metabolites. *Molecules*, 26, 5758. https://www.mdpi.com/1420-3049/26/19/5758/htm.

Bakermans, Corien, Alexandre I. Tsapin, Virginia Souza-Egipsy, David A. Gilichinsky, & Kenneth H. Nealson. 2003. Reproduction and metabolism at –10°C of bacteria isolated from Siberian permafrost. *Environmental Microbiology*, 5(4), 321–326.

Bakermans, Corien, Héctor L. Ayala-del-Río, Monica A. Ponder et al. 2006. Psychrobacter cryohalolentis sp. nov. and Psychrobacter arcticus sp. nov., isolated from Siberian permafrost. *International Journal of Systematic and Evolutionary Microbiology*, 56(6), 1285–1291.

Bakermans, Corien, & Mark L. Skidmore. 2011. Microbial metabolism in ice and brine at –5°C. *Environmental Microbiology*, 13(8), 2269–2278.

Bakermans, Corien. 2017. Determining the limits of microbial life at subzero temperatures. In: *Psychrophiles: From Biodiversity to Biotechnology*. Springer International Publishing, pp. 21–38.

Balcazar, Wilvis, Johnma Rondón, Marcos Rengifo, et al. 2015. Bioprospecting glacial ice for plant growth promoting bacteria. *Microbiological Research*, 177, 1–7.

Barraquio, W.L., Ladha, J.K., & I Watanabe. 1983. Isolation and identification of N2-fixing Pseudomonas associated with wetland rice. *Canadian Journal of Microbiology*, 29(8), 867–873.

Bergholz, Peter W., Corien Bakermans, & James M. Tiedje. 2009. Psychrobacter arcticus 273-4 uses resource efficiency and molecular motion adaptations for subzero temperature growth. *Journal of Bacteriology*, 191(7), 2340–2352.

Bhat, Archana, Syed Riyaz-Ul-Hassan, Nasier Ahmad, Nidhi Srivastava, & Sarojini Johri. 2013. Isolation of cold-active, acidic endocellulase from Ladakh soil by functional metagenomics. *Extremophiles*, 17, 229–239.

Bhattacharya, Pamela, Pankaj Tiwari, Gautam Talukdar, & Gopal S. Rawat. 2022. Shifts in bacterial community composition and functional traits at different time periods post-deglaciation of Gangotri Glacier, Himalaya. *Current Microbiology*, 79(3), 91.

Bhattacharya, Pamela, Pankaj Tiwari, Ishwari Datt Rai, Gautam Talukdar, & Gopal Singh Rawat. 2022. Edaphic factors override temperature in shaping soil bacterial diversity across an elevation-vegetation gradient in Himalaya. *Applied Soil Ecology*, 170, 104306.

Bisht, S., Pandey, P., Kaur, G., Aggarwal, H., & Sood, A. 2014. Utilization of endophytic strain Bacillus sp. SBER3 for biodegradation of polyaromatic hydrocarbons (PAH) in soil model system. *European Journal of Soil Biology*, 60, 67–76.

Blachinsky, Daphna, Jana Antonov, Amir Bercovitz, et al. 2007. *Commercial Applications of Shemer for the Control of Pre- and Post-harvest Diseases*. ResearchGate.

Bowman, Jeff S., & Jody W. Deming. 2014. Alkane hydroxylase genes in psychrophile genomes and the potential for cold active catalysis. *BMC Genomics*, 15(1), 1–11.

Bratchkova, Anna, & Veneta Ivanova. 2011. Bioactive metabolites produced by microorganisms collected in Antarctica and the Arctic. *Biotechnology and Biotechnological Equipment*, 25(sup1), 1–7.

Bruno, Stefano, Daniela Coppola, Guido di Prisco, Daniela Giordano, & Cinzia Verde. 2019. Enzymes from marine polar regions and their biotechnological applications. *Marine Drugs*, 17(10), 544.

Bryant, Jessica A., Christine Lamanna, Hélene Morlon, Andrew J. Kerkhoff, Brian J. Enquist, & Jessica L. Green. 2008. Microbes on mountainsides: Contrasting elevational patterns of bacterial and plant diversity. *Proceedings of the National Academy of Sciences*, 105(supplement_1), 11505–11511.

Buzzini, Pietro, & Rosa Margesin. 2014. Cold-adapted yeasts. In: *Miscellaneous Cold-Active Yeast Enzymes of Industrial Importance.* New York: Springer.

Cabrera, Ma Ángeles, Sebastián L. Márquez, & José M. Pérez-Donoso. 2022. Comparative genomic analysis of Antarctic pseudomonas isolates with 2,4,6-trinitrotoluene transformation capabilities reveals their unique features for xenobiotics degradation. *Genes*, 13(8),1354.

Cann, D.C., Barbara B. Wilson, Hobbs, G., & Shewan, J.M. 1965. The growth and toxin production of Clostridium botulinum type E in certain vacuum packed fish. *Journal of Applied Microbiology*, 28(3), 431–436.

Cavicchioli, R., Charlton, T., Ertan, H., S. Mohd Omar, Siddiqui, K.S., & Williams, T.J. 2011. Biotechnological uses of enzymes from psychrophiles. *Microbial Biotechnology*, 4(4), 449–460.

Chauhan, Mansi, Ayushi Kimothi, Avinash Sharma, & Anita Pandey. 2023. Cold adapted pseudomonas: Ecology to biotechnology. *Frontiers in Microbiology*, 14.

Chen, J. 2018. *Food Enzymes: Global Markets*. BCC Research.

Chica, Eduardo, Lenys Buela, Adrián Valdez, Paulina Villena, Denisse Peña, & Luis Andrés Yarzábal. 2019. Metagenomic survey of the bacterial communities in the rhizosphere of three Andean tuber crops. *Symbiosis*, 79(2), 141–150.

Chin, Jason P., Julianne Megaw, Caroline L. Magill et al. 2010. Solutes determine the temperature windows for microbial survival and growth. *Proceedings of the National Academy of Sciences*, 107(17), 7835–7840.

Chumpitaz-Segovia, Carolina, Débora Alvarado, Katty Ogata-Gutiérrez, & Doris Zúñiga-Dávila. 2020. Bioprospection of native psychrotolerant plant-growth-promoting rhizobacteria from Peruvian Andean Plateau soils associated with Chenopodium quinoa. *Canadian Journal of Microbiology*, 66(11), 641–652.

Collins, Tony, Georges Feller, Charles Gerday, & Marie-Alice Meuwis. 2012. *Family 8 Enzymes with Xylanolytic Activity*. United States Patent and Trademark Office (USPTO).

Collins, Tony, & Charles Gerday. 2017. Enzyme catalysis in psychrophiles. *Psychrophiles: From Biodiversity to Biotechnology*. Springer International Publishing, pp. 209–235.

Collins, Tony, & Rosa Margesin. 2019. Psychrophilic lifestyles: Mechanisms of adaptation and biotechnological tools. *Applied Microbiology and Biotechnology*, 103, 2857–2871.

Connell, L., Redman, R., Craig, S., Scorzetti, G., Iszard, M., & Rodriguez, R., 2008. Diversity of soil yeasts isolated from South Victoria Land, Antarctica. *Microbial Ecology*, 56, 448–459.

Craig, Kelly, Brant R. Johnson, & Amy Grunden. 2021. Leveraging Pseudomonas stress response mechanisms for industrial applications. *Frontiers in Microbiology*, 12, 660134.

Cui, Tingchen, Peng Wang, Jialu Li, YaoMing Su, Na Liu, & Mei Hong. 2023. Effects of temperature, ph, and salinity on the growth kinetics of Pseudomonas sp. Nb-1, a newly isolated cold-tolerant, alkali-resistant, and high-efficiency nitrobenzene-degrading bacterium. *Environmental Technology*, 44(14), 2171–2183.

D'Amico, Salvino, Paule Claverie, Tony Collins et al. 2002. Molecular basis of cold adaptation. *Philosophical Transactions of the Royal Society of London. Series B: Biological Sciences*, 357(1423), 917–925.

D'Amico, S., Collins, T., Marx, J.C., Feller, G., & Gerday, C. 2006. Psychrophilic microorganisms: Challenges for life. *EMBO Rep*, 7(4), 385–389.

Daranas, Núria, Gemma Roselló, Jordi Cabrefiga et al. 2019. Biological control of bacterial plant diseases with Lactobacillus plantarum strains selected for their broad-spectrum activity. *Annals of Applied Biology*, 174(1), 92–105.

Day, J.M., & Johanna Döbereiner. 1976. Physiological aspects of N2-fixation by a Spirillum from Digitaria roots. *Soil Biology and Biochemistry*, 8(1), 45–50.

De Maayer, Pieter, Dominique Anderson, Craig Cary, & Don A. Cowan. 2014. Some like it cold: Understanding the survival strategies of psychrophiles. *EMBO Reports*, 15(5), 508–517.

Dhakar, Kusum, & Anita Pandey. 2016. Wide pH range tolerance in extremophiles: Towards understanding an important phenomenon for future biotechnology. *Applied Microbiology and Biotechnology*, 100, 2499–2510.

Dortey, Miriam Dedeama, George Aboagye, & Bernardine Tuah. 2020. Effect of storage methods and duration of storage on the bacteriological quality of processed liquid milk post-opening. *Scientific African*, 10, e00555.

Doyle, S., Dieser, M., Broemsen, E., Christner, B., Miller, R.V., & Whyte, L.G. 2012. *Polar Microbiology: Life in a Deep Freeze*. AMS Press.

Du Jardin, Patrick. 2015. Plant biostimulants: Definition, concept, main categories, and regulation. *Scientia Horticulturae*, 196, 3–14.

Edgington, Steven, Emma Thompson, Dave Moore, Kevin A. Hughes, & Paul Bridge. 2014. Investigating the insecticidal potential of Geomyces (Myxotrichaceae: Helotiales) and Mortierella (Mortierellacea: Mortierellales) isolated from Antarctica. *SpringerPlus*, 3, 1–8.

Ehlers, Ralf-Udo, Jesko Oestergaard, Sonja Hollmer, Michael Wingen, & Olaf Strauch. 2005. Genetic selection for heat tolerance and low temperature activity of the entomopathogenic nematode–bacterium complex Heterorhabditis bacteriophora–Photorhabdus luminescens. *BioControl*, 50, 699–716.

Fairén, Alberto G., Alfonso F. Davila, Darlene Lim et al. 2010. Astrobiology through the ages of Mars: The study of terrestrial analogues to understand the habitability of Mars. *Astrobiology*, 10(8), 821–843.

Feller, Georges, Emmanuel Narinx, Jean Louis Arpigny et al. 1996. Enzymes from psychrophilic organisms. *FEMS Microbiology Reviews*, 18(2–3), 189–202.

Feller, Georges, & Charles Gerday. 2003. Psychrophilic enzymes: Hot topics in cold adaptation. *Nature Reviews Microbiology*, 1(3), 200–208.

Fenice, Massimiliano. 2016. The psychrotolerant Antarctic fungus Lecanicillium muscarium CCFEE 5003: A powerful producer of cold-tolerant chitinolytic enzymes. *Molecules*, 21(4), 447.

Florczak, Tomasz, Maurycy Daroch, Mark Charles Wilkinson et al. 2013. Purification, characterisation, and expression in Saccharomyces cerevisiae of LipG7 an enantioselective, cold-adapted lipase from the Antarctic filamentous fungus Geomyces sp. P7 with unusual thermostability characteristics. *Enzyme and Microbial Technology*, 53(1), 18–24.

Furhan, Junaid. 2020. Adaptation, production, and biotechnological potential of cold-adapted proteases from psychrophiles and psychrotrophs: Recent overview. *Journal of Genetic Engineering and Biotechnology*, 18, 1–13.

Giri, Anand, & Deepak Pant. 2019. CO_2 management using carbonic anhydrase producing microbes from western Indian Himalaya. *Bioresource Technology Reports*, 8, 100320.

Gomes, Joseph, & Walter Steiner. 2004. The biocatalytic potential of extremophiles and extremozymes. *Food Technology and Biotechnology*, 42(4), 223–225.

Goordial, Jackie. 2021. Cryomicrobial ecology: Still much to learn about life left out in the cold. *Msystems*, 6(5), e00852-21.

Gounot, Anne-Monique. 1991. Bacterial life at low temperature: Physiological aspects and biotechnological implications. *Journal of Applied Bacteriology*, 71(5), 386–397.

Gray, N.F., CHE Wyborn, & Ronald I.L. Smith. 1982. Nematophagous fungi from the maritime Antarctic. *Oikos*. Wiley-Blackwell, pp. 194–201.

Gray, N.F. 1985. Ecology of nematophagous fungi: Effect of soil moisture, organic matter, pH, and nematode density on distribution. *Soil Biology and Biochemistry*, 17(4), 499–507.

Gupta, Santosh K., Sampriti Kataki, Soumya Chatterjee et al. 2020. Cold adaptation in bacteria with special focus on cellulase production and its potential application. *Journal of Cleaner Production*, 258, 120351.

Gupta, Vijaya, Naveen Gupta, Neena Capalash, & Prince Sharma. 2017. Bio-prospecting bacterial diversity of hot springs in Northern Himalayan region of India for laccases. *Indian Journal of Microbiology*, 57, 285–291.

Hamid, B., Poonam Singh, Ravinder Singh Rana, & Sanjay Sahay. 2012. Isolation and identification of psychrophilic yeast from the soil of northern region of India. *Biotechnology*, 18, 103–107.

Hamid, B., Bashir, Z., Yatoo, A.M., Mohiddin, F., Majeed, N., Bansal, M. et al. (2022). Cold-active enzymes and their potential industrial applications: A review. *Molecules*, 27(18), 5885. https://doi.org/10.3390/molecules27185885.

Hamid, B., Baba, Z.A., Sheikh, T.A., Perveen, K., & Sayyed, R.Z. (2024). Investigating potential protease activity of psychrotrophic bacteria from a municipal landfill for solid waste management. *Biomass Conversion and Biorefinery*. https://doi.org/10.1007/s13399-024-05621-2.

Hu, Hao, Fujie Yan, Charles Wilson, Qing Shen, & Xiaodong Zheng. 2015. The ability of a cold-adapted Rhodotorula mucilaginosa strain from Tibet to control blue mold in pear fruit. *Antonie van Leeuwenhoek*, 108, 1391–1404.

Jain, Rahul, & Anita Pandey. 2016. A phenazine-1-carboxylic acid producing polyextremophilic Pseudomonas chlororaphis (MCC2693) strain, isolated from mountain ecosystem, possesses biocontrol and plant growth promotion abilities. *Microbiological Research*, 190, 63–71.

Jain, Rahul, Anita Pandey, Mukesh Pasupuleti, & Veena Pande. 2017. Prolonged production and aggregation complexity of cold-active lipase from Pseudomonas proteolytica (GBPI_Hb61) isolated from cold Desert Himalaya. *Molecular Biotechnology*, 59, 34–45.

Jeffrey, J.S., Hunter, A., & Atwill, E.R. 2000. A field-suitable, semisolid aerobic enrichment medium for isolation of Campylobacter jejuni in small numbers. *Journal of Clinical Microbiology*, 38(4), 1668–1669.

Joseph, Babu, Nitisha Shrivastava, & Pramod W. Ramteke. 2012. Extracellular cold-active lipase of Microbacterium luteolum isolated from Gangotri glacier, western Himalaya: Isolation, partial purification, and characterization. *Journal of Genetic Engineering and Biotechnology*, 10(1), 137–144.

Joshi, Swati, & Tulasi Satyanarayana. 2013. Biotechnology of cold-active proteases. *Biology*, 2(2), 755–783.

Junge, Karen, Hajo Eicken, & Jody W. Deming. 2004. Bacterial activity at –2° to –20°C in Arctic wintertime sea ice. *Applied and Environmental Microbiology*, 70(1), 550–557.

Kaira, Gaurav Singh, Kusum Dhakar, & Anita Pandey. 2015. A psychrotolerant strain of Serratia marcescens (MTCC 4822) produces laccase at wide temperature and pH range. *Amb. Express*, 5(1), 1–8.

Karlovsky, Petr. 2008. Secondary metabolites in soil ecology. In: *Secondary Metabolites in Soil Ecology*. Springer.

Kaur, Rajvinder, John Macleod, William Foley, & Murali Nayudu. 2006. Gluconic acid: An antifungal agent produced by Pseudomonas species in biological control of take-all. *Phytochemistry*, 67(6), 595–604.

Kaur, Ramandeep, Abhishake Saxena, Punesh Sangwan, Ajar Nath Yadav, Vinod Kumar, & Harcharan Singh Dhaliwal. 2017. Production and characterization of a neutral phytase of Penicillium oxalicum EUFR-3 isolated from Himalayan region. *Nusantara Bioscience*, 9(1), 68–76.

Kaur, Tanvir, Rubee Devi, Divjot Kour, Ashok Yadav, and Ajar Nath Yadav. 2021. Plant growth promotion of barley (Hordeum vulgare L.) by potassium solubilizing bacteria with multifarious plant growth promoting attributes. *Plant Science Today*, 8(sp1), 17–24.

Kim, Hak Jun, Jun Hyuck Lee, Young Baek Hur, Chang Woo Lee, Sun-Ha Park, & Bon-Won Koo. 2017. Marine antifreeze proteins: Structure, function, and application to cryopreservation as a potential cryoprotectant. *Marine Drugs*, 15(2), 27.

Köhl, Jürgen, Rogier Kolnaar, & Willem J. Ravensberg. 2019. Mode of action of microbial biological control agents against plant diseases: Relevance beyond efficacy. *Frontiers in Plant ZScience*, 845.

Kour, Divjot, Kusam Lata Rana, Imran Sheikh et al. 2020. Alleviation of drought stress and plant growth promotion by Pseudomonas libanensis EU-LWNA-33, a drought-adaptive phosphorus-solubilizing bacterium. *Proceedings of the National Academy of Sciences, India Section B: Biological Sciences*, 90, 785–795.

Kuddus, Mohammed. 2018. Cold-active enzymes in food biotechnology: An updated mini review. *Journal of Applied Biology and Biotechnology*, 6(3), 58–63.

Kumar, Bhavesh, Pankaj Trivedi, & Anita Pandey. 2007. Pseudomonas corrugata: A suitable bacterial inoculant for maize grown under rainfed conditions of Himalayan region. *Soil Biology and Biochemistry*, 39(12), 3093–3100.

Kumar, L., & Bharadvaja, N. 2019. *Smart Bioremediation Technologies*. Academic Press.

Kumar, V., Yadav, A.N., Saxena, A., Sangwan, P., & Dhaliwal, H.S. 2016. Unravelling rhizospheric diversity and potential of phytase producing microbes. *SM. J. Biol.*, 2(1), 1009.

Kumar, Vinod, Ajar Nath Yadav, Priyanka Verma et al. 2017. β-propeller phytases: Diversity, catalytic attributes, current developments, and potential biotechnological applications. *International Journal of Biological Macromolecules*, 98, 595–609.

Lin, Xuezheng, Baijuan Yang, Jihong Shen, & Ning Du. 2009. Biodegradation of crude oil by an Arctic psychrotrophic bacterium Pseudoalteromomas sp. P29. *Current Microbiology*, 59, 341–345.

Liu, Kezhen, Haitao Ding, Yong Yu, & Bo Chen. 2019. A cold-adapted chitinase-producing bacterium from Antarctica and its potential in biocontrol of plant pathogenic fungi. *Marine Drugs*, 17(12), 695.

Lo Giudice, Angelina, & Renato Fani. 2016. Antimicrobial potential of cold-adapted bacteria and fungi from polar regions. *Biotechnology of Extremophiles: Advances and Challenges*. Springer, pp. 83–115.

Lonhienne, Thierry, Charles Gerday, & Georges Feller. 2000. Psychrophilic enzymes: Revisiting the thermodynamic parameters of activation may explain local flexibility. *Biochimica et Biophysica Acta (BBA)-Protein Structure and Molecular Enzymology*, 1543(1), 1–10.

Mahajan, Rishi, Shalini Verma, Madhulika Kushwaha, Dharam Singh, Yusuf Akhter, & Subhankar Chatterjee. 2019. Biodegradation of di-n-butyl phthalate by psychrotolerant Sphingobium yanoikuyae strain P4 and protein structural analysis of carboxylesterase involved in the pathway. *International Journal of Biological Macromolecules*, 122, 806–816.

Mangiagalli, Marco, & Marina Lotti. 2021. Cold-active β-galactosidases: Insight into cold adaptation mechanisms and biotechnological exploitation. *Marine Drugs*, 19(1), 43.

Margesin, R., & Schinner, F. 2001. Biodegradation and bioremediation of hydrocarbons in extreme environments. *Applied Microbiology and Biotechnology*, 56, 650–663.

Margesin, Rosa, Silvia Gander, Gabriele Zacke, Anne Monique Gounot, & Franz Schinner. 2003. Hydrocarbon degradation and enzyme activities of cold-adapted bacteria and yeasts. *Extremophiles*, 7, 451–458.

Margesin, Rosa. 2007. Alpine microorganisms: Useful tools for low-temperature bioremediation. *The Journal of Microbiology*, 45(4), 281–285.

Margesin, Rosa, & Georges Feller. 2010. Biotechnological applications of psychrophiles. *Environmental Technology*, 31(8–9), 835–844.

Margesin, Rosa, & Vanya Miteva. 2011. Diversity and ecology of psychrophilic microorganisms. *Research in Microbiology*, 162(3), 346–361.

Margesin, Rosa, & Tony Collins. 2019. Microbial ecology of the cryosphere (glacial and permafrost habitats): Current knowledge. *Applied Microbiology and Biotechnology*, 103, 2537–2549.

Marizcurrena, Juan José, María Fernanda Cerdá, Diego Alem, & Susana Castro-Sowinski. 2019. Living with pigments: The colour palette of Antarctic life. *The Ecological Role of Micro-organisms in the Antarctic Environment*. Springer Polar Sciences, pp. 65–82.

Martínez-Rosales, Cecilia, & Susana Castro-Sowinski. 2011. Antarctic bacterial isolates that produce cold-active extracellular proteases at low temperature but are active and stable at high temperature. *Polar Research*, 30(1), 7123.

McKay, Christopher P., Nadia, C.S., Mykytczuk, & Lyle G. Whyte. 2011. Life in ice on other worlds. *Polar Microbiology: Life in a Deep Freeze*. ASM Press, pp. 290–304.

Melo, Itamar, Wallace Souza, Leonardo Silva et al. 2016. Antifungal activity of Pseudomonas frederiksbergensis CMAA 1323 isolated from the Antarctic hair grass Deschampsia antarctica. *British Microbiology Research Journal*, 14(3), 1–11.

Miri, Saba, Mitra Naghdi, Tarek Rouissi, Satinder Kaur Brar, & Richard Martel. 2019. Recent biotechnological advances in petroleum hydrocarbons degradation under cold climate conditions: A review. *Critical Reviews in Environmental Science and Technology*, 49(7), 553–586.

Mishra, Pankaj K., Smita Mishra, Govindan Selvakumar et al. 2008. Characterisation of a psychrotolerant plant growth promoting Pseudomonas sp. strain PGERs17 (MTCC 9000) isolated from North Western Indian Himalayas. *Annals of Microbiology*, 58, 561–568.

Moreno, Renata, and Fernando Rojo. 2014. Features of pseudomonads growing at low temperatures: Another facet of their versatility. *Environmental Microbiology Reports*, 6(5), 417–426.

Mukhopadhyay, Arka, Anjan Kr Dasgupta, & Krishanu Chakrabarti. 2015. Enhanced functionality and stabilization of a cold active laccase using nanotechnology based activation-immobilization. *Bioresource Technology*, 179, 573–584.

Muñoz, Patricio A., Sebastián L. Márquez, Fernando D. González-Nilo, Valeria Márquez-Miranda, & Jenny M. Blamey. 2017. Structure and application of antifreeze proteins from Antarctic bacteria. *Microbial Cell Factories*, 16, 1–13.

Mykytczuk, Nadia, Simon J. Foote, Chris R. Omelon, Gordon Southam, Charles W. Greer, & Lyle G. Whyte. 2013. Bacterial growth at –15 C; molecular insights from the permafrost bacterium Planococcus halocryophilus Or1. *The ISME Journal*, 7(6), 1211–1226.

Mykytczuk, N.C.S., Lawrence, J.R., Omelon, C.R., Southam, G., & Whyte, L.G. 2016. Microscopic characterization of the bacterial cell envelope of Planococcus halocryophilus Or1 during subzero growth at –15°C. *Polar Biology*, 39, 701–712.

Negi, Yogesh Kumar, Deepti Prabha, Satyendra K. Garg, & Kumar, J. 2011. Genetic diversity among cold-tolerant fluorescent Pseudomonas isolates from Indian Himalayas and their characterization for biocontrol and plant growth-promoting activities. *Journal of Plant Growth Regulation*, 30, 128–143.

Nieto-Penalver, Carlos G., María J. Savino, Elisa V. Bertini, Leandro A. Sánchez, & Lucía I.C. de Figueroa. 2014. Gluconic acid produced by Gluconacetobacter diazotrophicus Pal5 possesses antimicrobial properties. *Research in Microbiology*, 165(7), 549–558.

Nottingham, Andrew T., Noah Fierer, Benjamin L. Turner et al. 2018. Microbes follow Humboldt: Temperature drives plant and soil microbial diversity patterns from the Amazon to the Andes. *Ecology*, 99(11), 2455–2466.

Ogata-Gutiérrez, Katty, Carolina Chumpitaz-Segovia, Jesus Lirio-Paredes, Mariella M. Finetti-Sialer, & Doris Zúñiga-Dávila. 2017. Characterization and potential of plant growth promoting rhizobacteria isolated from native Andean crops. *World Journal of Microbiology and Biotechnology*, 33, 1–13.

Ohye, D.F., & Scott, W.J. 1957. Studies in the physiology of Clostridium botulinum type E. *Australian Journal of Biological Sciences*, 10(1), 85–94.
Ortiz-Ojeda, Paola, Katty Ogata-Gutiérrez, & Doris Zúñiga-Dávila. 2017. Evaluation of plant growth promoting activity and heavy metal tolerance of psychrotrophic bacteria associated with maca (Lepidium meyenii Walp.) rhizosphere. *AIMS Microbiology*, 3(2), 279.
Ortuño, Noel, Claudia Miranda, & Mayra Claros. 2013. Selección de cepas de Trichoderma spp. generadoras de metabolitos secundarios de interés para su uso como promotor de crecimiento en plantas cultivadas. *Journal of the Selva Andina Biosphere*, 1(1), 16–32.
Ortuño, Noel, José Antonio Castillo, Mayra Claros et al. 2013. Enhancing the sustainability of quinoa production and soil resilience by using bioproducts made with native microorganisms. *Agronomy*, 3(4), 732–746.
Pandey, A., & Palni, L.M.S. 1998. Isolation of Pseudomonas corrugata from Sikkim Himalaya. *World Journal of Microbiology and Biotechnology*, 14, 411–413.
Pandey, Anita, and Luis Andrés Yarzábal. 2019. Bioprospecting cold-adapted plant growth promoting microorganisms from mountain environments. *Applied Microbiology and Biotechnology*, 103(2), 643–657.
Pandey, Anita, Kusum Dhakar, Rahul Jain et al. 2019. Cold adapted fungi from Indian Himalaya: Untapped source for bioprospecting. *Proceedings of the National Academy of Sciences, India Section B: Biological Sciences*, 89, 1125–1132.
Pandey, K.D., Shukla, S.P., Shukla, P.N. et al. 2004. Cyanobacteria in Antarctica: Ecology, physiology, and cold adaptation. *Cellular and Molecular Biology-Paris-Wegmann*, 50(5), 575–584.
Pandey, Neha, Rahul Jain, Anita Pandey, & Sushma Tamta. 2018. Optimisation and characterisation of the orange pigment produced by a cold adapted strain of Penicillium sp. (GBPI_P155) isolated from mountain ecosystem. *Mycology*, 9(2), 81–92.
Panicker, Gitika, Jackie Aislabie, David Saul, & Asim K. Bej. 2002. Cold tolerance of Pseudomonas sp. 30-3 isolated from oil-contaminated soil, Antarctica. *Polar Biology*, 25, 5–11.
Panikov, Nicolai S., & Maria V. Sizova. 2007. Growth kinetics of microorganisms isolated from Alaskan soil and permafrost in solid media frozen down to –35°C. *FEMS Microbiology Ecology*, 59(2), 500–512.
Panwar, A.S., Molpa, D., & Joshi, G.K. 2019. Biotechnological potential of some cold-adapted bacteria isolated from North-Western Himalaya. *Microbiology*, 88, 343–352.
Pischedda, Alessandro, Kesava Priyan Ramasamy, Marco Mangiagalli et al. 2018. Antarctic marine ciliates under stress: Superoxide dismutases from the psychrophilic Euplotes focardii are cold-active yet heat tolerant enzymes. *Scientific Reports*, 8(1), 14721.
Rafeeq, Hamza, Nadia Afsheen, Sadia Rafique et al. 2023. Genetically engineered microorganisms for environmental remediation. *Chemosphere*, 310, 136751.
Ravi, Keerthikka, Nicole R. Falkowski, Brittan S. Scales, Volha D. Akulava, Leonid N. Valentovich, & Gary B. Huffnagle. 2022. The Psychrotrophic Pseudomonas lundensis, a non-aeruginosa pseudomonad, has a Type III secretion system of the Ysc family, which is transcriptionally active at 37°C. *Mbio*, 13(1), e03869-21.
Razzaq, Abdul, Sadia Shamsi, Arfan Ali et al. 2019. Microbial proteases applications. *Frontiers in Bioengineering and Biotechnology*, 7, 110.
Rehakova, Klara, Katerina Capkova, Pavel Hrouzek, Michal Koblizek, & Jiri Dolezal. 2019. Microbial photosynthetic and photoprotective pigments in Himalayan soils

originating from different elevations and successional stages. *Soil Biology and Biochemistry*, 132, 153–164.

Rivkina, E.M., Friedmann, E.I., McKay, C.P., & Gilichinsky, D.A. 2000. Metabolic activity of permafrost bacteria below the freezing point. *Applied and Environmental Microbiology*, 66(8), 3230–3233.

Rivkina, Elizaveta, Andrey Abramov, Elena Spirina et al. 2018. Earth's perennially frozen environments as a model of cryogenic planet ecosystems. *Permafrost and Periglacial Processes*, 29(4), 246–256.

Rizvi, Asfa, Bilal Ahmed, Mohammad Saghir Khan, Shahid Umar, & Jintae Lee. 2021. Psychrophilic bacterial phosphate-biofertilizers: A novel extremophile for sustainable crop production under cold environment. *Microorganisms*, 9(12), 2451.

Robinson, Clare H. 2001. Cold adaptation in Arctic and Antarctic fungi. *New Phytologist*, 151(2), 341–353.

Rodrigues, Debora F., Natalia Ivanova, Zhili He, Marianne Huebner, Jizhong Zhou, & James M. Tiedje. 2008. Architecture of thermal adaptation in an Exiguobacterium sibiricum strain isolated from 3 million year old permafrost: A genome and transcriptome approach. *BMC Genomics*, 9, 1–17.

Ruberto, Lucas A.M., Vazquez, S., Lobalbo, A., & Mac Cormack, W.P. 2005. Psychrotolerant hydrocarbon-degrading Rhodococcus strains isolated from polluted Antarctic soils. *Antarctic Science*, 17(1), 47–56.

Santiago, Margarita, César A. Ramírez-Sarmiento, Ricardo A. Zamora, & Loreto P. Parra. 2016. Discovery, molecular mechanisms, and industrial applications of cold-active enzymes. *Frontiers in Microbiology*, 7, 1408.

Sarmiento, Felipe, Rocío Peralta, & Jenny M. Blamey. 2015. Cold and hot extremozymes: Industrial relevance and current trends. *Frontiers in Bioengineering and Biotechnology*, 3, 148.

See-Too, Wah-Seng, Robson Ee, Munusamy Madhaiyan et al. 2017. Planococcus versutus sp. nov., isolated from soil. *International Journal of Systematic and Evolutionary Microbiology*, 67(4), 944–950.

See-Too, Wah Seng, Peter Convey, David A. Pearce, & Kok-Gan Chan. 2018. Characterization of a novel N-acylhomoserine lactonase, AidP, from Antarctic Planococcus sp. *Microbial Cell Factories*, 17, 1–14.

Segner, W.P., Schmidt, C.F., & Boltz, J.K. 1966. Effect of sodium chloride and pH on the outgrowth of spores of Type E Clostridium botulinum at optimal and suboptimal temperatures. *Applied Microbiology*, 14(1), 49–54.

Selvakumar, Govindan, Piyush Joshi, Pankaj K. Mishra, Jaideep K. Bisht, & Hari S. Gupta. 2009. Mountain aspect influences the genetic clustering of psychrotolerant phosphate solubilizing Pseudomonads in the Uttarakhand Himalayas. *Current Microbiology*, 59, 432–438.

Shanmugam, Veerubommu, Priya Chugh, & Pratibha Sharma. 2015. Cold-tolerant Trichoderma species for the management of Fusarium wilt of tomato plants. *Annals of Microbiology*, 65, 543–551.

Sharma, Sarika, Farrah Gul Khan, & Ghulam Nabi Qazi. 2010. Molecular cloning and characterization of amylase from soil metagenomic library derived from Northwestern Himalayas. *Applied Microbiology and Biotechnology*, 86, 1821–1828.

Sheikh T., Baba, Z., Hamid, B., Iqbal, S., Yatoo, A., Fatima, S. et al. (2022). Extracellular polymeric substances in psychrophilic cyanobacteria: A potential bioflocculant and carbon sink to mitigate cold stress. *Biocatalysis and Agricultural Biotechnology*, 42, 102375. https://doi.org/10.1016/j.bcab.2022.102375.

Singh, Brajesh K., Richard D. Bardgett, Pete Smith, & Dave S. Reay. 2010. Microorganisms and climate change: Terrestrial feedbacks and mitigation options. *Nature Reviews Microbiology*, 8(11), 779–790.

Spadaro, Davide, & Samir Droby. 2016. Development of biocontrol products for postharvest diseases of fruit: The importance of elucidating the mechanisms of action of yeast antagonists. *Trends in Food Science and Technology*, 47, 39–49.

LLC, BCC Research. 2025. Global Markets for Enzymes in Industrial Applications. BCC Research LLC https://www.bccresearch.com/market-research/biotechnology/global-markets-for-enzymes-in-industrial-applications.html?srsltid=AfmBOoq1JShSyvqsRVZlxrP9mjIYtX2PvdEXc0r7nZsol2J0Zs06Q-7Q.

Steven, Blaire, Richard Leveille, Wayne H. Pollard, & Lyle G. Whyte. 2006. Microbial ecology and biodiversity in permafrost. *Extremophiles*, 10, 259–267.

Steven, Blaire, Thomas D. Niederberger, Eric M. Bottos, Michael R. Dyen, & Lyle G. Whyte. 2007. Development of a sensitive radiorespiration method for detecting microbial activity at subzero temperatures. *Journal of Microbiological Methods*, 71(3), 275–280.

Stowers, Mark D. 1985. Carbon metabolism in Rhizobium species. *Annual Review of Microbiology*, 39(1), 89–108.

Tarfeen, Najeebul, Khair Ul Nisa, Burhan Hamid et al. 2022. Microbial remediation: A promising tool for reclamation of contaminated sites with special emphasis on heavy metal and pesticide pollution: A review. *Processes*, 10(7), 1358.

Tian, Yuan, Yan-Ling Li, & Feng-Chun Zhao. 2017. Secondary metabolites from polar organisms. *Marine Drugs*, 15(3), 28.

Torracchi, C., José Esteban, María A. Morel, Irán Tapia-Vázquez, Susana Castro-Sowinski, Ramón Alberto Batista-García, & Luis Andrés Yarzábal R. 2020. Fighting plant pathogens with cold-active microorganisms: Biopesticide development and agriculture intensification in cold climates. *Applied Microbiology and Biotechnology*, 104, 8243–8256.

Touchette, D., Altshuler, I., Gostinčar, C. et al. 2022. Novel Antarctic yeast adapts to cold by switching energy metabolism and increasing small RNA synthesis. *The ISME Journal*, 16(1), 221–232.

Tribelli, Paula M., & Nancy I. López. 2018. Reporting key features in cold-adapted bacteria. *Life*, 8(1), 8.

Trivedi, Pankaj, Anita Pandey, & Lok Man S. Palni. 2005. Carrier-based preparations of plant growth-promoting bacterial inoculants suitable for use in cooler regions. *World Journal of Microbiology and Biotechnology*, 21, 941–945.

Troncoso, Evelin, Salvador Barahona, Mario Carrasco et al. 2017. Identification and characterization of yeasts isolated from the South Shetland Islands and the Antarctic Peninsula. *Polar Biology*, 40, 649–658.

Vallesi, Adriana, Sandra Pucciarelli, Federico Buonanno, Angelo Fontana, & Marco Mangiagalli. 2020. Bioactive molecules from protists: Perspectives in biotechnology. *European Journal of Protistology*, 75, 125720.

Van, Vân Trân, Suong Ngôkê, Odile Berge et al. 1997. Isolation of Azospirillum lipoferum from the rhizosphere of rice by a new, simple method. *Canadian Journal of Microbiology*, 43(5), 486–490.

Vardharajula, Sandhya, Ali SkZ, Sai Shiva Krishna Prasad Vurukonda, & Manjari Shrivastava. 2017. Plant growth-promoting endophytes and their interaction with plants to alleviate abiotic stress. *Current Biotechnology*, 6(3), 252–263.

Velázquez, Encarna, Luis R. Silva, Martha-Helena Ramírez-Bahena, & Alvaro Peix. 2016. Diversity of potassium-solubilizing microorganisms and their interactions with plants. *Potassium Solubilizing Microorganisms for Sustainable Agriculture*. Springer, pp. 99–110.

Venketesh, S., & Dayananda, C. 2008. Properties, potentials, and prospects of antifreeze proteins. *Critical Reviews in Biotechnology*, 28(1), 57–82.

Verma, Priyanka, Ajar Nath Yadav, Livleen Shukla, Anil Kumar Saxena, & Archna Suman. 2015. Hydrolytic enzymes production by thermotolerant Bacillus altitudinis IARI-MB-9 and Gulbenkiania mobilis IARI-MB-18 isolated from Manikaran hot springs. *Int. J. Adv. Res.*, 3(9), 1241–1250.

Vero, Silvana, Gabriela Garmendia, M. Belén González, Oscar Bentancur, & Michael Wisniewski. 2013. Evaluation of yeasts obtained from Antarctic soil samples as biocontrol agents for the management of postharvest diseases of apple (Malus× domestica). *FEMS Yeast Research*, 13(2), 189–199.

Vester, Jan Kjølhede, Mikkel Andreas Glaring, & Peter Stougaard. 2015. Improved cultivation and metagenomics as new tools for bioprospecting in cold environments. *Extremophiles*, 19, 17–29.

Vishnivetskaya, Tatiana A., Susanne Liebner, Roland Wilhelm, & Dirk Wagner. 2011. Microbial carbon cycling in permafrost. *Polar Microbiology: Life in a Deep Freeze*. ASM Press, pp. 181–199.

Wackett, Lawrence P. 2019. Microbial industrial enzymes: An annotated selection of World Wide Web sites relevant to the topics in microbial biotechnology. *Microbial Biotechnology*, 12(5), 1090.

Wang, Manman, Jianqing Tian, Meichun Xiang, & Xingzhong Liu. 2017. Living strategy of cold-adapted fungi with the reference to several representative species. *Mycology*, 8(3), 178–188.

Yadav, Ajar Nath. 2021. Biodiversity and bioprospecting of extremophilic microbiomes for agro-environmental sustainability. *Journal of Applied Biology and Biotechnology*, 9(3), 1–6.

Yadav, F.D., Ranjit Kumar, Anjali Yadav, & Madhurima Yadav. 2019. Effective management of cercospora leaf spot disease of brinjal by fungicides in Ayodhya UP. *Journal of Medicinal Plants*, 7(3), 134–135.

Yang, Guang Li, Shu Gui Hou, Ri Le Baoge et al. 2016. Differences in bacterial diversity and communities between glacial snow and glacial soil on the Chongce Ice Cap, West Kunlun Mountains. *Scientific Reports*, 6(1), 36548.

Yarzábal, Luis Andrés, Lorena Monserrate, Lenys Buela, & Eduardo Chica. 2018. Antarctic Pseudomonas spp. promote wheat germination and growth at low temperatures. *Polar Biology*, 41, 2343–2354.

Zajc, Janja, Anja Cernosa, Alessandra Di Francesco et al. 2020. Characterization of Aureobasidium pullulans isolates selected as biocontrol agents against fruit decay pathogens. *Fungal Genomics and Biology*, 10(163).

Zhang, Yi, Chen Li, Timothy Geary, & Benjamin Kofi Simpson. 2020. Contribution of special structural features to high thermal stability of a cold-active transglutaminase. *Journal of Agricultural and Food Chemistry*, 68(30), 7935–7945.

CHAPTER
12

Applications of Psychrophiles in Molecular Biology as an Innovative Tool

Anisa Tariq[1], Ayesha Aftab[1], Sania Tabassum[1], Taliha Ahmed[1], Syeda Marriam Bakhtiar[1*] and F.A. Mohiddin[2]

[1] Genetic and Molecular Epidemiology Research Group, Department of Bioinformatics and Biosciences, Capital University of Science and Technology, Islamabad, Pakistan

[2] Section of Plant Pathology, MRCFC, Khudwani, SKUAST-K, 192101, India

1. Introduction

Psychrophiles have evolved to thrive in cold environments, such as the Arctic, Antarctic deep sea, permafrost, and alpine regions, demonstrating exceptional resilience to cold (Feller, 2013). They include a diverse array of organisms spanning Eukarya, Archaea, and Prokaryotes, which comprise plants, animals, and various other life forms (Bhatia et al., 2021), yeast (Buzzini et al., 2012), algae (Cvetkovska et al., 2022), and bacteria (Jin et al., 2022). These psychrophiles carry out their cellular activities at very low temperatures, such as metabolism and reproduction in a permafrost bacteria reported at –20°C and –12°C (Cavicchioli, 2016), photosynthesis in lichen *Umbilicaria aprina* reported at –17°C and food spoilage by yeast *Rhodotorula glutinis* reported at –18°C (Hassan et al., 2016).

Psychrophilic organisms have evolved unique strategies to confront the obstacles of cold environments. They modify the composition of their cell envelopes, suppress primary metabolic processes, and synthesize specific small molecules and proteins serving as antifreeze agents, crucial for sustaining osmotic balance (Collins & Margesin, 2019). This molecular adaptation has sparked significant interest among contemporary researchers due to these enzymes' considerable biotechnological potential. However, it remains crucial to delve into

*Corresponding author: marriam@cust.edu.pk

DOI: 10.1201/9781003717003-12

various aspects of psychrophilic physiology, ecology, and molecular adaptation and their contribution to the biogeochemical cycle. Because of their distinct qualities, cold-adapted enzymes effectively catalyze reactions at lower temperatures, serving as essential components in the detergent industry (Al-Ghanayem & Joseph, 2020), food industry, and molecular industry (Kumar et al., 2021; Kuddus et al., 2024) by reducing energy consumption and mitigating undesirable reactions, meeting the increasing demand of novel biocatalysts. Research has also outlined the potential applications of cold-active enzymes across various industries, including detergent and fabric, food and brewing, molecular biology, pharmaceuticals, medicine, bioremediation, and cosmetics (Kumar et al., 2021; Hamid et al., 2022).

The global enzyme market in 2022 is estimated to be worth USD 12.27 billion and is projected to grow at a Compound Annual Growth Rate (CAGR) of 6.5% until 2030. This growth trajectory indicates a significant shift, with enzymatic reactions expected to replace up to 40% of industrial chemical reactions that currently utilize toxic solvents, underscoring the substantial industrial potential of these enzymes (Xu et al., 2024). Hence, this chapter delves into the utilization of these enzymes adapted to cold in the realm of molecular biology, as well as strategies geared towards improving their expression. Additionally, it explores tools and methodologies for mining and screening to identify novel psychrophilic enzymes.

Adaptive Features of Psychrophiles

In cold climates characterized by harsh and unusual weather conditions, the normal functioning of cells and the maintenance of structural integrity depend significantly on managing intracellular water content (De Maayer et al., 2014). At the metabolic level, psychrophiles undergo reprogramming which is crucial for long-term survival and also reduces oxidative stress in a cold environment (Kostadinova et al., 2017). They increase the contents of the Long-Chain Polyunsaturated Fatty Acids (LCPUFAs) that act as a barrier against reactive oxygen species and H_2O_2, short-chain, cis isomeric, and methyl branch fatty acids to achieve home viscous adaptation (Wang et al., 2020). At the cellular level, psychrophiles alter the fatty acid composition of the lipid bilayer (Siddiqui et al., 2013). Some species have thickened peptidoglycan layers by upregulation of the peptidoglycan biosynthesis gene that protects the cell disruption in cold conditions (Mykytczuk et al., 2013; Pandey et al., 2018). Pigments, particularly carotenoids, influence the cell membrane fluidity. They downregulate certain oxidative metabolic processes like the Glycolysis, Pentose Phosphate pathway, TCA cycle, and electron transport chain. Some activate the alternative pathways such as acetate metabolism, 2-methyl citrate, glyoxylate cycle, propionyl-CoA catabolism, and ethanol oxidation pathway (Ayala-Del-Río et al., 2010). These organisms utilize various novel mechanisms such as protein and nucleic acid chaperones, Extracellular Polymeric Substance (EPS), Ice Nucleating Proteins (INPs), Antifreeze Proteins (AFPs), bio-surfactants, and some compatible solute to cope with cold challenges (Collins & Margesin, 2019). Protein chaperones are essential for ensuring efficient replication, translation, and transcription processes, while also preventing them from misfolding and aggregating. For example, the first known Chaperone Trigger Factor (TF) catalyzes the proline isomerization (Godin-Roulling et al., 2014), a rate-limiting step of protein folding to prevent protein misfolding and cold (Baldwin, 2008). AFPs function to inhibit recrystallization and ice growth (Bar Dolev et al., 2016), while INPs are membrane-bound proteins that facilitate ice formation (Pandey et al., 2016). Additionally, biocompatible solutes such as sorbitol, glycerin, mannitol, trehalose, and sucrose contribute to restoring osmotic balance and counteracting water loss (Goordial et al., 2016). EPS is the high molecular weight biopolymer complex attached to the outer surface of the cell that server various functions such as protein protection, ROS scavenging, and cell adhesion (Deming & Young, 2017) while biosurfactants regulate the biofilm development and bacterial pathogenicity (Perfumo et al., 2017).

2. Psychrophiles: A Reservoir of Enzymes Adapted to Cold Environments

Life in cold environments presents significant challenges due to severe physiochemical constraints imposed by extremely low temperatures (De Maayer et al., 2014). These challenges include slow biochemical reactions, disruption of weak interactions crucial for molecular recognition, reduced molecular diffusion rates, high water viscosity, stabilization of toxic metabolites, and low fluidity of cell membranes. Psychrophiles have evolved to produce cold-adapted enzymes characterized by enhanced flexibility and weakened intermolecular interactions within proteins. The adaptive strategies of these enzymes involve reduced substrate affinity, increased flexibility, intrinsic structural instability, and structural modifications to enhance catalytic rates (Parvizpour et al., 2021). A short description of the adaptive features of psychrophiles and their enzymes is shown (Figure 1).

Adaptive Features of Psychrophilic Enzymes

Psychrophilic enzymes are cold-adapted and maintain the high specific activity required for optimal metabolism in a cold environment (Collins & Margesin, 2019). Their increased flexibility is achieved by weak intermolecular interactions necessary for protein folding and stability of nascent polypeptide chains. This adaptation involves reducing the arginine/lysine ratio, hydrophobic interaction, disulfide bonds, hydrogen bonds, and salt bridges and increases the aromatic residues and contents of glycine (Kumar et al., 2021). Additionally, these enzymes feature a significant abundance of loops with fewer secondary structures, resulting in elevated entropy (Al-Ghanayem & Joseph, 2020). The study of 2011 highlights the key structural difference between the psychrophilic enzymes

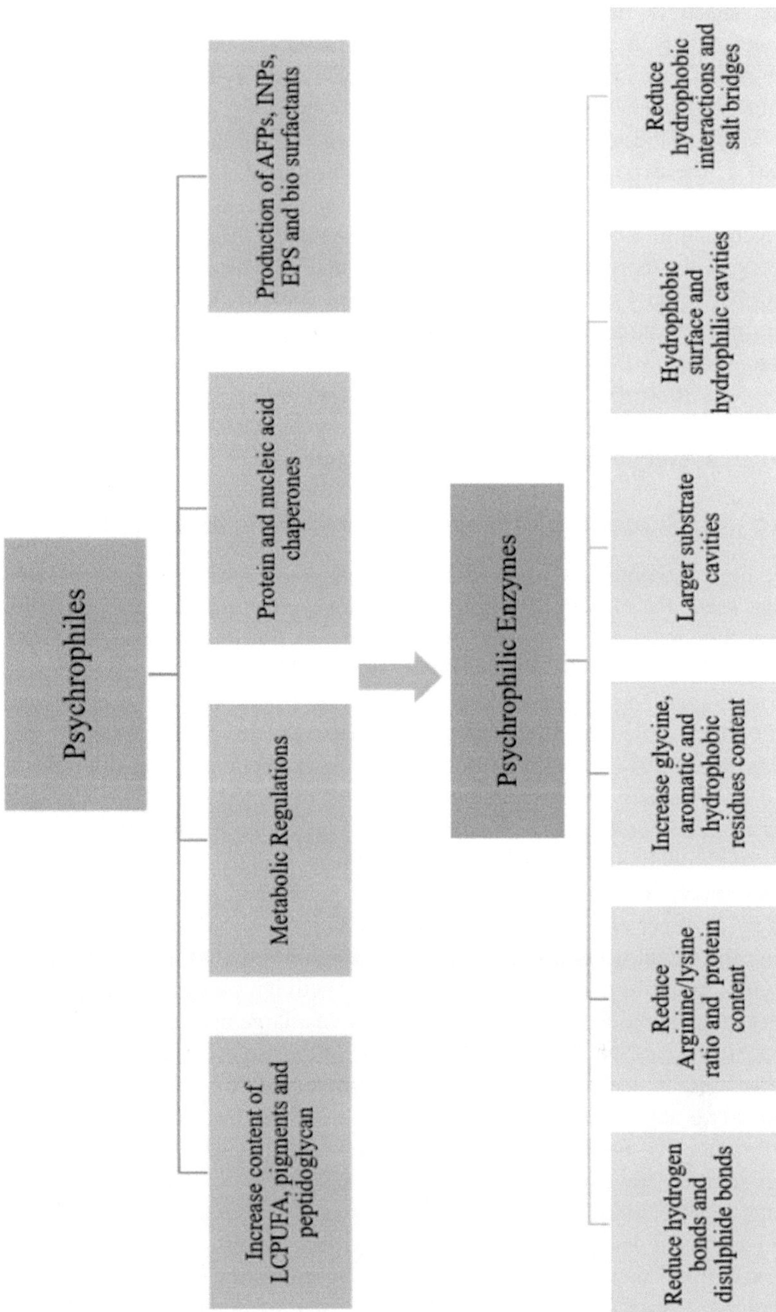

Figure 1: Key adaptive features of psychrophiles and their enzymes.

and their mesophilic counterparts (Paredes et al., 2011). These adaptations enhance substrate accessibility and facilitate conformational modifications at a lower energy cost, thereby reducing the activation energy needed for the enzyme-substrate complex. Because of their advanced dynamics of the active site, substrates bind less tightly, resulting in elevated K_m (Michaelis constant) values. Many of these enzymes compensate by increasing the K_{cat} (catalytic rate constant) at the expense of K_m, leading to sub-optimal values of the K_{cat}/K_m ratio (Feller, 2013). Zanphorlin et al. (2016) emphasized is the oligomerization of a psychrophilic GH1 β-glucosidase from *Exiguobacterium antarcticum* B7, which forms the cornerstone of its cold adaptation. These enzymes have 10 times higher activity as they stabilize the natural confirmation through tetramer structure than disassembled monomer (Zanphorlin et al., 2016). The adaptation of β-glucosidase to freezing temperature in the Antarctic archaeon *Halorubrum lacusprofundi* is attributed to mutations in only six amino acid residuals (Karan et al., 2020). To thrive in cold habitats, psychrophiles undergo intricate adaptations at every level of their cellular structure.

Mining and Screening of Novel Psychrophilic Enzymes

Indeed, the vast potential of psychrophilic enzymes across various research fields has made the mining and utilization of novel psychrophilic enzymes with enhanced functionality a prominent area of research. For this purpose, the direct approach is to produce the psychrophilic enzyme by growing bacteria on the culture medium. In a study, two extracellular alkaline phosphatases were isolated from the culture medium in which *Arthrobacter* isolate, D10, was cultivated (Orellana et al., 2018). Although this method finds applications in the field of Applied Biochemistry and Biotechnology but is restricted by the insignificant number of microbes that can be cultured, i.e., only 1%. Another limitation in mining psychrophilic enzymes is the reliance on specialized temperature control systems (Fenice et al., 2021). Therefore, alternative methods are necessary to effectively discover and utilize these enzymes.

With the recent development in the field of bioinformatics, a window opens for the mining of novel enzymes with enhanced activity. Some researchers have used these technologies in combination to retrieve a large dataset of genomes of psychrophilic organisms (Liu et al., 2022). One of the bioinformatics techniques, metagenomics, is used for isolating any biomolecules from samples. In this technique, two major methods exist, sequence-based mining and function-based dataset mining. By employing this bioinformatics approach, resources typically involved in microbial culture and streamlining the process of mining our protein of interest can be conserved. In the sequence-based dataset mining method, the first step involves designing oligonucleotide probes specific to the target gene. These probes are then utilized in the colony hybridization method to retrieve metagenomic clones. Afterward, the target gene can be amplified through PCR (Polymerase Chain Reaction), and then cloning is facilitated using expression vectors (Madhavan et al., 2017). It has been reported that with the help of this technique, a bacterial laccase gene lac21 is retrieved from a microbial metagenomics library of the South China Sea (Fang et al., 2012). Another novel psychrophilic enzyme, type I pullulanase, is isolated from the metagenome analysis of an aquatic habitat characterized by thermal conditions (Thakur et al., 2021). After this step, the resultant proteome datasets are derived from the genome databases. In the end, researchers perform sequence similarity and prediction of protein structures along with catalytic sites that resemble already known enzymes. To retrieve a target protein from proteome datasets by using the primary sequence of target protein, the Multiple Sequence Alignment (MSA) method is used. In MSA, the candidates with a high identity score can be obtained from the datasets. This method is also quite worthy in protein function annotation by applying strict similarity criteria like sequence identity of at least 40% and query coverage at least 90% (Profiti et al., 2017).

There is also an observation that enzymes having entirely different sequences may have the same function. The sequence similarity mining approach faces significant constraints, including its limitation to matching the query sequence solely to known genes. Additionally, achieving a high query coverage value is necessary to successfully mine the target gene, which may represent only a small subset of genes within the databases (Jin et al., 2022). Whereas the function-based mining approach does not need a sequence of genes, there are different strategies in the field of metagenomics such as phenotypic detection, enzyme activity-based mining, induced gene expression, heterologous complementarity to host strains, and Substrate-Induced Gene Expression (SIGEX) (Madhavan et al., 2017). This approach is best for finding new genes as gene sequence is not required. Over the years, nine genes of DNA polymerase I were obtained from the metagenomics library of glacier ice by using the method of heterologous complementarity to host strains to get viability (Simon et al., 2009).

Based on the flexible properties of the psychrophilic enzymes, and less number of hydrogen bonds, the RMSD value changes significantly. With the aid of these properties, the identification of potential psychrophilic enzymes through structural prediction and molecular dynamics simulation becomes more straightforward. For instance, by using the techniques of computational prediction and molecular dynamics simulation, a cold-adapted cellobiose 2-epimerase is retrieved (Chen et al., 2020). Consequently, these bioinformatics approaches aid the uncovering of novel psychrophilic enzymes that are not possible by using traditional approaches (Figure 2).

Strategies for Promoting Enzyme Expression

Expressing proteins solubilized in a host bacterium is feasible, but extracting and purifying them for characterization can be challenging. The purification process, dictated by the properties of the protein, often involves intricate and cumbersome steps (Javed & Qazi, 2016). Usually, the expression plasmid

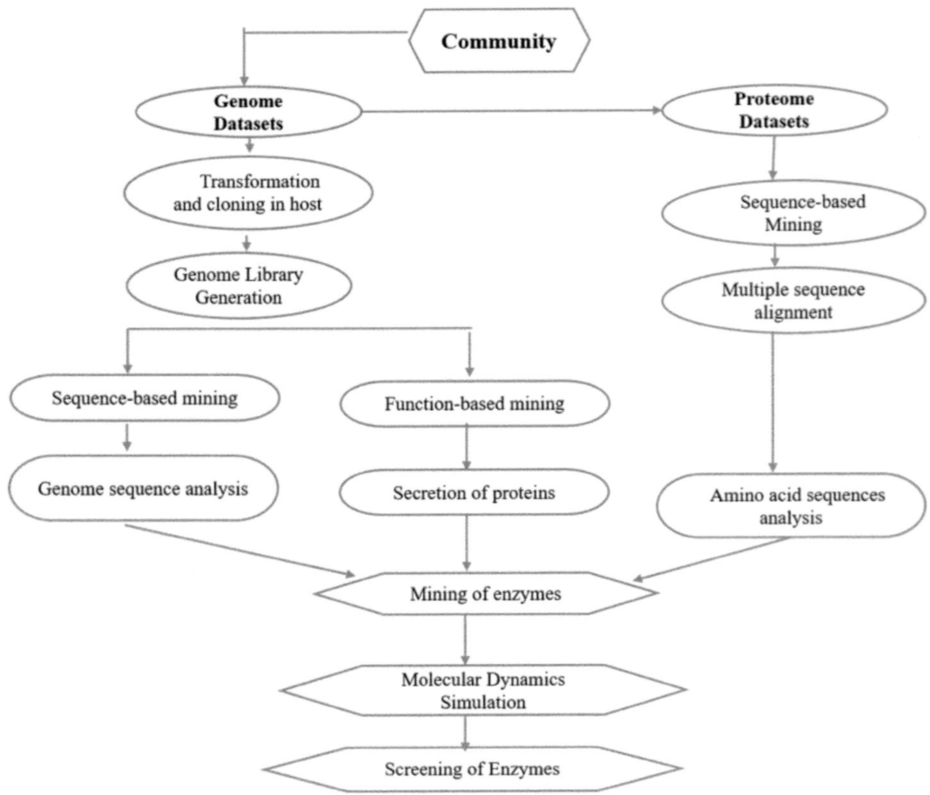

Figure 2: Mining enzymes via metagenomics coupled with molecular dynamics simulation-based screening.

containing the targeted gene and a T7 promoter is introduced into *E. coli* cells. Upon induction with IPTG (Isopropyl β-D-1-thiogalactopyranoside) at low temperatures, protein expression is initiated. When protein expression rises by 2% of all cellular proteins, the formation of inclusion bodies occurs (Santiago et al., 2016) (Figure 3).

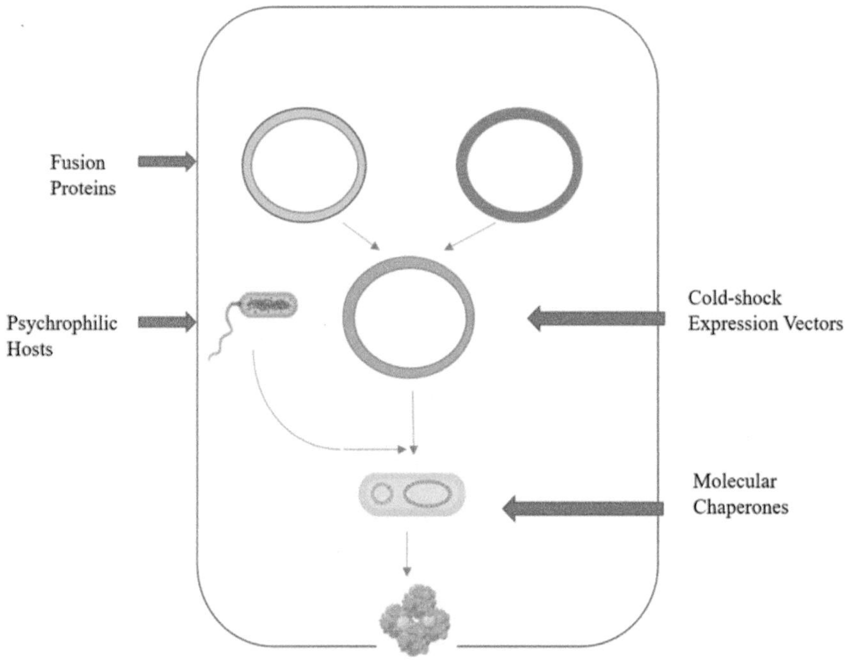

Figure 3: Strategies to promote protein expression.

The solubility and expression of recombinant enzymes can be enhanced by potent tags and fusion proteins located at either the N or C end of the target proteins. In the purification process, the fusion protein is cleaved from the target protein at a precise cleavage location found between them (Xu et al., 2019). Commonly used fusion proteins are new chain chaperone Trigger Factor (TF), G protein Gβ-1 domain (Gb1), thioredoxin (Trx), N-utilization substance, ubiquitin-like modifier (SUMO), Glutathione-S-transferase (GST), and maltose binding protein (MBP) (Santiago et al., 2016; Xu et al., 2019). Six N terminal fusion proteins from the psychrophilic pathogen of the Pisces family, *Vibrio salmomicida* are compared in expressing a protein in *E. coli*, MBP, and NusA and were found to perform best. However, in the case of MBP, a fusion protein utilized for the soluble expression of lipase from Antarctic seawater bacteria *Psychrobacter* sp., it forms inclusion bodies even in the absence of MBP assistance (Bjerga et al., 2016). The chaperone family, also known as Heat Shock Proteins (HSP), assists newly formed polypeptides and denatured proteins in attaining their normal shape. These proteins are widely distributed in all types of life forms. The genes responsible for molecular chaperones and their expression are affected under stress, including hypoxia, ultraviolet radiation, chemical challenges, and heat shock (Collins & Margesin, 2019). Molecular chaperones are classified into HSP40 to HSP100 and small HSPs, categorized according to their molecular weight (Hartl et al., 2011). The GroELS and the DnaK/DnaJ system of *E. coli* are commonly used molecular chaperones (Kumar et al., 2020b). An esterase, highly heat-labile yet boasting 180 times greater specific activity, is expressed in *E. coli* and cultured at 37°C. This expression is facilitated by molecular chaperones Cpn60 and Cpn10 sourced from the bacterium *Oleispira antarctica* RB8T, a psychrophile (Ferrer et al., 2004). Kim et al. (2015) emphasized that the enzyme activity of a psychrophilic esterase is more than 70% higher when co-expressed with PsyGroELS from *Psychrobacter* spp. PAMC21119 in *E. coli* at 10°C compared to that of OleGroELS (Kim et al., 2015). Qing et al. (2004) developed a cold-shock expression vector pCold that induces the expression of multiple genes upon cold shock, in conjunction with the cspA promoter (Qing & Mal, 2004). The Antarctic bacteria *Psychrobacter* sp. successfully cloned a cold-active lipase gene, Lip-948, by ligating it into the plasmid pColdI. Total obtained proteins account for 39% and most of them are the inclusion bodies. The soluble protein is improved when co-expressed with different molecular chaperons (Shuo-shuo et al., 2011). *E. coli* allows the accurate folding of proteins and lowers the inclusion of body formation at low expression temperatures. It is worth noting that protein expression systems utilizing psychrophiles have been established, with the commonly studied host being the Antarctic Gram-negative bacterium *Pseudoalteromonas haloplanktis* TAC125. A psychrophilic α-amylase from *P. haloplanktis* TAB23 is expressed using this host. There is a correct formation of disulfide bonds of chimera components and secreted proteins can easily be purified without contamination with other proteins (Ji & Wei, 2020). A psychrophilic bacterial strain, *Pseudoalteromonas* spp. SM20429, was utilized, with Puc19 ligated with pSW429, derived from the psychrophilic bacterium *Pseudoalteromonas* spp. BSi20429, to construct a shuttle vector known as pWD (Zhao et al., 2011). A eukaryotic expression system based on *Aspergillus oryzae* is used to express α-amylase from the psychrophilic fungus *Geomyces pannorum*, and has also been developed (Mao et al., 2015).

3. Applications of Psychrophilic Enzymes in Molecular Biology and Other Industries

Psychrophilic enzymes have various applications in the food industry, detergents, molecular biology, and other sectors (Hamid & Mohiddin, 2018; Hamid, 2021). Particularly, in industries related to food, the formation of products often occurs at low temperatures to avoid food spoiling due to microbial activity at higher temperatures. The primary function of these enzymes at lower temperatures is to reduce food spoilage. α-Amylase enzyme helps in degrading complex sugar into simpler forms and is thus utilized in food manufacturing for the production of bread, beverages, and dairy (Joseph et al., 2018). β-Galactocidases are involved in the hydrolysis of dairy to obtain lactose-free products (Czyzewska & Trusek, 2021). Pectinase hydrolysis pectin, which makes juice less viscous and clearer (Mangiagalli et al., 2020). Cold-adopted protease enzymes help improve the tenderness and preservation of meat at lower temperatures (Mageswari et al., 2017). In the food sector, lipases are used to enhance the flavor of certain food items. They facilitate the synthesis of butyl butyrate, ethyl lactate, and ethyl caprylate, which is used for flavoring (Cong et al., 2019).

Psychrophilic enzymes also hold significant importance in the detergent and fabric industry as these enzymes are used to enhance washing quality. The most frequently used cold-adapted enzymes in the detergent sector are lipases, proteases, and amylases. These enzymes have been studied for their effectiveness and recommended for detergent formulation (Kumar et al., 2020a).

Cold-adapted microbial enzymes are economical and effective in bioremediation, removing contaminants from the environment. Lactase enzyme is involved in catalyzing the oxidation of aromatic amines, phenolic compounds, and other organic compounds (Bhandari et al., 2021). A study revealed that bioremediation of xylene-impacted environments at cooler temperatures could be achieved by using xylene monooxygenase and catechol 2,3-dioxygenase to destroy p-xylene (Miri et al., 2021). In molecular biology, psychrophilic have been used since 1980, and due to their high specificity and heat liability psychrophilic find a wide range of applications in molecular biology. Cold-adapted enzyme applications in various industrial sectors are summarized in Table 1 and overviewed in Figure 4.

Alkaline Phosphates

Alkaline phosphatases (APs) are zinc-containing dimeric enzymes that act as nonspecific phosphor-monoesterases found in various organisms, ranging from bacteria to mammals. Alkaline phosphatases are widely used enzymes in molecular biology. Their primary function is to dephosphorylate DNA vectors before cloning, a process that reduces self-ligation and improves cloning efficiency. Alternatively, they can be used to dephosphorylate the 5′-nucleic acid terminal before labeling with polynucleotide kinase at the 5′-terminal (Xu et al., 2024). Psychrophilic enzymes exhibit enhanced catalytic efficiencies at lower and moderate temperatures, coupled with notable thermolability. Therefore, an AP sourced from psychrophiles possessing these attributes presents significant potential for biotechnological applications. Additionally, it offers an intriguing avenue for investigating protein adaptations to cold environments (Siddiqui et al., 2012).

The characteristics of psychrophilic enzymes were formerly attributed to their increased structural flexibility, which allows them to make better use of

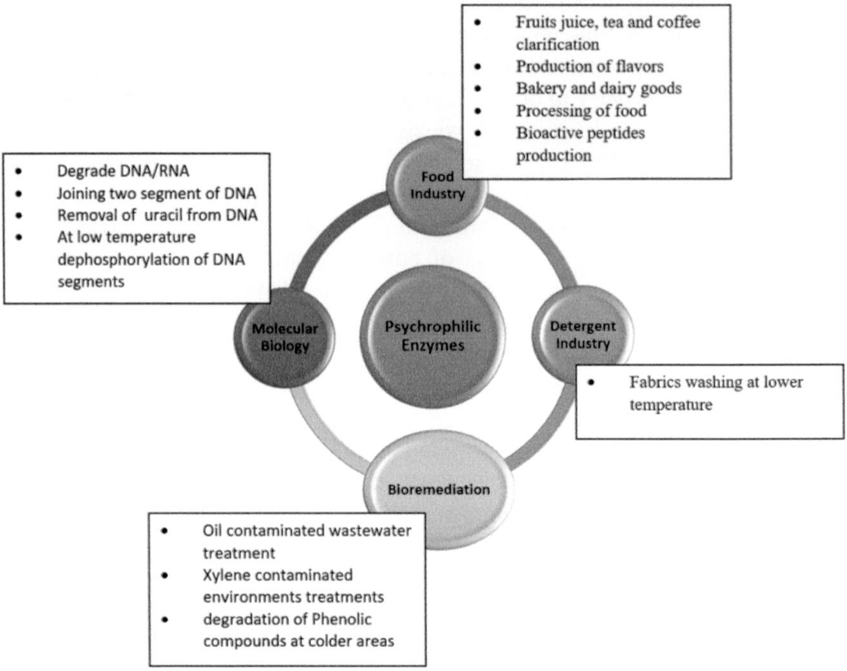

Figure 4: A summary of applications of psychrophilic enzymes in different industries.

the reduced thermal energy present in their surroundings. Extensive analysis of sequences has led to the proposal of various adaptations in sequence that are connected with increased flexibility in structure. These modifications include the grouping of glycine residues close to catalytic and metal sites for binding, a reduction in the sequence content of inflexible proline residues, and/or a decrease in interactions between hydrogen bonds and electrostatic forces. Furthermore, favorable surface charge distributions that facilitate substrate and cofactor binding as well as enhanced active site accessibility are hypothesized to be adaptations of psychrophilic enzymes based on their crystallographic structures. Previous research has demonstrated that these enzymes with shorter surface loops may increase access of substrate to the catalytic site, hence increasing catalysis (Xu et al., 2024). The AP derived from the calf intestine, which functions optimally at moderate temperatures, presents significant obstacles due to its heat resistance. This resistance necessitates thorough or time-intensive methods for deactivation, such as the application of detergents or exposure to high temperatures, before advancing to subsequent procedures. However, these treatments have the potential to affect subsequent processes adversely. Conversely, psychrophilic APs, which are sensitive to heat, can be easily deactivated with slight heat treatment. The first psychrophilic AP was isolated from Antarctic bacterium HK47 in 1984, and it irreversibly lost its activity at 65°C. Additionally, it notably decreases the

Table 1: Applications of psychrophilic enzymes in various industries

Industry	Psychrophilic Enzymes	Applications	Origin
Detergent	Amylase	Fabrics washing at a lower temperature	*Bacillus subtilis* N8 (Arabacı and Arıkan, 2018) *Pseudoalteromonas* spp. 2-3 (Sanchez et al., 2019)
	Protease		*Bacillus subtilis* WLCP1 (Furhan et al., 2019)
	Lipase		*Chryseobacterium polytrichastri* EMR1:04 (Kumar et al., 2020a) *Anoxybacillus* spp. ARS-1 (Sahoo et al., 2020) *Pseudomonas helmanticensis* HS6 (Phukon et al., 2020)
Foods	Pectinase	Natural oil extraction, wine, coffee, fruit juice clarification, and tea preparation	*Guehomyces pullulans*, *Cystofilobasidium* *Cryptococcus adeliensis* (Cavello et al., 2017) *Tetracladium* spp. (Carrasco et al., 2019) *Aureobasidium pullulans* (Merín et al., 2020)
	Xylanase	Baking and bread industries help to enhance the bread's tenderness	*Luteimonas* sp. (Han et al., 2018) *Bacillus pumilus* (Liu et al., 2019) *Bifidobacterium longum* (Li et al., 2020) *Echinicola rosea* sp. Nov. JL3085^{T} (He et al., 2020) *Acinetobacter* sp. HC4 (Zang et al., 2020)
	Protease	Peptides with bioactive properties and food properties (meat tenderization at low temperatures)	*Chryseobacterium* sp. (Mageswari et al., 2017)
	Lipase	Enhancing the flavors of various food products	*Candida Antarctica* (Souza et al,. 2017) *Pseudomonas* AMS8 (Musa et al., 2018) *Aspergillus* (Cong et al., 2019)

(Contd.)

(*Contd.*)

Industry	Psychrophilic Enzymes	Applications	Origin
	β-galactosidase	Enhancing the sweetness of milk products and manufacturing them as lactose-free products	*Bacillus velezensis* (Liu et al., 2021`) *Bifidobacterium adolescentis* BqaC (Mulualem et al., 2021)
	α-amylase	Bread, beverages, and different types of dairy products	*Bacillus subtilis* N8 (Arabacı et al., 2018) *Pseudoalteromonas* sp. 2-3 (Sanchez et al., 2019) *Shewanella* sp. ISTPL2 (Rathour et al., 2020)
Bioremediation	Laccase	Aromatic amines, phenolic compounds, and other organic compounds breakdown in colder areas	*Aspergillus nidulans* TTF6 (Sahay et al., 2020) *Bacillus* sp. (Thomas et al., 2023) Enterobacter sp. (Thomas et al., 2020) *Bacillus* sp. (Deepa Thomas et al., 2020)
	Xylene monoxygenase, catechol 2,3-dioxygenase	Xylene environmental contamination treated by low-temperature bioremediation	*Pseudomonas synxantha* S2TR-26 & *Pseudomonas mandelii* S2TR-08 (Miri et al., 2021)
	Lipase	Wastewater contaminated with oils in cold areas can be remediated	*Pseudomonas aeruginosa* KC1 (Das et al., 2018)
Molecular Biology	Proteases	To remove the proteins and create the active surface by degrading it into peptides	*Gadus morhua* (Clarsund et al., 2016)
	DNA ligases	To make temperature-sensitive strains by joining two segments of DNA	*Colwellia psychrerythraea* (Duplantis et al., 2010) *Pseudoalteromonas haloplanktis* (Georlette et al., 2000b) *Aliivibrio salmonicida* (Williamson and Pedersen, 2014)

	Alkaline phosphatases	Dephosphorylation of the 5′ end of DNA	*Pandalus borealis* (Nooralabettu, 2015) *Colwellia psychrerythraea, Arthrobacter* genus D10 *Shewanella* sp. (Rathour et al., 2020) *Vibrio* sp. (Markússon et al., 2022) *Gadus morhua* (Pires, 2017)
	Uracil DNA glycosylase	Recognize and remove DNA from uracil	*Bacillus* sp. HJ171 (Lee et al., 2009) *Psychrobacter* sp. HJ147 (Lee et al., 2009)
	RNA polymerase	Transcription (*in vitro*)	*Pseudomonas fluorescens bacteriophage* VSW-3 (Zhu et al., 2013) *Synechococcus* (Wang et al., 2022)
	DNA polymerase	Replication of DNA	*Psychromonas ingrahamii* (Xue et al., 2021)
	Nuclease	To remove the DNA/RNA by degrading them	*Psychromonas ingrahamii* 37 (Maciejewska et al., 2019) *Psychrobacter* spp. ANT206 (Wang et al., 2019) *Pseudomonas syringae* (Sulthana et al., 2011b) *Chlamys islandica* (Øverbø and Myrnes, 2006)

occurrence of non-recombinants in cloning experiments to approximately 5% of the transformed colonies when treating BamHI-digested pBR322. Later on, in 1991, another psychrophilic AP, derived from the Northern shrimp *Pandalus borealis*, was found. In 1993, ArcticZymes released this enzyme as the first psychrophilic AP that was available for purchase. A modified version was then released in 2010 (Collins & Margesin, 2019).

Uracil-DNA Glycosylases

Uracil-DNA glycosylase (UDG) plays a vital role in the repair pathway for base excision, addressing uracil generated by the hydrolytic deamination of cytosine. Consequently, its presence is essential for preserving the integrity of the genome (Chang et al., 2022). Uracil is a common non-standard base found in the human genome, emerging either through the natural breakdown of cytosine or due to errors in incorporating deoxyuridine monophosphate (dUMP) by DNA polymerases. The concentration of uracil within the genome is regulated by the effectiveness of DNA repair mechanisms, primarily through base excision repair (BER) with mismatch repair (MMR) serving as a secondary option. If left unaddressed, uracil within DNA can lead to mutations, specifically causing G:C to A:T transitions during the process of DNA replication (Perkins & Zhao, 2021).

Within cellular mechanisms aimed at preserving genetic stability, BER stands out as a fundamental process for rectifying DNA damage, such as that caused by uracil. BER is evolutionarily conserved and operates through two primary pathways: the short-patch pathway, resulting in the repair of a single nucleotide tract, and the long-patch pathway, which repairs a tract of at least two nucleotides. BER is a meticulously coordinated process, proceeding through several sequential steps. Initially, the damaged nucleotide base undergoes enzymatic hydrolysis by a DNA glycosylase which is a damage-specific enzyme, leading to the development of an Apurinic/Apyrimidinic (AP) intermediate site. Subsequently, an endonuclease cleaves the sugar-phosphate backbone at the AP site, followed by DNA end cleanup facilitated by a lyase enzyme. Gap filling is then carried out by a DNA polymerase, and finally, the nick in the DNA strand is sealed by a ligase enzyme (Chang et al., 2022).

Examples of Uracil-DNA Glycosylases from Psychrophiles

In a research study, the DNA sequence responsible for encoding Psp HJ147 UDG (*Psychrobacter* sp. HJ147 uracil-DNA glycosylase) was isolated, cloned, and analyzed. This gene comprises 735 base pairs, encoding a protein consisting of 244 amino acid residues. Comparison of the deduced amino acid sequence of Psp HJ147 UDG revealed significant similarity to those found in *Psychrobacter arcticus*, *Psychrobacter cryohalolentis* K5, and *Psychrobacter* sp. PRwf-1 (Bakermans, 2018). The psychrophilic, Gram-positive marine strain (BMTU3346) yielded the heat-labile UDG gene, which has been successfully extracted, sequenced, and translated into *E. coli*. This UDG exhibits cold-active properties, demonstrating a seeming optimal temperature of 35°C and a half-life of 2 minutes at 40°C. By comparing its amino acid sequence to that of UDGs from mesophilic bacteria, identities are also revealed to range from 39.1% to 46.2%. Investigation into the primary structure aimed to identify features associated with the enzyme's thermolability. The heat-labile UDG's amino acid sequence differs from the consensus sequence of bacterial UDGs by 22 amino acid variations, but it lacks characteristic features of cold-active enzymes, like longer surface loops or fewer arginine or proline residues within loops. The study suggests that more subtle alterations in conformation likely underlie the thermolability of the UDG from BMTU3346, as no distinct characteristics directly related to its thermolability were identified (Mangiagalli et al., 2020).

Properties of Uracil-DNA Glycolase

The catalytic competence of cold-adapted enzymes is increased, their thermostability is reduced, and they are more flexible in structural regions that are important for catalytic activity. As temperatures drop from 37 to 0°C, the rate of reaction typically decreases significantly, ranging from 30-80 fold. Compared to their mesophilic counterparts, these enzymes frequently have fewer proline residues, more glycine residues, a lower relative arginine content (arg/(arg + lys)), and fewer pairs of ions, aromatic contacts, and/or hydrogen bonds. For instance, the cold-adapted uracil DNA glycosylase from cod (cUDG) demonstrates up to tenfold higher catalytic efficiency (k_{cat}/K_m) within the temperature range of 15–37°C compared to the warm-active human counterpart (hUDG). UDG is the first enzyme in the BER pathway, which helps single- or double-stranded DNA's promutagenic uracil residues to be hydrolyzed (Åqvist et al., 2017).

Lipases

Cold-adapted lipases find extensive application in various industries. In the detergent sector, they contribute to energy savings and minimize textile fiber damage through cold-washing processes. Similarly, in the food industry, their ability to be inactivated at relatively low temperatures helps preserve the nutritional integrity of food products. Moreover, these enzymes play a crucial role in environmental endeavors, such as wastewater treatment and the remediation of cold environments contaminated with fats. Ganasen et al. (2016) reported that the thermolabile nature of the cold-adapted alkaline AMS8 lipase is particularly desirable in the latest advancements within biotechnological applications. The Antarctic soil bacterium *Streptomyces* sp. So3.2 can synthesize a biosurfactant, potentially valuable for hydrocarbon bioremediation purposes. Psychrophiles exhibit inherent adaptability to handle diverse heavy metal concentrations in their environment. This adaptive characteristic is likely attributed to the prevalence of heavy metals in Antarctica's sedimentary regions. Furthermore, Antarctic bacteria are frequently associated with a high incidence of plasmids, likely facilitating the transference of genes conferring resistance or tolerance to heavy metals like mercury, tellurium, cadmium, copper, chromium, and lead (Orellana et al., 2018). DNA ligases derived from psychrophilic organisms like *Pseudoalteromonas haloplanktis* and the psychrophilic fish-pathogen *Aliivibrio salmonicida* exhibit activity at low temperatures, yet they are not currently available for commercial use. These ligases could potentially serve as essential genes to replace counterparts in mammalian pathogens, rendering them temperature-sensitive strains. Such strains could then be utilized in the development of a distinct class of bacterial temperature-sensitive vaccines targeting significant human pathogens. Similarly, RNA ligases aid in the creation of phosphodiester linkages between RNA molecules by going through similar chemical stages as DNA ligases. Although there is an urgent need to address their relatively low efficiency, the functional diversity of RNA ligases has led to a variety of uses, including chemical misacylation, connecting single-stranded adenylated primers, RNA radioactive labeling, and miRNA cloning. ArcticZymes has recently introduced a novel ligase called ArcticZymes R2D ligase. This ligase stands out as the first of its kind on the market capable of ligating DNA to 5′-phosphorylated ends of RNA in the presence of a DNA template that positions the joinable ends (Xu et al., 2024).

Nucleases

The enzymes that perform cleavage of the phosphodiester bonds of nucleic acids are called nucleases and are categorized based on their function as DNases, RNases, endo-nucleases, exo-nucleases, recombinases, ribozymes, topoisomerases, and RNA splicing enzymes. The mechanism of action of nucleases in general can be divided into three phases, i.e., nucleophilic attack, formation of a highly negative penta-covalent intermediate, followed by the breakage of a phosphodiester bond. The dissociation of phosphodiester bonds is assumed to be catalyzed by a general acid-base reaction in which the general acid protonates the leaving group to aid in product formation and the general base activates the nucleophile (Yang, 2011).

There are two categories of nucleases based on the number of strands of nucleic acid, one is double-strand nucleases which are specialized for the breakdown of double-stranded nucleic acids without affecting any single-stranded nucleic acids whereas single-stranded nucleases work exactly in the reverse way. Whenever nucleases are involved they must be inactivated quickly after their required action, so psychrophilic nucleases are quite significant. Many types of deoxyribonucleases II are characterized in crown-of-thorns starfish (*Acanthaster planci*) and Icelandic scallops (*Chlamys islandica*) (Shiomi et al., 2004; Maciejewska et al., 2019). The biochemical properties of a psychrophilic RNase R present in Antarctic bacterium *Pseudomonas syringae* Lz4W have been reported for the first time (Sulthana et al., 2011a). In another study, an unusual salt-tolerant RNase R was obtained from Antarctic bacterium *Psychrobacter* spp. ANT206 has been reported. There is also a first report on endonuclease I retrieved from *Psychromonas ingrahamii* 37 which is an extreme psychrophilic microorganism (Maciejewska et al., 2019; Wang et al., 2019).

Currently, many companies have introduced psychrophilic nucleases in the market, like HL-ExoI and the dsDNase from ArcticZymes, Affymetrix USB, and Thermo Fisher Scientific (MA, USA), whereas Takara-Clontech (Mountain View, CA, USA) has introduced the Cryonase enzyme (Sarmiento et al., 2015; Mangiagalli et al., 2020). It is a useful enzyme produced by using *E. coli* and is isolated from a psychrophile *Shewanella* spp. It is effective with ice-cold materials and may break down every type of DNA or RNA, including circularized, single-stranded, double-stranded, and linear DNA (Google Patents, 2006).

Ligases

DNA ligases are the enzymes that perform catalysis of a phosphodiester bond formation at single-strand breaks in double-stranded DNA (Zheng & Shabek, 2017). The mechanism of this catalysis involves three steps. First, in the absence of DNA, the ligase forms a ligase-AMP linkage at a conserved lysine residue. The second step involves the formation of an adenylated DNA intermediate by the adenylated ligase binding to a nicked DNA. To create a phosphodiester bond and release AMP, ligase catalyzes the final step, which involves activating AMP for the nucleophilic assault of 3'-OH in the DNA nick on the adenylated 5′-P (Williamson & Leiros, 2020).

When it comes to annealing short DNA protrusion ends, the ideal temperature for ligase to work with DNA T_{opt}, is typically relatively low. Due to this, high-activity enzymes that can work under low temperatures are needed. There comes the need for psychrophilic ligases, such as the ligases from psychrophilic species *Pseudoalteromonas haloplanktis* (Georlette et al., 2000) and *Aliivibrio salmonicida* (a psychrophilic fish-pathogen) (Williamson & Pedersen, 2014). These two psychrophilic ligases work best at low temperatures but currently are not being synthesized commercially. The essential gene DNA ligase can be used to replace counterparts in the genomes of mammalian pathogens, turning them into temperature-sensitive pathogens. These pathogens can then be used to create a unique class of temperature-sensitive bacterial vaccines that can be used to protect against a variety of human pathogens (Duplantis et al., 2010).

In the same way, the RNA ligases do the catalysis of phosphodiester bond formation in RNA. The mechanism of working of RNA ligase is somehow similar to that of DNA ligases (Unciuleac et al., 2015). One of the ligases is T4 RNA ligase 1, which joins single-stranded 5′-P and 3′-OH nucleic acids to ligate single-stranded RNA, whereas T4 RNA ligase 2 prefers to work on double-stranded nucleic acids to catalyze the ligation of 3′-OH of RNA to 5′-P of RNA or DNA (Nichols et al., 2008). There exists a significant functional diversity of RNA ligases that have paved the way for its various applications, for example, chemical misacylation, radioactive labeling of RNA, joining a single-stranded adenylated primer, and miRNA cloning. But there is a need to resolve low efficiency. The first ligase has been launched in the market by ArcticZymes, i.e., R2D ligase. It can catalyze the joining of DNA to 5′-P ends of RNA when a DNA template is present for the positioning of the joinable ends. But they have not mentioned the details such as the source and inactivation temperature of the enzyme on their official website. Apart from this, it has been reported that two new ligases named AZ Prototype L18 and L19 can be used as prototypes in the near future (Xu et al., 2024). This research field on the functional characterization of RNA ligases needs much more exploration.

Proteases

The hydrolysis reactions are catalyzed by proteases, which break down proteins into peptides and ultimately amino acids. These are categorized into six major groups based on their mechanism of action i.e. serine, cysteine, threonine, aspartic, glutamic acid proteases, and metalloproteases (Zhu et al., 2011; Mohiddin & Hamid, 2018). In molecular biology, the protease that finds most applications is Protease K, which is a member of the alkaline serine protease family. This enzyme is mostly used in nucleic acid extraction because of its extensive cutting activity, *in situ* hybridization, and detection of proteins (Xu et al., 2024). It works by using its substrate recognition site (peptides 99–104 and 132–136), after substrate recognition it uses an active serine to break a peptide bond (Li et al., 2020).

The protease K that is generally used remains stable under 95°C, and the temperature required for its inactivation has a great impact on the integrity of nucleic acids exclusively in the case of RNA. ArcticZymes has introduced a valuable alternative proteinase i.e. a cold-adapted and heat-labile endopeptidase isolated from an arctic marine microbe. Nucleic acid destruction is reduced since this proteinase completes deactivation at 65°C for 30 minutes (Mangiagalli et al., 2020). Another valuable product has been launched, i.e., ColdZyme (Enzymatica AB, Lund, Sweden) that is playing a vital role in the management of viral infections. Trypsin extracted from cod that has acclimated to the cold is present in this spray. ColdZyme spray treats the common cold by forming an active surface barrier having proteolytic activity when sprayed in the oral and throat cavity (ArticZymes, 2020). From this, it is inferred that psychrophilic enzymes can be worthwhile in medical treatment.

4. Conclusion

In this chapter, we have highlighted how psychrophilic enzymes provide valuable opportunities for energy-efficient processes and improved product quality in diverse industries, driving advancements in biotechnology and environmental sustainability. Despite the expanding research in molecular biology, the exploration of psychrophilic enzymes remains relatively limited compared to other enzyme types. However, these enzymes offer valuable tools for various molecular biology techniques, including DNA manipulation, gene expression, and protein analysis. This suggests untapped potential for psychrophilic enzymes to contribute significantly to advancements in molecular biology research. Challenges persist in purification, mining, and expression methods for cold-tolerant enzymes. Further comprehensive studies are necessary to fully understand the diversity of psychrophiles and their enzymes. Innovative research into the applications of psychrophilic enzymes across industries holds promising potential for future developments.

References

Al-Ghanayem, A.A., & Joseph, B. 2020. Current prospective in using cold-active enzymes as eco-friendly detergent additive. *Appl. Microbiol. Biotechnol.*, 104, 2871–2882.

Åqvist, J., Isaksen, G.V., & Brandsdal, B.O. 2017. Computation of enzyme cold adaptation. *Nat. Rev. Chem.*, 1, 51.

Arabacı, N., & Arıkan, B. 2018. Isolation and characterization of a cold-active, alkaline, detergent stable α-amylase from a novel bacterium Bacillus subtilis N8. *Preparative Biochem Biotechnol.*, 48(9). https://doi.org/10.1080/10826068.2018.1452256.

ArticZymes. 2020. ArcticZymes Proteinase is an Unspecific Endopeptidase. https://www.arcticzymes.com/product-details/arcticzymes-proteinase.

Ayala-Del-Río, H.L., Chain, P.S., Grzymski, J.J., Ponder, M.A., Ivanova, N., Bergholz, P.W. et al. 2010. The genome sequence of Psychrobacter arcticus 273-4, a Psychroactive Siberian permafrost bacterium, reveals mechanisms for adaptation to low-temperature growth. *Applied Environ. Microbiol.*, 76, 2304–2312. https://doi.org/10.1128/AEM.02101-09.

Bakermans, C. 2018. Adaptations to marine versus terrestrial low temperature environments as revealed by comparative genomic analyses of the genus Psychrobacter. *FEMS Microbiol. Ecol.*, 94, fiy102.

Baldwin, R.L. 2008. The search for folding intermediates and the mechanism of protein folding. *Annu. Rev. Biophys.*, 37, 1–21. https://doi.org/10.1146/ANNUREV.BIOPHYS.37.032807.125948.

Bar Dolev, M., Braslavsky, I., & Davies, P.L. 2016. Ice-binding proteins and their function. *Annu. Rev. Biochem.*, 85, 515–542. https://doi.org/10.1146/ANNUREV-BIOCHEM-060815-014546.

Bhandari, S., Poudel, D.K., Marahatha, R. et al. 2021. Microbial enzymes used in bioremediation. *J. Chem.* https://doi.org/10.1155/2021/8849512.

Bhatia, R.K., Ullah, S., Hoque, M.Z. et al. 2021. Psychrophiles: A source of cold-adapted enzymes for energy efficient biotechnological industrial processes. *J. Environ. Chem. Eng.*, 9, 104607.

Bjerga, G.E.K., Lale, R., & Williamson, A.K. 2016. Engineering low-temperature expression systems for heterologous production of cold-adapted enzymes. *Bioengineered*, 7, 33–38.

Buzzini, P., Branda, E., Goretti, M., & Turchetti, B. 2012. Psychrophilic yeasts from worldwide glacial habitats: Diversity, adaptation strategies, and biotechnological potential. *FEMS Microbiol. Ecol.*, 82, 217–241.

Carrasco, M., Rozas, J.M., Alcaíno, J. et al. 2019. Pectinase secreted by psychrotolerant fungi: Identification, molecular characterization, and heterologous expression of a cold-active polygalacturonase from Tetracladium sp. *Microb. Cell. Fact.*, 18. https://doi.org/10.1186/S12934-019-1092-2.

Cavello, I., Albanesi, A., Fratebianchi, D. et al. 2017. Pectinolytic yeasts from cold environments: Novel findings of Guehomyces pullulans, Cystofilobasidium infirmominiatum, and Cryptococcus adeliensis producing pectinases. *Extremophiles*, 21, 319–329. https://doi.org/10.1007/S00792-016-0904-0.

Cavicchioli, R. 2016. On the concept of a psychrophile. *ISME J.*, 10, 793–795.

Chang, H-L., Su, K-Y., Goodman, S.D. et al. 2022. Uracil-DNA glycosylase assay by matrix-assisted laser desorption/ionization time-of-flight mass spectrometry analysis. *JoVE* (*Journal Vis Exp*), e63089.

Chen, Q., Xiao, Y., Zhang, W. et al. 2020. Computer-aided search for a cold-active cellobiose 2-epimerase. *J. Dairy Sci.*, 103, 7730–7741. https://doi.org/10.3168/JDS.2020-18153.

Clarsund, M., Blom, U., & Gardulf, A. 2016. Evaluation of ColdZyme® Mouth Spray on prevention of upper respiratory tract infections in a boy with primary immunodeficiency: A case report. *J. Med. Case Rep.*, 10, 1–5. https://doi.org/10.1186/S13256-016-1085-2.

Collins, T., & Margesin, R.2019. Psychrophilic lifestyles: Mechanisms of adaptation and biotechnological tools. *Appl. Microbiol. Biotechnol.*, 103, 2857–2871.

Cong, S., Tian, K., Zhang, X. et al. 2019. Synthesis of flavor esters by a novel lipase from Aspergillus niger in a soybean-solvent system. *3 Biotech*, 9. https://doi.org/10.1007/S13205-019-1778-5.

Cvetkovska, M., Vakulenko, G., Smith, D.R. et al. 2022. Temperature stress in psychrophilic green microalgae: Minireview. *Physiol. Plant*, 174, e13811.

Czyzewska, K., & Trusek, A. 2021. Encapsulated NOLA™ Fit 5500 Lactase: An economically beneficial way to obtain lactose-free milk at low temperature. *Catalysts*, 11, 527.

Das, A. 2018. *A Cold Tolerant Lipase Develops Enhanced Activity, Thermal Tolerance, and Solvent Stability in the Presence of Calcium Nanoparticles: AN Alternative Approach to Genetic*. Elsevier.

De Maayer, P., Anderson, D., Cary, C., & Cowan, D.A. 2014. Some like it cold: Understanding the survival strategies of psychrophiles. *EMBO Rep.*, 15, 508–517. https://doi.org/10.1002/EMBR.201338170.

Deming, J.W., & Young, J.N. 2017. The role of exopolysaccharides in microbial adaptation to cold habitats. In: *Psychrophiles: From Biodivers to Biotechnology* (2nd Edn.). Springer, pp. 259–284. https://doi.org/10.1007/978-3-319-57057-0_12.

Duplantis, B.N., Osusky, M., Schmerk, C.L. et al. 2010. Essential genes from Arctic bacteria used to construct stable, temperature-sensitive bacterial vaccines. *Proc. Natl. Acad. Sci. USA*, 107, 13456–13460. https://doi.org/10.1073/PNAS.1004119107/SUPPL_FILE/PNAS.201004119SI.PDF.

Fang, Z.M., Li, T.L., Chang, F. et al. 2012. A new marine bacterial laccase with chloride-enhancing, alkaline-dependent activity and dye decolorization ability. *Bioresour. Technol.*, 11(1), 36–41. https://doi.org/10.1016/J.BIORTECH.2012.01.172.

Feller, G. 2013. Psychrophilic enzymes: From folding to function and biotechnology. *Scientifica* (*Cairo*) 2013.

Fenice, M., Khare, S.K., & Gorrasi, S. 2021. Mining, designing, mechanisms, and applications of extremophilic enzymes. *Front. Microbiol.*, 12, 709377.

Ferrer, M., Chernikova, T.N., Timmis, K.N., & Golyshin, P.N. 2004. Expression of a temperature-sensitive esterase in a novel chaperone-based Escherichia coli strain. 70(8), 4499–4504. https://doi.org/10.1128/AEM.70.8.4499-4504.2004.

Furhan, J., & Awasthi, P. 2019. *Biochemical Characterization and Homology Modelling of Cold-Active Alkophilic Protease from Northwestern Himalayas and Its Application in Detergent Industry*. Elsevier.

Ganasen, M., Yaacob, N., Abd Rahman, R.N.Z.R., Leow, A.T.C., Basri, M., Salleh, A.B., & Ali, M.S.M. (2016). Cold-adapted organic solvent tolerant alkalophilic family I. 3 lipase from an Antarctic Pseudomonas. *International Journal of Biological Macromolecules*, 92, 1266–1276.

Georlette, D., Jónsson, Z.O., Van Petegem, F. et al. 2000a. A DNA ligase from the psychrophile Pseudoalteromonas haloplanktis gives insights into the adaptation of proteins to low temperatures. *Eur. J. Biochem.*, 267, 3502–3512. https://doi.org/10.1046/J.1432-1327.2000.01377.X.

Godin-Roulling. A., Philipp A.M. Schmidpeter, Schmid, F.X., & Feller, G. 2015. Functional adaptations of the bacterial chaperone trigger factor to extreme environmental temperatures. *Environmental Microbiol.*, 17, 2407–2420. https://doi.org/10.1111/1462-2920.12707.

Google Patents. 2006. US8034597B2: Microorganism-Derived Psychrophilic Endonuclease. https://patents.google.com/patent/US8034597B2/en.

Goordial, J., Raymond-Bouchard, I., Zolotarov, Y., Bethencourt, L., Ronholm, J., Shapiro, N. et al. 2016. Cold adaptive traits revealed by comparative genomic analysis of the eurypsychrophile Rhodococcus sp. JG3 isolated from high elevation McMurdo Dry Valley permafrost, Antarctica. *FEMS Microbiol Ecol.*, 92(2), fv154.

Hamid, B., & Mohiddin, F.A. 2018. Cold-active enzymes in food processing. In: *Enzymes in Food Technology*, Springer, Singapore, pp. 383–400.

Hamid, B. (2021). Potent biotechnological applications of psychrozymes. In: *Microbiomes and the Global Climate Change*, Springer, pp. 349–364.

Hamid, B., Bashir, Z., Yatoo, A.M. et al. 2022. Cold-active enzymes and their potential industrial applications: A review. *Molecules*, 27, 5885.

Hamid, B., Baba, Z.A., Sheikh, T.A., Perveen, K., & Sayyed, RZ (2025). Investigating potential protease activity of psychrotrophic bacteria from a municipal landfill for solid waste management. *Biomass Conversion and Biorefinery*, 15(15), 22615–22624. https://doi.org/10.1007/s13399-024-05621-2

Han, Z., Shang-guan, F., & Yang, J. 2018. Characterization of a novel cold-active xylanase from Luteimonas species. *World J. Microbiol. Biotechnol.*, *34*. https://doi.org/10.1007/S11274-018-2505-9.

Hartl, F., Bracher, A., & M. Hayer-Hartl. 2011. Molecular chaperones in protein folding and proteostasis. *Nature*. https://doi.org/10.1038/nature10317.

Hassan, N., Rafiq, M., Hayat, M. et al. 2016. Psychrophilic and psychrotrophic fungi: A comprehensive review. *Rev. Environ. Sci. Bio/Technology*, 15, 147–172.

He, J., Liu, L., Liu, X. et al. 2020. Isolation and characterization of a novel cold-active, Halotolerant Endoxylanase from Echinicola rosea sp. nov. JL3085T. *Marine Drugs*, 2020.

Javed, A., & Qazi, J.I. 2016. Psychrophilic microbial enzymes implications in coming biotechnological processes. *Am. Sci. Res. J. Eng. Technol. Sci.*, 23, 103–120.

Ji, X., & Wei, Y. 2020. Mechanism of microbial adaptation and survival within psychrophilic habitat. In: *Microbial in Versatility Varied Environments: Microbes in Sensitive Environments*. Springer Verlag, Singapore, pp. 67–84.

Jin, S., Wang, Y., & Zhao, X. 2022. Cold-adaptive mechanism of psychrophilic bacteria in food and its application. *Microb. Pathog.*, 169, 105652.

Joseph, B., & Kumar, V. 2019. *Biotechnology PR-E in F, Undefined Psychrophilic Enzymes: Potential Biocatalysts for Food Processing*. Elsevier.

Karan, R., Mathew, S., Muhammad, R., Bautista, D.B., Vogler, M., Eppinger, J. et al. 2020. Understanding high-salt and cold adaptation of a polyextremophilic enzyme. *Microorganisms*, 2020.

Kim, H.W., Wi, A.R., Jeon, B.W. et al. 2015. Cold adaptation of a psychrophilic chaperonin from Psychrobacter sp. and its application for heterologous protein expression. *Biotechnol. Lett.*, 37, 1887–1893. https://doi.org/10.1007/S10529-015-1860-Y.

Kostadinova, N., Tosi, S., Spassova, B., & Angelova, M. 2017. Comparison of the oxidative stress response of two Antarctic fungi to different growth temperatures. *Polish Polar Res.*, 393–408.

Kuddus, M., Roohi, Bano, N., Sheik, G.B., Joseph, B., Hamid, B., Raveendran Sindhu, & Aravind Madhavan. 2024. Cold-active microbial enzymes and their biotechnological applications. *Microbial. Biotechnology*, 17, e14467.

Kumar, A., Mukhia, S., Kumar, N. et al. 2020a. A broad temperature active lipase purified from a psychrotrophic bacterium of Sikkim Himalaya with potential application in detergent formulation. *Front. Bioeng. Biotechnol.*, 8. https://doi.org/10.3389/FBIOE.2020.00642/FULL.

Kumar, A., Mukhia, S., & Kumar, R. 2021. Industrial applications of cold-adapted enzymes: Challenges, innovations and future perspective. *3 Biotech*, 11, 426.

Kumar, S., Suyal, D.C., Yadav, A. et al. 2020b. Psychrophilic Pseudomonas helmanticensis proteome under simulated cold stress. *Cell Stress Chaperones*, 25, 1025–1032.

Lee, M.S., Kim, G.A., Seo, M.S. et al. 2009. Characterization of heat-labile uracil-DNA glycosylase from Psychrobacter sp. HJ147 and its application to the polymerase chain reaction. *Biotechnol. Appl. Biochem.*, 52, 167–175.

Li, D., Fei, T., Wang, Y. et al. 2020. A cold-active 1,4-α-glucan branching enzyme from Bifidobacterium longum reduces the retrogradation and enhances the slow digestibility of wheat starch. *Food Chem.*, 2020. Elsevier.

Li, W., Wang, J.J., Qian, H. et al. 2020. Insights into the role of extracellular DNA and extracellular proteins in biofilm formation of Vibrio parahaemolyticus. *Front. Microbiol.*, 11, 813.

Liu, Y., Huang, L., Zheng, D. et al. 2019. Biochemical characterization of a novel GH43 family β-xylosidase from Bacillus pumilus. *Food Chemistry*. Elsevier.

Liu, Y., Wu, Z., Zeng, X. et al. 2021. A novel cold-adapted phospho-beta-galactosidase from Bacillus velezensis and its potential application for lactose hydrolysis in milk. *International J. Biol. Macromol.*, 164(4).

Liu, Y., Ji, M., Yu, T. et al. 2022. A genome and gene catalog of glacier microbiomes. *Nat. Biotechnol.*, 40. https://doi.org/10.1038/S41587-022-01367-2.

Maciejewska, N., Walkusz, R., Olszewski, M., & Szymańska, A. 2019. New nuclease from extremely psychrophilic microorganism Psychromonas ingrahamii 37: Identification and characterization. *Mol. Biotechnol.*, 61, 122–133.

Madhavan, A., Sindhu, R., Parameswaran, B. et al. 2017. Metagenome analysis: A powerful tool for enzyme bioprospecting. *Appl. Biochem. Biotechnol.*, 183, 636–651. https://doi.org/10.1007/S12010-017-2568-3.

Mageswari, A., Subramanian, P., Chandrasekaran, S. et al. 2017. Systematic functional analysis and application of a cold-active serine protease from a novel

Chryseobacterium sp. *Food Chem.*, 217, 18–27. https://doi.org/10.1016/J.FOODCHEM.2016.08.064.

Mangiagalli, M., Brocca, S., Orlando, M., & Lotti, M. 2020. The "cold revolution": Present and future applications of cold-active enzymes and ice-binding proteins. *N. Biotechnol.*, 55, 5–11.

Mao, Y., Yin, Y., Zhang, L. et al. 2015. Development of a novel Aspergillus uracil deficient expression system and its application in expressing a cold-adapted α-amylase gene from Antarctic fungi Genomes panorum. *Process Biochemistry*, 50(10).

Markússon, S., Hjorleifsson, J.G., Kursula, P., & Ásgeirsson, B. 2022. Structural characterization of functionally important chloride binding sites in the marine Vibrio alkaline phosphatase. *Biochemistry*, 61, 2248–2260.

Merín, M., & Morata Ambrosini, V.I. 2020. Application of a grape surface majority pectinolytic species, Aureobasidium pullulans, to low-temperature red winemaking: Development and stability of wine colour. *Journal Wine Res.*, 31, 218–239. https://doi.org/10.1080/09571264.2020.1816534.

Miri, S., Davoodi, S., Brar, S. et al. 2021. Psychrozymes as novel tools to biodegrade p-xylene and potential use for contaminated groundwater in the cold climate. *Bioresource Technol.*, 321(6), 124464.

Mohiddin, F.A.., & Hamid, B. 2018. Cold-active yeast proteases as a versatile tool for industry: A review. *SKUAST Journal of Research*, 20(2), 127–134.

Mulualem, D.M., Agbavwe, C., Ogilvie, L.A. et al. 2021. Metagenomic identification, purification, and characterisation of the Bifidobacterium adolescentis BgaC β-galactosidase. *Appl. Microbiol. Biotechnol.*, 105, 1063–1078. https://doi.org/10.1007/S00253-020-11084-Y.

Musa, N., Latip, W., Rahman, R.A. et al. 2018. Immobilization of an Antarctic Pseudomonas AMS8 lipase for low temperature ethyl hexanoate synthesis. *Catalysts*, 8(6), 234.

Mykytczuk, N.C., Foote, S.J., Omelon, C.R. et al. 2013. Bacterial growth at –15 °C; molecular insights from the permafrost bacterium Planococcus halocryophilus Or1. *The ISME Journal*, 7, 1211–1226. https://doi.org/10.1038/ismej.2013.8.

Nichols, N.M., Tabor, S., & McReynolds, L.A. 2008. RNA ligases. *Curr. Protoc. Mol. Biol.* https://doi.org/10.1002/0471142727.MB0315S84.

Nooralabettu, K.P. 2015. Characteristics of alkaline phosphatase isolated from the hepatopancreas of Indian white shrimp(Penaeus indicus). *Int. J. Anal. Bio-Sci.*, 3.

Orellana, R., Macaya, C., Bravo, G. et al. 2018. Living at the frontiers of life: Extremophiles in Chile and their potential for bioremediation. *Front. Microbiol.*, 9, 2309.

Øverbø, K., & Bjornar, M. 2006. Deoxyribonuclease II from the Icelandic scallop (Chlamys islandica): Isolation and partial characterization. *Comp. Biochem. and Physiology Part B Biochem. and Mol. Bio*l., 143(3), 315–318.

Pandey, N., Jain, R., Pandey, A., & Tamta, S. 2018. Optimisation and characterisation of the orange pigment produced by a cold adapted strain of Penicillium sp. (GBPI_P155) isolated from mountain ecosystem. *Mycology*, 9, 81–92.

Pandey, R., Usui, K., Livingstone, R.A. et al. 2016. Ice-nucleating bacteria control the order and dynamics of interfacial water. *Sci. Adv*. 2. https://doi.org/10.1126/SCIADV.1501630.

Paredes. D.I., Watters, K., Pitman, D.J. et al. 2011. Comparative void-volume analysis of psychrophilic and mesophilic enzymes: Structural bioinformatics of psychrophilic enzymes reveals sources of core flexibility. *BMC Struct. Biol.*, 11. https://doi.org/10.1186/1472-6807-11-42.

Parvizpour, S., Hussin, N., Shamsir, M.S., & Razmara, J. 2021. Psychrophilic enzymes: Structural adaptation, pharmaceutical and industrial applications. *Appl. Microbiol. Biotechnol.*, 105, 899–907.

Perfumo, A., Banat, I.M., & Marchant, R. 2017. Going green and cold: Biosurfactants from low-temperature environments to biotechnology applications. *Trends in Biotechnol.*, 36(3). https://doi.org/10.1016/j.tibtech.2017.10.016.

Perkins, J.L., & Zhao, L. 2021. The N-terminal domain of uracil-DNA glycosylase: Roles for disordered regions. *DNA Repair* (*Amst*), 101, 103077.

Phukon, L., Chourasia, R., Kumari, M. et al. 2020. Production and characterisation of lipase for application in detergent industry from a novel Pseudomonas helmanticensis HS6. *Bioresource Technol.*, 2020.

Pires, N.M. 2017. *Dietary Taurine Requirements of European Sea Bass* (*Dicentrarchus Labrax*, L.) *Juveniles*. Porto.

Profiti, G., Martelli, P.L., & Casadio, R. 2017. The Bologna Annotation Resource (BAR 3.0): Improving protein functional annotation. *Nucleic Acids Res.*, 45, W285–W290. https://doi.org/10.1093/NAR/GKX330.

Qing, G., Mal, T.K. et al. 2004. Cold-shock induced high-yield protein production in Escherichia coli. *Nature Biotechnol.*, 2004. https://doi.org/10.1038/nbt984.

Rathour, R., Gupta, J., Tyagi, B., & Thakur, I.S. 2020. Production and characterization of psychrophilic α-amylase from a psychrophilic bacterium, Shewanella sp. *ISTPL2 Amylase*, 4, 1–10. https://doi.org/10.1515/AMYLASE-2020-0001/PDF.

Sahay, S., Chaurse, V., & Chauhan, D. 2020. Laccase from Aspergillus nidulans TTF6 showing Pb activation for smaller substrates and dyes remediation in all climates. *Proc. Natl. Acad. Sci. India Sec. B Biol. Sci.*, 90, 143–150.

Sahoo, R.K., Das, A., Gaur, M. et al. 2020. Parameter optimization for thermostable lipase production and performance evaluation as prospective detergent additive. *Prep. Biochem. Biotechnol.*, 50, 578–584. https://doi.org/10.1080/10826068.2020.1719513.

Sanchez, A., & Ravanal, M. 2019. Heterologous expression and biochemical characterization of a novel cold-active α-amylase from the Antarctic bacteria Pseudoalteromonas sp. 2-3. *Protein Expression and Purification*, 155.

Santiago, M., Ramírez-Sarmiento, C.A., Zamora, R.A., & Parra, L.P. 2016. Discovery, molecular mechanisms, and industrial applications of cold-active enzymes. *Front. Microbiol.*, 7. https://doi.org/10.3389/FMICB.2016.01408.

Sarmiento, F., Peralta, R., & Blamey, J.M. 2015. Cold and hot extremozymes: Industrial relevance and current trends. *Front. Bioeng. Biotechnol.*, 3, 148.

Shiomi, K., Midorikawa, S., Ishida, M. et al. 2004. Plancitoxins, lethal factors from the crown-of-thorns starfish Acanthaster planci, are deoxyribonucleases II. *Toxicon*, 44, 499–506.

Shuo-shuo C., & Xue-zheng, L. 2011. Effects of co-expression of molecular chaperones on heterologous soluble expression of the cold-active lipase Lip-948. *Protein Expression and Purification*, 77(2), 166–172. https://doi.org/10.1016/j.pep.2011.01.009.

Siddiqui, K.S., Williams, T.J., Wilkins, D. et al. 2013. Psychrophiles. *Annu. Rev. Earth Planet Sci.*, 41, 87–115. https://doi.org/10.1146/annurev-earth-040610-133514.

Souza, M. De, Santos K. Dos, et al. 2017. Production of flavor esters catalyzed by lipase B from Candida Antarctica immobilized on magnetic nanoparticles. *Brazilian J. Chem. Eng.*, 34, 147–159.

Thakur, M., Sharma, N., Rai, A.K., & Singh, S.P. 2021. A novel cold-active type I pullulanase from a hot-spring metagenome for effective debranching and production of resistant starch. *Bioresour. Technol.*, 320. https://doi.org/10.1016/J.BIORTECH.2020.124288.

Thomas, D., Gangwane, A., Sayyed, R.Z., Sapna, Enshasy H.E., & Zaidel, D.N.A. 2020. UV induced mutagenesis elevates the production of laccase in *Enterobacter cloacae*. *Journal of Scientific and Industrial Research*, 79, 442–448. http://nopr.niscpr.res.in/handle/123456789/54695.

Thomas, D., Gangwane, A., Sayyed, R.Z., Jadhav, H.P., & Mehjabeen. 2020. Optimization and scale-up of laccase production by Bacillus sp. isolated from the soap industry waste site. *Environmental Sustainability*, 3(4), 471–479. https://doi.org/10.1007/s42398-020-00135-9.

Thomas, D., Gangwane, A.K., Khan, S., Sayyed, R.Z., Jadhav, H.P., Khan, M. et al. 2023. Optimization of artificial neural network-genetic algorithm simulated laccase production from *Bacillus* sp. BAB-4151. *Biomass Conversion and Biorefinery*. https://doi.org/10.1007/s13399-023-04815-4.

Unciuleac, M.C., Goldgur, Y., & Shuman, S. 2015. Structure and two-metal mechanism of a eukaryal nick-sealing RNA ligase. *Proc. Natl. Acad. Sci. USA*, 112, 13868–13873. https://doi.org/10.1073/PNAS.1516536112/SUPPL_FILE/PNAS.201516536SI.PDF.

Wang, G., Cheng, R., Chen, Q. et al. 2022. mRNA produced by VSW-3 RNAP has high-level translation efficiency with low inflammatory stimulation. *Cell Insight*, 1(4), 100056.

Wang, Y., Hou, Y., Nie, P. et al. 2019. A novel cold-adapted and salt-tolerant RNase R from Antarctic sea-ice bacterium Psychrobacter sp. ANT206. *Molecules*, 24, 2229.

Wang, Y., Wang, Q., & Hou, Y. 2020. A new cold-adapted and salt-tolerant glutathione reductase from Antarctic psychrophilic bacterium Psychrobacter sp. and its resistance to oxidation. *Int. J. Mol. Sci.*, 21, 420.

Williamson, A., & Pedersen, H. 2014. Recombinant expression and purification of an ATP-dependent DNA ligase from Aliivibrio salmonicida. *Protein Expr. Purif.*, 97, 29–36. https://doi.org/10.1016/J.PEP.2014.02.008.

Williamson, A., & Leiros, H.K.S. 2020. Structural insight into DNA joining: From conserved mechanisms to diverse scaffolds. *Nucleic Acids Res.*, 48, 8225–8242. https://doi.org/10.1093/NAR/GKAA307.

Xu, H., Wang, Q., Zhang, Z. et al. 2019. A simplified method to remove fusion tags from a xylanase of Bacillus sp. HBP8 with HRV 3C protease. *Enzyme and Microbial Tech.*, 123.

Xu, H., Xu, D., & Liu, Y. 2024. Molecular biology applications of psychrophilic enzymes: Adaptations, advantages, expression, and prospective. *Appl. Biochem. Biotechnol.*, 1–25.

Xue, Y., & Braslavsky, I. 2021. Temperature effect on polymerase fidelity. *ASBMB*, 297, 20–24.

Yang, W. 2011. Nucleases: Diversity of structure, function, and mechanism. *Q. Rev. Biophys.*, 44, 1–93.

Zang, H., Du, X., Wang, J. et al. 2020. Insight into cold-active xylanase production and xylan degradation pathways in psychrotrophic Acinetobacter sp. HC4 from the cold region of China. *Cellulose*, 27, 7575–7589. https://doi.org/10.1007/S10570-020-03286-4.

Zanphorlin, L., Giuseppe, P.O. de et al. 2016. Oligomerization as a strategy for cold adaptation: Structure and dynamics of the GH1 β-glucosidase from Exiguobacterium antarcticum B7. *Scientific Reports*, 6, Article No. 23776.

Zhao, D.L., Yu, Z.C., Li, P.Y. et al. 2011. Characterization of a cryptic plasmid pSM429 and its application for heterologous expression in psychrophilic Pseudoalteromonas. *Microb. Cell Fact.*, 10, https://doi.org/10.1186/1475-2859-10-30.

Zheng, N., & Shabek, N. 2017. Ubiquitin ligases: Structure, function, and regulation. *Annu. Rev. Biochem.*, 86, 129–157.

Zhu, B., Tabor, S., Raytcheva, D. et al. 2012. The RNA polymerase of marine cyanophage Syn5. *J. Biological Chem.*, 228(5).

Zhu, D., Wu, Q., & Wang, N., 2011. Industrial enzymes. In: *Compr. Biotechnology* (2nd Edn.), 3, 3–13. https://doi.org/10.1016/B978-0-08-088504-9.00182-3.

CHAPTER

13

Psychrophiles and Food Spoilage: Challenges and Control Strategies

Fatima Bibi[1], Noshin Ilyas[2*], Maimona Saeed[1], Sumera Shabir[3], Sohail[3] and Muhammad Izhar Shafi[4]

[1] Department of Botany, Rawalpindi Women University, 6th Road, Satellite Town, Rawalpindi, 46300, Pakistan

[2] Department of Botany, PMAS Arid Agriculture University, Rawalpindi 46300, Pakistan

[3] Department of Agriculture, Forestry and Bioresources, Forest Ecophysiology Lab, Seoul National University, Seoul, South Korea

[4] Department of Soil & Environmental Sciences, Faculty of Crop Production Sciences, University of Agriculture, Peshawar Pakistan

1. Introduction

The term 'psychrophiles' is derived from the Greek words 'psukhros' (cool) and 'philein' (loving). Psychrophiles are microorganisms that thrive in cold conditions, including the deep sea, high heights, and polar regions. They can endure temperatures ranging from 0 to 20°C, with an ideal growth temperature of 5°C (Cavicchioli, 2016; Salwan & Sharma, 2020). Food microbiologists are particularly interested in psychrophile/psychrotrophic bacteria, which can adapt to stressful living environments (Afrouz et al., 2023; Vasut & Robeci, 2009; Baba et al., 2021; Sheikh et al., 2022; Baba et al., 2021), *Vibrio marinus* and *V. psychroerythrus* were the first species of psychrophiles to be identified and characterized (Moyer & Morita, 2007). Microorganisms known as psychrophilic microorganisms thrive in adverse conditions and exhibit excellent growth and reproduction at cold temperatures, this makes them potential sources of cold-active enzymes (Kumari et al., 2021; Hamid et al., 2022). The ability of microorganisms to withstand low temperatures is a major factor in the quick decomposition of some perishable goods, even when they are kept in cold

*Corresponding author: noshinilyas@yahoo.com

DOI: 10.1201/9781003717003-13

storage (Hamid et al., 2024; Yi & Xie, 2021). Like the two sides of a coin, there has been much talk and reporting in recent decades about the promising potential application of cold-adapted enzymes from microorganisms across a variety of industries, including the production of bio-diesel, detergents, food, dairy, and pharmaceuticals (Hamid & Mohiddin, 2018; Furhan, 2020).

2. Psychrophiles in Extreme Environment

In contrast, microbes can thrive at high temperatures. Bacteria thrive at temperatures below 15°C and inhabit almost all of the Earth's surface. They can be found in high-altitude alpine and arctic soils, ocean depths, arctic ice, ice ages, and snow-covered regions (Vasut & Robeci, 2009). Furthermore, these microbes have high enzyme potential and can maintain considerable activity even in cold temperatures, lowering the temperature dependence of reactions (Yayanos, 2009; Cavicchioli et al., 2011; Irwin, 2020; Rai & Kumar, 2015). Psychrophiles use multiple metabolic processes, including photosynthesis, chemoautotrophism, heterotrophism, and potent communities. Psychrophile bacteria have fat cell walls, are chemically resistant to extreme cold, and create 'antifreeze' proteins to safeguard their DNA when water temperatures approach the frost threshold (Siegel, 2001). Psychrophilic yeasts produce extremozymes with high stability and activity in cold temperatures. These psychrozymes have diverse applications due to their optimal performance at low temperatures (Hamid, 2021; Kuddus et al., 2024).

3. Food Spoilage and Its Impact on Consumer Acceptance

Food spoiling is a natural phenomenon that causes unwanted changes in food's organoleptic features such as appearance, shape, and color, as well as flavor, odor, and texture (Figure 1). Spoiled food has a significant impact on consumer acceptance and is unfit for consumption. Perishable foods with a high protein content (meat production, seafood, poultry, milk, and other dairy products) are more vulnerable to food spoiling bacteria invasion, whereas non-perishable foods are frequently spoiled by yeast and mold (Figure 2). Several factors that influence the decay of food include indigenous microbiota, pH, temperature, the activity of water, people who handle food and food transportation (Figure 3) (Khalili et al., 2023; Iulietto et al., 2015; Qian et al., 2023; Odeyemi et al., 2020).

Spoilage microorganisms can proliferate and accumulate organic acids and ammonia during the production of food. These by-products are a factor in the development of organoleptic rejection, which is a sign that the product is past its expiration date (Amit et al., 2017; Boziaris & Parlapani, 2017) prevent adverse responses and microbiological contamination at low temperatures (Kuddus & Ramteke, 2012; Javed & Qazi, 2016).

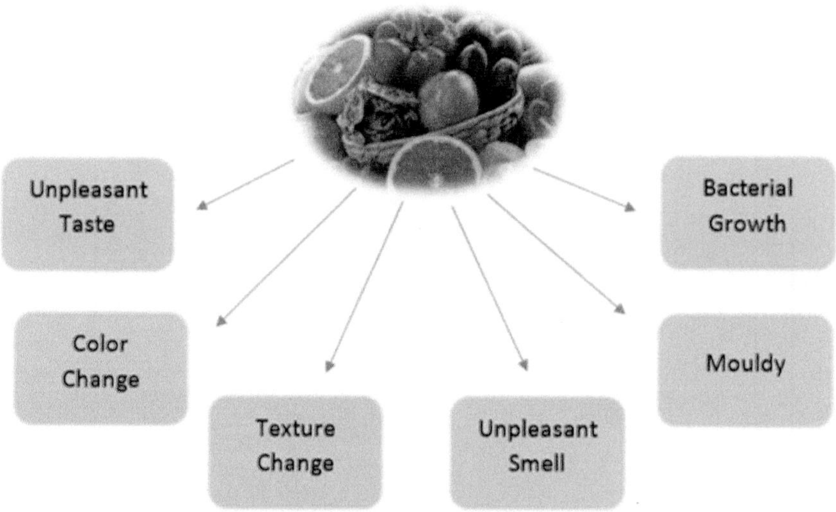

Figure 1: Effect of spoilage on food items.

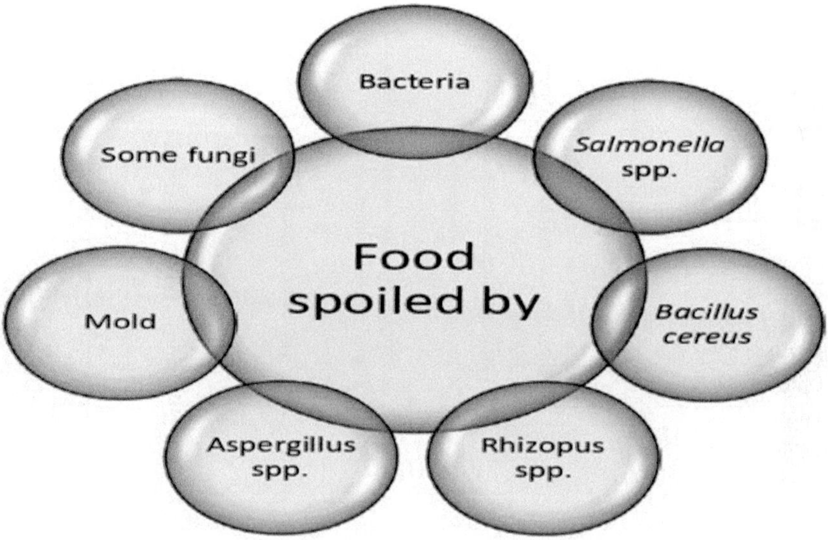

Figure 2: Microorganisms responsible for food spoilage.

Table 1: Enlisting of food items spoiled by various microorganisms

Food Item	Microorganisms	Spoilage Type	References
Bread	*Rhizopus nigricans*; Penicillium; *Bacillus subtilus*	Moldy	Jovanovic et al., 2022; De la Cruz Quiroz et al., 2021
Eggs	*Alcaligenes*; *Proteus*; *Pseudomonas*; *P. fluorescens*	Colorless and green rot	Wu et al., 2024
Milk/cream	*Alcaligenes viscolactis*, *Micrococcus*, *Enterobacter*, *Lactobacillus*, *Streptococcus*, *Bacillus* Yersinia enterocolitica	Ropy	Reguillo-Granados et al., 2018
Poultry	*Alcaligenes*, *Pseudomonas*	Slime, odor	Wu et al., 2024; Powell et al., 2020
Fresh meat	*Alcaligenes*, *Clostridium*, *Proteus vulgaris*, *Pseudomonasfluorescens*	Putrefaction	Andritsos et al., 2021
Cured meats	*Rhizopus*, *Aspergillus*, *Penicillium*	Moldy	Wu et al., 2024
Fish	*Alcaligenes*, *Pseudomonas*, *Flavobacterium*	Discolorations	De la Cruz Quiroz et al., 2021; Powell et al., 2020; Tsironi et al., 2019
Concentrated juices	*Acetobacter*, *Lactobacillus*, *Leuconostoc*	Off flavor	Wu et al., 2024
Fresh fruits and vegetables	*Rhizopus*, *Erwinia*	Soft rot	Garcia Cortes et al., 2022; Getahun et al., 2017

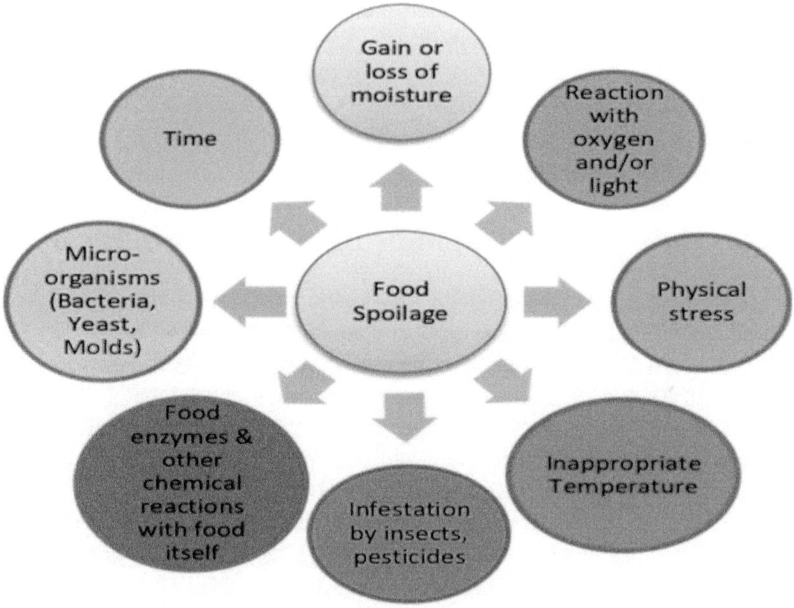

Figure 3: Factors affecting the quality of food.

4. Psychrophilic Enzymes Advantages

Psychrophilic enzymes are ideal for industrial uses due to their cost-effectiveness, ability to catalyze reactions without extra heat, and selective inactivation with light heat. Cold-active enzymes are employed in biotechnology to minimize unwanted reactions and reduce the loss of volatile components (Gerday, 2013; Kuddus & Bano, 2022). Cold-active enzymes are likely to have significant opportunities for expansion in the future due to their widespread applications (Santiago et al., 2016; Al-Ghanayem & Joseph, 2020).

5. Impacts of Psychrophiles on Food Quality and Safety

As the world's population grows, so does the demand for high-quality, nutritious food. However, the advent of food spoilage microorganisms remains a challenge in meeting the requirements of society in the present-day food industry. Food deterioration microorganisms can enter the food supply chain at any stage, from farm to fork, and cause a significant decline in food quality. Food that has been impacted by spoiling may no longer be fit for human consumption. Even if customers take contaminated food by accident, most food spoilage bacteria do not cause severe sickness. Chilling and freezing are often employed to prevent microbial development and affect food quality. However, cooling and freezing are not successful for psychrophilic spoilage bacteria, respectively due to their ability to withstand colder or freezing temperatures and cause rotting in refrigerated food. *Brochothrix thermosphacta*, *Pseudomonas fluorescens*, and *Shewanella putrefaciens* cause spoiling in dairy products, meat, and seafood, correspondingly (Wei et al., 2019).

6. Economic Impact of Microbial Contamination

Food decomposing is regarded as recyclable waste products, resulting in financial losses for producers as well as customers. According to the FAO and WHO, one-third of the food produced annually is wasted (FAO, 2011; Odeyemi et al., 2020). As per the European Commission (EC), approximately 88 tons of food are misused each year in Europe, costing 143 billion euros (EC, 2018). Despite advances in food processing and preservation technology, food rotting continues to be a major universal problem. It is assumed that over 30% of produced food products experience microbial deterioration, and approximately 25% of food worldwide is wasted during or after harvest or post-slaughter operations. However, the true global loss due to food spoiling is impossible to determine precisely (Bourdichon & Ronzeau, 2012; Iulietto et al., 2015; Odeyemi et al., 2020).

Understanding Bacterial Spoilage in Vacuum-Packed and Mapped Meat Preservation

Vacuum packing and MAP procedures are used to preserve meat and extend its shelf-life while inhibiting aerobic bacteria development. Though, the mixing of carbon dioxide and nitrogen encourages a higher occurrence of *B. thermosphacta* (Iulietto et al., 2015). Meat deterioration caused by *B. thermosphacta* can be classified into aerobic and anaerobic types. *B. thermosphacta* can withstand up to 50% carbon dioxide concentration in anaerobic environments, which leads to a shift away from the aerobic bacteria that were once dominant (such as lactic acid bacteria) and toward replacing them. Meat deterioration is considered by meat souring instead of meat putrefaction (Betts, 2006; Gribble & Brightwell, 2013; EFSA, 2016). *B. thermosphacta's* primary metabolism produces ethanol and lactic acid but not acetoin. Furthermore, *B. thermosphacta* spoiling is caused by the generation of volatile fatty acids. However, little is now known about the precise functions that lipolysis and the generation of volatile fatty acid play in meat rotting (Casaburi et al., 2015).

Control Strategies for Microbial Contamination in Food Preservation

Microbial contamination that causes food spoiling cannot be properly handled without suitable cleaning measures, especially during the storage process and processing (Scott, 2003; Doyle & Erickson, 2006). Chilling and freezing are widely used food preservation technologies that prevent microbial infiltration (Qian et al., 2023). However, since these microbes normally need temperatures between 0 and 5°C to grow and propagate, the dependability of psychrotolerant invasion is undermined (Walker, 2003). Psychrotolerant bacteria, also known as specialized spoilage microorganisms (SSO), are hardy enough to withstand low temperatures, which makes them a potential hazard to food quality when stored in cold environments. SSO are the main cause of food quality deterioration since they can outcompete and even eclipse other spoilage bacteria (Walker, 2003; Qian et al., 2023) (Table 2).

Table 2: Basic approaches used in food preservation

Food Preservation Approach	Examples	References
Low temperature	Refrigeration, freezing	Lemos et al., 2023
High temperature	Heat inactivation partially or completely, i.e., pasteurization and canning	Brumfitt et al., 2002; Mehling, 2023
Eradication of Microorganisms	Removal of microbial contamination, centrifugation, physical filtration	Manzur et al., 2023
Preservation through chemicals	Use of inhibitory compounds (i.e., organic acids, sulphur dioxide, nitrates)	Zhang et al., 2024
Reduction of available water	By lyophilization or freeze-drying use of dryer sprays, use of heating drums, and addition of solutes such as salts or sugars	Shen et al., 2024; Mehling, 2023
Use of microbicides	Use of microbial killing products such as bacteriocins to food	Barut Gök and Erdoğdu, 2024
Radiations	Use of non-ionizing radiation (i.e., UV) and ionizing radiation (i.e., gamma rays)	Karamalakova et al., 2024

7. Conclusion and Future Perspective

Psychrophilic and psychrotrophic microorganisms are found throughout the world. They adjust to harsh climatic conditions by changing the ratio of saturated to unsaturated fatty acids in their cell membranes. Certain bacterial species, such as *Pseudomonas, Moraxella*, and *Acinetobacter*, can form partnerships to utilize substrates. Changes in the smell and taste of food can be complex and difficult to identify due to the expansion of microorganisms with diverse actions and consequences. Products with such flaws are often pulled from the market. Ensuring food safety throughout the whole supply chain is crucial. When food-spoiling bacteria are present, the food becomes inappropriate for human consumption. This not only decreases the quality of the food but also dramatically shortens its shelf-life, resulting in food waste. As a result, it is critical to carefully monitor multiple factors such as pH, temperature, water activity, and environmental conditions utilizing a well-established food safety system. By doing so, we can effectively reduce the threat of microbial contamination while ensuring the quality of the food supply cycle and maintaining the safety of the public, reducing food waste.

References

Afrouz, M., Sayyed, R.Z., Fazeli-Nasab, B., Piri, R., Almalki, W.H., & Fitriatin, B.N. 2023. Seed biopriming with beneficial *Trichoderma harzianum* alleviates cold stress in maize. *PeerJ.*, 11, e15644. http://doi.org/10.7717/peerj.15644.

Al-Ghanayem, A.A., & Joseph, B. 2020. Current prospective in using cold-active enzymes as eco-friendly detergent additive. *Appl. Microbiol. Biotechnol.*, 104(7), 2871–2882.

Amit, S.K., Uddin, M.M., Rahman, R., Islam, S.M.R., & Khan, M.S. 2017. A review on mechanisms and commercial aspects of food preservation and processing. *Agric. Food Secur.*, 6, 51.

Andritsos, N.D., Stasinou, V., Tserolas, D., & Giaouris, E. 2021. Temperature distribution and hygienic status of domestic refrigerators in Lemnos island, Greece. *Food Control*, 127(108121), 1–7.

Baba, Z.A., Hamid, B., Sheikh, T.A., Alotaibi, S., Enshasy, H.E., Ansari, M.J. et al. 2021. Psychrotolerant *Mesorhizobium* sp. isolated from temperate and cold desert regions solubilize potassium and produces multiple plant growth promoting metabolites, *Molecules*, 26, 5758. https://www.mdpi.com/1420-3049/26/19/5758/htm.

Barut Gök, S., & Erdoğdu, Y. 2024. Chemical composition and antimicrobial activity of essential oils from six lavenders (Lavandula angustifolia Mill.). *Plant Soil Environ.*, 70, 111–123.

Betts, G. 2006. *Food Spoilage Microorganisms*. Cambridge: Woodhead Publishing.

Bourdichon, F., & Rouzeau, K. 2012. Microbial food spoilage: A major concern for food business operators. *New Food*, 15(3), 54.

Boziaris, I.S., & Parlapani, F.F. 2017. Specific spoilage organisms (SSOs) in fish. In: *The Microbiol. Quality of Food*. Woodhead Publishing, pp. 61–98.

Brumfitt, W., Salton, M.R., & Hamilton-Miller, J.M. 2002. Nisin, alone and combined with peptidoglycan-modulating antibiotics: Activity against methicillin-resistant Staphylococcus aureus and vancomycin-resistant enterococci. *J. Antimicrob. Chemother.*, 50(5), 731–734.

Casaburi, A., Piombino, P., Nychas, G.J., Villani, F. & Ercolini, D. 2015. Bacterial populations and the volatilome associated to meat spoilage. *Food Microbiol.*, 45(A), 83–102.

Cavicchioli, R., Charlton, T., Ertan, H., Omar, S.M., Siddiqui, K.S., & Williams, T. 2011.

Biotechnological uses of enzymes from psychrophiles. *Microb. Biotechnol.*, 4(4), 449–460.

Cavicchioli, R. 2016. On the concept of a psychrophile. *ISME J.*, 10(4), 793–795.

de la Cruz Quiroz, R., Fagotti, F., Welti Chanes, J., & Torres, J.A.. 2021. Food preservation performance of residential refrigerators: Pasteurized milk and ground beef as animal food models. *Food Eng. Rev.*, 13(1), 104–114.

Doyle, M., & Erickson, M. 2006. Emerging microbiological food safety issues related to meat. *Meat Sci.*, 74(1), 98–112.

EFSA Panel on Biological Hazards (BIOHAZ). 2016. Growth of spoilage bacteria during storage and transport of meat. *EFSA J.*, 14(6), e04523.

European Commission (EC). 2018. *Food Waste*. European Union: DG Health and Food Safety.

Food and Agriculture Organization (FAO). 2011. *Global Food Losses and Food Waste: Extent, Causes, and Prevention*. Rome: Food and Agriculture Organization Publication.

Furhan, J. 2020. Adaptation, production, and biotechnological potential of cold-adapted proteases from psychrophiles and psychrotrophs: Recent overview. *J. Genet. Eng. Biotechnol.*, 18, 36.

Garcia Cortes, A., Martinez-Martinez, E., Fernandez Villanueva, G., Torres-Valdez, A.Y., Fagotti, F., Garcia Garcia, R.R. et al. 2022. Absolute preservation indicator for the assessment of refrigeration performance based on dynamic temperature measurements and predictive microbiology. *Food Control.*, 136(108891), 1–9.

Gerday, C. 2013. *Catalysis and Protein Folding in Psychrophiles*. Caister Academic Press: Norfolk, pp. 137–160.

Getahun, S., Ambaw, A., Delele, M., Meyer, C.J. & Opara, U.L. 2017. Analysis of airflow and heat transfer inside fruit packed refrigerated shipping container: Part I – Model development and validation. *J. Food Eng.*, 203, 58–68.

Gribble, A., & Brightwell, G. 2013. Spoilage characteristics of Brochothrix thermosphacta and campestris in chilled vacuum packaged lamb, and their detection and identification by real time PCR. *Meat Sci.*, 94(1), 361–368.

Hamid, B., & Mohiddin, F.A. 2018. Cold-active enzymes in food processing. In: *Enzymes in Food Technology*. Springer, Singapore, pp. 383–400.

Hamid, B. 2021. Potent biotechnological applications of psychrozymes. In: *Microbiomes and the Global Climate Change*. Springer, pp. 349–364.

Hamid, B., Bashir, Z., Yatoo, A.M., Mohiddin, F., Majeed, N., Bansal, M. et al. 2022. Cold-active enzymes and their potential industrial applications: A review. *Molecules*, 27, 5885. https://doi.org/10.3390/molecules27185885.

Hamid, B., Baba, Z.A., Sheikh, T.A., Perveen, K., & Sayyed, R.Z. 2024. Investigating potential protease activity of psychrotrophic bacteria from a municipal landfill for solid waste management. *Biomass Conversion and Biorefinery*. https://doi.org/10.1007/s13399-024-05621-2.

Irwin, J.A. 2020. Overview of extremophiles and their food and medical applications. In: *Physiological and Biotechnological Aspects of Extremophiles*. Academic Press, pp. 65–87.

Iulietto, M.F., Paola, S., Beniamino, E., & Cenci-Goga, T. 2015. Meat spoilage: A critical review of a neglected alteration due to ropy slime producing bacteria. *Ital. J. Anim. Sci.*, 14(3), 4011.

Javed, A., & Qazi, J.I. 2016. Psychrophilic microbial enzymes implications in coming biotechnological processes. *Am. Sci. Res. J. Eng. Technol. Sci.*, 23(1), 103–120.

Jovanovic, J., Djekic, I., Smigic, N., Tomic, N., & Rajkovic, A. 2022. Temperature profile and hygiene in household refrigerators in Belgrade, Serbia and their relation to consumers food safety knowledge and characteristics of the refrigerators. *Food Control*, 136(108813), 1–8.

Karamalakova, Y., Petkova-Parlapanska, K., Hinkov, I., Vetskov, N. Georgieva, E., & Nikolova, G. 2024. The effect of radiation preservation on macromolecule content, antimicrobial activity and radical-scavenging potential of Portulaca oleracea L. leaves extract. *J. Chem. Technol. Metall.*, 59(2), 243–249.

Khalili, L., Bright, J.B., & Sayyed, R.Z. 2023, Dec. 10. Genetically modified, functional, and organic foods and consumers' approach: a critical review. *J. of Consumer Protection and Food Safety*, 2023. https://doi.org/10.1007/s00003-023-01466-5.

Kuddus, M., & Ramteke, P.W. 2012. Recent developments in production and biotechnological applications of cold-active microbial proteases. *Crit. Rev. Microbiol.*, 38(4), 330–338.

Kumari, M., Padhi, S., Sharma, S., Chiring, L., Sudhir, P., Amit, P.S., & Rai, K. 2021. Biotechnological potential of psychrophilic microorganisms as the source of cold-active enzymes in food processing applications. *Biotech.*, 11(11), 1–18.

Kuddus, M., Roohi, N., Bano, N., Sheik, G.B., Joseph, B., Hamid, B. et al. 2024. Cold-active microbial enzymes and their biotechnological applications. *Microb. Biotechnol.*, 17, e14467.

Lemos, Á.T., Martins, A.P., Delgadillo, I., & Saraiva, J.A. 2023. A quasi-energetically costless novel food preservation methodology: Moderate pressure pasteurisation followed by hyperbaric storage, both at room temperature. *Appl. Food Res.*, 3(1), 100251.

Manzur, M., Luciardi, M.C., Blázquez, M.A., Alberto, M.R., Cartagena, E., & Arena, M.E. 2023. Citrus sinensis essential oils: An innovative antioxidant and antipathogenic dual strategy in food preservation against spoilage bacteria. *Antioxidants*, 12(2), 246.

Mehling, H. 2023. Use of phase change materials for food applications: State of the art in 2022. *Appl. Sci.*, 13(5), 3354.

Moyer, C.L., & Morita, R.Y. 2007. Psychrophiles and psychrotrophs. *Encycl. Life Sci.*, 1(6).

Odeyemi, O.A., Alegbeleye, O.O., Strateva, M., & Stratev, D. 2020. Understanding spoilage microbial community and spoilage mechanisms in foods of animal origin. *Compr. Rev. Food Sci. Food Saf.*, 19(2), 311–331.

Powell, C.D., López, S., & France, J. 2020. New insights into modelling bacterial growth with reference to the fish pathogen Flavobacterium psychrophilum. *Animals*, 10(3), 435.

Qian, Y.F., Yu, J.Y., Xie, J., & Yang, S.P. 2023. A mini-review on cold-adapted enzymes from psychrotrophic microorganisms in foods: Benefits and challenges. *Curr. Res. Biotechnol.*, 100162.

Rai, A.K., & Kumar, R. 2015. Potential of microbial bio-resources of Sikkim Himalayan region. *Envis Bull. Himalayan Ecol.*, 23, 99–105.

Reguillo-Granados, L., Pérez, F., & Valero, A. 2018. Food quality management systems in the dairy industry: A case study on the application of predictive microbiology in the microbial quality of milk. In: *Technological Approaches for Novel Applications in Dairy Processing*. Intech Open, London, UK.

Salwan, R., & Sharma, V. 2020. Physiology of extremophiles. In: *Physiological and Biotechnological Aspects of Extremophiles*. Academic Press, pp. 13–22.

Santiago, M., Ramírez-Sarmiento, C.A., Zamora, R.A., & Parra, L.P. 2016. Discovery, molecular mechanisms, and industrial applications of cold-active enzymes. *Front. Microbiol.*, 7, 1408.

Scott, E. 2003. Food safety and foodborne disease in 21st century homes. *Can. J. Infect. Dis.*, 14(5), 277–280.

Sheikh, T., Baba, Z., Hamid, B., Iqbal, S., Yatoo, A., Fatima, S. et al. 2022. Extracellular polymeric substances in psychrophilic cyanobacteria: A potential bioflocculant and carbon sink to mitigate cold stress. *Biocatalysis and Agricultural Biotechnology*, 42, 102375. https://doi.org/10.1016/j.bcab.2022.102375.

Shen, B., Yan, Z., Yang, T., Zhu, L., Wang, Y., & Jiang, L. 2024. Waste citrus pectin/garlic bionanohybrids for edible food preservation. *J. Food Eng.*, 364, 111800.

Siegel, L.J. 2001. Life on ice. *Electron. Astrobiol. Newsl.*, 8, 30.

Tsironi, T., Lougovois, V., Simou, V.N., Mexi, A., Koussissis, S., Tsakali, E. et al. 2019. Next generation sequencing (NGS) for the determination of fish fresh microbiota. *J. Food Res.*, 8(4), 101–110.

Vasut, R.G., & Robeci, M.D. 2009. Food contamination with psychrophilic bacteria. *Lucr St Med Vet Timisoara*, 13(2), 325–330.

Walker, S.J. 2003. *Chilled Storage*. Amsterdam: Elsevier Science Ltd.

Wei, Q., Wang, X., Sun, D.W., & Pu, H. 2019. Rapid detection and control of psychrotrophic microorganisms in cold storage foods: A review. *Trends Food Sci. Technol.*, 86, 453–464.

Wu, S., Zhang, M., Yu, Q., Mujumdar, A.S., & Yang, C. 2024. Fresh food quality deterioration detection and labeling: A review of recent research and application in supply chain. *Food and Bioprocess Technology*, 17(7), 1706–1726.

Yayanos, A.A. 2009. *High-pressure Habitats*. Wiley Online Library, pp. 228–239.

Yi, Z., & Xie, J. 2021. Comparative proteomics reveals the spoilage-related factors of Shewanella putrefaciens under refrigerated condition. *Front. Microbiol.*, 12, 740482.

Zhang, Q., Kong, B., Liu, H., Du, X., Sun, F., & Xia, X. 2024. Nanoscale Pickering emulsion food preservative films/coatings: Compositions, preparations, influencing factors, and applications. *Compr. Rev. Food Sci. Food Saf.*, 23(1), e13279.

CHAPTER

14

Microbial Interactions and Symbioses in Cold Environments: Ecological and Applied Perspectives

Younes Rezaee Danesh[1*], Mina Najafi[2], Marika Pellegrini[3] and Nayereh Akbari[4]

[1] Department of Plant Protection, Faculty of Agriculture, Van Yuzuncu Yil University, Van – Türkiye

[2] Department of Soil Science and Plant Nutrition, Faculty of Agriculture, Ege University, Izmir – Türkiye

[3] Department of Life, Health and Environmental Sciences, University of L'Aquila, 67100 L'Aquila, Italy

[4] Department of Plant Protection, Faculty of Agriculture, Urmia University, Urmia – Iran

1. Introduction

Microorganisms, abbreviated as microbes, are microscopic organisms that belong to various species, including bacteria, archaea, protozoa, microalgae, fungi, and viruses. Within the microbial population, many interactions occur, mostly co-evolutionary and age-old, and some more recent and specific ones. Understanding microbial interactions is vital for microbial ecologists and researchers applying microbial taxa in biotechnology because interactions significantly affect cooperating microbes' activities, development, and evolution. Considerable research has focused on microbes' taxonomy, activity, and physiology for bioprospecting and biocontrol; however, only a few studies have addressed microbial interactions. Ongoing research aims to elucidate the network of interactions among microbes, leading to a comprehensive understanding of their bioprospecting potential (Weiland-Bräuer, 2021). Microbial interactions can occur through a variety of microbially-derived mechanisms, including changes in intracellular signaling; the exchange of small, chemically-modifiable metabolite signaling biomolecules; direct signaling by cell-to-cell contacts or

*Corresponding author: y.rdanesh@yahoo.com

DOI: 10.1201/9781003717003-14

capillary action; and shared production of insoluble, continuously secreted, polymeric biopolymers (Nawaz et al., 2022). Interactions can either be two-bodied or many-bodied and are usually not static but dynamic and subject to change during the co-evolution of microbes. Those interactions may be specific and promote mutualistic behavior or degrade species-to-species recognition, triggering a competitive or parasitic situation. A growing body of evidence indicates that microbial interactions were established co-evolutionarily. Microbial interactions usually lead to strategies stably maintained over long periods of physicochemical environments and evolutionary histories. However, currently known interactions mainly concern model organisms that can easily be cultivated in laboratories. A broader understanding is expected from evolving freshwater and marine microbial communities, which may have interesting and unknown interactions due to their evolutionary history and biodiversity.

2. Cold Environments: An Overview

Cold environments represent diverse habitats characterized by a low temperature, a high availability of nutrients such as phosphorus and iron, a low availability of light and salinity, and an extended competition for nutrients. A polar ecosystem is usually organized by three main compartments: water column, sea ice, and sediments, which host different microbial communities (Lauritano et al., 2020). Polar oceans are characterized by a strong stratification, leading to extreme light availability, salinity, and temperature conditions. Sea ice represents a cold, ice-covered, carbon-limited ecosystem surrounding the Arctic and Antarctic continents, where microalgae bloom occurs during spring-summer. However, the metabolic adaptations of polar prokaryotes to maintain bioactivity in this extreme ecosystem and their role in 'first-pioneer' symbiosis are still poorly understood. The other antifreeze mechanisms studied in cold-adapted bacteria, microorganisms assumingly enriched during priority effects in polar environments, consisting of the production and the exopolymer alginate. Microbial communities on the sea-ice water column and sediment samples from the Kongsfjorden fjord were revealed in 454 pyrosequencing sequence reads, and descriptions of novel taxa in the bacteria, archaea, and diatom kingdoms. The field sampling approach revealed the potential for accurately observing various situations and conditions. Cold-adapted ecosystems generally had short food chains and low impact on organic matter decomposition due to strong temperature changes and prey character selection driven by cell size (Gutt et al., 2021).

3. Types of Microbial Interactions

Microbes are species of microscopic scale (~1–100 μm) with diverse morphology and physiology (Weiland-Bräuer, 2021). They inhabit all environments—land, sea, ice, volcanoes, hot springs, oil, and animal or human guts—and colonize tissues of macroscopic hosts such as plants, animals, insects, and fungi. Microorganisms form single or mixed-species communities, displaying emergent properties where interactions occur at intra- or interspecific levels, including bacteria–bacteria, bacteria–fungus, bacteria–virus, and bacteria–host. Understanding these interactions is essential to utilize their benefits. Outcomes range from mutualism (win–win), commensalism (neutral for one partner), to parasitism (benefiting one while harming the other). Pathogens are well studied, including *Bacillus anthracis*, *Borrelia* spp., *Campylobacter jejuni*, *Haemophilus influenzae*, *Enterovirus B*, and *Ebolavirus* (Nawaz et al., 2022) (Figure 1).

Figure 1: Types of microbial interactions in cold environments. Mutualism: Both organisms benefit; e.g., algae and cold-tolerant fungi forming cryo-lichens. Commensalism: One organism benefits without affecting the other; e.g., bacteria residing in fungal biofilms. Parasitism: One organism benefits at the expense of the other; e.g., bacteriophages targeting cold-adapted bacterial hosts.

Mutualism

Although mutualism is considered common, direct bacterial–bacterial cases are rarely reported. Abiotic stress is suggested as a promoter, as shown in composting systems where temperature shifts favored positive interactions. For instance, *Thermobifida fusca* and *Saccharomonospora viridis* shared cobalamin, enhancing growth in co-culture (Kazamia et al., 2012). Sampling across six time points revealed higher diversity and activity at elevated temperatures, with networks showing competitive interactions at low temperatures and mutualistic interactions at high temperatures. Laboratory studies confirmed that thermal dynamics facilitated thermophilic mutualism. Such findings suggest bacterial interactions drive compositional and functional stability in communities. Cross-feeding is a typical mutualistic mechanism. However, many microbes harm others by competing for nutrients or releasing toxins. Few parasitoid bacterial relationships are known, and even honey bee gut microbiota lacked predatory bacteria. Experiments on culturable isolates from tree holes and guts showed positive pairings were rare, leading to debate over whether bacterial mutualism is as common as in multicellular organisms (Zhao et al., 2023).

Commensalism

Commensalism benefits one partner without harming the other. On human skin, commensals colonize epidermal flakes, keratinous cells, and non-infectious strains, exploiting tunnels and ducts where pathogens cannot thrive. This creates a protective veil, preventing pathogen colonization (Ranjan Ghosh, 2013). Commensals also support immune defense through surface colonization. In the gut, commensals such as *E. coli* can switch to pathogens under stress, while reservoirs of plasmids spread pathogenicity. Encounters with heterologous strains may disrupt tolerance and trigger disease. Studies of chemotaxis show that microbial communities sense light, aroma, quorum signals, surfactants, gravity, metals, and redox states (Zhao et al., 2023). In ectotherm guts, *Chitinophagaceae* dominates, though molecular mechanisms remain unknown. Comparative omics revealed evolution of polysaccharide deconstruction, adhesion, and chemotaxis. Big-data-driven approaches now investigate underground and network-like microbial communities, fostering resilience against gut diseases.

Paratism

Parasitism occurs when one life form exploits another as its ecological niche (Araújo et al., 2003). Niches may range from specific organs (micro-parasitism) to entire hosts (macro-parasitism). Parasitism encompasses a wide fraction of biodiversity, including configurations where the host provides food, shelter, or defense. Parasites and symbionts represent extremes of interspecific interactions; the same species may shift roles depending on environmental or genetic constraints. Symbiosis relies on stable host–symbiont associations, whereas parasitism often arises from opportunism. Definitions vary along a gradient of interdependence, where commensalism, mutualism, and parasitism overlap. The host–parasite–environment system is dynamic, shaped by genetic, biotic, and abiotic factors, with multiple equilibrium states. This complexity makes it difficult to set thresholds between benefit and harm, so parasitism may encompass other forms of symbiosis (Overstreet & Lotz, 2016).

4. Microbial Diversity in Cold Habitats

Cold environments host microbial communities that are able to withstand stresses such as low temperature, desiccation, and high UV radiation, offering significant biotechnological potential (Table 1). Bioprospection in Polar Regions faces logistical and technical challenges, leaving them less explored than temperate regions. These environments serve as evolutionary laboratories

Table 1: Dominant microbial taxa in cold ecosystems

Microbial Group	Habitat	Functional Traits
Bacteria	Cryo-soils, glaciers	Cold-active metabolism, antifreeze proteins
Archaea	Permafrost, saline ice	Methanogenesis, salt tolerance
Fungi	Ice cores, mosses	Pigment production, desiccation resistance
Protists	Meltwater pools	Photosynthesis, motility at low temperature

where low temperatures drive unique adaptations. Novel taxa have already been reported, and future exploration is expected to reveal more (Gallo & Aulitto, 2024). Snow and ice store most of Earth's freshwater, and though frozen, brief thawing enables microbial activity. Biogeochemical cycles continue, albeit slower, and recent studies reveal high microbial abundance and diversity. Eukaryotic algae and fungi thrive in snowpacks, shifting from inactive to active states with warming, while bacteria remain metabolically active in ice. Despite global studies of bacterial biogeography, frozen environments, particularly the Polar Regions, remain poorly understood (Larose et al., 2013).

Bacterial Diversity

In glaciers and cryospheres, microbial adaptation has long been studied, linking to changes in albedo, water balance, permafrost thaw, and nutrient release (Larose et al., 2013). Warming may increase bioavailable nitrogen and microbial activity, with global biogeochemical implications, though debated. Periglacial zones, more reactive to climate change than ice sheets, remain understudied due to difficulties identifying microbes in frozen soils and limited sequencing resolution. Advances in sequencing, phenotyping, and igloo-style laboratories now uncover permafrost diversity and interactions, shifting geomicrobiology towards earth system science. Snow-pit analyses provide valuable data, though isolating older snow layers is difficult. Synthetic snowpacks modeled in laboratories allow experimental validation of natural processes (Bischoff et al., 2013).

Fungal Diversity

Cold-adapted fungi grow near or below 0°C, classified as stenothermal psychrophiles (~10°C) or eurythermal psychrophiles (20–40°C). They evolved biochemical adaptations such as cold-active enzymes to sustain metabolism at freezing points. Facultative cold-adapted microbes are abundant in such environments (Jodłowska & Białkowska, 2024). Cold ecosystems host highly diverse fungi, with active communities representing >10% of biodiversity. Besides decomposition roles, fungal taxa include epiphytic yeasts, planktonic yeasts, and symbionts with algae. Some psychrophiles are plant and animal pathogens; mycovirus-infected strains reduce virulence in agriculture or insects. Cold fungi span Ascomycota, Basidiomycota, and Zygomycetes, colonizing various substrates (Wang et al., 2017). Paleomycological studies show fungi in Antarctica since the Permian, with soils dominated by ascomycetes and yeasts. Over 400 genera and 1,000 species occur in Antarctica, and more than 4,350 species are reported in the Arctic.

Archael Diversity

Archaea from Doñana National Park revealed dominance of Euryarchaeota and Thaumarchaeota (82.9%), shaping community variability. In Argentina's Bahía Blanca Estuary, Thaumarchaeota (41%), Crenarchaeota (26.4%), and Euryarchaeota (19.5%) predominated. In hypersaline alkaline lakes of Patagonia, salinity chiefly determined archaeal communities dominated by Euryarchaeota and Halobacterota (89.9%). This first characterization of these lakes links halophilic archaea with the evolutionary history of halophilicity (Jahnke et al., 2008; Wang et al., 2020). Microbial group distribution in cold ecosystems is shown in Figure 2.

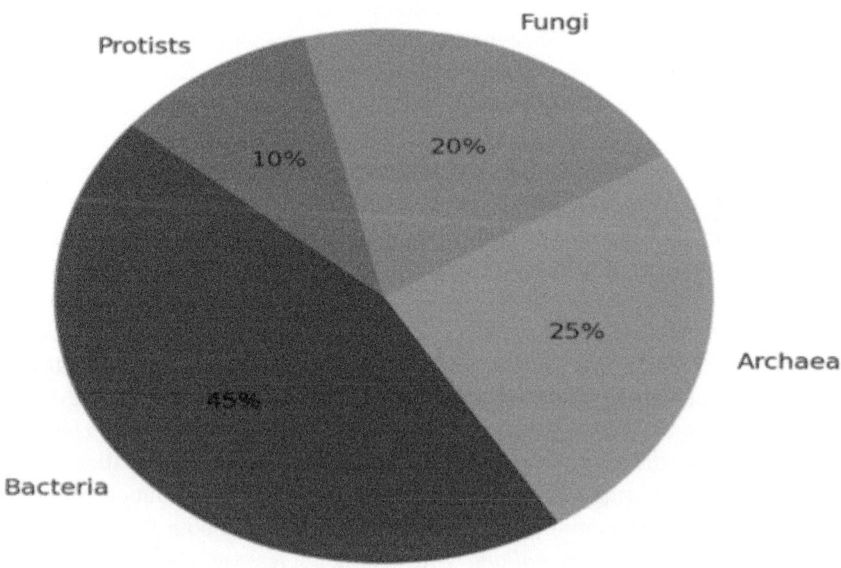

Figure 2: Microbial diversity in cold ecosystems.

Symbiotic Relationships in Cold Environments

Symbioses in cold environments include mutualism, commensalism, and parasitism. At least one partner is often prokaryotic, but most research emphasizes eukaryotic associations, especially plant–microbial symbioses. Invertebrate–

microbe interactions in aquatic cold systems remain underexplored, though marine invertebrates like *Riftia* tubeworms with sulfur-oxidizing bacteria and hydrothermal mussels with methanotrophic symbionts are well studied. Sponges (Porifera) host diverse bacterial partners (Overstreet & Lotz, 2016). Exploring marine invertebrate associations may uncover new applications. Fossil evidence shows ancient sponge–bacteria symbioses, while modern cases such as Eidothea with Gammaproteobacteria demonstrate ongoing colonization. Filamentous bacteria in fossil sponges (Chondrosia) influenced reef facies formation, suggesting long-term biotechnological potential (Lo Giudice & Rizzo, 2022).

Plant-Microbe Interactions

Sustainable food production is a global challenge requiring approaches that utilize a combination of biotic and environmental resources to increase plant productivity while promoting ecosystem health. Applying microbial inoculant to the seed, soil, or plant has received considerable attention as a cost-effective and environmentally benign means of improving plant productivity, with beneficial species increasingly discovered among the cold clime microbiota (Eswaran et al., 2024; Basu et al., 2021; Ferioun et al., 2025; Chaithra et al., 2025; Bright et al., 2025). The plant-microbiome partnership is responsive to stress, with the details of the signaling between Eubacteria, Fungi, and Planta only beginning to be investigated (Al-Turki et al., 2023; Rai et al., 2023; Juurakko et al., 2021). Sub-zero temperatures challenge plants due to their lack of internal heating. As with desert plants, this arid environment causes cellular dehydration, increased osmolality, and increased vulnerability to pathogens such as snow mold. However, some cold-acclimating perennials respond to non-freezing temperatures (4°C) in cold acclimation. This developmental change is a signaling response to temperature variation, not the result of damage from sub-zero temperatures (Hamid et al., 2021). During this signaling response, the plant alters the expression levels of diverse genes, guiding changes in metabolism to reach cold-hardened states, such as the production of antifreeze proteins, phosphorylated sugars, and plant sugars to mitigate freeze events. Cold acclimation is a complex stress response that extends well beyond changes in metabolic biochemistry to structural alterations, defense modifications, and community restructuring of associated agents, such as pathogenic fungi. Changes in the leaf microbiome community profile were observed in both *Brachypodium distachyon* and *Arabidopsis thaliana*, with similar but less substantial community shifts observed in the rhizosphere. While *Brachypodium* plants were moved to low-temperature conditions immediately after germination, changes in the rhizosphere community after the shift to low temperatures were less drastic, with overlapping PCoA groupings. The *Brachypodium* rhizosphere communities contained many taxa reported inbound soils of a similarly arid region associated with this host and those similar to taxa reported in association with wheat (Xin & Browse, 2000).

Animal-Microbe Interactions

Many animal taxa are home to complex microbial communities that benefit their hosts through detoxification, nutrition, and immunity. Examples of such animal-microbe interactions span phylogeny, illustrated by aquatic sponges housing diverse bacterial communities. These interactions generally begin soon after the habitats' settlement but require a complex contact, defense, and tolerance pattern. Investigating the structure, diversity, and composition of microbiomes and their initiation, development, and structure within the goby-fungus mutualism reveals the importance of microbial dispersal, competition, and chemical and physical interactions between partners in establishing good and stable relationships. The framework extends previous studies of the dynamics of the onset, development, and stability of different types of animal-microbe interactions, including manipulative interactions that may lead to some of the most devastating diseases (Braga et al., 2016). Microbiota is essential for host survival, metabolism, and health and protection of many multicellular organisms. However, ensuring stable microbiota colonization in birth and early life is one of the most important factors for developing a healthy organism. Aquatic laboratories utilizing aquaria usually mimic the natural habitats with low microbial diversity. Moreover, sustainable and manageable alternatives to the traditional laboratory microhabitats, generally colonies of the same animal species, can be readily employed in animal-microbe interaction studies. An animal in this environment initially lacks the microbiota and acquires it from poly-specific sources. Two paradigms governing initial aquaria colonization were discovered: priority effects, in which an early colonizing bacteria species prevents later colonization, and assemblage-induced ecological selection, whereby introduced bacteria change environmental conditions and preclude other taxa colonization. Resolving these mechanisms on the species level and understanding the environment's role in establishing healthy holobionts provides a blueprint for animal-microbe interaction studies across several environments (McFall-Ngai et al., 2013).

Ecological Roles of Microbial Symbionts

Recent studies have demonstrated that the genes and enzymes of cold-adapted microbes and their enzymes are of significant interest. The mechanisms underlying the specific ecological roles of symbiotic microbes in various habitats remain unknown. Anthropogenic activities have increased the introduction of alien species, which need to be investigated regarding their ecological effects. This review aims to promote further research into the ecological roles of microbial symbionts living in cold environments. The ecological interactions of microbes have recently received intensive attention because of their important roles in regulating the biodiversity, stability, and functioning of microbial communities. These microbes can interact with other microorganisms, plants, and animals. Such ecological interactions can produce positive or negative effects on microbes and can be classified into mutualistic, parasitic, predatory, commensal, neutral, or competitive relationships (Nawaz et al., 2022). To date, most studies on microbial interactions have been limited to microbial terrestrial or marine environments. Symbiotic microbes contribute considerably to their hosts' function, diversity, stability, and resilience. In return, their hosts can provide a variety of metabolites, nutrients, substrates, and niches to their symbionts. Microbial symbiotic relations can be classified into mutualism, commensalism, parasitism, and others based on their advantages to the symbionts and hosts. Microcrustacean hosts provide symbiotic ecto- and endo-symbiotic bacteria with nutrients, habitats, and protection. The ectosymbionts of its host species, which are adapted to grow at 4°C and can use and degrade polysaccharides and polysulfur compounds, have been discovered. Microbes represent a huge and unexplored resource of low-temperature-adapted enzymes, biopolymers, drugs, and other high-value bioproducts, which are of plentiful interest in academia and industries. Cold habitats have been strictly protected as conservation areas to reduce the risks of pollution and other potential ecological effects arising from the transportation of cold biotic resources. Given the serious consequences of ichthyological invasion and the emerging significance of microbial biodiversity, a research focus on the ecological roles of symbiotic microorganisms in cold environments is becoming increasingly essential and relevant.

Nutrient Cycling

Microbial interactions are a key aspect of ecosystem functioning, and nutrient availability is an important determinant for different types of microbial interactions. This last parameter, by affecting the metabolic profile of the interactants, can promote or diminish different types of interactions among microorganisms in diverse environments. However, the variability of microbial interactions under different nutrient regimes and the potential impacts on microbiome composition and functioning are poorly understood. Cross-kingdom interactions among fungi and bacteria from a phosphorous oligotrophic aquatic system were explored to study these patterns. Using a nutrient-dependent approach, potential interactions were characterized by a network analysis of fungal-bacterial interactions in low and high-nutrient media. To understand the nutrient-driven shifts in types of interactions, metabolome profiling of specific taxa was performed to evaluate the organic compounds produced and consumed during the interactions. The nutrient-dependent patterns in this fungal-bacterial system were estimated using a metabolic network approach. Under oligotrophic conditions, beneficial interactions dominated, while under enriched-nutrient conditions, antagonistic interactions prevailed. Beneficial interactions were promoted by nutrient-dependent shifts in the metabolic profile of the interactants, leading to nutrient exchange from bacteria to fungi. Most taxa exhibited the ability to benefit from the interactions, with bacteria exhibiting a more diverse array of benefits. Moreover, microbial interactions shifted from competition to cooperation as environmental stress increased. The results agree with the stress-gradient hypothesis, as predation favored competition and maintained enhanced decomposition rates under predation. These findings represent specific case studies of interactions that promote growth under oligotrophic conditions, an adaptive trait in extreme environments (Qiao et al., 2024).

Soil Formation

Soil formation is the process that transforms a thin layer of material on the Earth's surface (the regolith) into soil. The regolith is either newly weathered or formed from the alteration of pre-existing material and, subsequently, deformed or sediment-loaded material deposited on earlier weathered substrates. The conversion of regolith to soil is enhanced by the progressive incorporation of organic matter (OM) and biological activity in the soil (Kalam et al., 2020). Well-developed soils have intricate horizons or layers that differ in physical, chemical, and mineralogical attributes due to the abovementioned processes. Soils are diverse and influenced by both climate and biota. Cold regions like Antarctica, mountainous, and polar regions comprise one-fifth of the land surface. Cold soils, often called desert soils, have limited development, sparse vegetation, and a low abundance of soil mesobenthos. Lack of bioturbation means that they are heterogeneous and spatially chaotic (Graham et al., 2010). Polar soils and montane soils, on the other hand, are less spatially variable but have low organic carbon stocks and nitrogen availability. They generally possess higher nutrient contents than the polar 'desert' soils and greater clay contents and cation exchange capacity than montane soils. In the last few decades, attention has turned to more focused investigations into how these environmental factors can affect terrestrial community structure, emphasizing testing predictions from classical ecological theory. The fledgling task of understanding microorganisms' spatial and temporal co-occurrence patterns in extreme environments has produced valuable and entirely novel insights into microbial dispersion and assembly processes in cold deserts. Beyond autoecological considerations, where the physiological capacities of organismal types dictate their distributions, the emphasis is strongly on understanding ecological interactions at the community level (Malard et al., 2023). Diverse soil-forming processes contribute to pedogenesis and soil functioning. The mechanistic understanding of these processes in polar and montane soils has advanced remarkably over the past few decades. The congruence of chemical weathering theory is secured through modeling and is supported by advances in quantifying chemical weathering. Although knowledge of the underlying geomorphological controls (e.g. organic matter accumulation) has improved, much remains to be elucidated. Studying soil formation, functional processes, and biodiversity in extreme environments could yield significant advances in soil science (Carrero et al., 2023).

Ecosystem Stability

The effects of temperature and nutrient enrichments can cascade through ecological and evolutionary processes to hinder ecosystem stability, and impacts include shifts in ecosystem functioning, alterations in species composition and diversity, and extinctions. Microbial communities are assumed to offer ecosystem functioning advantages because of their diversity and redundancy. However, their extremely high diversity raises questions about how they maintain it and possibly compete at an ecosystem scale. Resistance to biotic homogenization via income and dispersal to preserve nonlocal microbes has been reported in contrived communities of cultured bacteria. On a broader scale, residents may resist immigrants due to the biotic filtering of existing microbes and cross-feeding. However, interactions among existing species that diminish resource availability or interfere with the metabolism of immigrants have been investigated empirically (Song et al., 2019). Resistance and resilience are broader concepts and include all stabilizing mechanisms, including abiotic ones. Stability is composed of multiple components. Resistance limits the severity of change under the impact of disturbance, whereas resilience is the rate at which the system recovers after it occurs. There are numerous possible contributions to resistance and resilience. However, the traditional approaches to ecosystem stability ignore the potentially strong effect of temperature, an environmentally and phytosanitary relevant disturbance driver. Microbes comprise the bulk of Earth's biodiversity, realizing important ecosystem services in soil bioremediation and human health. Soil microbial communities undertake a wealth of essential ecosystem functions, but this wealth is hard to understand regarding species richness. Ecological models can be used to explore the effects of temperature and nutrient enrichment on ecosystems, and they usually include all the mechanisms that affect resistance and the effects of a disturbance. Microbial assemblages can regulate temperature sensitivity via adaptations of resource use and strategies to deal with temperature and/or nutrient perturbations without even major compositional shifts, affecting ecosystems' temperature and nitrogen sensitivity (Mori, 2016).

5. Adaptations of Microbes to Cold Stress

Microorganisms inhabiting cold terrestrial, freshwater, and marine environments, defined as having constant temperatures below 5°C (psychrophilic) or 0°C (bacterioplankton), must face several stressor conditions, primarily related to low/freezing temperatures (Baeza et al., 2022). Because of their low temperature, these environments are usually devoid of macromolecules. Thus, microorganisms assimilating carbon from organic molecules are subjected to low diffusion and metabolic rates, anoxia, and cryopreservation. Lows of the temperature are typically concomitant with increases in osmolality, influencing cell turgidity and cytoplasm viscosity and modifying the solubility of free gases and metals. Conservative genes and more specific ones were not the only mechanisms acting on growth at 4°C. Microorganisms living in cold environments have evolved to function efficiently at low temperatures by molecular (gene complement) and metabolic adaptations (induction or repression of metabolic pathways, metabolic literacy) (Raymond-Bouchard et al., 2018). Cryoprotection is crucial for the effectiveness of life at subzero temperatures, and common freezing strategies include membrane remodeling, synthesis of cryoprotectant molecules, cold-active enzymes, and antifreeze proteins (Hamid et al., 2022, 2024) (Figure 3). The freezing point of solutions mostly depends on the osmolality, which usually increases in polar areas, even exceeding 2 M in specific cases. Mechanisms to counteract it include membrane remodeling, synthesis of protective/cryoprotective compounds, and lyophilization. Microbial adaptation to low temperatures is a complex topic. After the studies addressing the adaptation of visible eukaryote organisms/phenotypes to cold and cryogenic preservation, several pieces of evidence supporting the molecular adaptation of microorganisms to low temperatures have accumulated. These examples include comparative genomics/proteomics/transcriptomics and transcriptomic studies developed with new-generation sequencing techniques. Low-temperature stress (LTS) and its reciprocal stress appear important even to terrestrial biomes, with variability depending on their latitude and altitude. Most studies addressing the molecular adaptation of microorganisms to low temperatures have used mesophilic models. As a consequence, LTS investigations resulted in a variety of criteria, including negative ones (Jahed et al., 2023). The key molecular mechanisms that allow microorganisms to survive and function under freezing conditions are summarized in Table 2.

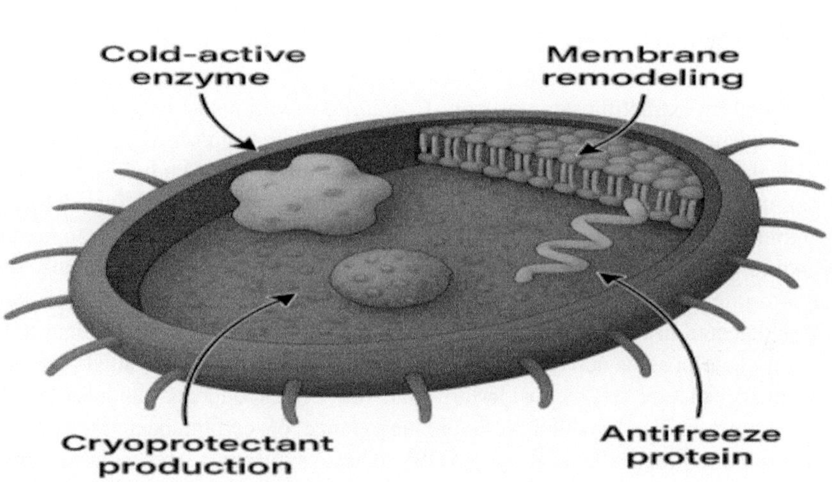

Figure 3: 3D illustration of cold-adaptation mechanisms in a microbial cell: This cross-sectional 3D illustration of a microbial cell highlights cold-adaptive components such as cold-active enzymes, cryoprotectant production, antifreeze proteins, and membrane remodeling.

Table 2: Cold adaptation mechanisms and their biological functions

Adaptation Strategy	Molecular Feature	Biological Function
Cold-active enzymes	Increased surface hydrophilicity	Metabolic activity at low temperature
Membrane remodeling	Unsaturated fatty acids	Maintains membrane fluidity
Cryoprotectants	Trehalose, glycine betaine	Prevents ice formation, stabilizes proteins
Antifreeze proteins	Ice-binding domains	Inhibits ice recrystallization
Stress gene activation	Cold-shock proteins, chaperones	Stabilizes RNA, prevents misfolding

Physiological Adaptations

Cold environments such as polar regions and high altitudes are not as inhospitable as they seem. Indeed, several organisms thrive in these areas: free-living microorganisms and sociable macroorganisms harbor various bacteria in their symbiotic environments to facilitate their survival in these harsh ecosystems. Since polar regions experience highly seasonal temperature and nutrient input changes, bacterial symbionts in these habitats are expected to be adapted to various ecological conditions (Raymond-Bouchard et al., 2018). This chapter summarizes the current knowledge regarding cold-adapted microbial symbiotic partners with a focus on the Antarctic biota: the bacterial partners of Antarctic brown algae and sponges and the cold-adapted symbionts of Antarctic and Arctic invertebrates. Bacterial symbionts are embedded in organic polymer matrices of silicified tissue in the basal holdfast region of the brown alga. Possible source microorganisms of the Antarctic *L. digitata*'s bacterial community were examined by extracting samples from *L. digitata* of various geographical origins and sequencing their amplicons. Despite the significant presence of bacteria in the Icelandic and Norwegian algal samples, those of M. Skudra's free-living community, no matches for P. antifraud were observed, implying obligate symbiotic lifestyles of the bacteria within these parameters. Various environmental conditions were examined for the algal models by cloning axenic cells. Only the growth rates of the axenic cultures under low light conditions were slower than those detected in wild-type algal cells. This study illustrates that some temperate brown algae possess active bacterial endosymbionts that degrade algal-released secondary metabolites, considering that the correspondence addressed the temperature-induced reduction in the abundance of certain bacterial taxa in algal communities (Ihua et al., 2020). These symbionts are isolated from temperate waters, presumably to temporally prevent the eruption of greenhouse gas-releasing sulfide by healing bacterial symbiont eviction. Ecosystems were heavily influenced as the world entered the Cryogenian, with major changes resulting from the accompanying climatic cooling. Microbial mats dominated by psychrophiles evolved due to the pervasive cooling in the oceans. Such microbial ecosystems were suggested to have enhanced the proliferation of multicellular parazoans: sponges. Sponges disrupt and reintegrate their metagenomic contents by external stimuli as evolvable ecosystems as multidimensional entities. During this ecological process, sponge-associated psychrophilic bacteria are expected to have evolved cold-adapted phenotypes to briefly cope with short-term temperature changes in polar oceans and high-altitude environments worldwide (Cavicchioli et al., 2019).

Genetic Adaptations

Cold-adapted psychrophilic organisms have successfully colonized various cold ecosystems on Earth, including polar regions, remote high-altitude locales, hydrocarbon seeps from the seafloor, sea-ice, subglacial lakes, and the continental deep biosphere. They have been shown to play crucial roles in biogeochemical cycling in these ecosystems, as no life can thrive without microbes. This review presents a largely overlooked perspective of these unique microbes by showcasing their often subtle yet pivotal microbial interactions and symbioses along the ecological continuum of cold to ultra-cold environments. Examples are presented on how such interactions may be truly symbiotic but also mutually harmful and can comprise microbial communities populating the same habitat and/or even including very different organisms. Some of these interactions may have never been documented before, or rarely, and offer novel perspectives that may inspire similar investigations in other environments. Furthermore, studies using omics approaches in tandem with cultivation-independent approaches from these extreme environments will advance the field of cold ecological microbiology (Feller et al., 2017). Mechanistic details and applied aspects are also provided, including the potential use of cold-adapted enzymes in agricultural/geothermal processes, biomolecules in bioprospecting, and new technologies that surpass current global high-volume bio-exploration paradigms. Cold ecosystems are some of Earth's least explored and most complex ecological niches. Studying these environments may open doors to discoveries with no current predictive analogue outside cold habitats. Among environmental genomics studies of psychrophiles, the genomic insights of the very first psychrophilic archaeon, *Methanogenium frigidum*, recently isolated from permafrost, have revealed several important features of their cold-adaptive metabolism (Raymond-Bouchard et al., 2018). Metagenomic studies of remote cold ecosystems, including the ancient blood from a mammoth and, in other pioneering work, an ice core from ancient Antarctic subglacial lakes, have brought to light the ecology and evolutionary history of psychrophilic communities in extremely cold environments. By reconstructing nearly complete genomes of psychrophiles from metagenomes, studies also present intriguing perspectives on their ecophysiology, particularly insights that may be uniquely applicable to research on marine psychrophiles where most cultivation-independent environmental genomics studies have been conducted. Psychrophilic microbes, which thrive below 20°C, bring new levels of complexity to the study of microbial interactions and symbioses. While there is growing ecological data on the former, genomic insights into how the incredible metabolic diversity of microbes allows them to adapt to such extreme environments have been rare. Prior to these studies, endeavors have primarily focused on free-living microbes. *Cryobacterium* 09-b14, a Gram-positive cold-adapted bacterium, was isolated from the glacial ice of an Antarctic glacier (Liu et al., 2020).

6. Microbial Interactions in Polar Regions

Cold-adapted microorganisms also known as Psychrophiles, thriving in polar regions, alpine glaciers, and deep-sea ice, showcase the remarkable ability to survive and function at temperatures near or below freezing. These microorganisms have evolved different biochemical mechanisms that not only demonstrates fundamental aspects of life in extreme conditions but also anchorage significant potential for biotechnological applications. Their research and utilization, particularly regarding cold-active enzymes and antifreeze proteins, hold promise for improving various industrial processes in cold conditions. One of the essential characteristics of psychrophiles is their production of cold-active enzymes such as proteases, lipases, amylases, and cellulases. For example, some species of the psychrobacter are identified to produce proteases that remain active at 4°C, providing anticipation for applications in energy efficient industrial processes, including detergent formulation, food processing, and biomedical diagnostics (Ayala-del-Río et al., 2010; Gault et al., 2021). These Psychrophilic enzymes maintain their activity at low temperatures, which is an efficient feature on reducing energy costs compared to traditional processes that need heat (Tajima et al., 2015; Gault et al., 2021). Alterations in these enzymes demonstrate that they keep stable and active even in cold conditions, making them useful for different biotechnological applications (Yang et al., 2021; Huang et al., 2023). Psychrophiles synthesize antifreeze proteins (AFPs) that prevent ice formation in their cells by making alterations to the pattern of ice crystals. This trait of these proteins holds substantial relevance in biotechnology, especially in cryopreservation techniques for long term storage of cell, tissues, and organs storage (Muñoz-Torres et al., 2017; Liu et al., 2020). Also, these proteins are crucial in agriculture by helping crops withstanding frost thereby increasing their growing season and boosting food production (Xiao et al., 2010; Majewska et al., 2025). Besides, these antifreeze proteins are functioning effectively as cryoprotectants that can extend foods shelf-life (Majewska et al., 2025). The production of extracellular polymeric substances (EPS) by *psychrobacter* like *Pseudoalteromonas*, plays a crucial part in how they are pertinent for both ecological and industrial uses. These substances are helping in biofilms formation, moisture retention, and surviving tough environmental conditions (Mykytczuk et al., 2013; Muñoz-Torres et al., 2017). EPS derived from these microbes have impressive applications in different fields, like cosmetics, medical dressings, and eco-friendly packaging, thanks to their unique texture and hydration properties in cold habitats (Alcaíno et al., 2015; Azman et al., 2024). The ability of these microorganisms in surviving extreme conditions divulges their complex adaptations including cell membrane modifications in fluidity and metabolic processes that strengthens nutrient fusion in cold conditions (Firdaus-Raih et al., 2018; Majewska et al., 2025). Ongoing research is diving into the genes and proteins of psychrophiles, which could lead to some new uses in different industries thanks to biotech tools like genetic engineering and machine learning (Huang et al., 2023). But ethical issues come with these discoveries, especially since these discoveries affect delicate cold ecosystems that are already struggling with climate change. It's crucial to innovate responsibly and take care of these unique environments while benefiting from what biotechnology has to offer (Buzzini et al., 2012; Maayer et al., 2014). In conclusion, the special traits of these psychrophile microbes open up interesting possibilities in biotechnology, especially with their enzymes and antifreeze proteins. As scientists keep exploring these cold places, they're finding new ways to improve industrial processes and make progress, all while considering the environmental impact and the need for resource conservation.

Antarctic Ecosystems

Antarctica is a vast and pristine ecosystem characterized by low temperatures, low water availability, and dynamic geochemistry. Most of Antarctica's remote environments are covered by snow and ice for most of the year; little is known regarding these systems' biodiversity and ecological processes. Microbial communities in Antarctic ecosystems play a central role in biogeochemical cycles. Understanding cyanobacterial symbioses in these ecosystems will further our understanding of their responses to global change. Studying microbial habitats in Antarctica has broad ramifications for understanding how microorganisms can reproduce and evolve in extreme environments. Approaching questions in a new habitat, such as the Antarctic biogeographical realm, can provide insights into the earliest stages of community assembly and evolution, as this area has been isolated moderately recently. Unique terrestrial environments in Antarctica have been shaped by geological episodes, resulting in quiescent and somewhat sheltered results of over a million years. Understanding how these habitats affect microbial diversity and adaptations is fascinating and challenging. Morphology, physiology, and low-temperature genomic alterations are mostly conserved among cyanobacteria, while some proteins related to protein synthesis are found to be exclusively in psychrophilic representatives. However, only a few studies have examined the genetic basis of traits relevant to the success of bacterioplankton in the cold. Some extreme habitats likely have a unique and interesting evolutionary history or highly conserved core inhabitants. Although most studies have focused on genetic adaptations, a multifaceted approach will likely reveal novel diversity and evolution of Antarctic cyanobacteria. We present the crustose lichen *Trebouxia* sp. as a model organism to study cold habitats. Initially focusing on phylogenomic inference, approaches from systematics and biogeography will be applied to explore its population genetics, dispersal, and adaptation in the cold (Malard et al., 2019).

Arctic Ecosystems

The Arctic is an ocean surrounded by land with a maximum depth of about 6,000 m. Classical ocean circulation patterns overprint it, but current states are affected by local influences such as shelf/slope glacial melting and riverine input of organic matter (Lauritano et al., 2020). Based on global estimates of human populations, the Arctic region had a population density of 2–7 individuals per sq. km in the mid-18th century. Summer months provide an ice-free ocean for about 1–4 months, but daylight is 24 hours for about 6 months in the early stage, forming a thermal stratification of Atlantic waters. Strong seasonality in photosynthesis provides an opportunity for long-term observation of seasonally sampled microbial communities. The clam and amphipod communities have been disrupted due to rising temperatures and a decadal shift in community composition away from amphipod dominance. Climate change can result in bio-physical coupling of the microbial community structure and processes at longer time scales (weeks to months) by reducing sea-ice cover and increasing stratification. Melting sea ice can be a secondary source of nutrients, preventing nutrient limitation in bloom scenarios. During the warmest three months in the Arctic summer, it has been suggested that two benthic community types exist: shallow, high diversity, and productive communities on diverse substrates and deep, low diversity and productivity on sand (Larose et al., 2013). Due to a slow active layer freeze-thaw, permafrost can be an archive of ancient bacteria, but the active fraction may still carry possible pathogen risks at warm temperatures. Highly concentrated solutes, contaminants, and microorganisms can be transported during snowmelt under environmental change. Biogeography of glacial and polar environments is rich and relatively unexplored.

Microbial Symbiosis in Extreme Cold

Symbioses are prevalent in all organisms' disciplines of life and phylogenetic scales with different interests of study and questions addressed. Invertebrate hosts, from cnidarians to molluscs and even tunicates, are often suitable models to investigate these associations and have received attention for the description of novel symbiotic associations, the comprehension of mechanisms involved in establishing and maintaining these contexts and their potential biotechnological applications. Some of the described associations are species-specific, while many others may encompass a large diversity of symbionts or be common to different *Gentania*. Associations are also diverse in terms of symbionts' hosting preferences and of 'multilayering', whereby different levels of cyanobacterial hosts can be simultaneously present in the holobiont. Symbiosis in extreme environments has consequently only recently yet successfully been investigated, bringing a new understanding of microbial ecology and evolution, biogeography of extremophiles and potential exploitation pitfalls and prospects (Fierer et al., 2007). Cold environments are characterized by substantially different temporal and spatial distributions of physical and biochemical properties compared to other ecosystems, which impose specific stresses on the resident microbial communities. These differences appear to be reflected at all levels of biodiversity, from the physical environment to community structure, metabolite utilization spectra and genome organization. Accordingly, cold habitats can be highly diversified geomorphologically, and a sharp freezing of the environment can cause a selective elution of the microbial communities. Different survival strategies at low temperatures have been observed in Antarctic ice and permafrost bacteria. Cold is also linked to large-scale glaciation and ice cover, which lead to the transport and deposition of microbial cells containing laboratory-viable cells in otherwise inhospitable settings. Thriving cells upon the modifications of ecological constraints and conditions of growth have been observed. A further environment of deposition is on snow, an ice-covered landscape and a living habitat with prevalent bacteria with different morphology, survival strategies, and physiological traits in their concerted resort to various discrete ecological niches. Bacteria associated with benthic invertebrates from extreme marine environments are promising but underexplored sources of biotechnologically relevant molecules. From a microbial perspective, symbiosis can be considered an ancient survival strategy significantly tailored by evolution at multiple levels. It furnishes microorganisms with advantages for their proliferation in spatial segregation from competitors, enhanced nutrient acquisition and shielding from host predation and stress. The four major symbiont-harboring benthic invertebrate groups colonizing extreme marine environments are emphasized because they are poorly explored and represent uncommonly enriched sources of biotechnologically relevant molecules. Symbiotic partners' astonishingly close structural, metabolic, and even molecular integrations show the breadth of the symbiotic specificity that can result from the fusion of two or more different organisms (Margesin & Collins, 2019).

7. Psychrophilic Microorganisms

Microbial life in glaciers, ice-rich permafrost, sea ice, glacial and ice meltwater permafrost, and ice-covered lakes has been shown to be present over wide ranges of temperature and moisture. Limits to microbial viability and growth temperatures are being evaluated to understand better how the limits to microbial life are possible. There is some microbial activity at –20°C down to –30°C, and it is anticipated that microbes are alive at –50°C due to non-freezing salt concentrations in brines and the rediscovery of halobacteria in permafrost (Figure 4). Psychrophilic, psychrotolerant, or cryosensitive microbes are also of immense biotechnological value. Activism with cold-active

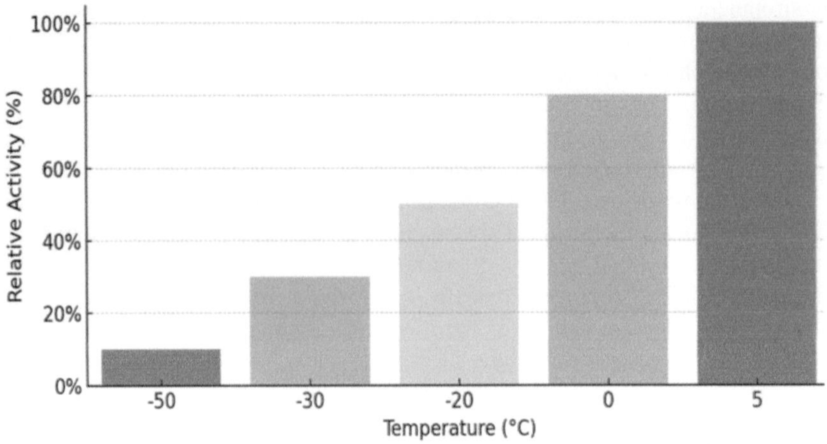

Figure 4: Psychrophilic microbial activity vs. temperature. This bar graph illustrates the relative metabolic activity of psychrophilic microorganisms across a temperature gradient from –50°C to +5°C. Despite extreme subzero conditions, measurable microbial activity persists, highlighting the survival strategies of cold-adapted microbes. Activity increases significantly as temperatures rise toward the freezing point, indicating potential for biotechnological exploitation in cold environments and implications for astrobiology on icy planets.

enzymes and biopolymers, biofertilizers, biosensors, biocontrol agents against pests and bioremediation biocatalysts are reviewed. Understanding extreme conditions such as cryogenic temperate and frozen states within the Earth is vital to understanding life elsewhere on icy bodies like Mars, Europa, Titan, and Enceladus. Cold environments, snow, and ice cover about 75% of Earth's surface. The coldest temperatures recorded are between –84.5°C and –89.2°C at the high altitudes of the Antarctic Plateau in July 2022 (Milojevic et al., 2022). However, independent scientists have reported microbial life widely in polar habitats using several methodologies and knowledge: culture-dependent, metagenomics 16SrRNA amplicon sequencing, and microscopy-based. The observation of microorganisms is surprising without photosynthetically assimilable light or organic carbon. It is proposed that they are part of a glacial subsurface biosphere. Microbial life in glaciers, ice-rich permafrost, sea ice, glacial and ice melt water permafrost, and ice-covered lakes has been shown to be present or detectable at least over wide ranges of temperature and moisture. How limits to microbial viability and growth temperatures become possible have been researched concerning microbial physiology, methodology, nucleic acid, and protein preservation. There is some microbial activity at –20°C down to –30°C, and it is anticipated that microbes are alive at least at –50°C due to non-freezing salt concentrations in brines, viscous actions restricting proton transfer of psychoactive metabolites, and the rediscovery of halobacteria in permafrost (Muthappa et al., 2022).

Biofilm Formation

Biofilm provides a unique living environment for most organisms. Biofilm is defined as microbial cells embedded in a self-produced EPS, forming a structure on living or inert surfaces. The EPS matrix of biofilm consists predominantly of polysaccharides but also contains proteins, nucleic acids, lipids, and other biopolymers. The number of microorganisms growing in biofilm communities dramatically exceeds the number of suspended growth conditions. The biofilm process commences when individual microbial cells attach to a solid surface and multiply to form microcolonies. The biofilm attachment consists of two steps: passive (reversible) and active (irreversible). Bacterial cells establish attachment by weak electrostatic interactions, such as van der Waals and hydrophobic interactions. Then, some cells secrete EPS to facilitate strong attachment, resulting in a more substantial bond between microbial cells and surfaces (Yin et al., 2019). Biofilm is responsible for causing economic loss and environmental pollution by contaminating industrial water systems and drinking water distribution systems, obstructing shipping lanes and pipelines, and fouling contact lenses, medical implants, and membrane bioreactors. Biofilm formation by beneficial microorganisms can be exploited for removing pollutants from wastewater, bioremediation of contaminated areas, and removing metal ions from industrial effluents. Biofilm can be found in various ecosystems, from the human body to industrial systems. Biofilm has a unique physical structure, including a complex architecture with various shapes and sizes, channels connecting the microcolonies that allow water and solutes to penetrate the biofilm matrix, and a heterogeneous population of microorganisms with various species combinations and metabolic states. The biofilm provides a unique living environment for microorganisms in aquatic habitats by allowing them to attach to surfaces and aggregate together. Biofilm formation provides survival advantages in the competition for nutrients with planktonic organisms. Biofilm formation is essential for colonizing the gut and for beneficial species. Biofilms can serve as source populations for planktonic dissemination in biofilm-based inoculation. Biofilm formation of biocontrol microorganisms may enhance biocontrol. Biofilm formation and disassembly can influence population structure, competition, and recruitment processes in complex communities for a long time. Biofilms may also significantly alter the biogeochemical processing of materials absorbed by surfaces in aquatic habitats, affecting the fate of these materials (Figure 5) (Erdei-Tombor et al., 2024).

Microbial Interactions and Climate Change

The situation of the microbial communities in cold environments brings new challenges to understanding the responses of biological processes, functions and interactions to climate changes and alternative environmental conditions. Deconfinement on environmental conditions and unveiling the geology of other

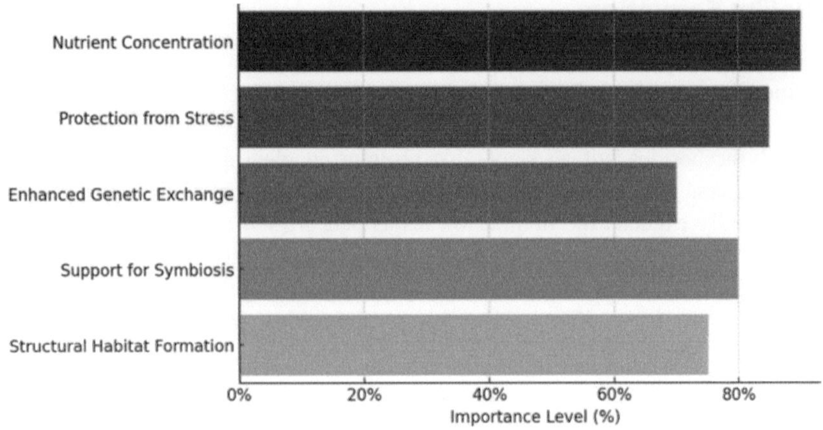

Figure 5: Functional importance of biofilms in cold environments. This horizontal bar chart highlights the key ecological functions of microbial biofilms in frigid habitats. Biofilms enable microorganisms to concentrate nutrients, resist environmental stress, enhance genetic exchange, facilitate symbiotic relationships, and form protective structural habitats. These features are crucial for microbial survival and community resilience in extreme cold, suggesting strategic targets for future biotechnological and environmental applications.

planets or satellite bodies with surface water will also consider cold ecosystems. One significant challenge is that the taxonomy of microorganisms, notably bacteria, is extremely rich in those environments, with many new lineages, orders, and families still to be assigned at the phyla level. In the cold regions of Earth, the unavailability of sampled ice cores with the proper size, size-fractionation, and preservation would significantly impact biological response and history reconstructions (Larose et al., 2013). This is coupled to understanding bacterial functions, whether autochthonous or not. The typical timescale of the processes may also influence the interpretation of observations. On Earth, many anaerobic bacterial strains were detected as active despite their patchy availability, raising questions about their source inoculum and limitations. Their persistence, survival, genetic drift and niche access should be revisited in interaction with the extent of alternation and confinement of environments. Identification of the strain level performed for a large number of strains up to now only in the viable populations should be followed by metabolic studies, including possible predation on key strains by DNS or viral infection. When and how did those scavenging and predator strains enter and adapt to those environments? It is very unlikely that they originate from space after detection of viral genomes. Beyond the mere identification of taxa, the extent and the time scale of activities, resilience and adaptability, syntrophy and predation, and the sequencing of genome-encoded metabolic functions are today necessary building bricks to reconstruct microbial interactions and subsequent propagated condensate evolution. In this context, key strains should be submitted to meta-'omics' exhaustive analyses within each defined biogeochemical context (Musat et al., 2006).

Impact on Ecosystems

Microbial cognition, or the ability of microbes to acquire, process, and respond to information from their surroundings, is implicated in the success of many microbial communities. It has recently been recognized that microbes benefit from forming multispecies biofilms by engaging in intercellular signaling, energetically favorable metabolic niche differentiation and interspecies symbioses. In particular, the local exchange of molecular signals, through the secretion of diverse compounds ranging from small organic molecules and gaseous signals to peptides, is thought to influence interspecies interactions, leading to social cooperation, altruism, metabolic niche differentiation, or antagonism, thereby playing a significant role in structuring multispecies biofilms. Biofilm growth mode protects against environmental stressors and promotes biocompound formation and bioavailability. Thus, biofilm growth mode heightens the ecological importance of microbial interactions in nutrient-rich environments. Such a consortia-wide capacity that emerges from the collective behavior of an individual population is often referred to as emergent behavior. Some well-studied emergent behaviors include dimerization, pattern formation, swarm crawling, and meditation (Sharma et al., 2023). From industrial and medical perspectives, the detrimental side of the above-mentioned processes that strengthen biofilms and favor population existence is significant, and control mechanisms involve molecular-level understanding. There is a limited research base on microbial interactions and symbioses in cold environments primarily because of inaccessibility to such extreme habitats and methodological difficulty in laboratory cultivation of psychrophilic microbes, unless for simple model systems bearing high similarity with well-studied systems. Microbial interactions with viral and protistan grazers have been studied in snow, glacial ice, and other cryospheric systems. Lichen microflora with high microbial diversity within a single sample, predominantly comprising filamentous fungi with more than 90% species recovery, were recently found in polyextreme high-altitude environments. Microbial communities in rock glaciers and cold meteorites have also been analyzed. From industrial and environmental perspectives, biofouling due to psychrophilic biofilms and bioremediation by bio-sedimentary formations of cold environments is ecologically significant but has received only modest attention. Despite the vastness of cold environments and the widely accepted presence of microbial populations with different physiological classifications and ecological lifestyles, there is a limited understanding of their diversity, distribution, biogeochemistry, and interactions. In particular, experiential knowledge of microbial interactions and symbioses in cold environments concerning their ecological and applied implications is mostly absent. Therefore, this chapter aims to review and discuss current knowledge on microbial interactions and symbioses in cold environments and put together gaps and future perspectives regarding investigations of both biotic and abiotic interactions of cold-adapted microbes (Larose et al., 2013).

8. Adaptation Mechanisms

Cold environments on Earth, such as the polar regions and high-altitude mountains, harbor diverse biota with microorganisms as the primary initiators of food webs. Low temperatures significantly lower the kinetic energy in aquatic systems, causing the different biochemical and biophysical properties of water, which can challenge the basic life functions of microorganisms. Nevertheless, many microorganisms, such as bacteria, archaea, fungi, algae, and rotifers, can survive and thrive in cold environments. Among the many strategies proposed by researchers, only a handful have been directly associated with the cold adaptation of the organisms in the study. Here, upon recent findings from two diverse and unique cold-environment microorganisms, *Psychrobacter arcticus* 273-4 and *Methanogenium tensions*, examples of new adaptations associated with the cold growth of this eury- and steno-psychrophiles are described. As more genomes of cold-adapted organisms are sequenced and analyzed, novel features will be discovered which are likely to be more biogeochemical significant for understanding the origin and evolution of life on Earth. A cold-active beta-galactosidase of a marine psychrophilic bacterium, *Microbulbifer* sp. JY07 was identified. This enzyme is the first example of a psychrophilic beta-galactosidase and exhibits catalytic activity at 0°C. Catalytic efficiency at low temperatures is a common cold-adaptation mechanism (Raymond-Bouchard et al., 2018). Growing in the cold, the sorption of organic solutes to less-hydrophobic surfaces is diminished. One possible strategy is to increase the adsorption mechanism by expanding the size of the whole sorbent. Molecular dynamics (MD) simulations show that phages Ps-83 and Ps-3, isolated from cold water environments, adopt more extended conformations. A novel microbial-peptide prediction algorithm was developed via MS analysis marPro was evaluated to help analyze and predict the peptide libraries generated by mass spectrometry. A sea ice brine metagenome sequence was obtained (Rapp et al., 2021).

9. Biotechnological Applications of Cold-Adapted Microbes

The polar regions of our planet represent the most extreme environment on Earth. Due to their large extension, complexity, and insensitivity to climate change, they are often considered natural laboratories for researching the evolution of life on our planet. At the same time, polar areas are considered as unique resources for biotechnological applications. In particular, they represent highly unexplored environments that are effectively colonized by microorganisms, enabling them to thrive at low temperatures (–2°C to 20°C), low nutrient concentration, and high salinity characteristics of the polar marine environment. Cold-adapted enzymes are of biotechnological interest because they are more efficient than their mesophilic counterparts at low temperatures and in extreme environments. Low mesophilic/psychrophilic temperature and high salt concentrations common to marine environments reduce the kinetic barriers to catalysis, resulting in high enzyme efficiency for psychrophilic biocatalysts (Rizzo & Lo Giudice, 2020). Moreover, cold-adapted enzymes are highly flexible, and their use at suboptimal temperatures reduces the risk of thermal denaturation in biotechnological processes. As a consequence, there is an increasing interest in cold-adapted enzymes that can be usefully employed in biotechnology as alternatives to mesophilic and thermophilic biocatalysts, in biocatalysis and organic synthesis, in food and agriculture, and in the bioremediation of contaminated marine areas. Cold-adapted enzymatic activities of bacteria have been extensively investigated in polar regions. Enzymatic activities of cold-adapted bacteria have been extensively investigated in polar areas, resulting in many different cold enzymes that have been identified, most of them suitable for industrial applications. Cold enzymes possess high catalytic efficiency at low temperatures and great molecular flexibility, making them highly attractive for industrial and biotechnological applications. The relatively low activity of one of these enzymes at higher temperatures and the tight regulation of its expression could be used to engineer the thermal stability of alternative oxidases for biotechnological applications at temperatures where other enzymes in these pathways are inactive. Other aspects of microbial physiology that could be engineered could be those that improve solvent tolerance, knocking out genes/pathways that produce toxic intermediates (Huston, 2008).

Enzyme Production

The enzyme production lab deals with cold-active enzymes, which catalyze biochemical reactions with an increase in temperature up to their optimum. Low-temperature-active enzymes investigate all types of enzymes which exhibit typical physiological features in corresponding temperatures. Low-temperature active microbial processes are the processes at temperatures inferior to 5°C. Cold-active enzymes have great potential as green biocatalysts in several applications, such as the food sector, biotechnology, and biopolymers, attending to society's demands for sustainability and eco-friendliness. The biopharmaceutical industry has a growing interest in low-temperature-active enzymes. Ecology and pharmacology applications are other sectors, as cold-active organisms are of interest for their polysaccharides and secondary metabolites. Marine environments have a strong, still not well understood, potential biodiversity that is attractive for novel biotechnology. Cold habitats are still poorly explored, and biocatalysts and strains with novel abilities can be expected. Biocatalysts from cold Antarctic environments have proved more effective in bioplex and oil detoxification than their classical mesophilic counterparts. Marine enzymes such as cold-active amylases and proteases are used in the food industry. The porcine pancreas lipase is similar to a psychrophilic lipase from a marine microbe, with a potential application in the detergent industry. Although the industrial potential of cold-active enzymes is broadly recognized, only a few marine enzymes are marketed. This is mainly due to the technical difficulties in studying and engineering such enzymes. The optimization of solid-state fermentation for biocatalyst production from marine microorganisms and using high-throughput sequencing will provide the strain diversity and bioprocesses needed to explore the potential biodiversity from those underexplored environments and their applied knowledge (Kuddus et al., 2024).

Bioremediation

Bioremediation is a natural bio-process in which microorganisms degrade hydrocarbons into harmless products (such as CO_2, H_2O, or microbial cells) (Yap et al., 2021). Since hydrocarbons are highly potent ecosystems, bioremediation can effectively detoxify hydrocarbon-polluted soils. Soil nitrogen and phosphorus are nutrients necessary for bioremediation treatment. Hydrocarbon-polluted soil must be initially assessed for nutrient availability to identify if fertilizers need to be amended. Nutrients are optimally applied in low quantities to stimulate the growth of autochthonous microorganisms. Adding high amounts of nutrients can lead to biostimulation, but this treatment strategy is not strongly recommended under cold conditions, which slow down biodegradation processes and can cause nutrient leaching. Surfactants greatly increase the partition of hydrophobic compounds into the aqueous phase and desorb them from soil micropores. The addition of surfactants is beneficial for bioremediation treatment, but their dosage should be optimally controlled to prevent a toxic effect on microorganisms. Biosurfactants are preferred over chemical surfactants due to their eco-friendly property and, therefore, do not represent a major concern during detoxification. The bio-stimulant addition is another important factor to enhance the biodegradation rate. A combination of municipal compost and organic sludge is highly suitable for soil bioremediation. These substrates are easily degradable compounds and represent a primary nutrient source to promote the growth of autochthonous microbes. Spraying a foam mixture of these compounds is a very simple and effective technique in activating hydrocarbon removal treatment. The poor mixing of nutrients in contaminated soils limits their bioavailability to microbes, diminishing pollutant removal efficiency. Under cold climates, nutrient amendments in bioremediation treatment must be optimally designed. Continuous geological frozen soils inhibit pollutant removal processes due to a lack of microbial activity. Implementing bioremediation treatment in colder regions requires an assessment of the frozen depth of soils and predicting their de-freezing periods. It is highly recommended that bioremediation implementation in cold regions be performed when soil temperature rises above the limit needed for effective biological treatments (Kuppan et al., 2024).

10. Future Directions in Research

Microbes from cold environments are a promising resource for biotechnology but are still poorly investigated. They are in extreme environments that constrain the chemiosmotic proton potential the cellular bioenergetics is based on (from –300 to –100 mV). Their survival requires specialized biosynthetic pathways to seize energy and build cellular materials. Cold-adapted microorganisms usually display unique biological and biochemical properties that allow them to thrive at extremely low temperatures. Since physiological properties are linked to genomic contents, cold-adapted microbes are of interest for studies on genome evolution. Omics approaches are useful tools for understanding cellular machinery and fate; they produce enormous amounts of information, and the bottleneck understands the functional implications of such information (Rizzo & Lo Giudice, 2020). New genomic and cultivation techniques improve knowledge of polar microbiota structure and physiology. Environmental shotgun sequencing has revealed the biodiversity of many biomes. Among these, polar environments have been recognized as a source of biotech products. Experimental evolution approaches suggest that adaptation to low temperatures is associated with consistent genomic changes. Nevertheless, most of these results are derived from the same isolate and methodology. Thus, new approaches addressing different strains/populations would be insightful for a more general understanding of adaptation to cold (Nawaz et al., 2022). New strains from unexplored environments still reveal novel bioproperties and biocatalysts with relevant AD and substituent-chain lengths for biodiesel production. Diversified screening methods would reveal new entities and include strains in different phylogenetic groups that may produce novel natural products. New bioinformatic approaches and increasing genomic databases offer novel opportunities to prospect new biosynthetic gene clusters in culturable and unculturable bacteria. Future research directions may benefit from combining the results of diverse approaches.

Emerging Technologies

As cold environments, such as the polar and deep-sea regions, are the repositories of many microorganisms, the knowledge about their biology and ecology is increasing. Moreover, knowing that these microorganisms can also produce biomolecules of biotechnological interest, discovered or unknown, new habitats are continually being prospected. In this scenario, any discovery has a low arsenal of tools and knowledge. All investigations upon these new promising niches, both ecological and biotechnologically oriented, need their development, from screening strategies to bioprospecting methods. The last years have seen spectacular breakthroughs in the availability of complete genomes of cold microorganisms. This is explained by the progress made in genome sequencing technologies and the associated drop in costs made possible with the emergence of the next-generation sequencing (NGS) methods. The genome sequencing effort devoted to polar and cold waters or soil organisms rapidly increases. The analysis of these genomes provides a wealth of information about gene content, function, and evolution of adaptation to extreme conditions, metabolic and biosynthetic pathways, and gene expression regulators (Rampelotto, 2010). Many new sequencing methods today focus on population structures and dynamics. In virtually all of the new marine field studies, obtaining a sequence/access number from the collected samples is common. Attention has all focused on 'genomics' targeted analysis. Within the overall inclination, however, a particular branch of "functional metagenomics" has been investigated much less extensively. This approach first aims at detecting new biochemical functions, and thus the genetic markers giving rise to them, and only afterwards doing environmental population bioprospection to identify the organisms or multiple organisms able to execute the new functions (Rizzo & Lo Giudice, 2020). In metagenomics approaches, the habitat is typically marine and the target function concerns bioluminescence. Metagenomic studies have used similar approaches, selecting different marine environments as PCR targets for sequencing rare bioluminescent genes and organisms. Translating the bioinformatics approach into a functional biochemistry register is still a challenge. The elements of chance will always play a large role when searching for biomolecules of interest among hundreds or thousands of candidate sequences. Because of these elements of chance and the wide array of molecular outcomes of the abundant unknown 'LOS' genes, rigorous characterization of the promising new gene sequences as applied to carefully selected model organisms is needed.

Interdisciplinary Approaches

Microbial interactions are central to community structure, biogeochemical processes, and the maintenance of ecosystem resilience. Interactions influence how organisms cope with environmental perturbation and guide the design of microbial communities to obtain desired functions. The focus on interactions among microbial species has increased in recent years, leading to the development of many new approaches to elucidate microbial interactions. These approaches harness technique advancements, pooled experimental designs, and interdisciplinary efforts to characterize the interaction patterns of microbes in various habitats. For example, understanding interactions is critical for inferring species persistence, community succession, and the design of synthetic microbial communities. However, inferring the interaction patterns of microbial communities is technically challenging due to the coupling of innumerous organisms, the need for high spatiotemporal sampling depths, and the co-occurring abiotic factors and indirect interactions (Nawaz et al., 2022). The development of technologies such as community sequencing, large-scale experiments, and modeling approaches, as well as the integration of multi-omic techniques across disciplines, have provided new tools for asking interaction-related questions in the context of high-throughput and big-data analysis. The centre stage in these integrative efforts is process-based modeling tools, capable of distilling long-term dynamics of microbial communities across spatial and temporal scales, thereby revealing hidden interaction patterns. A recent modeling effort highlighted current progress in this era of genome-centered microbial ecology, described how modeling, synthetic ecology, and ecological modeling and inference techniques can be integrated with feedback on each other, and excised avenues for future exploration in these expansive fields (Kodera et al., 2022). Understanding patterns of association and how they vary over time can shed light on the interactions among microbial species within complex communities. Work in this area also opens up many challenging questions on modeling, as these relatively simple but powerful tools can reveal new insights into interaction patterns that may remain hidden in detailed mechanistic models. A modeling framework based on stochastic differential equations (SDE) can describe how the states of the entire system influence the growth of one population. Further proof-of-concept analyses that adopt this modeling framework will likely lead to Bayesian methods combined with perturbation assays that formally test interactions in synthetic systems of microbes. Extensive environmental datasets paired with interaction characterizations provide a great opportunity to investigate association patterns at the community scale in a specific environment; variable partitioning tools and semi-supervised methods are branches in this direction that merit further exploration (Zuñiga et al., 2017).

Challenges in Studying Cold Microbial Interactions

Microbial communities are exposed to extreme environments and cascading population and community ecology constraints. Both abiotic and biotic factors shape local microbial communities and their functions (Larose et al., 2013). However, only a few studies have explored micro-aggregates' micro-environment within snow-pack. Extreme topography, environmental stressors, and wide and varied microhabitats characterize polar environments. Recognizing that natural environments include a hierarchical organization of physical and biological structures of varying scales is critical for assessing functioning and microbial life in extreme environments. This is because coarse-scale structures can harbor diverse biota that thrive under specific selective pressures. Some species occupy narrow niches within these tolerable ranges while others live in a broader tolerance zone and, hence, together help maintain ecosystem function. This hierarchical architecture offers clues and tools for studies, with insights into their biophysical or biogeochemical behavior gained from the systematic characterization of their macro to micro-inhabiting structures. Developing creative sampling strategies that can efficiently capture a representative set of the many different-scale structures can dramatically improve the quality of empirical studies and open many previously inaccessible environments for exploration. Similarly, biogeographically inspired modeling platforms can create parametric scales by mixing algorithms or coalescing diverse ecological models into one. However, techniques and tools to disentangle the complex network of interactions between the abiotic factors and biotic factors are still scarce (Loreau et al., 2001). Using microbial communities from bulk environments as model systems, sequencing techniques and bioinformatic tools have been developed to assess and annotate population and community structures systematically. Subsequently, these statistical modeling platforms have been applied to molecular profiles from highly replicated environmental samples to discover the predictive power of metabolic activity or community structure for local conditions. Soon, it is expected that genomic or transcriptomic clustering techniques will become available to assess the functional outputs of the ecological architectures. These studies are truly exciting because they involved and successfully conquered many of the latest challenges in microbial ecology. Whether the insights and tools will apply to compositional studies of the cold microbial communities remains to be seen. Negative results from applying the tools straightforwardly may provide empirically important insights into the necessity of using tailored techniques and tools (Leite & Kuramae, 2020).

Field Studies

Field studies of cold microbial communities provide the opportunity to directly study the interactions and influences of environmental variation in situ. Iceland has diverse cold environments, and three field studies will be presented that span various environments and techniques. The first study focuses on cold and cryo-ecosystems, vital to life on Earth, and microbial environments with unique properties, including permanent frigid temperatures and prolonged darkness. Field sites include a glacial pond, a permanent snowfield stream, and a subglacial cave. These sites enable research addressing how cold temperatures and darkness impact the physiology and biogeochemistry of still poorly understood ecosystems, as well as fundamental principles, such as how temperature plays a role in controlling microbial community structure and activity. This understanding will contribute to current initiatives in glacier monitoring and ice-caps research and provide a foundation for comparative studies across similar environments. The results of a first field campaign that initially took measurements of multiple biogeochemical and microbial variables are detailed (Guðmundsdóttir et al., 2019). Laboratory studies with frozen samples have been implemented to gain insights unapproachable in the field, and future directions and opportunities will be presented. The second study focuses on the microbial ecology of glacial ecosystems and glaciers in Iceland. Glaciers hold approximately 10% of the world's freshwater and rapidly melt due to climate change. Glacial retreat exposes new subglacial environments, and understanding the microbial ecosystems of glaciers will help understand how they evolve and impact the environments they emerge into. Another focus area is the sediments of the glacial rivers that emerge from the glaciers, which have complex and extreme environments that vary spatially and temporally. Understanding the microbial ecology of such rivers is gaining increasing attention because of their role in biogeochemical cycles and microbial ecology (Edwards & Cook, 2015). Work with geobiology, sedimentology, and microbial genomics is highlighted to look at the evolution of the glacial ecosystems and their impact on the aquatic environments they affect. The final field study highlights Iceland's diverse and unique soils, from the high-altitude lava fields on Iceland's central plateau to agricultural soils around the lowland areas. Active soil production occurs in Iceland, with fresh tephra, volcanic ash, and lava produced by recent eruptions providing a new substrate for colonization. Despite Iceland's young geologic history, soils possess a rich microbial diversity that has evolved unique adaptations to the specific conditions of the soils.

Laboratory Cultivation

Laboratory cultivation of cold microbial communities, in isolation and/or manipulation of co-presence with eukaryotic microorganisms and/or a solid matrix, has led to almost all the advancements, alterations and detection of novel metabolite production, initial understanding of long-term co-evolutions, inventions of new cultivation techniques, and have doubly-triple confirmed their isolation motives. Growth conditions of interest for the cold environments, e.g., low/negative temperature and/or high pressure. In isolation cultivation of cold bacteria and archaea, firstly, 55°C above growth temperature was optimized to apply another cold 35°C/38°C for cultivation. *De novo* detection of cold enzymes and novel microorganisms. Metabolomics of novel *Cryoglomus* species and psychrophilic methanotroph-involved glycolic soil biogeochemical cycles. Cultivation in co-presence environments of microorganisms gifted them long-term co-evolution motivations and new knowledge, including how fungi farm/disband for bacteria protection and adjustment of growth environments. Newly designed wet-surface solid modulates supply more nutrients and growth spaces to isolate many previously uncultured microorganisms through harsh environments. Further understanding has led to large changes, such as motivations for its knowledge and discovering in advanced knowledge, as examples introductions of liquid co-culturing techniques and ripened potato womb and wet-surface solid genesis, illustrated their invents and wide applications of applicable levels (Yan et al., 2023).

Case Studies of Microbial Symbiosis

Microbial interactions and symbiosis are examples of such interactions that have acquired notoriety in the last decades. These relationships are intimate, persistent, and often evolutionarily stable, and they are becoming the center of attention in many scientific fields, including ecology, evolutionary biology, and biotechnology. For many microorganisms, symbiosis is essential to their lifecycle, enhancing their evolutionary success. However, despite recognizing the importance of microbial symbionts in shaping the diversity and functioning of their hosts, in comparative studies, scientists still rely primarily on descriptors of microbial taxonomy and single microbial features (Lo Giudice & Rizzo,

2022). In extremely cold environments, where microbial resources are scarce and physicochemical parameters are outside the tolerance ranges of many known species, microbial interactions and symbiosis constitute important undiscovered territories from an ecological but also biotechnological point of view. A few accounts of microbial symbiosis in cold, extreme environments suggest this phenomenon could be highly prevalent in such ecosystems (Khomutovska et al., 2020). The coupling between the ecology of cold environments and symbiosis suggests that more studies of this kind should be conducted to better understand the existence, consequence, and efficiency of microbial symbiosis in extreme conditions. The structure of microbial consortia colonizing different types of cryogenic rocks in cold deserts shows the formation of unique and different communities. For one of the ecosystems under study, the frequency of the endolithic activity was four orders of magnitude higher than that of the other two cold deserts. This suggests the possibility of microbiota exchange between these extreme locations. Primary habitats, resources, cooperation, and synergy in communities might be common factors for forming endolithic communities in different bio-climatic zones.

Lichens in Cold Climates

Lichens are photosynthetic symbioses between a fungus, the mycobiont, and a green alga and/or a *Cyanobacterium*, the photobiont. Lichens grow in many habitats, including cold habitats, where important biogeochemical processes occur. Yet, little is known about how these ecosystems function, particularly in the microbiome, the structure of the primary producer consortium, and the species-level occurrence and distribution of its members. The high-latitude lichen community in Greenland was studied, including the dominant lichen species (e.g., *Cetraria*, *Cladina*, and *Stereocaulon*), the associated bacteria and archaea, and the cyanobacterial endosymbionts. Genomic loci and phylogenetic markers were chosen that could be amplified with previously published barcoding primers. Sampling sites along a latitude gradient with different environmental conditions adopted a phylogenetically broad approach. The cryptobiotic soil crusts around two prominent mountains in Utah were investigated to determine communities of soil-dwelling, cyanobacterial lichen-forming fungi and to characterize the environmental conditions favoring their growth. In Iceland, gut metagenomes of ground beetles containing Cradle lichen were analyzed to see whether and/or how long Cradle can survive inside a gut. Lichens are symbiotic systems composed of mycobionts (fungi) and photobionts (algae and/or cyanobacteria) (Aschenbrenner et al., 2016). The dominant mycobionts of Lichens are ascomycetes, while photobionts include green algae from the genera *Trebouxia*, *Coccomyxa*, and *Trentepohlia*, as well as cyanobacteria from the genera *Nostoc* and *Scytonema*. Several groups of lichenized fungi have been described from Iceland since the 1700s, and key areas for inventories of lichen biota are northern latitudes like Iceland, Nunavut, and Svalbard. On islands like Svalbard, large natural lichen herbivores exert top-down control on the structure and dynamics of lichen communities at different spatial scales. The lichen association does not represent a strictly specific relationship between the partners; thus, promiscuity is common. However, the choice of partner, especially the lichen fungus, is limited to some extent by ecological and phylogenetic constraints, such as climatic factors and soil chemistry. A network based on literature data visualizes these limits by separating partners into four modules, distinguished by their preferences for soil chemistry and bioclimatic conditions (Škvorová et al., 2022).

Corals and Symbiotic Algae

Coral reefs are a diverse ecosystem built upon rigid calcium carbonate frameworks. These reefs are formed by symbiotic relationships between crustacean coral animals and symbiotic dinoflagellates (genus *Symbiodinium*). Therefore, the ecologies of coral reefs, which support enormous marine biodiversity, rely upon the coral-dinoflagellate symbiosis, where dinoflagellates benefit from nutrient-rich environments in coral tissues, and corals receive a wide range of energy sources produced by photosynthesis and the absorption of dissolved nutrients. Invaded by stressors such as rising seawater temperature due to climate change and the runoff of terrestrial reefs, many corals lose their dinoflagellate symbionts. The increase in the density of light-absorbing and heat-radiating zooxanthellae in the endodermal tissues is thought to confer resilience against coral bleaching. Consequently, corals with more heat-resilient symbionts may benefit from an opportunity to survive under rising seawater temperatures. However, the ecological significance of symbiont variability and switching in the context of continual warming remains elusive because the risk of thermal expulsion was similar among symbiotic species (Pernice & Levy, 2014). In addition to environmental temperature, a host-induced symbiont switching that affects coral physiology can be triggered by nutrient enrichment. This new avenue of research gained traction due to mounting evidence that changes in nutrient loads affected by coral death occur globally. Meanwhile, given that the loss of mutualisms can have disastrous socio-economic implications for biodiversity, much research has focused on the detrimental effects upon corals and dinoflagellates due to elevated temperatures. However, research addressing the implications of nutrient enrichment and herbivore loss to the breakdown of the coral-dinoflagellate association remains limited (Takagi et al., 2023). This situation constitutes research opportunities to fill critical knowledge gaps regarding coral-dinoflagellate symbiosis, at least temporally and spatially.

Ethical Considerations in Microbial Research

As the study of environmental microbes has transitioned from culture-dependent and taxonomic studies to rapid genomic approaches, it has become necessary to consider new ethical challenges in the field. Ethical issues are not unique to environmental microbiomes, and many of them have already been addressed in biomedical microbiome research, albeit often in more discrete case studies. As in other fields of microbiome research, the focus on environmental microbes has commercialized quickly, and while it is important to consider the potential beneficial uses of engineered environmental microbes, it is equally essential to consider their unintended consequences. Synthetic biology approaches will enable the engineering of environmental microbes, but these microbes differ from those in biomedical studies in key ways that pose unique challenges for regulatory frameworks. One important difference is that while biomedical microbes are most often managed in laboratory conditions, environmental microbes are mobilized into complex, co-evolved systems that interact poorly (Shamarina et al., 2017). Beyond consideration of regulatory frameworks for environmental microbes, deeper ethical considerations are also warranted in order to avoid opening ecological Pandora's boxes. How can ethical considerations in studying environmental microbes' best be anticipated when such studies are so encompassing? These microbes are studied in a range of environments, including urban areas and their impacts on infrastructure, aquifers, oceans, and climate change, and in all manner of experiments, from the monitoring of pathogens in drinking water to the design of corroding bacteria in pipelines and from the engineering of methane-eating microbes in landfills to climate-friendly microbes in the oceans. The movement to consider the microbiome as a whole has necessitated the consideration of new and responsible ethical frameworks for research (Robinson et al., 2022).

Biodiversity Conservation

As dense, dry, and cold environments seem to resist biological invasion compared to other extreme ecosystems, it is widely accepted that microbial life in extreme arctic and antarctic localities is less diverse than in milder habitats, such as temperate regions. However, the sparseness of data and the extent of unexplored territories lead to some doubts. Despite low species density, niche evolution in these extreme environments can lead to highly diverse and/ or endemic communities. Thus, it might not be valid to assume a priori that microbial communities in polar environments should be less diverse (Lauritano et al., 2020). During warming trends, ice cover may become patchy, opening up territory to opportunistic and potentially invasive organisms. Melting glaciers will release nutrient-rich materials that may allow blooms of microbes and phytoplankton to occur wherever light can penetrate. Such blooms can reduce the ambient pH, exposing organisms that cannot survive extinction. More gradual changes that allow changes in microbes' community structure will result in unknown biogeochemical consequences under previous environments. Such changes are an important aspect of conservation efforts, which are important for monitoring and protecting polar habitats. These can be rapid or long-term, but in both cases, stages can be easily overlooked when they occur longer than human observation periods. Current habitats in areas of protective significance need to be identified and protected so that any shifts in distributions of microbial assemblages can be detected at an early stage. Model regions should be selected for sites where additional studies can improve understanding of polar microbial driven biogeochemistry dynamics. By taking a long-term observational and experimental approach, mechanisms of change in areas of special interest can be uncovered. Enhanced monitoring of physical conditions and their microbiological response should be developed. To address the issue, a combination of modeling, ecological surveys, molecular techniques, and laboratory studies is better used to locate connections between microbes and the physical environment (Venegas et al., 2023).

Bioprospecting Ethics

The necessity of bioprospecting and searching for new ecophysiological strategies and biochemical products in polar and cold-adapted bacteria is beyond question. However, some ABPs were also discovered in Hyperboreal and temperate environments (Rizzo & Lo Giudice, 2020). It must be borne in mind that scientific and melodic knowledge could be an advantage, an opportunity for different operators, or something neutral that brings calamities. It is diligently requested that those involved in bioprospecting initiatives, especially in setting up bio observatories, apply ethical practices concerning the local, Indigenous nations and traditional knowledge holders involved in and affected by such initiatives. Cultural exchanges and science should not become instruments for disenfranchisement, oppression and colonization. The Humboldt penguins, an emblematic species of the coast of Peru, are regarded as a national treasure. Over a recent symposium, scientists expressed their apprehension over the damage and risks that globalization and new capitalism could incur upon marine life and coastal zones in Peru, and public markers were erected as warnings not to pursue toxic mining and fishing-technology-based jobs that would compromise marine life (Loayza-Aguilar et al., 2024). Bio-colonization occurs when scientists become owners of the "genetic resources", beholders of the "intellectual property" on the means to access cold-adapted bacteria and transmitters of a technology which can render them their royalties while depriving other nations of these bioprospecting resources and revenues. It is necessary to put into practice processes to jointly monitor, validate, and patent interested scientific work, generations and technologies concerning all local nations and traditional heirs involved in these matters. Bio-colonization occurs when scientists invest the benefits of their scientific work entirely for their states or institutions without considering Indigenous nations, traditional and local holders, and the communities involved in and affected by such activities. Untamed and unexchanged knowledge should keep the nature of neutral bio-resources, but when crossed with melodic, industrial, and epistemic perspectives, this knowledge must be regarded as intellectual property. Bioprospecting either in cold climates or temperate seas can be regarded as a chance for the cooperative betterment of the personal, social, and scientific quality of life of different people and nations because this option is sustainable and congruent (Millum, 2010).

11. Conclusion

Studying microbial interactions is important in elucidating ecological gradients and evolutionary transitions. An increasingly complex autocatalysis-based evolutionary state that builds upon earlier analysis has been applied to ocean planktonic assemblages from coastal sites, coastal mesocosm experiments, and the entire ocean surface. These approaches have demonstrated promising outcomes for guiding future biological discovery and assessing ecosystem responses to environmental change, congruent with ecosystem assembly observations at various spatial scales. The growing availability of genomic and environmental information on the deepest reaches of Earth's ocean and the assemblages of microbial organisms present new horizons for understanding the mechanisms regulating ecological structure and function in cold addiction. Much basic mechanistic insight is still needed in many areas of inquiry, and current analytical capabilities come with restrictions that should be addressed theoretically and experimentally if future progress is to be effective and meaningful. As some initial directions for future research, an integrated approach employing tracer studies and modeling would be a fruitful avenue for assessing modes of predation in soft sediment systems. As another suggestion, the auto-inoculation hypothesis holds great potential for elucidating the role of sediments as a source of microbial cells to the planktonic community in the open ocean. Exploring this concept would entail novel experimental designs and challenges for new measurement systems to assess the vertical transport of solids into suspension and the fate of the attached microbes. As cornerstone ecological models point out, the richness and composition of a community are regulated by the interplay between immigration and extinction, where environmental filtering can play an important role in forcing the outcomes of this balance in favor or against certain taxa. At the mesopelagic level, an improved understanding of the regulation of microbial diversity would enable comparison with low-diversity systems like the solid-phase benthic sediments, and the account of gradients in environmental conditions and microbial diversity would expose potential uncertainties for present-day biogeochemical models.

References

Alcaíno, J., Cifuentes, V., & Baeza, M. 2015. Physiological adaptations of yeasts living in cold environments and their potential applications. *World Journal of Microbiology and Biotechnology*, 31(10), 1467–1473.

Al-Turki, A., Murali, M., Omar, A.F., Rehan, M., & Sayyed, R.Z. 2023. Exploring the recent advances in PGPR mediated resilience towards interactive effects of drought and salt stress in plants. *Frontiers in Microbiology*, 14, 1214845. https://doi.org/10.3389/fmicb.2023.1214845.

Araújo, A., Reinhard, K., Ferreira, L.F., & Gardner, S.L. 2003. Parasites as probes for prehistoric human migration? *Trends in Parasitology*, 19(2), 94–96.

Aschenbrenner, I.A., Cernava, T., Berg, G., & Grube, M. 2016. Understanding microbial multi-species symbioses. *Frontiers in Microbiology*, 7, 180.

Ayala-del-Río, H.L., Chain, P., Grzymski, J.J., Ponder, M.A., Ivanova, N., Bergholz, P.W. et al. 2010. The genome sequence of *Psychrobacter arcticus* 273-4, a psychroactive siberian permafrost bacterium, reveals mechanisms for adaptation to low-temperature growth. *Applied and Environmental Microbiology,* 76(7), 2304–2312.

Azman, A.N.A., Kamaruddin, S., Quay, D.H.X., Bakar, F.D.A., Hashim, N.H.F., Murad, A.M.A. et al. 2024. Profiling the growth conditions and persistent organic pollutants (pops) tolerance of *Phenoliferia glacialis* usm-psy62. *Malaysian Applied Biology,* 53(4), 103–113.

Baeza, M., Sanhueza, L., García, K., & Alberdi, M. 2022. Cold stress response of plants with emphasis on Chilean species: Insights from the molecular, biochemical, and physiological perspectives. *Plants,* 11(21), 2863.

Basu, A., Prasad, P., Subha Narayan Das, Sadaf Kalam, Sayyed, R.Z., Reddy, M.S., & Hesham El Enshasy. 2021. Plant growth promoting rhizobacteria (PGPR) as green bioinoculants: Recent developments, constraints, and prospects. *Sustainability*, 13, 1140. https://www.mdpi.com/2071-1050/13/3/1140.

Bischoff, J., Mangelsdorf, K., Gattinger, A., Schloter, M., Kurchatova, A.N., Herzschuh, U. et al. 2013. Response of methanogenic archaea to Late Pleistocene and Holocene climate changes in the Siberian Arctic. *Global Biogeochemical Cycles*, 27(2), 305–317.

Braga, R.M., Dourado, M.N., & Araújo, W.L. 2016. Microbial interactions: Ecology in a molecular perspective. *Brazilian Journal of Microbiology,* 47(Suppl. 1), 86–98.

Bright, J.P., Maheshwari, H.S., Thangappan, S., Perveeen, K., Bukhari, N.A., Mitra, D. et al. 2025. Biofilmed-PGPR: A next-generation bioinoculant for plant growth promotion in rice (*Oryza sativa* L.) under changing climate. *Rice Science*, 2025(32), https://doi.org/10.1016/j.rsci.2024.08.008.

Buzzini, P., Branda, E., Goretti, M., & Turchetti, B. 2012. Psychrophilic yeasts from worldwide glacial habitats: Diversity, adaptation strategies, and biotechnological potential. *FEMS Microbiology Ecology,* 82(2), 217–241.

Carrero, S., Slotznick, S.P., Fakra, S.C., Sitar, M.C., Bone, S.E., Mauk, J.L. et al. 2023. Mineralogical, magnetic, and geochemical data constrain the pathways and extent of weathering of mineralized sedimentary rocks. *Geochimica et Cosmochimica Acta*, 343, 180–195.

Cavicchioli, R., Ripple, W.J., Timmis, K.N., Azam, F., Bakken, L.R., Baylis, M. et al. 2019. Scientists' warning to humanity: Microorganisms and climate change. *Nature Reviews Microbiology*, 17(9), 569–586.

Chaithra, M., Nishmitha, K., Patel, P.S., Mitra, D., Gangaraj, R., & Sayyed, R.Z. 2025. Fungal bio-stimulants: Cutting-edge bioinoculants for sustainable agriculture. In: Mathur, P., & Roy, S. (Eds.), *Plant Microbiome and Biological Control. Sustainability in Plant and Crop Protection* (Vol 20). Springer, Cham, pp. 289–307. https://doi.org/10.1007/978-3-031-75845-4_13.

Edwards, A., & Cook, S. 2015. Microbial dynamics in glacier forefield soils show succession is not just skin deep. *Molecular Ecology*, 24(5), 963–966.

Erdei-Tombor, P., Kiskó, G., & Taczman-Brückner, A. (2024). Biofilm formation in water distribution systems. *Processes*, 12(2), 280.

Eswaran, S.U.D., Sundaram, L., Perveen, K., Bukhari, N.A., & Sayyed, R.Z. 2024. Osmolyte-producing microbial biostimulants regulate the growth of *Arachis hypogaea* L. under drought stress. *BMC Microbiology*, 4, 165. https://doi.org/10.1186/s12866-024-03320-6.

Feller, G. 2017. Cryosphere and psychrophiles: Insights into a cold origin of life? *Life*, 7(2), 25.

Ferioun, M., Zouitane, I., Bouhraoua, S., Elouattassi, Y., Belahcen, D., Errabbani, A. et al. 2025. Applying microbial biostimulants and drought-tolerant genotypes to enhance barley growth and yield under drought stress. *Frontiers in Plant Sciences*, 15, 1494987. https://doi.org/10.3389/fpls.2024.1494987.

Fierer, N., Bradford, M.A., & Jackson, R.B. 2007. Toward an ecological classification of soil bacteria. *Proceedings of the National Academy of Sciences*, 104(3), 896–903.

Firdaus-Raih, M., Hashim, N.H.F., Bharudin, I., Bakar, M.F.A., Huang, K.K., Alias, H. et al. 2018. The *Glaciozyma antarctica* genome reveals an array of systems that provide sustained responses towards temperature variations in a persistently cold habitat. *PLOS One*, 13(1), e0189947.

Gallo, G., & Aulitto, M. 2024. Advances in extremophile research: Biotechnological applications through isolation and identification techniques. *Life*, 14(9), 1205.

Gault, S., Higgins, P.M., Cockell, C.S., & Gillies, K. 2021. A meta-analysis of the activity, stability, and mutational characteristics of temperature-adapted enzymes. *Bioscience Reports,* 41(4), BSR20210336.

Ghosh, A.R. 2013. Appraisal of microbial evolution to commensalism and pathogenicity in humans. *Clinical Medicine Insights: Gastroenterology*, 6, 1–12.

Graham, R.C., Rossi, A.M., & Hubbert, K.R. 2010. Rock to regolith conversion: Producing hospitable substrates for terrestrial ecosystems. *GSA Today*, 20(2), 4–9.

Guðmundsdóttir, R., Kreiling, A.-K., Kristjánsson, B.K., Marteinsson, V.Þ., & Pálsson, S. 2019. Bacterial diversity in Icelandic cold spring sources and in relation to the groundwater amphipod *Crangonyx islandicus*. *PLOS One*, 14(10), e0222527.

Gutt, J., Isla, E., Xavier, J.C., Adams, B.J., Ahn, I.Y., Cheng, C.H.C. et al. 2021. Antarctic ecosystems in transition: Life between stresses and opportunities. *Biological Reviews*, 96(3), 798–821.

Hamid, B., Zaman, M., Farooq, S., Fatima, S., Sayyed, R.Z., Baba, Z.A. et al. 2021. Bacterial plant biostimulants: A sustainable way towards improving growth, productivity, and health of crops. *Sustainability*, 21(13), 2856. https://www.mdpi.com/2071-1050/13/5/2856

Hamid, B., Bashir, Z., Yatoo, A.M., Mohiddin, F., Majeed, N., Bansal, M. et al. 2022. Cold-active enzymes and their potential industrial applications: A review. *Molecules*, 27, 5885. https://doi.org/10.3390/molecules27185885.

Hamid, B., Baba, Z.A., Sheikh, T.A., Perveen, K., & Sayyed, R.Z. 2024. Investigating potential protease activity of psychrotrophic bacteria from a municipal landfill for solid waste management. *Biomass Conversion and Biorefinery*. https://doi.org/10.1007/s13399-024-05621-2.

Huang, A., Lu, F., & Liu, F. 2023. Discrimination of psychrophilic enzymes using machine learning algorithms with amino acid composition descriptor. *Frontiers in Microbiology*, 14, 1130594.

Huston, A.L. 2008. Biotechnological aspects of cold-adapted enzymes. In: R. Margesin, F. Schinner, J.-C. Marx, & C. Gerday (Eds.), *Psychrophiles: From Biodiversity to Biotechnology*. Springer, pp. 347–363.

Ihua, M.W., FitzGerald, J.A., Guihéneuf, F., Jackson, S.A., Claesson, M.J., Stengel, D.B. et al. 2020. Diversity of bacteria populations associated with different thallus regions of the brown alga *Laminaria digitata*. *PLOS One*, 15(11), e0242675.

Jahed, K.R., Saini, A.K., & Sherif, S.M. 2023. Coping with the cold: Unveiling cryoprotectants, molecular signaling pathways, and strategies for cold stress resilience. *Frontiers in Plant Science*, 14, 1246093.

Jahnke, L.L., Orphan, V.J., Embaye, T., Turk, K.A., Kubo, M.D., Summons, R.E. et al. 2008. Lipid biomarker and phylogenetic analyses to reveal archaeal biodiversity and distribution in hypersaline microbial mat and underlying sediment. *Geobiology*, 6(4), 394–410.

Jodłowska, I., & Białkowska, A.M. (2024). Cold-adapted fungi: Goldmine of biomolecules applicable in industry. *Applied Sciences*, 14(24), 11950.

Juurakko, C.L., diCenzo, G.C., & Walker, V.K. 2021. Cold acclimation in *Brachypodium* is accompanied by changes in above-ground bacterial and fungal communities. *Plants*, 10(12), 2824.

Kalam, S., Basu, A., Ahmad, I., Sayyed, R.Z., Enshasy, H.E., Dailin, D.J. et al. 2020. Recent understanding of soil Acidobacteria and their ecological significance: A critical review. *Frontiers in Microbiology*, 11, 580024. https://www.frontiersin.org/articles/10.3389/fmicb.2020.580024.

Kazamia, E., Czesnick, H., Nguyen, T.T., Croft, M.T., Sherwood, E., Sasso, S. et al. 2012. Mutualistic interactions between vitamin B_{12}-dependent algae and heterotrophic bacteria exhibit regulation. *Environmental Microbiology*, 14(6), 1466–1476.

Khomutovska, N., de Los Ríos, A., & Jasser, I. 2020. Diversity and colonization strategies of endolithic cyanobacteria in the cold mountain desert of Pamir. *Microorganisms*, 9(1), 6.

Kodera, S.M., Das, P., Gilbert, J.A., & Lutz, H.L. 2022. Conceptual strategies for characterizing interactions in microbial communities. *iScience*, 25(2), 103775.

Kuddus, M., Roohi, Bano, N., Sheik, G.B., Joseph, B., Hamid, B., Sindhu, R. et al. 2024. Cold-active microbial enzymes and their biotechnological applications. *Microbial Biotechnology*, 17(4), e14467.

Kuppan, N., Padman, M., Mahadeva, M., Srinivasan, S., & Devarajan, R. 2024. A comprehensive review of sustainable bioremediation techniques: Eco-friendly solutions for waste and pollution management. *Waste Management Bulletin*, 2(1), 154–171.

Larose, C., Dommergue, A., & Vogel, T.M. 2013. The dynamic Arctic snow pack: An unexplored environment for microbial diversity and activity. *Biology*, 2(1), 317–330.

Lauritano, C., Rizzo, C., Lo Giudice, A., & Saggiomo, M. 2020. Physiological and molecular responses to main environmental stressors of microalgae and bacteria in polar marine environments. *Microorganisms*, 8(12), 1957.

Leite, M.F.A., & Kuramae, E. 2020. You must choose, but choose wisely: Model-based approaches for microbial community analysis. *Soil Biology and Biochemistry*, 151, 108042.

Liu, Y., Shen, L., Zeng, Y., Xing, T., Xu, B., & Wang, N. 2020. Genomic insights of *Cryobacterium* isolated from ice core reveal genome dynamics for adaptation in glacier. *Frontiers in Microbiology*, 11, 1530.

Lo Giudice, A., & Rizzo, C. 2022. Bacteria associated with benthic invertebrates from extreme marine environments: Promising but underexplored sources of biotechnologically relevant molecules. *Marine Drugs*, 20(10), 617.

Loayza-Aguilar, R.E., Medina-Quezada, H., Merino, F., Saldaña-Rojas, G.B., Mendoza, S., & Olivos-Ramirez, G.E. 2024. Occurrence of the Humboldt Penguin (*Spheniscus humboldti*) in El Ferrol Bay, Chimbote, Peru, and strategies for its conservation. *bioRxiv*.

Loreau, M., Naeem, S., Inchausti, P., Bengtsson, J., Grime, J.P., Hector, A. et al. 2001. Biodiversity and ecosystem functioning: Current knowledge and future challenges. *Science*, 294(5543), 804–808.

Maayer, P.D., Anderson, D., Cary, S.C., & Cowan, D.A. 2014. Some like it cold: Understanding the survival strategies of psychrophiles. *EMBO Reports*, 15(5), 508–517.

Majewska, E., Twarda-Clapa, A., Jędrzejczak-Krzepkowska, M., Kamińska-Dwórznicka, A., Zakłos-Szyda, M., & Białkowska, A. 2025. Antifreeze proteins produced by antarctic yeast from the genus *Glaciozyma* as cryoprotectants in food storage. *PLOS One*, 20(3), e0318459.

Malard, L.A., Šabacká, M., Magiopoulos, I., Mowlem, M., Hodson, A., Tranter, M. et al. 2019. Spatial variability of Antarctic surface snow bacterial communities. *Frontiers in Microbiology*, 10, 461.

Malard, L.A., Bergk-Pinto, B., Layton, R., Vogel, T.M., Larose, C., & Pearce, D.A. 2023. Snow microorganisms colonise Arctic soils following snow melt. *Microbial Ecology*, 86(3), 1661–1675.

Margesin, R., & Collins, T. 2019. Microbial ecology of the cryosphere (glacial and permafrost habitats): Current knowledge. *Applied Microbiology and Biotechnology*, 103(6), 2537–2549.

McFall-Ngai, M., Hadfield, M.G., Bosch, T.C.G., Carey, H.V., Domazet-Lošo, T., Douglas, A.E. et al. 2013. Animals in a bacterial world, a new imperative for the life sciences. *Proceedings of the National Academy of Sciences of the United States of America*, 110(9), 3229–3236.

Millum, J. (2010). How should the benefits of bioprospecting be shared? *Hastings Center Report*, 40(1), 24–33.

Milojević, T., Cramm, M.A., Hubert, C.R.J., & Westall, F. 2022. “Freezing” thermophiles: From one temperature extreme to another. *Microorganisms*, 10(12), 2417.

Mori, A.S. (2016). Resilience in the studies of biodiversity–ecosystem functioning. *Trends in Ecology and Evolution*, 31(2), 87–89.

Muñoz-Torres, P., Márquez, S., González-Nilo, F.D., Márquez-Miranda, V., & Blamey, J.M. 2017. Structure and application of antifreeze proteins from antarctic bacteria. *Microbial Cell Factories*, 16, 138.

Musat, N., Werner, U., Knittel, K., Kolb, S., Dodenhof, T., van Beusekom, J.E.E. et al. 2006. Microbial community structure of sandy intertidal sediments in the North Sea, Sylt-Rømø Basin, Wadden Sea. *Systematic and Applied Microbiology*, 29(4), 333–348.

Muthappa, D.M., Lamba, S., Sivasankaran, S.K., Naithani, A., Rogers, N., Srikumar, S. et al. 2022. 16S rRNA-based profiling of bacterial communities colonizing bakery-production environments. *Foodborne Pathogens and Disease*, 19(7), 485–494.

Mykytczuk, N., Foote, S., Omelon, C.R., Southam, G., Greer, C.W., & Whyte, L.G. 2013. Bacterial growth at –15°C; molecular insights from the permafrost bacterium *Planococcus halocryophilus* Or1. *The ISME Journal*, 7(6), 1211–1226.

Nawaz, M.Z., Subin Sasidharan, R., Alghamdi, H.A., & Dang, H. 2022. Understanding interaction patterns within deep-sea microbial communities and their potential applications. *Marine Drugs*, 20(2), 108.

Oktarina, T.N., & Yulianti, A. 2022. The role of women in sustainable development and environmental protection: A discourse of ecofeminisme in Indonesia. *Indonesian Journal of Environmental Law and Sustainable Development*, 1(2), 107–138. missing

Overstreet, R.M., & Lotz, J.M. 2016. Host–symbiont relationships: Understanding the change from guest to pest. In: *The Rasputin Effect: When Commensals and Symbionts Become Parasitic* (Vol. 3). Academic Press, pp. 27–64.

Pernice, M., & Levy, O. 2014. Novel tools integrating metabolic and gene function to study the impact of the environment on coral symbiosis. *Frontiers in Microbiology*, 5, 448.

Qiao, Y., Huang, Q., Guo, H., Qi, M., Zhang, H., Xu, Q. et al. 2024. Nutrient status changes bacterial interactions in a synthetic community. *Applied and Environmental Microbiology*, 90(1), e01566-23.

Rai, S., Omar, A.F., Rehan, M., Al-Turki, A., Sagar, A., Ilyas, N. et al. 2023. Crop microbiome: Their role and advances in molecular and omic techniques for the sustenance of agriculture. *Planta*, 257, 27. https://doi.org/10.1007/s00425-022-04052-5.

Rampelotto, P.H. 2010. Resistance of microorganisms to extreme environmental conditions and its contribution to astrobiology. *Sustainability*, 2(6), 1602–1623.

Rapp, J.Z., Sullivan, M.B., & Deming, J.W. 2021. Divergent genomic adaptations in the microbiomes of Arctic subzero sea-ice and cryopeg brines. *Frontiers in Microbiology*, 12, 701186.

Raymond-Bouchard, I., Tremblay, J., Altshuler, I., Greer, C.W., & Whyte, L.G. 2018. Comparative transcriptomics of cold growth and adaptive features of a eury- and steno-psychrophile. *Frontiers in Microbiology*, 9, 1565.

Rizzo, C., & Lo Giudice, A. 2020. The variety and inscrutability of polar environments as a resource of biotechnologically relevant molecules. *Microorganisms*, 8(9), 1422.

Robinson, J.M., Redvers, N., Camargo, A., Bosch, C.A., Breed, M.F., Brenner, L.A. et al. 2022. Twenty important research questions in microbial exposure and social equity. *mSystems*, 7(1), e01240-21.

Shamarina, D., Stoyantcheva, I., Mason, C.E., Bibby, K., & Elhaik, E. 2017. Communicating the promise, risks, and ethics of large-scale, open space microbiome, and metagenome research. *Microbiome*, 5(1), 132.

Sharma, S., Mohler, J., Mahajan, S.D., Schwartz, S.A., Bruggemann, L., & Aalinkeel, R. 2023. Microbial biofilm: A review on formation, infection, antibiotic resistance, control measures, and innovative treatment. *Microorganisms*, 11(6), 1614.

Škvorová, Z., Černajová, I., Steinová, J., Peksa, O., Moya, P., & Škaloud, P. 2022. Promiscuity in lichens follows clear rules: Partner switching in *Cladonia* is regulated by climatic factors and soil chemistry. *Frontiers in Microbiology*, 12, 781585.

Song, M.H., Zong, N., Jiang, J., Shi, P.L., Zhang, X.Z., Gao, J.Q. et al. 2019. Nutrient-induced shifts of dominant species reduce ecosystem stability via increases in species synchrony and population variability. *Science of the Total Environment*, 692, 441–449.

Tajima, T., Hamada, M., Nakashimada, Y., & Kato, J. 2015. Efficient aspartic acid production by a psychrophile-based simple biocatalyst. *Journal of Industrial Microbiology and Biotechnology*, 42(10), 1319–1324.

Takagi, T., Aoyama, K., Motone, K., Aburaya, S., Yamashiro, H., Miura, N., & Inoue, K. 2023. Mutualistic interactions between dinoflagellates and pigmented bacteria mitigate environmental stress. *Microbiology Spectrum*, 11(1), e0246422.

Venegas, R.M., Acevedo, J., & Treml, E.A. 2023. Three decades of ocean warming impacts on marine ecosystems: A review and perspective. *Deep-Sea Research Part II: Topical Studies in Oceanography*, 212, Article 105318.

Wang, H., Bier, R., Zgleszewski, L., Peipoch, M., Omondi, E., Mukherjee, A. et al. 2020. Distinct distribution of Archaea from soil to freshwater to estuary: Implications of archaeal composition and function in different environments. *Frontiers in Microbiology*, 11, 576661.

Wang, M., Tian, J., Xiang, M., & Liu, X. 2017. Living strategy of cold-adapted fungi with reference to several representative species. *Mycology*, 8(3), 178–188.

Weiland-Bräuer, N. 2021. Friends or foes: Microbial interactions in nature. *Biology*, 10(6), 496.

Xiao, N., Suzuki, K., Nishimiya, Y., Kondo, H., Miura, A., Tsuda, S. et al. 2010. Comparison of functional properties of two fungal antifreeze proteins from *Antarctomyces psychrotrophicus* and *Typhula ishikariensis*. *The FEBS Journal*, 277(2), 394–403.

Xin, Z., & Browse, J. 2000. Cold comfort farm: The acclimation of plants to freezing temperatures. *Plant, Cell, and Environment*, 23(9), 893–902.

Yan, C., Owen, J.S., Seo, E.Y., Jung, D., & He, S. 2023. Microbial interaction is among the key factors for isolation of previously uncultured microbes. *Journal of Microbiology*, 61(7), 655–662.

Yang, G., Mozzicafreddo, M., Ballarini, P., Pucciarelli, S., & Miceli, C. (2021). An *in silico* comparative study of lipases from the antarctic psychrophilic ciliate euplotes focardii and the mesophilic congeneric species *Euplotes crassus*: Insight into molecular cold-adaptation. *Marine Drugs*, 19(2), 67.

Yap, H.S., Zakaria, N.N., Zulkharnain, A., Sabri, S., Gomez-Fuentes, C., & Ahmad, S.A. 2021. Bibliometric analysis of hydrocarbon bioremediation in cold regions and a review on enhanced soil bioremediation. *Biology*, 10(5), 354.

Yin, W., Wang, Y., Liu, L., & He, J. 2019. Biofilms: The microbial "protective clothing" in extreme environments. *International Journal of Molecular Sciences*, 20(14), 3423.

Zhao, Y., Liu, Z., Zhang, B., Cai, J., Yao, X., Zhang, M. et al. 2023. Inter-bacterial mutualism promoted by public goods in a system characterized by deterministic temperature variation. *Nature Communications*, 14(1), 5394.

Zuñiga, C., Zaramela, L., & Zengler, K. (2017). Elucidation of complexity and prediction of interactions in microbial communities. *Microbial Biotechnology*, 10(6), 1500–1522.

CHAPTER

15

Psychrophilic Microbiomes in Cold Climate Agriculture: Prospects and Applications

Ayse Feyza Tufan Dulger[1], Hasan Murat Aksoy[2] and Yilmaz Kaya[1*]

[1] Department of Agricultural Biotechnology, Faculty of Agriculture, Ondokuz Mayis University, Samsun, Türkiye

[2] Plant Protection Department, Faculty of Agriculture, Ondokuz Mayis University, Samsun, Türkiye

1. Introduction

Agricultural production in cold climate regions is constrained by abiotic stressors such as low soil temperatures, shortened growing seasons, and frost-induced plant injuries, all of which pose serious threats to global food security (Sharma et al., 2005). As the global population grows and arable land becomes increasingly limited, expanding cultivation into high-latitude and high-altitude regions is considered a strategic necessity (Su et al., 2024). However, traditional methods often fall short in coping with these environmental extremes. In this context, psychrophilic and psychrotolerant microbiomes, particularly those associated with plant roots and rhizospheres, have gained attention due to their ability to enhance nutrient uptake, stimulate plant growth, and improve cold tolerance (Vega-Celedon et al., 2021; Abdullah et al., 2023; Hou et al., 2024). These cold-adapted microbes offer ecologically sustainable solutions to support crop resilience under harsh conditions (Arvas et al., 2025). Moreover, as modern biotechnology seeks environmentally friendly alternatives to chemical inputs, the role of microbiomes in supporting agricultural biodiversity has become increasingly central (Arvas & Kaya, 2019).

Psychrophilic microorganisms, adapted to permanently cold environments such as polar regions, alpine soils, and permafrost zones, exhibit unique

*Corresponding author: yilmaz.kaya@omu.edu.tr

DOI: 10.1201/9781003717003-15

physiological and molecular traits that enable them to maintain metabolic activity at sub-zero temperatures. Their cell membranes remain fluid due to increased unsaturated fatty acids, and they produce cold-active enzymes, antifreeze proteins, and exopolysaccharides that confer cryoprotection and maintain cellular integrity (Collins & Margesin, 2019). These adaptations support survival and facilitate critical ecological functions such as nutrient cycling and organic matter degradation under cold stress (Sheikh et al., 2022). Moreover, low-temperature ecosystems—particularly those undergoing rapid transformation due to climate change—are now recognized as hotspots of microbial diversity and resilience (Perfumo & Lo Giudice, 2023). Understanding the functionality and ecological roles of these microbes is essential, especially as permafrost thaw and hydrological changes alter biogeochemical cycles and microbial community structure in cold terrestrial and aquatic ecosystems (Miner et al., 2021; Picazo et al., 2021).

Harnessing the plant-growth-promoting (PGP) potential of psychrophilic microbes offers a promising path toward climate-resilient agriculture. Cold-tolerant plant-associated bacteria—especially phosphate-solubilizing strains—can significantly enhance nutrient availability, root development, and overall crop vigor under suboptimal temperature conditions (Rizvi et al., 2021). Integrating such microorganisms into seed biopriming technologies is gaining attention as an eco-friendly strategy to improve seed germination, stress resistance, and field establishment in cold environments (Sarkar et al., 2021; Nasab et al., 2021; Afrouz et al., 2023; Moghaddam et al., 2024). In this regard, psychrophilic microbiomes contribute to sustainable crop production and support biodiversity conservation by reducing reliance on chemical inputs (Arvas & Kaya, 2019). As low-temperature microbiology increasingly intersects with biotechnological innovation, using these extremophiles in agriculture represents a valuable step toward developing circular bioeconomy models suited to marginal environments (Perfumo & Lo Giudice, 2023).

Psychrophilic microorganisms occupy a central role in cold agroecosystems by their metabolic versatility and adaptation to low-temperature environments. They are essential for functioning nutrient cycles, such as nitrogen fixation, phosphorus solubilization, and organic matter decomposition (Suyal et al., 2023). Genera like *Arthrobacter*, *Pseudomonas*, *Paenibacillus*, and *Bacillus* have been isolated from cold soils and have demonstrated strong PGP traits under suboptimal thermal conditions (Dasila et al., 2023). These organisms act as biological alternatives to chemical fertilizers in high-altitude and high-latitude agriculture, thereby enhancing sustainability and reducing ecological footprints.

One of the remarkable traits of psychrophilic microbes is their ability to produce cold-adaptive proteins such as antifreeze proteins (AFPs), ice-binding proteins (IBPs), and cold shock proteins (CSPs), which are crucial not only for microbial survival but also for plant protection in freezing environments. AFPs reduce the freezing point of water and inhibit ice recrystallization, providing cryoprotection to both microbes and host plant tissues. These proteins also help stabilize membranes and prevent frost injury, especially during sudden freeze-thaw cycles. Such biological frost control mechanisms offer promising applications in frost-prone agriculture by reducing structural soil damage and plant tissue necrosis (Białkowska et al., 2020; Jin et al., 2022; Rahman et al., 2024).

Psychrophiles display a range of molecular strategies for coping with cold stress, including alterations in cell membrane lipid composition, production of cold-active enzymes, and increased expression of cold-inducible genes (Hassan et al., 2020; Collins & Feller, 2023; Wu et al., 2023). These adaptations facilitate efficient metabolic functioning at temperatures below 5°C. Their cold-active enzymes—such as amylases, proteases, and lipases—exhibit high catalytic activity at low temperatures, making them suitable for biotechnological applications in low-input agriculture and cold-chain bioprocessing (Yusof et al., 2021; Purwar & Srivastava, 2024). Furthermore, these enzymes and their gene clusters present in psychrophilic genomes can be harnessed for transgenic approaches or microbial inoculants designed for cold agroecosystems (Arvas & Kaya, 2019; Jin et al., 2022). To better understand these unique features of psychrophilic microorganisms, it is first necessary to examine their fundamental characteristics and adaptation mechanisms.

2. Characteristics of Psychrophilic Microbiomes

Psychrophilic microorganisms thrive in environments characterized by extremely low temperatures, including polar regions, high-altitude ecosystems, and permanently frozen soils. Antarctica, the coldest and most remote continent on Earth, presents one of the most pristine and extreme habitats for these organisms. Despite its harsh conditions—such as strong winds, low precipitation, and intense UV radiation—Antarctica harbors a surprisingly rich microbial diversity (Yusof et al., 2021). These microorganisms adapt to survive and play essential ecological roles in nutrient cycling, organic matter decomposition, and biological interactions within cold ecosystems (Kırkıncı et al., 2021). The microbial communities isolated from Antarctic soil, ice, freshwater, and permafrost include diverse bacterial genera such as *Psychrobacter*, *Pseudomonas*, *Streptomyces*, *Arthrobacter*, and *Planococcus*, many of which have demonstrated potential applications in biotechnology, including bioremediation and cryoprotection (Kırkıncı et al., 2021; Marcoleta et al., 2022; Ramasamy et al., 2023). Understanding the adaptive features of these microbes is fundamental for exploring their functional capacities in agroecological contexts and addressing environmental challenges linked to cold climate resilience.

Survival in cold environments necessitates complex biochemical and physiological strategies among psychrophilic microorganisms. These adaptations primarily target membrane structure and function, where an increase in unsaturated fatty acids enhances membrane fluidity, enabling proper nutrient transport and cellular communication under low temperatures (Purwar & Srivastava, 2024).

CSPs and AFPs, including IBPs, are produced to stabilize nucleic acids, inhibit ice crystal formation, and maintain the structural integrity of cellular components during freeze–thaw cycles. Additionally, pigments such as carotenoids and secondary metabolites like violacein and indigoidine are produced to protect cells from oxidative stress and UV radiation in cold ecosystems (Sajjad et al., 2020; Garcia-Lopez et al., 2021; Lauritano & Coppola, 2023). These mechanisms collectively allow psychrophiles to survive and thrive in environments where temperatures frequently drop below 0°C.

The genomic architecture of psychrophilic microorganisms reflects a high degree of metabolic flexibility and adaptation to cold. Comparative genomic analyses have revealed the presence of specific gene clusters responsible for cryoprotection, membrane modification, cold-active enzyme synthesis, and transcriptional regulation under low temperatures (Perfumo & Lo Giudice, 2023). Next-generation sequencing has uncovered more than 180 complete or draft psychrophile genomes, with a significant proportion originating from marine and polar habitats (Orellana-Saez et al., 2019). These organisms possess a suite of cold-adapted enzymes such as amylases, proteases, lipases, and DNA polymerases that remain catalytically active at near-freezing temperatures, often exhibiting greater structural flexibility and lower activation energies than their mesophilic counterparts (Xu et al., 2024). Such enzymes support survival in natural ecosystems and offer significant biotechnological potential for industrial applications in low-energy and low-temperature processes.

Psychrophilic microorganisms demonstrate remarkable modifications to their cell membranes, which are essential for maintaining functional integrity under freezing conditions. To preserve membrane fluidity at subzero temperatures, these microbes increase the proportion of unsaturated and branched-chain fatty acids in their phospholipid bilayers, reducing the melting point and enabling continued transport and signal transduction (Hassan et al., 2020; Jin et al., 2022). Additionally, the expression of extramembrane proteins and transporters is upregulated in response to cold stress, helping compensate for the slower passive diffusion at low temperatures (Mykytczuk et al., 2011). Some cold-adapted microbes also synthesize pigments such as carotenoids that contribute to membrane stability and antioxidant defense, especially under high UV exposure common in polar regions (Kırkıncı et al., 2021). These structural and molecular adaptations distinguish psychrophiles from mesophiles and are critical for survival in cryospheric habitats.

Metabolically, psychrophilic microbes exhibit extraordinary flexibility, allowing them to thrive in oligotrophic, ice-covered, and permafrost ecosystems. Enzymes produced by these organisms, such as cold-active amylases, lipases, and proteases, are adapted to function efficiently at low temperatures with high catalytic rates and low activation energies (Collins & Feller, 2023; Perfumo & Lo Giudice, 2023). In addition to carbon metabolism, several psychrophilic taxa are capable of cycling sulfur, nitrogen, iron, and even methane, enabling them to maintain energy production in diverse geochemical niches (Margesin & Collins, 2019; Patil et al., 2024). These metabolic traits, often encoded in cold-adapted operons and gene clusters, support microbial survival and stabilize ecosystem functions by sustaining nutrient turnover in cold agroecosystems. As global warming alters cryospheric boundaries and permafrost regions, the ecological significance of these cold-adapted communities in maintaining biogeochemical balance is becoming increasingly evident (Jin et al., 2022; Kırkıncı et al., 2021).

Comparative genomic studies have revealed that psychrophilic bacteria possess distinct genetic regions absent in their mesophilic counterparts, which are crucial for adaptation to low-temperature environments. For instance, the genome of *Alteromonas* sp. SN2, isolated from polar marine habitats, includes more than 15 unique cold-adaptation-related loci, including those regulating membrane transport, protein folding, and cryoprotection systems. These genes are often organized in operons or genomic islands that are horizontally transferred and subject to intense selective pressure in cryospheric ecosystems (Math et al., 2012). Such genomic rearrangements facilitate efficient nutrient assimilation, resistance to freeze–thaw cycles, and enhanced cellular stability at low temperatures—traits indispensable for survival in extreme cold environments. Moreover, many of these psychrophiles exhibit gene duplications in pathways related to oxidative stress defense and lipid desaturation, reflecting evolutionary pressure for redundancy and robustness in cold conditions.

In archaeal species such as *Halobacterium* and *Halorubrum* isolated from Antarctic environments, genomic analyses have identified cold-specific genetic traits including gas vesicle synthesis proteins, photoreceptor-linked light-harvesting proteins, and AFPs. These elements help stabilize intracellular structures under low temperatures and support energy acquisition and buoyancy regulation in low-light polar habitats. Across broader databases, at least 83 fully sequenced or draft genomes of psychrophilic microorganisms have been cataloged, providing rich comparative data. These genomes frequently contain adaptations in DNA replication fidelity, DNA repair enzymes, and RNA helicases—factors that ensure proper gene expression despite the rigid physical conditions of icy ecosystems. Additionally, alterations in lipid biosynthesis genes allow dynamic membrane remodeling, while modifications in translation and ribosome-associated proteins further enhance protein synthesis efficiency at sub-zero temperatures. These adaptations collectively highlight the genomic plasticity underpinning microbial survival in the cold biosphere (Cavicchioli et al., 2002; Bowman, 2017; Luo et al., 2024).

In cold environments, psychrophilic microorganisms are often forced to reconfigure their central metabolic pathways to maintain energy production under thermodynamically constrained conditions. Studies have shown that classical glycolytic routes are frequently downregulated or even absent in many cold-adapted species. Instead, alternative pathways such as gluconeogenesis, glyoxylate, and 2-methylcitrate cycles are activated (Purwar & Srivastava, 2024). These routes bypass the energetically unfavorable steps of glycolysis and enable efficient utilization of reduced carbon substrates. For instance, the glyoxylate shunt permits the conservation of carbon skeletons by avoiding decarboxylation steps, which is particularly beneficial in oligotrophic, carbon-limited polar habitats. These metabolic rewirings reflect the evolutionary flexibility of psychrophiles and are supported by cold-regulated transcriptional programs that activate enzymes optimized for low-temperature kinetics.

Psychrophilic bacteria such as *Psychrobacter arcticus* have been shown to lack classical glycolytic genes and instead rely on oxidized substrates such as acetate, ethanol, or fatty acids, utilizing them via modified pathways tailored to cold environments (Ayala-del-Río et al., 2010). Similarly, in *Pseudomonas putida* KT2440, cold exposure induces the upregulation of genes involved in the 2-methylcitrate pathway, which channels propionate and propionyl-CoA into the TCA cycle—effectively linking amino acid catabolism with central carbon metabolism (Fonseca et al., 2011). Such organisms also express cold-active isozymes with broader substrate ranges and lower activation energies, which help maintain catalytic efficiency even near the freezing point. These strategies enable psychrophiles not only to persist but also to perform essential ecological functions such as nutrient turnover and organic matter degradation in cold agroecosystems (Ayala-del-Río et al., 2010; Fonseca et al., 2011; Purwar & Srivastava, 2024).

Transcriptomic analyses have become indispensable in understanding how psychrophilic microorganisms reprogram their metabolism in response to cold stress. In *Psychrobacter* sp. PAMC 21119, high-throughput RNA sequencing revealed that genes involved in classical energy-yielding pathways, including electron transport chain components and ATP synthase complexes, are significantly downregulated at low temperatures. Instead, there is a notable upregulation of acetyl-CoA-associated metabolic pathways, suggesting a shift towards fermentation-type or acetyl-CoA-driven metabolic modules that are more efficient under low-oxygen, energy-limited cryoenvironments (Koh et al., 2017). This transcriptional reconfiguration indicates an adaptive strategy to conserve energy and maintain cellular viability in cold and nutrient-poor conditions.

Proteomic and metabolomic studies further support the transcriptomic findings by confirming the elevated expression of enzymes involved in cold-active metabolic circuits. In psychrophilic species such as *Nesterenkonia* sp. AN1 and *Exiguobacterium sibiricum*, key enzymes of the glyoxylate shunt and β-oxidation pathways, are found to be significantly upregulated, particularly during prolonged cold exposure (Purwar & Srivastava, 2024). These metabolic shifts enhance the use of fatty acids and alternative carbon sources, bypassing the carbon loss typical in the TCA cycle and supporting biomass maintenance in carbon-limited soils. Integration of omics datasets thus provides a holistic understanding of how psychrophiles restructure their metabolic networks, revealing novel targets for biotechnological applications in low-temperature processes (Koh et al., 2017; Purwar & Srivastava, 2024). These biochemical and molecular adaptations are crucial in understanding how plant–microbe interactions are shaped and function under cold climate conditions.

3. Plant–Microbe Interactions Under Low Temperatures

Microbiomes refer to the entire community of microorganisms present in a specific environment (such as soil, the rhizosphere, leaf surfaces, or internal plant tissues), and the mutual relationships these microorganisms establish with plants are studied under *plant–microbe interactions* (Aksoy et al., 2018; Aksoy et al., 2021). These interactions are critically important for plant production systems, as microorganisms promote plant growth, enhance resistance to biotic and abiotic stresses, facilitate the mobilization of nutrients, and regulate soil health. For instance, rhizobacteria can fix atmospheric nitrogen, while mycorrhizal fungi help plants absorb phosphorus more efficiently. Additionally, certain endophytic microbes directly influence plant development by producing phytohormones. These symbiotic relationships offer ecological solutions for sustainable agriculture and hold potential to reduce the use of chemical fertilizers and pesticides. Recent research has shown that engineering the microbiome to form targeted microbial communities represents a promising strategy for cultivating high-yield, stress-resilient plant varieties (Mendes et al., 2013; Philippot et al., 2013; Berg et al., 2020; Triverdi et al., 2020).

Cold soils pose significant challenges to plant growth due to reduced microbial activity, limited nutrient availability, and low organic matter decomposition rates. However, psychrophilic and psychrotolerant microorganisms in the rhizosphere of cold environments have evolved adaptive strategies that allow them to maintain functional interactions with plant roots. These microbes contribute to nutrient cycling, enhance soil structure, and mediate stress responses, promoting plant resilience under harsh climatic conditions (Perfumo & Lo Giudice, 2023). Particularly, phosphate-solubilizing and nitrogen-fixing bacteria native to cold soils play a vital role in sustaining plant productivity without synthetic inputs, highlighting their potential as biofertilizers (Suyal et al., 2023). Moreover, cold-adapted microbial communities are characterized by high taxonomic and functional diversity, which ensures ecosystem stability despite thermal limitations. Classic studies by Margesin and Miteva (2011) have shown that such communities are metabolically active at low temperatures and capable of forming complex networks with plants and other soil organisms. Understanding these rhizospheric dynamics is essential for developing sustainable agricultural systems in alpine, subpolar, and other cold climate regions (Margesin & Miteva, 2011; Perfumo & Lo Giudice, 2023; Suyal et al., 2023).

Psychrophilic endophytic microorganisms establish highly specialized symbiotic relationships with their plant hosts in cold and sub-zero environments, contributing significantly to cold stress tolerance and overall plant fitness. These endophytes colonize internal plant tissues without causing harm and can influence key physiological processes such as nutrient uptake, hormone regulation, and antioxidant defense. Studies have demonstrated that psychrotolerant bacterial and fungal endophytes enhance cold tolerance by modulating plant stress responses, often through the production of antifreeze proteins, exopolysaccharides, and phytohormones like indole acetic acid (IAA) and abscisic acid (ABA). Furthermore, they can trigger the expression of cold-responsive genes in plants, reinforcing the plant's ability to cope with chilling injuries. These endophyte-mediated interactions are crucial for maintaining plant growth under limited water and nutrient conditions, especially in high-altitude and polar agriculture. Such findings underscore the biotechnological potential of psychrophilic endophytes as sustainable tools for crop improvement in cold climate regions (See-Too et al., 2016; Muñoz et al., 2017; Tapia-Vázquez et al., 2020; Bhardwaj et al., 2025).

Cold-adapted microbial communities, particularly psychrophilic plant growth-promoting rhizobacteria (PGPR), play a vital role in mitigating cold-induced stress in plants. These microorganisms enhance cold tolerance through multiple mechanisms such as phosphate solubilization, production of antifreeze compounds, and synthesis of stress-related phytohormones (Rizvi et al., 2023). For instance, phosphorus-solubilizing psychrotolerant strains improve root growth and nutrient acquisition, allowing plants to maintain metabolic function under chilling stress. Moreover, the excretion of exopolysaccharides by these microbes forms protective biofilms around root surfaces, reducing desiccation and insulating the rhizosphere. These physiological adaptations are particularly beneficial for crops cultivated in high-altitude and subpolar agricultural systems (Table 1), where early-season frost events and cold soil conditions can severely hamper crop establishment (Figure 1) (Purwar & Srivastava, 2024; Rahman et al., 2024).

At the molecular level, psychrophilic microbes interact with plant signaling pathways to induce the expression of cold-responsive genes, including those encoding late embryogenesis abundant (LEA) proteins, dehydrins, and C-repeat binding factors (CBFs) that collectively enhance cellular protection against freezing damage (Liu et al., 2011). Furthermore, Rahman et al. (2024) demonstrated that certain psychrophiles can influence soil thermal properties by producing antifreeze proteins and modifying ice–water phase dynamics around plant roots. This microbial regulation of soil thermodynamics effectively buffers root tissues from sudden freezing events, creating a microenvironment more conducive to seed germination and root elongation. These findings collectively suggest that psychrophilic microbes support plant metabolism and act as biological climate modulators, offering promising solutions for frost-prone agricultural zones.

The mechanisms by which these microbial interactions mitigate cold stress are also closely related to their fundamental traits promoting plant growth. The mechanisms by which these microbial interactions mitigate cold stress are closely associated with their fundamental characteristics, promoting plant growth. It becomes particularly important to develop innovative techniques such as seed biopriming to utilize these plant growth-promoting traits effectively.

Table 1: Cold-adapted microorganisms and their agricultural applications

Microorganism	**Origin**	**Key Agricultural Application**	**References**
Pseudomonas lurida M2RH3 (MTCC 9245)	High-altitude rhizospheric soil, Uttarakhand Himalayas	Phosphate solubilization and enhancement of wheat seedling growth under cold conditions (4–30°C)	Selvakumar et al., 2011
Bacillus megaterium M3, *Bacillus subtilis* OSU 142, *Azospirillum brasilense* Sp 245, *Raoultella terrigena*	Greenhouse experiment (Türkiye)	Reduced freezing injury, improved antioxidant enzyme (SOD, POD, CAT) activity, enhanced wheat and barley growth	Turan et al., 2013
Pseudomonas sp. L3, *Pseudomonas* sp. P2, *Streptomyces* sp. T3, *Streptococcus* sp. T4	Agroforestry soils (field trials in India)	Phosphate solubilization at low temperature, increased wheat grain yield and soil phosphatase activity	Dasila et al., 2023
Acinetobacter rhizosphaerae EU-KL44	Wheat rhizosphere, Keylong region (Himachal Pradesh, India)	Alleviates cold stress and enhances nutrient uptake in wheat through multiple PGP traits (K, Zn solubilization, IAA, siderophore)	Kour and Yadav, 2023
Pseudomonas palleroniana N26	Himalayan cold soil	Diazotrophic psychrotroph exhibiting nitrogen fixation and stress adaptation under low-temperature conditions	Soni et al., 2015
Psychrobacter arcticus 273-4	Siberian permafrost soil (Kolyma region)	Genomic adaptation to subzero growth; cold shock proteins, membrane modifications, and flexible proteome aiding survival	Ayala-del-Río et al., 2010
Mesorhizobium sp. (K-solubilizing strain)	Rhizosphere soils of temperate and cold desert regions, northern India	Potassium solubilization and production of multiple PGP metabolites (IAA, cellulase, protease, chitinase)	Baba et al., 2021

(Contd.)

(*Contd.*)

Microorganism	Origin	Key Agricultural Application	References
Naganishia liquefaciens ARY7 (yeast endophyte)	Roots of cold-desert plant Arnebia euchroma	Enhances cold tolerance in Arabidopsis thaliana through auxin and salicylic acid signaling, EPS accumulation	Bhardwaj et al., 2025
Trichoderma harzianum	Laboratory and greenhouse, University of Mohaghegh Ardabili (Iran)	Seed biopriming improves cold tolerance in maize, enhances chlorophyll, enzyme activity, and seedling emergence	Afrouz et al., 2023
Pseudomonas sp. ANT_H4	Antarctic soil	Produces pyomelanin pigment with antioxidant, UV-protective, and plant-priming effects (e.g., Calendula officinalis hairy roots)	Styczynski et al., 2022
Rhodococcus sp. JG3, *Polaromonas* sp. Eur3 1.2.1	Permafrost subzero environments	Transcriptomic adaptation to cold; improved nutrient transport, oxidative stress response, and capsule synthesis	Raymond-Bouchard et al., 2018
Colwellia psychrerythraea 34H	Marine cold ecosystem	Cold-adapted enzymes, cryotolerance mechanisms, and membrane fluidity regulation; relevance to carbon and nutrient cycling	Methé et al., 2005

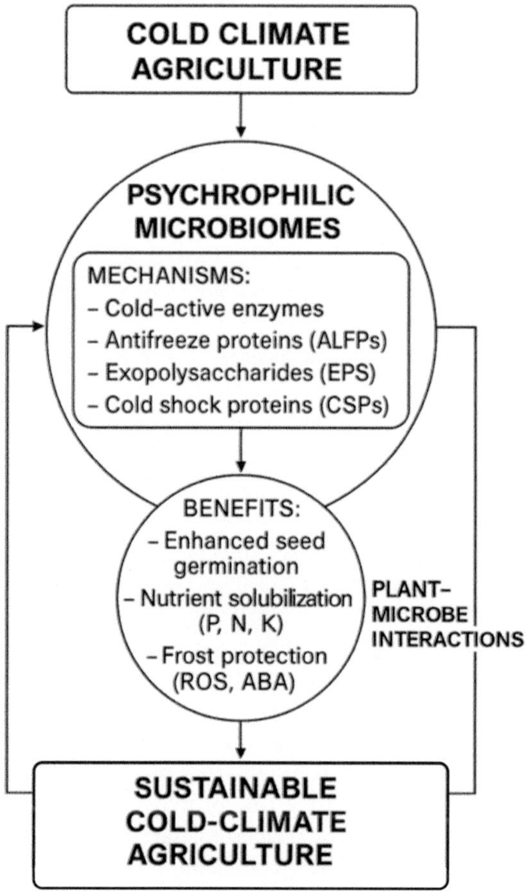

Figure 1: Conceptual diagram illustrating the role of psychrophilic microbiomes in cold-climate agriculture.

4. Biopriming with Psychrophilic Microorganisms

Biopriming is a biological seed treatment technique involving soaking seeds in a moist environment with beneficial microorganisms—such as *Trichoderma*, *Bacillus*, or *Pseudomonas*—for a specific period. This allows microbial inoculants to adhere to the seed surface, colonize the embryo region, and support seedlings' healthy germination and early growth. Unlike conventional seed coating, biopriming is not only protective but also growth-promoting. It enhances germination rate, stimulates root system development, and optimizes the microecological balance around the seed in favor of the plant (Singh et al., 2016; Mahmood & Kataoka, 2018).

Biopriming is highly valuable in plant production as it enables the development of stress-tolerant and vigorously growing seedlings. The microorganisms residing near the seed help activate plant defense mechanisms against both biotic (e.g., pathogens) and abiotic (e.g., drought, salinity, heavy metals) stressors. These microbes also contribute to nutrient cycles through actions like nitrogen fixation and phosphorus solubilization, and they produce phytohormones such as auxins and cytokinins. As a biofertilizer and biopesticide alternative, biopriming aligns with sustainable agriculture goals by reducing dependency on synthetic inputs (Kumar et al., 2020; Dubey et al., 2024). Particularly in low-input systems, it presents a promising biotechnological strategy to boost agricultural productivity while maintaining ecological integrity.

Biopriming is a seed treatment strategy that combines hydration with the application of beneficial microbes, allowing early metabolic activation of both seeds and microbial inoculants. Under cold stress conditions, where germination and early seedling growth are typically delayed, biopriming with psychrotolerant microbes offers a promising solution. Suyal et al. (2023) demonstrated that inoculation of seeds with microbial isolates from Himalayan soils—rich in cold-adapted bacteria—promotes early rhizoplane colonization and accelerates the establishment of mutualistic interactions. These microbes initiate metabolic activity even at low temperatures, enhancing nutrient mobilization and improving seed vigor. Moreover, biopriming facilitates a pre-emptive physiological readiness in the seed, including enhanced antioxidant defense and osmolyte accumulation, which helps plants cope with early-season abiotic stress.

In addition to microbial colonization, the metabolites produced during biopriming play a critical role in modulating plant responses to cold stress. Styczynski et al. (2022) explored the role of pyomelanin, a dark pigment with antioxidant properties, produced by psychrotolerant *Pseudomonas* strains during priming. Their findings showed that pyomelanin improved root architecture and conferred cellular protection against reactive oxygen species (ROS) accumulation under cold exposure. This indicates that microbial metabolites can act as signaling molecules or protective agents during early plant development. Such mechanisms underscore the multifaceted nature of biopriming, which goes beyond simple inoculation by leveraging the biochemical arsenal of cold-adapted microbes to enhance plant resilience from the earliest growth stages.

Seed biopriming using psychrotolerant microbes has emerged as a viable approach to enhance germination efficiency and early vigor, particularly under cold soil conditions. In a study by Kour and Yadav (2023), wheat seeds treated with *Acinetobacter rhizosphaerae* exhibited significantly improved germination rates and early root elongation at suboptimal temperatures. The treated seeds demonstrated higher metabolic activity, as evidenced by increased α-amylase activity and faster mobilization of storage reserves. Additionally, the bioprimed seedlings displayed enhanced expression of antioxidant enzymes such as superoxide dismutase (SOD) and catalase (CAT), suggesting that microbial priming contributes to cold stress mitigation through physiological protection mechanisms. These results highlight the dual benefit of microbial biopriming—both as a stimulant of developmental processes and as a buffer against abiotic stress.

Psychrotolerant PGPR have been increasingly explored in cereal crops cultivated under cold climate conditions. Studies demonstrated that microbial inoculation with strains such as Pseudomonas fluorescens and Bacillus subtilis enhanced shoot length, chlorophyll content, root biomass, and early tiller development in wheat and barley grown under suboptimal temperatures. These effects were linked to increased indole-3-acetic acid (IAA) production, siderophores, and cold-active enzymes facilitating nutrient uptake and protecting root tissues. Furthermore, inoculated plants exhibited lower electrolyte leakage and higher relative water content, indicating improved membrane stability and stress resilience. Such findings highlight the utility of cold-adapted microbial strains as bio-inputs for temperate cereal production systems (Mishra et al., 2011; Selvakumar et al., 2011; Turan et al., 2013; Verma et al., 2015).

Low-temperature soil conditions in tuber crops like potato can severely impede root expansion and nutrient absorption, especially during early vegetative stages. Rahman et al. (2024) investigated the application of psychrophilic microbial consortia in frost-prone soils. They found that strains producing AFPs were critical in modifying the thermal environment surrounding the root zone. The inoculated soils exhibited altered ice–water phase dynamics, with a measurable reduction in ice crystal formation and surface soil freezing. As a result, inoculated potato plants maintained more stable root zone temperatures, leading to improved germination rates, greater stolon development, and increased tuber mass. These outcomes demonstrate the potential of psychrophilic inoculants as biofertilizers and biological frost protectants in cold-region agriculture.

Integrating psychrophilic microorganisms into seed coating formulations is gaining attention as a next-generation approach to enhance crop establishment in cold and marginal environments. Cold-active extremozymes produced by psychrotolerant bacteria could be embedded into seed coatings to facilitate early metabolic activation at low soil temperatures. These enzymes—such as cold-active amylases, cellulases, and phosphatases—could assist in seed coat weakening, starch mobilization, and early root penetration even under suboptimal thermal conditions (Mishra et al., 2011; Yadav et al., 2016). Encapsulating psychrophilic spores or metabolites in biodegradable polymers may offer a controlled release system, ensuring microbial viability and functionality throughout early seedling development. This represents a promising biotechnological innovation for sustainable agriculture in high-altitude and high-latitude regions.

Recent molecular studies on psychrotolerant diazotrophic strains have further informed the selection of microbial candidates for functional seed coatings. Soni et al. (2015) reported that *Pseudomonas palleroniana*, a nitrogen-fixing bacterium isolated from cold soils, exhibits enhanced expression of cold-adaptive proteins, including those related to membrane transport and oxidative stress mitigation. These traits suggest that such microbes can maintain metabolic activity at low temperatures, supporting nitrogen assimilation and early plant vigor. Embedding these strains in polymer-based coatings or biochar matrices may offer dual benefits: microbe-mediated plant growth promotion and protection from early-season frost damage. As cold climate agriculture expands, integrating functional genomics into coating design will be pivotal for tailoring microbe–seed interactions to specific environmental challenges.

5. Biotechnological Applications in Cold Climate Agriculture

Biofertilizers are formulations containing living microorganisms that enhance plant nutrient acquisition through processes such as nitrogen fixation, phosphorus solubilization, and production of plant growth-promoting substances. Biopesticides, on the other hand, are biological agents—derived from bacteria, fungi, viruses, or plant extracts—that suppress pests and diseases through ecological and physiological mechanisms. Genetically engineered organisms (GEOs), including genetically modified (GM) crops and microbial inoculants, are organisms whose genetic material has been modified to enhance traits such as disease resistance, drought tolerance, or nutrient use efficiency. Together, these tools provide alternatives to synthetic agrochemicals and contribute to environmentally friendly and sustainable agricultural systems (Ali et al., 2020; Hernández-Fernández et al., 2021).

Applying biofertilizers and biopesticides is pivotal in improving soil fertility, enhancing crop productivity, and reducing the dependency on chemical fertilizers and pesticides. These agents improve plant health and growth under optimal and stress conditions, while preserving soil microbiota and ecological balance. Genetically engineered microbes and plants further support agricultural efficiency by introducing precise traits that enhance resource utilization and resistance to biotic or abiotic stress. These biotechnologies are fundamental in advancing sustainable agriculture, supporting climate-resilient farming, and ensuring food security under resource-limited conditions (Sudheer et al., 2020; Ortiz & Sansinenea, 2023).

Psychrophilic and psychrotolerant microbes have demonstrated remarkable plant growth-promoting capabilities, making them valuable candidates for biofertilizer formulations in cold and high-altitude agroecosystems. According to Suyal et al. (2023), many cold-adapted bacterial strains isolated from Himalayan and Arctic soils exhibit traits such as nitrogen fixation, phosphate solubilization, and potassium mobilization, which are essential for plant nutrition in nutrient-limited cold soils (Baba et al., 2021). For example, strains of *Pseudomonas*, *Bacillus*, and *Arthrobacter* have been shown to increase the availability of phosphorus and nitrogen in the rhizosphere, leading to improved plant biomass and early growth. Purwar and Srivastava (2024) further emphasize that these microbes can remain metabolically active at temperatures below 10°C, maintaining their beneficial functions even when mesophilic inoculants fail. Thus, psychrophilic biofertilizers represent a promising solution for reducing chemical inputs in sustainable cold climate agriculture.

Beyond nutrient mobilization, psychrophilic microbes also effectively control phytopathogens, making them suitable for development as biopesticides. Kaushal et al. (2024) reported that several cold-tolerant strains produce antifungal compounds such as siderophores, lytic enzymes (e.g., chitinases and β-1,3-glucanases), and volatile organic compounds that inhibit the growth of pathogens like *Fusarium oxysporum* and *Rhizoctonia solani* at low temperatures. These microbes exert both direct antagonistic effects and indirect mechanisms, such as induced systemic resistance (ISR), contributing to enhanced plant defense responses. The ability of psychrophilic microbes to function under suboptimal conditions gives them a competitive edge in integrated pest management strategies in cold regions. Their incorporation into bioformulations can reduce reliance on synthetic fungicides while enhancing crop resilience and soil health.

Cold tolerance in microorganisms is driven by a suite of structural and regulatory genes that allow metabolic activity at low temperatures. Advances in omics technologies have enabled the identification of these cold-adaptive genes, which can be harnessed through genetic engineering to develop cold-tolerant microbial strains for agricultural use. One notable example is *Pseudomonas lurida* NPRp15, whose proteomic profile revealed upregulation of multiple cold-responsive proteins, including cold shock proteins (Csps), chaperonins, and enzymes involved in membrane lipid modification (Jin et al., 2022). These proteins enhance survival at subzero temperatures and help maintain membrane integrity, enzyme functionality, and nutrient transport—all critical for plant-microbe interactions in cold soils. The characterization of such proteins provides valuable genetic targets for synthetic biology approaches aimed at constructing designer microbes for cold-region agriculture (Margesin et al., 2017).

Raymond-Bouchard et al. (2018) performed comparative genomics on *Shewanella* species from polar marine environments and revealed significant genome-level adaptations to cold stress. These included gene expansions in signal transduction, DNA repair, and oxidative stress resistance pathways. Notably, cold-environment *Shewanella* strains contained abundant genes encoding unsaturated fatty acid synthases and modified respiratory enzymes adapted to low oxygen availability in icy ecosystems. These adaptations suggest that genes from extremophiles such as *Shewanella* can be helpful to donors in horizontal gene transfer or recombinant DNA strategies to construct novel cold-tolerant chassis organisms. Such engineered strains may prove highly effective as inoculants in frozen soils or microbial factories in low-temperature bioprocesses.

The ability to engineer cold-tolerant strains using well-characterized psychrophilic genes offers exciting prospects for biofertilizer development, bioremediation in alpine zones, and low-energy industrial fermentation. With advances in CRISPR/Cas genome editing and gene circuit design, it is now feasible to insert cold-adaptive gene clusters into agriculturally relevant microbial hosts such as *Bacillus*, *Azospirillum*, or *Rhizobium*. Furthermore, identifying active promoters at low temperatures could improve expression control in engineered strains. Combining systems biology, evolutionary genomics, and synthetic biology will ultimately accelerate the development of tailor-made microorganisms capable of performing in cold and fluctuating environments where traditional microbial inoculants underperform.

Psychrophilic microorganisms such as *Psychrobacter cryohalolentis* F5-6 have demonstrated significant potential for industrial bioprocessing under low-temperature conditions. These organisms can maintain enzymatic activity and metabolic functions at temperatures as low as 4°C, making them attractive candidates for cold-chain biomanufacturing and energy-efficient fermentation systems. Specifically, *P. cryohalolentis* has been explored for its ability to ferment sugars into bioethanol in cold environments, which can be advantageous in reducing the need for external heating in traditional fermentation reactors. Additionally, the strain's robust stress tolerance and ability to metabolize diverse carbon sources make it suitable for bioconversion processes using agricultural residues or industrial effluents in temperate climates (Methé et al., 2005; Margesin & Feller, 2010).

Other cold-adapted marine bacteria, such as *Pseudoalteromonas* and *Colwellia psychrerythraea*, are increasingly recognized for producing high-value proteins, pigments, and secondary metabolites at low temperatures. These species exhibit cold-active enzymes—including proteases, lipases, and esterases—functioning efficiently under refrigeration or near-freezing conditions. Their biotechnological applications range from enzyme production for the food and pharmaceutical industries to biosurfactants and antimicrobial compounds used in agriculture and aquaculture. Low-temperature fermentation with these organisms reduces energy consumption and minimizes contamination risks, as most mesophilic spoilage organisms cannot compete under such conditions. This makes psychrophilic fermentation a promising avenue for sustainable and safe bioprocessing in cold supply chains.

6. Limitations and Challenges

One of the major limitations in the field application of psychrophilic microbial inoculants is ensuring their survivability under fluctuating environmental conditions. In cold and often moisture-limited soils, microbial activity is naturally reduced, and without adequate formulation, inoculants may fail to establish themselves or remain viable for long periods. Dasila et al. (2023) noted that the viability of cold-adapted PGPRs can be drastically affected by soil structure, freeze–thaw cycles, and nutrient availability. Furthermore, the absence of robust carrier materials tailored for cold environments limits the efficiency of delivery systems. Inoculants applied as powders or liquid suspensions may desiccate, freeze, or degrade, reducing microbial counts before plant interaction begins.

The formulation of microbial products for cold climates requires novel encapsulation techniques to protect microbes from freezing injury and desiccation while maintaining their metabolic readiness. Rizvi et al. (2023) emphasized the need for cryo-compatible matrices, such as alginate beads enriched with cryoprotectants or biopolymer coatings infused with antifreeze proteins. However, these advanced formulations remain in early research stages and are rarely available in commercial biofertilizer products. Until scalable, shelf-stable, and cold-resistant delivery systems are developed, field-scale success with psychrophilic inoculants will remain inconsistent and site-specific.

Introducing non-native psychrophilic strains into soil microbiomes can unintentionally disrupt existing microbial communities. Studies such as those reported in *Microorganisms* (2023) show that even beneficial inoculants may alter microbial diversity and functional guilds in the rhizosphere, particularly under stressful conditions where competition is intensified. Psychrophilic microbes with high resource efficiency or antimicrobial activity may outcompete indigenous strains, leading to shifts in nitrogen, carbon, or phosphorus cycling. These community-level disturbances could impair long-term soil health or alter ecosystem services beyond the cropping season.

In alpine or subpolar soils with low microbial diversity, the introduction of psychrophilic inoculants may amplify selective pressures and reduce ecosystem resilience.

Some scientists observed that even mild inputs such as lignin-rich residues caused shifts in community composition and enzymatic activities in cold soils. The competitive imbalance may be even greater when psychrophilic inoculants with engineered traits or strong metabolic capabilities are introduced. These findings underscore the need for long-term field studies evaluating not just short-term plant responses, but also the legacy effects on microbial networks (Dasila et al., 2023).

From a regulatory perspective, psychrophilic bioinoculants raise concerns regarding ecological containment, horizontal gene transfer, and unintended interactions with native biota. Some scientists argue that despite their natural origin, cold-adapted microbes used in biotechnology should undergo rigorous biosafety evaluation, especially when used in open environments like farms. Moreover, Rahman et al. (2024) warned that microbial production of antifreeze proteins in soil could interfere with natural freeze–thaw patterns and soil hydrology, potentially affecting non-target organisms. Currently, few regulatory frameworks specifically address these cold-climate scenarios, highlighting a need for updated guidelines and targeted risk assessment models. To overcome existing challenges, it is essential to define forward-looking research perspectives to ensure sustainable progress in psychrophilic microbiomes.

7. Future Prospects and Research Directions

Synthetic biology is revolutionizing the development of microbial inoculants through precision genome engineering, enabling the design of microbial strains customized for cold climate functionality. This includes inserting cold-responsive regulatory elements, antifreeze protein genes, and metabolic circuits that remain active at low temperatures. Bowman (2017) highlighted how synthetic approaches transfer cold-adaptive genes from extremophiles into agriculturally relevant microbial hosts, such as *Bacillus* or *Pseudomonas* spp. These engineered strains can function as consistent biofertilizers or biopesticides in suboptimal soil conditions, thereby overcoming limitations that naturally isolated strains face in field applications.

Moreover, synthetic biology paves the way for constructing synthetic microbial consortia or "designer microbiomes" where microbial members are selected based on complementary traits, such as nitrogen fixation, phosphate solubilization, and ROS detoxification—all active in cold stress conditions. Gene editing tools like CRISPR/Cas9 and genome-scale metabolic models facilitate removing unwanted traits and enhancing cooperative metabolic functions. This level of microbial control can potentially reduce field variability and create reproducible outcomes in cold climate agriculture, moving microbial biotechnology from trial-and-error to predictable system design.

Metagenomics has emerged as a powerful tool to explore the vast, uncultured majority of psychrophilic microbes in extreme environments. By sequencing environmental DNA (eDNA) from permafrost, glacial soils, or alpine rhizospheres, researchers can infer microbial diversity and detect genes responsible for cold adaptation. However, linking genetic potential to function remains challenging. Here, machine learning and AI-based trait prediction play a crucial role. Mocali et al. (2017) integrated genomic and phenomic data to predict cold-active enzymatic activities and metabolic pathways relevant to plant-microbe interactions, enabling more targeted bioinoculant development.

Artificial intelligence (AI) tools are increasingly used to screen large metagenomic datasets for traits such as osmolyte production, antifreeze protein genes, and stress-related regulatory motifs. These computational approaches accelerate the discovery of novel gene candidates and reduce experimental costs. For instance, neural network-based models can predict microbial resilience under different thermal conditions, aiding in the pre-selection of strains most likely to succeed in specific cold-field conditions. This fusion of big data analytics with microbiology heralds a new era in microbiome engineering, where inoculant design is driven by predictive precision rather than empirical screening alone.

Climate change is dramatically altering psychrophilic microbiome distribution, diversity, and function, particularly in sensitive ecosystems like the Arctic, Antarctic, and high-altitude regions. Warming temperatures, increased freeze–thaw events, and altered snowmelt patterns lead to shifts in microbial community composition and disrupt long-established symbiotic relationships. Perfumo and Lo Giudice (2023) emphasize that such changes not only affect microbial diversity. Still, they may also influence nutrient cycling, greenhouse gas emissions, and plant–microbe interactions—factors critical to natural and agricultural cold-climate systems (Bowman, 2017).

These disruptions demand a proactive research agenda focused on microbial resilience, adaptation, and ecological feedback. Longitudinal monitoring of microbial populations, functional redundancy assessments, and integrating microbiome shifts into climate models will be key. In addition, breeding or engineering cold-tolerant crops that form stable associations with shifting microbiomes will help buffer agricultural systems against ecological instability. Cold-climate microbiome research must now align more closely with climate mitigation and adaptation strategies, transforming these microbial communities from passive indicators to active agents in sustainable ecosystem management. Future research perspectives will help summarize the importance of psychrophilic microbiomes in sustainable cold climate agriculture and enable a comprehensive evaluation of the findings obtained.

8. Conclusions

Psychrophilic microorganisms have emerged as key players in pursuing sustainable solutions for cold climate agriculture. This chapter highlighted their unique adaptations—such as cold-active enzymes, antifreeze proteins, and membrane modifications—that enable metabolic activity under subzero temperatures. Through their roles in nutrient cycling, biocontrol, stress mitigation, and biostimulation, these microbes offer alternative strategies to chemical fertilizers and pesticides, especially in high-altitude or high-latitude regions. Research has demonstrated their effectiveness in enhancing seed germination, promoting plant growth, and providing resilience against cold-induced abiotic stress in crops like wheat, barley, and potato (Kırkıncı et al., 2021; Perfumo & Lo Giudice, 2023).

Beyond their immediate agricultural functions, psychrophiles hold broader implications for biotechnology and climate adaptation. Recent metagenomic and omic studies demonstrate that psychrophilic communities harbor vast, untapped genetic diversity with industrial applications ranging from fermentation to bioremediation (Hamid et al., 2024). Their ecological importance is also growing in the context of climate change, where they may buffer ecosystems against warming and permafrost degradation effects. However, to fully realize their promise, integrated research combining synthetic biology, long-term ecological monitoring, and cold-environment formulation technologies is essential. With proper stewardship, psychrophilic microbiomes could become cornerstone tools for resilient and low-impact agriculture in the 21st century.

Funding

The authors acknowledged Ondokuz Mayis University Grant Number BAP05-2025-5836, BAP05-2025-6009 and BAP05-2025-6080 for partly sponsoring this research.

References

Abdullah, M., Tariq, M., Zahra, S.T., Ahmad, A., Zafar, M., & Ali, S. 2023. Potential of psychrotolerant rhizobacteria for the growth promotion of wheat (Triticum aestivum L.). *PeerJ.*, 11, e16399.

Afrouz, M., Sayyed, R.Z., Fazeli-Nasab, B., Piri, R., Almalki, W.H., & Fitriatin, B.N. 2023. Seed bio-priming with beneficial *Trichoderma harzianum* alleviates cold stress in maize. *PeerJ.*, 11, e15644. http://doi.org/10.7717/peerj.15644.

Ali, R., Zulaykha, K.D., & Sajjad, N. 2020. Genetically modified microbes as biofertilizers. In: *Bioremediation and Biotechnology* (Vol. 4). *Techniques for Noxious Substances Remediation*. Springer Nature, pp. 275–293.

Aksoy, H.M., Arslanoglu, S.F., Edbeib, M.F., Kaya, Y., & Marakli, S. 2021. Antibacterial activity of Calendula officinalis and Echinacea purpurea extracts against the causal agent of tomatoes' bacterial canker: Clavibacter michiganensis subsp. michiganensis. *Bol. Latinoam. Caribe Plantas Med.*, 20, 496–502. https://doi.org/10.37360/blacpma.21.20.5.36.

Aksoy, H.M., Saruhan, İ., Kaya, Y., & Ozturk, M. 2018. Morphological changes caused by Bacillus megaterium on Adult Emergence of Fall Webworm's Pupa, Hyphantria cunea (Drury) (Lepidoptera: Erebidae). *Journal of Agricultural Sciences*, 24(4), 539–546. https://doi.org/10.15832/ankutbd.349474.

Arvas, Y.E., Devlet, A., & Kaya, Y. 2025.. Cultivation of new crops under changing climate. In: Faiz, S., Ashraf, U., Attia, K.A., & Amir, R.M. (Eds.), *Climate Smart Agriculture for Future Food Security*. Springer, Singapore. https://doi.org/10.1007/978-981-96-4499-5_3.

Arvas, Y.E., & Kaya, Y. 2019. Genetiği değiştirilmiş bitkilerin biyolojik çeşitliliğe potansiyel etkileri. *YYÜ Tarım Bilimleri Dergisi*, 29(1), 168–177. https://doi.org/10.29133/yyutbd.468218.

Ayala-del-Río, H.L., Chain, P.S., Grzymski, J.J., Ponder, M.A., Ivanova, N., Bergholz, P.W. et al. 2010. The genome sequence of Psychrobacter arcticus 273-4, a psychroactive Siberian permafrost bacterium, reveals mechanisms for adaptation to low-temperature growth. *Applied and Environmental Microbiology*, 76(7), 2304–2312.

Baba, Z.A., Hamid, B., Sheikh, T.A., Alotaibi, S., Enshasy, H.E., Ansari, M.J. et al. 2021. Psychrotolerant *Mesorhizobium* sp. isolated from temperate and cold desert regions solubilize Potassium and produces multiple plant growth promoting metabolites, *Molecules*, 26, 5758. https://www.mdpi.com/1420-3049/26/19/5758/htm.

Berg, G., Rybakova, D., Fischer, D., & Cernava, T. 2020. Microbiome definition re-visited: Old concepts and new challenges. *Microbiome*, 8(1), 103. https://doi.org/10.1186/s40168-020-00875-0.

Bhardwaj, P., Jain, R., Rawat, N., Joshi, R., Kumar, A., Pandey, S.S., & Kumar, S. 2025. Naganishia liquefaciens ARY7, a psychrophilic yeast endophyte ımproves plant low temperature acclimation through auxin and salicylic acid signaling. *Physiologia Plantarum*, 177(3), e70267.

Białkowska, A., Majewska, E., Olczak, A., & Twarda-Clapa, A. 2020. Ice binding proteins: Diverse biological roles and applications in different types of industry. *Biomolecules*, 10(2), 274.

Bowman, J.P. 2017. Genomics of psychrophilic bacteria and archaea. In: R. Margesin (Ed.), *Psychrophiles: From Biodiversity to Biotechnology*. Springer, pp. 345–387. https://doi.org/10.1007/978-3-319-57057-0_13.

Cavicchioli, R., Siddiqui, K.S., Andrews, D., & Sowers, K.R. 2002. Low-temperature extremophiles and their applications. *Current Opinion in Biotechnology*, 13(3), 253–261. https://doi.org/10.1016/S0958-1669(02)00317-8.

Collins, T., & Margesin, R. 2019. Psychrophilic lifestyles: Mechanisms of adaptation and biotechnological tools. *Applied Microbiology and Biotechnology*, 103, 2857–2871. https://doi.org/10.1007/s00253-019-09659-5.

Collins, T., & Feller, G. 2023. Psychrophilic enzymes: Strategies for cold-adaptation. *Essays in Biochemistry*, 67(4), 701–713.

Dasila, H., Sah, V.K., Jaggi, V., Kumar, A., Tewari, L., Taj, G. et al. 2023. Cold-tolerant phosphate-solubilizing Pseudomonas strains promote wheat growth and yield by improving soil phosphorous (P) nutrition status. *Frontiers in Microbiology*, 14, 1135693.

Dubey, A., Malla, M.A., Kumar, A., Khan, M.L., & Kumari, S. 2024. Seed bio-priming with ACC deaminase-producing bacterial strains alleviates impact of drought stress in soybean (Glycine max). *Rhizosphere.*_https://www.sciencedirect.com/science/article/pii/S2452219824000260.

Fonseca, P., Moreno, R., & Rojo, F. 2011. Growth of Pseudomonas putida at low temperature: Global transcriptomic and proteomic analyses. *Environmental Microbiology Reports*, 3(3), 329–339.

Garcia-Lopez, E., Alcazar, P., & Cid, C. 2021. Identification of biomolecules involved in the adaptation to the environment of cold-loving microorganisms and metabolic pathways for their production. *Biomolecules*, 11(8), 1155.

Hamid, B., Baba, Z.A., Sheikh, T.A., Perveen, K., & Sayyed, R.Z. 2024. Investigating potential protease activity of psychrotrophic bacteria from a municipal landfill for solid waste management. *Biomass Conversion and Biorefinery*. https://doi.org/10.1007/s13399-024-05621-2,

Hassan, N., Anesio, A.M., Rafiq, M., Holtvoeth, J., Bull, I., Haleem, A. et al. 2020. Temperature driven membrane lipid adaptation in glacial psychrophilic bacteria. *Frontiers in Microbiology*, 11, 824.

Hernández-Fernández, M., Cordero-Bueso, G., Ruiz-Muñoz, M., & Cantoral, J.M. 2021. Culturable yeasts as biofertilizers and biopesticides for a sustainable agriculture: A comprehensive review. *Plants*, 10(5), 822.

Hou, M., Leng, C., Zhu, J., Yang, M., Yin, Y., Xing, Y., & Chen, J. 2024. Alpine and subalpine plant microbiome mediated plants adapt to the cold environment: A systematic review. *Environmental Microbiome*, 19(1), 82.

Jin, S., Wang, Y., & Zhao, X. 2022. Cold-adaptive mechanism of psychrophilic bacteria in food and its application. *Microbial Pathogenesis*, 169, 105652. https://doi.org/10.1016/j.micpath.2022.105652.

Kaushal M., Devi, S., Kumawat, K.C., & Kumar, A. 2024. Microbial consortium: A boon for a sustainable agriculture. In: *Climate Change and Microbiome Dynamics, Climate Change Management*. Springer, Cham. (2023), pp. 15–31, 10.1007/978-3-031-21079-2_2.

Kırkıncı, S.F., Maraklı, S., Aksoy, H.M., Özçimen, D., & Kaya, Y. (2021). Antarktika: Yaşam bilimleri ve biyoteknoloji araştırmalarının gözden geçirilmesi. *International Journal of Life Sciences and Biotechnology*, 4(1), 158–177. https://doi.org/10.38001/ijlsb.853472.

Koh, H.Y., Park, H., Lee, J.H., Han, S.J., Sohn, Y.C., & Lee, S.G. 2017, Feb. Proteomic and transcriptomic investigations on cold-responsive properties of the psychrophilic

Antarctic bacterium Psychrobacter sp. PAMC 21119 at subzero temperatures. *Environ Microbiol.*, 19(2), 628–644. doi: 10.1111/1462-2920.13578. Epub 2016 Dec 8. PMID: 27750393.

Kour, D., & Yadav, A.N. 2023. Alleviation of cold stress in wheat with psychrotrophic phosphorus solubilizing Acinetobacter rhizosphaerae EU-KL44. *Brazilian Journal of Microbiology*, 54(1), 371–383.

Kumar, A., Droby, S., White, J.F., Singh, V.K., & Singh, S.K. (2020). Endophytes and seed priming: Agricultural applications and future prospects. In: *Microbial Endophytes*. Elsevier, pp. 123–141. https://www.sciencedirect.com/science/article/pii/B9780128196540000053.

Lauritano, C., & Coppola, D. 2023. Biotechnological applications of products released by marine microorganisms for cold adaptation strategies: Polyunsaturated fatty acids, antioxidants, and antifreeze proteins. *Journal of Marine Science and Engineering*, 11(7), 1399.

Liu, X., Wang, Y., Gao, H., & Xu, X. 2011. Identification and characterization of genes encoding two novel LEA proteins in Antarctic and temperate strains of Chlorella vulgaris. *Gene*, 482(1–2), 51–58.

Luo, B., Su, J.Y., Zhang, Y.F. et al. 2024. Alteromonas arenosi sp. nov., a novel bioflocculant-producing bacterium, isolated from intertidal sand. *Antonie van Leeuwenhoek*, 117, 28. https://doi.org/10.1007/s10482-023-01926-w.

Mahmood, A., & Kataoka, R. 2018. Potential of biopriming in enhancing crop productivity and stress tolerance. In: *Advances in Seed Priming*. Springer, pp. 133–150. https://link.springer.com/chapter/10.1007/978-981-13-0032-5_9.

Marcoleta, A.E., Arros, P., Varas, M.A., Costa, J., Rojas-Salgado, J., Berríos-Pastén, C. et al. 2022. The highly diverse Antarctic peninsula soil microbiota as a source of novel resistance genes. *Sci. Total Environ.*, 810, 152003. doi: 10.1016/j.scitotenv.2021.152003.

Margesin, R., & Collins, T. 2019. Microbial ecology of the cryosphere (glacial and permafrost habitats): Current knowledge. *Applied Microbiology and Biotechnology*, 103, 2537–2549.

Margesin, R., & Feller, G. 2010. Biotechnological applications of psychrophiles. *Environmental Technology*, 31(8–9), 835–844.

Margesin, R., & Miteva, V. 2011. Diversity and ecology of psychrophilic microorganisms. *Research in Microbiology*, 162(3), 346–361. https://doi.org/10.1016/j.resmic.2010.12.004.

Margesin, R., Schinner, F., Marx, J.C., & Gerday, C. (Eds.). 2017. *Psychrophiles: From Biodiversity to Biotechnology*. Berlin Heidelberg: Springer International Publishing, 450 pp.

Math, R.K., Jin, H.M., Kim, J.M., Hahn, Y., Park, W., Madsen, E.L., & Jeon, C.O. 2012. Comparative genomics reveals adaptation by Alteromonas sp. SN2 to marine tidal-flat conditions: Cold tolerance and aromatic hydrocarbon metabolism. *PLOS One*, 7(4), e35784.

Mendes, R., Garbeva, P., & Raaijmakers, J.M. 2013. The rhizosphere microbiome: Significance of plant beneficial, plant pathogenic, and human pathogenic microorganisms. *FEMS Microbiology Reviews*, 37(5), 634–663. https://doi.org/10.1111/1574-6976.12028.

Methé, B.A., Nelson, K.E., Deming, J.W., Momen, B., Melamud, E., Zhang, X. et al. 2005. The psychrophilic lifestyle as revealed by the genome sequence of Colwellia psychrerythraea 34H through genomic and proteomic analyses. *Proceedings of the National Academy of Sciences*, 102(31), 10913–10918.

Miner, K.R., D'Andrilli, J., Mackelprang, R., Edwards, A., Malaska, M.J., Waldrop, M.P., & Miller, C.E. 2021. Emergent biogeochemical risks from Arctic permafrost degradation. *Nature Climate Change*, 11(10), 809–819.

Mishra, P.K., Bisht, S.C., Ruwari, P., Selvakumar, G., Joshi, G.K., Bisht, J.K. et al. 2011. Alleviation of cold stress in inoculated wheat (Triticum aestivum L.) seedlings with psychrotolerant Pseudomonads from NW Himalayas. *Archives of Microbiology*, 193, 497–513.

Mocali, S., Chiellini, C., Fabiani, A., Decuzzi, S., de Pascale, D., Parrilli, E. et al. 2017. Ecology of cold environments: New insights of bacterial metabolic adaptation through an integrated genomic-phenomic approach. *Scientific Reports*, 7(1), 839.

Moghaddam, M.H., Moradi, A., Piri, R., Glick, B.R., Fazeli-Nasab, B., & Sayyed, R.Z. 2024. Seed coating with minerals and plant growth-promoting bacteria enhances drought tolerance in fennel (*Foeniculum vulgare* L.). *Biocatalysis and Agricultural Biotechnology*, 58, 103202. https://doi.org/10.1016/j.bcab.2024.103202.

Muñoz, P.A., Márquez, S.L., González-Nilo, F.D., Márquez-Miranda, V., & Blamey, J.M. 2017. Structure and application of antifreeze proteins from Antarctic bacteria. *Microbial Cell Factories*, 16, 1–13.

Mykytczuk, N.C., Trevors, J.T., Foote, S.J., Leduc, L.G., Ferroni, G.D., & Twine, S.M. 2011. Proteomic insights into cold adaptation of psychrotrophic and mesophilic Acidithiobacillus ferrooxidans strains. *Antonie Van Leeuwenhoek*, 100, 259–277.

Nasab, B.F., Sayyed, R.Z., Ahmad, R.P., & Rahmani, F. 2021. Biopriming and nanopriming: Green revolution wings to ıncrease plant yield, growth, and development under stress condition and forward dimensions. In: Singh, H.B., Anukool, V., Sayyed, R.Z. (Eds.), *Antioxidants in Plant-Microbe Interaction*. Springer, Singapore, pp. 623–655. https://doi.org/10.1007/978-981-16-1350-0_29.

Orellana-Saez, M., Pacheco, N., Costa, J.I., Mendez, K.N., Miossec, M.J., Meneses, C. et al. 2019. In-depth genomic and phenotypic characterization of the antarctic psychrotolerant strain Pseudomonas sp. MPC6 reveals unique metabolic features, plasticity, and biotechnological potential. *Frontiers in Microbiology*, 10, 1154.

Ortiz, A., & Sansinenea, E. 2023. Genetically modified plants based on Bacillus genes and commercial Bacillus-based biopesticides for sustainable agriculture. *Horticulturae*, 9(9), 963.

Patil, S.K., Islam, T., Tveit, A., Hodson, A., & Øvreås, L. 2024. Targeting methanotrophs and isolation of a novel psychrophilic Methylobacter species from a terrestrial Arctic alkaline methane seep in Lagoon Pingo, Central Spitsbergen (78° N). *Antonie van Leeuwenhoek*, 117(1), 60.

Perfumo, A., & Lo Giudice, A. 2023. Low-temperature microbiology meets the global challenges of our time. *Microorganisms*, 11(5), 1217. https://doi.org/10.3390/microorganisms11051217.

Philippot, L., Raaijmakers, J.M., Lemanceau, P., & van der Putten, W.H. 2013. Going back to the roots: The microbial ecology of the rhizosphere. *Nature Reviews Microbiology*, 11(11), 789–799. https://doi.org/10.1038/nrmicro3109.

Picazo, A., Villaescusa, J.A., Rochera, C., Miralles-Lorenzo, J., Quesada, A., & Camacho, A. 2021. Functional metabolic diversity of bacterioplankton in maritime Antarctic lakes. *Microorganisms*, 9(10), 2077.

Purwar, S., & Srivastava, S. 2024. Adaptations of psychrophilic microorganisms to low-temperature environments. *Applied Microbiology: Theory and Technology*, 5(2), 76–83. https://doi.org/10.37256/amtt.5220244537.

Rahman, R., Bheemasetti, T.V., Govil, T., & Sani, R. 2024. Psychrophiles to control ice–water phase changes in frost-susceptible soils. *Scientific Reports*, 14(477). https://doi.org/10.1038/s41598-023-51060-w.

Ramasamy, K.P., Mahawar, L., Rajasabapathy, R., Rajeshwari, K., Miceli, C., & Pucciarelli, S. 2023. Comprehensive insights on environmental adaptation strategies in Antarctic bacteria and biotechnological applications of cold adapted molecules. *Frontiers in Microbiology*, 14, 1197797.

Raymond-Bouchard, I., Tremblay, J., Altshuler, I., Greer, C.W., & Whyte, L.G. 2018. Comparative transcriptomics of cold growth and adaptive features of a Eury- and Steno-Psychrophile. *Front. Microbiol.*, 9, 1565. doi: 10.3389/fmicb.2018.01565.

Rizvi, A., Ahmed, B., Khan, M.S., Umar, S., & Lee, J. 2021. Psychrophilic bacterial phosphate-biofertilizers: A novel extremophile for sustainable crop production under cold environment. *Microorganisms*, 9(12), 2451.

Sajjad, W., Din, G., Rafiq, M., Iqbal, A., Khan, S., Zada, S. et al. 2020. Pigment production by cold-adapted bacteria and fungi: Colorful tale of cryosphere with wide range applications. *Extremophiles*, 24, 447–473.

Sarkar, D., Sarkar, A., Devika, O.S., Singh, S., Parihar, M., Rakshit, A. et al. 2021. Optimizing nutrient use efficiency, productivity, energetics, and economics of red cabbage following mineral fertilization and biopriming with compatible rhizosphere microbes. *Scientific Reports*, 11, 15680. https://www.nature.com/articles/s41598-021-95092-6.

See-Too, W.S., Lim, Y.L., Ee, R., Convey, P., Pearce, D.A., Yin, W.F., & Chan, K.G. 2016. Complete genome of Pseudomonas sp. strain L10. 10, a psychrotolerant biofertilizer that could promote plant growth. *Journal of Biotechnology*, 222, 84–85.

Selvakumar, G., Joshi, P., Suyal, P., Mishra, P.K., Joshi, G.K., Bisht, J.K. et al. 2011. Pseudomonas lurida M2RH3 (MTCC 9245), a psychrotolerant bacterium from the Uttarakhand Himalayas, solubilizes phosphate and promotes wheat seedling growth. *World Journal of Microbiology and Biotechnology*, 27, 1129–1135.

Sharma, P., Sharma, N., & Deswal, R. 2005. The molecular biology of the low-temperature response in plants. *Bioessays*, 27, 1048–1059. https://doi.org/10.1002/bies.20307.

Sheikh, T., Baba, Z., Hamid, B., Iqbal, S., Yatoo, A., Fatima, S. et al. 2022. Extracellular polymeric substances in psychrophilic cyanobacteria: A potential bioflocculant and carbon sink to mitigate cold stress. *Biocatalysis and Agricultural Biotechnology*, 42, 102375. https://doi.org/10.1016/j.bcab.2022.102375.

Singh, H.B., Upadhyay, R.S., & Sarma, B.K. 2016. Seed biopriming: A comprehensive approach towards agricultural sustainability. *Indian Phytopathology*, 69(2), 101–110. https://www.researchgate.net/publication/308723632.

Soni, R., Suyal, D.C., Agrawal, K., Yadav, A., Souche, Y., & Goel, R. 2015. Differential proteomic analysis of Himalayan psychrotolerant diazotroph Pseudomonas palleroniana N26 strain under low temperature diazotrophic conditions. *CryoLetters*, 36, 74–82.

Styczynski, M., Rogowska, A., Nyabayo, C., Decewicz, P., Romaniuk, F., Pączkowski, C., & Dziewit, L. 2022. Heterologous production and characterization of a pyomelanin of Antarctic Pseudomonas sp. ANT_H4: A metabolite protecting against UV and free radicals, interacting with iron from minerals and exhibiting priming properties toward plant hairy roots. *Microbial Cell Factories*, 21(1), 261.

Su, Z.E., Zhao, J., Zhuang, M., Liu, Z., Zhao, C., Pullens, J.W. et al. 2024. Climate-adaptive crop distribution can feed food demand, improve water scarcity, and reduce greenhouse gas emissions. *Science of the Total Environment*, 944, 173819.

Sudheer, S., Bai, R.G., Usmani, Z., & Sharma, M. 2020. Insights on engineered microbes in sustainable agriculture: Biotechnological developments and future prospects. *Current Genomics*, 21(5), 321–333.

Suyal, D.C., Yadav, A.N., El Enshasy, H.A., & Soni, R. 2023. Exploration of cold-adapted microorganisms for sustainable development. *Frontiers in Microbiology*, 14, 1191673. https://doi.org/10.3389/fmicb.2023.1191673.

Tapia-Vázquez, I., Sánchez-Cruz, R., Arroyo-Domínguez, M., Lira-Ruan, V., Sánchez-Reyes, A., del Rayo Sánchez-Carbente, M. et al. 2020. Isolation and characterization of psychrophilic and psychrotolerant plant-growth promoting microorganisms from a high-altitude volcano crater in Mexico. *Microbiological Research*, 232, 126394.

Trivedi, P., Leach, J.E., Tringe, S.G., Sa, T., & Singh, B.K. (2020). Plant–microbiome interactions: From community assembly to plant health. *Nature Reviews Microbiology*, 18(11), 607–621. https://doi.org/10.1038/s41579-020-0412-1.

Turan, M., Güllüce, M., Çakmak, R., & Şahin, F. 2013. Effect of plant growth-promoting rhizobacteria strain on freezing injury and antioxidant enzyme activity of wheat and barley. *Journal of Plant Nutrition*, 36(5), 731–748.

Vega-Celedon, P., Bravo, G., Velasquez, A., Cid, F.P., Valenzuela, M., Ramirez, I. et al. (2021). Microbial diversity of psychrotolerant bacteria isolated from wild flora of Andes Mountains and Patagonia of Chile towards the selection of plant growth-promoting bacterial consortia to alleviate cold stress in plants. *Microorganisms*, 9(3), 538.

Verma, P., Yadav, A.N., Khannam, K.S., Panjiar, N., Kumar, S., Saxena, A.K., & Suman, A. (2015). Assessment of genetic diversity and plant growth promoting attributes of psychrotolerant bacteria allied with wheat (Triticum aestivum) from the northern hills zone of India. *Annals of Microbiology*, 65, 1885–1899.

Wu, G., Baumeister, R., & Heimbucher, T. 2023. Molecular mechanisms of lipid-based metabolic adaptation strategies in response to cold. *Cells*, 12(10), 1353.

Xu, H., Xu, D. & Liu, Y. 2024. Molecular biology applications of psychrophilic enzymes: Adaptations, advantages, expression, and prospective. *Appl. Biochem. Biotechnol.*, 196, 5765–5789. https://doi.org/10.1007/s12010-023-04810-5.

Yadav, A.N., Sachan, S.G., Verma, P., Kaushik, R., & Saxena, A.K. 2016. Cold active hydrolytic enzymes production by psychrotrophic bacilli isolated from three sub-glacial lakes of NW Indian Himalayas. *J. Basic Microbiol.*, 56, 294–307.

Yusof, N.A., Hashim, N.H.F., & Bharudin, I. 2021. Cold adaptation strategies and the potential of psychrophilic enzymes from the antarctic yeast, Glaciozyma Antarctica PI12. *Journal of Fungi*, 7(7), 528.

Index

About the Editors

Burhan Hamid, PhD, is an Assistant Professor at Centre of Research for Development (CORD) & P.G. Programme in Microbiology, University of Kashmir, Srinagar, India since 2019. He has 12 years of teaching and research experience. He has worked as researcher in Regional Research Station (RRS) for biofertilizers production at Sher-e-Kashmir University of Agricultural Sciences and Technology of Kashmir (SKUAST-K), India. His area of interest in teaching and research are environmental microbiology, psychrophiles, cold-active enzymes, agricultural microbiology, food and industrial microbiology. Dr. Hamid is a reviewer for various national and international journals and has published more than 58 research papers, review papers and book chapters with reputed publishers such as Springer, Elsevier, Wiley, MDPI, Taylor & Francis. Moreover, he has worked as Senior Microbiologist in ISO/IEC-17025 accredited laboratory at Baddi, India. He is also supervising M.Sc. Microbiology students in their dissertation and project works.

Riyaz Sayyed is a Professor in the Department of Microbiology, PSGVP Mandal's ASC College, Shahada, India, with over two decades of experience in teaching, research, and academic leadership. He is a Fellow of the Royal Society of Biology (FRSB), the Indian Phytopathological Society (FIPS), and the Association of Agricultural Technology in South East Asia (FAATSEA); a Member of the American Chemical Society (ACS); and a Global Council Member of the World Agriculture Forum (WAF). He is also a Visiting Scholar at INTI International University, Malaysia, and an Associate Researcher at the Bashan Institute of Science, USA. Prof. Sayyed currently serves as Executive Editor of *Biocatalysis and Agricultural Biotechnology* (Elsevier), Editor for *Plant Pathology* (Wiley), and Associate Editor of *Environmental Sustainability* and *Vegetos* (Springer). His prolific research career includes over 380 research publications, the editing of 35 books with leading publishers including Springer, Elsevier, Wiley, and CRC Press, and the supervision of more than 100 graduate and doctoral students. Recognized globally for his scholarly impact, Elsevier and Clarivate have consistently listed him among the Top 2% most-cited researchers worldwide for three consecutive years and among the Top 1% of Highly Cited Researchers globally. Prof. Sayyed has successfully led several international R&D projects, organized major global scientific conferences, and received numerous prestigious awards for his outstanding contributions to science and research leadership.

For Product Safety Concerns and Information please contact our EU representative GPSR@taylorandfrancis.com Taylor & Francis Verlag GmbH, Kaufingerstraße 24, 80331 München, Germany